ENCYCLOPAEDIA OF TEACHING OF AGRICULTURE

PREFACE

If 'food' is a must for survival of man besides the air and water, it is AGRICULTURE which provides FOOD. Since the cave-dwelling stage to the present day journey of mankind Agriculture has seen many changes, ups and downs. Modern agricultural science include agriculture, horticulture, floriculture, pisciculture, animal husbandry, sericulture, veterinary science, soil science, irrigation and water technology, paultry, dairying, forestry, food processing and technology etc. etc. Thus from an occupation it has grown into an industry in modern sense of the term. Hence Agriculture education.

Agriculture, art, science, and industry of managing the growth of plants and animals for human use. In a broad sense agriculture includes cultivation of the soil, growing and harvesting crops, breeding and raising of livestock, dairying, and forestry. Modern agriculture depends heavily on engineering and technology and on the biological and physical sciences. Irrigation, drainage, soil conservation, and sanitation each of which is important in successful farming—are some of the fields requiring the specialized knowledge of agricultural engineers. Agricultural chemistry deals with other vital farm problems, such as uses of fertilizers, insecticide, and fungicide, soil makeup, analysis of agricultural products, and nutritional needs of farm animals. Plant breeding and genetics contribute immeasurably to farm productivity. Genetics has also placed livestock breeding on a scientific basis. Hydroponics, a method of soilless gardening in which plants are grown in chemical nutrient solutions, may solve additional agricultural problems.

Many universities, institutes and colleges are there imparting education of Agriculture in many of its specialized branches.

The present work (in two volume) is an effort to bring together encyclopaedic information on the subject. All these issues in the context of teaching Agriculture have been discussed in the present book in detail and therefore, it is hoped that it will be a very useful work for all concerned.

The authoritative information is organised in thirty seven chapters. To make the work more authentic, opinions and research of many scholars in the field are liberally cited and borrowed. We are deeply beholden to all those scholars of eminence and agricultural scientists. We are also thankful to all those who have assisted/guided us while preparing the draft of the work. Our colleagues and friends at the ISD deserve our appreciation. Finally we owe a deep sense of gratitude to Mr. J.L. Kumar, Managing Director, Anmol Publications Pvt. Ltd., New Delhi for his constant moral support and also for bringing out this work.

—Editor-in-Chief

1

WORLD AGRICULTURE: A PANAROMIC VIEW

Agriculture, art, science, and industry of managing the growth of plants and animals for human use. In a broad sense agriculture includes cultivation of the soil, growing and harvesting crops, breeding and raising of livestock, dairying, and forestry. Regional and national agriculture are covered in more detail in individual continent and country articles.

Modern agriculture depends heavily on engineering and *technology* and on the biological and physical sciences. *Irrigation, drainage, conservation,* and sanitation each of which is important in successful farming are some of the fields requiring the specialized knowledge of agricultural engineers.

Agricultural chemistry deals with other vital farm problems, such as uses of fertilizer, insecticide, and fungicide, soil makeup, analysis of agricultural products, and nutritional needs of farm animals.

Plant breeding and *genetics* contribute immeasurably to farm productivity. Genetics has also placed livestock breeding on a scientific basis. *Hydroponics,* a method of soilless gardening in which plants are grown in chemical nutrient solutions, may solve additional agricultural problems.

The packing, processing, and marketing of agricultural products are closely related activities also influenced by science. Methods of quick-freezing and dehydration have increased the markets for farm products. Mechanization, the outstanding characteristic of late 19th- and 20th-century agriculture, has eased much of the backbreaking toil of the farmer. More significantly, mechanization has enormously increased farm efficiency and productivity.

Airplanes and helicopters are employed in agriculture for such purposes as seeding, transporting perishable products, and fighting forest fires, and in spraying operations involved in insect and disease control. Radio and television disseminate vital weather reports and other information that is of concern to farmers.

WORLD AGRICULTURE

Greenhouse Fanning

Over the 10,000 years since agriculture began to be developed, peoples everywhere have discovered the food value of wild plants and animals and domesticated and bred them. The most important are cereals such as wheat, rice, barley, corn, and rye; sugarcane and sugar beets; meat animals such as sheep, cattle, goats, and pigs or swine; poultry such as chickens, ducks, and turkeys; and such products as milk, cheese, eggs, nuts, and oils. Fruits, vegetables, and olives are also major foods for people; feed grains for animals include soybeans, field corn, and sorghum.

Agricultural income is also derived from nonfood commercial crops such as rubber, fiber plants, tobacco, and oilseeds used in synthetic chemical compounds, as well as raising animals for pelt.

The conditions that determine what will be raised in an area include climate, water supply, and terrain. Nearly 50 per cent of the world's labor force is employed in agriculture. The distribution in the late 1980s ranged from 64 per cent of the economically active population in Africa to less than 4 per cent in the U.S. and Canada. For Asia the figure was 61 per cent; in South America, 24 per cent; in Eastern Europe and the Soviet Union, 15 per cent, and in Western Europe, 7 per cent.

Farm size varies widely from region to region. In the late 1980s the average for Canadian farms was about 230 ha (about 570 acres) per farm; for U.S. farms, about 185 ha (about 460 acres). The average size of a single landholding in the Philippines, however, may be somewhat less than 3.6 ha (less than 9 acres), and in Indonesia, a little less than 1.2 ha (less than 3 acres).

Size also depends on the purpose of the farm. Commercial farming, or production for cash, is usually on large holdings. The latifundia of Latin America are large, privately owned estates worked by tenant labor. Single-crop plantations produce tea, rubber, and cocoa. Wheat farms are most efficient when they comprise some thousands of hectares and can be worked by teams of people and machines. Australian sheep stations and other livestock farms must be large to provide grazing for thousands of animals. The agricultural plots of Chinese communes and the cooperative farms held by Peruvian communities are other necessarily large agricultural units, as were the collective farms that were owned and operated by state employees in the former Soviet Union.

Individual subsistence farms or small-family mixed-farm operations are decreasing in number in developed countries but are still numerous in the developing countries of Africa and Asia. A "back-to-the-land" movement in the U.S. reversed the decline of small farms in New England and Alaska in the decade from 1970 to 1980.

Nomadic herders range over large areas in sub-Saharan Africa, Afghanistan, and Lapland; and herding is a major part of agriculture in such areas as Mongolia.

Much of the foreign exchange earned by a country may be derived from a single commodity; for example, Sri Lanka depends on tea, Denmark specializes in dairy products, Australia in wool, and New Zealand and Argentina in meat products. In the U.S., wheat, corn, and soybeans have become major foreign exchange commodities in recent decades.

The importance of an individual country as an exporter of agricultural products depends on many variables. Among them is the possibility that the country is too little developed industrially to produce manufactured goods in sufficient quantity or technical sophistication. Such agricultural exporters include Ghana, with cocoa, and Burma (Myanmar), with rice. On the other hand, an exceptionally well-developed country may produce surpluses that are not needed by its own population; such has been the case of the U.S., Canada, and some of the Western European countries.

Because nations depend on agriculture not only for food but for national income and raw materials for industry as well, trade in agriculture is a constant international concern. It is regulated by international agreements such as the General Agreement on Tariffs and Trade and by trading areas such as the European Union.

The Food and Agricultural Organization (FAO) of the UN directs much attention to agricultural trade and policies. According to the FAO, world agricultural production, stimulated by improving technology, reached a record high in the late 1980s. Further, agricultural output in developing nations increased 41 per cent during the 1977-88 period, as compared to a rise of 9 per cent in developed countries. On a per capita basis, however, food production rose by only 12 per cent in developing nations, and less than 1 per cent in developed countries.

HISTORY

Scientific Agriculture

The history of agriculture may be divided into four broad periods of unequal length, differing widely in date according to region: prehistoric; historic through the Roman period; feudal; and scientific.

Prehistoric Agriculture

Early agriculturists were, it is agreed, largely of Neolithic culture. Sites occupied by such people are located in southwestern Asia, in what are now Iran, Iraq, Israel, Jordan, Syria, and Turkey; in southeastern Asia, in what is now Thailand; in Africa, along the Nile River in Egypt; and in Europe, along the Danube River and in Macedonia, Thrace, and Thessaly. Early centers of agriculture have also been identified in the Huang Ho (Yellow River) area of China; the Indus River valley of India and Pakistan; and the Tehuacan Valley of Mexico, northwest of the Isthmus of Tehuantepec.

The dates of domesticated plants and animals vary with the regions, but most predate the 6th millennium BC, and the earliest may date from 10,000 BC. Scientists have carried out carbon-14 testing of animal and plant remains (fossils) and have dated finds of domesticated sheep at 9000 BC in northern Iraq; cattle in the 6th millennium BC in northeastern Iran; goats at 8000 BC in central Iran; pigs at 8000 BC in Thailand and 7000 BC in Thessaly; onagers, or asses, at 7000 BC in Jarmo, Iraq; and horses at 4350 BC in Ukraine. The llama and alpaca were domesticated in the Andean regions of South America by the middle of the 3d millennium BC.

According to carbon dating, wheat and barley were domesticated in the Middle East in the 8th millennium BC; millet and rice in China and southeastern Asia by 5500 BC; and squash in Mexico about 8000 BC. Legumes found in Thessaly and Macedonia are dated as early as 6000 BC, Flax was grown and apparently woven into textiles early in the Neolithic period.

The fanner began, most probably, by noting which of the wild plants were edible or otherwise useful and learned to save the seed and to replant it in cleared land. Long cultivation of the most prolific and hardiest plants yielded a stable strain. Herds of goats and sheep were assembled from captured young wild animals, and those with the most useful traits—such as small horns and high milk yield—were bred. The aurochs seems to have been the ancestor of European cattle, and an Asian wild ox of the zebu, the humped cattle of Asia. The cat, dog, and chicken were domesticated very early. The transition from hunting and food gathering to a dependence on food production was gradual, and in a few isolated parts of the world has not yet been accomplished. Crops and domestic meat supplies were augmented by fish and wildfowl as well as by the meat of wild animals.

The Neolithic farmers lived in simple dwellings-in caves and in small houses of sunbaked mud brick or of reed and wood. These homes were grouped into small villages or existed as single farmsteads surrounded by fields, sheltering animals and humans in adjacent or joined buildings. In the Neolithic period, the growth of cities such as Jericho (founded c. 9000 BC) was stimulated by the production of surplus crops.

Pastoralism may have been a later development. Evidence indicates that mixed farming, combining cultivation of crops and stock raising, was the most common Neolithic pattern. Nomadic herders, however, roamed the steppes of Europe and Asia, where the horse and camel were domesticated.

The earliest tools of the farmer were made of wood and stone. They included the stone adz; the sickle or reaping knife with sharpened stone blades, used to gather grain; the digging stick, used to plant seeds, and, with later adaptations, as a spade or hoe; and a rudimentary plow, a modified tree branch used to scratch the surface of the soil and prepare it for planting. The plow was later adapted for pulling by oxen.

The hilly areas of southwestern Asia and the forests of Europe had enough rain to sustain agriculture, but Egypt depended on the annual floods of the Nile to replenish soil moisture and fertility. The inhabitants of the so-called Fertile Crescent, around the Tigris and Euphrates rivers, also depended on annual floods to supply irrigation water. Drainage was necessary to prevent the carrying off of land from the hillsides through which the rivers ran. The farmers who lived in the area near the Huang Ho developed a system of irrigation and drainage to control the damage caused to their fields in the floodplain of the Meandering river.

Although the Neolithic settlements were more permanent than the camps of hunting populations, villages had to be moved periodically in some areas, as the fields lost their fertility from continuous cropping. This was most necessary in northern Europe, where fields were produced by the slash-and-burn method of clearing. The settlements along the Nile, however, were more permanent, because the river deposited fertile silt annually.

HISTORICAL AGRICULTURE THROUGH THE ROMAN PERIOD

With the close of the Neolithic period and the introduction of metals, the age of innovation in agriculture was largely over. The historical period-known through written and pictured materials, including the Bible, Near Eastern records and monuments, and Chinese, Greek, and Roman writings-was devoted to improvement. A few high points must serve to outline the development of worldwide agriculture in this era, roughly defined as 2500 BC to AD 500. For a similar period of development in Central and South America, somewhat later in date, Some plants became newly prominent. Grapes and wine were mentioned in Egyptian records about 2900 BC, and trade in olive oil and wine was widespread in the Mediterranean area in the 1st millennium BC. Rye and oats were cultivated in northern Europe about 1000 BC.

Many vegetables and fruits, including onions, melons, and cucumbers, were grown by the 3d millennium BC in Ur. Dates and figs were an important source of sugar in the Near East, and apples, pomegranates, peaches, and mulberries were grown in the Mediterranean area. Cotton was grown and spun in India about 2000 BC, and linen and silk were used extensively in 2d-millennium China. Felt was made from the wool of sheep in Central Asia and the Russian steppes.

The horse, introduced to Egypt about 1600 BC, was already known in Mesopotamia and Asia Minor. The ox-drawn four-wheeled cart for farm work and two-wheeled chariots drawn by horses were familiar in northern India in the 2d millennium BC.

Improvements in tools and implements were particularly important. Metal tools were longer lasting and more efficient, and cultivation was greatly improved by such aids as the ox-drawn plow fitted with an iron-tipped point, noted in the 10th century BC in Palestine. In Mesopotamia in the 3d millennium BC a funnellike device was attached to the plow to aid in seeding, and other early forms of drills were used in China. Threshing was done with animal power in Palestine and Mesopotamia, although reaping, binding, and winnowing were still done by hand. Egypt retained hand seeding through this period, on individual farm plots and large estates alike. Storage methods for oil and grain were improved. Granaries-jars, dry cisterns, silos, and bins of one sort or another containing stored grain-supported city populations. Indeed, without adequate food supplies and trade in food and nonfood items, the high civilizations of Mesopotamia, northern India, Egypt, and Rome would not have been possible.

Irrigation systems in China, Egypt, and the Near East were elaborated, putting more land into cultivation. The forced labor of peasants and the bureaucracy built up to plan and supervise the work of irrigation were probably basic in the development of the city-states of Sumer. Windmills and water mills, developed toward the end of the Roman period, increased control over the many uncertainties of weather. The introduction of fertilizer, mostly animal manures, and the rotation of fallow and crop land made agriculture more productive.

Mixed farming and stock raising were flourishing in the British Isles and on the continent of Europe as far north as Scandinavia at the beginning of the historical period, already displaying a pattern that persisted throughout the next 3000 years. According to region, fishing and hunting supplemented the food grown by agriculturists.

Shortly after the time of Julius Caesar, the Roman historian Cornelius Tacitus described the "Germans" as a tribal society of free peasant warriors, who cultivated their own lands or left them to fight. About 500 years later, a characteristic European village had a cluster of houses in the middle, surrounded by rudely cultivated fields comprising individually owned farmlands; and meadows, woods, and wasteland were used by the entire community. Oxen and plow were passed from one field to another, and harvesting was a cooperative effort.

Rome appears to have started as a rural agricultural society of independent farmers. In the 1st millennium BC, after the city was established, however, agriculture started a capitalistic development that reached a peak in the Christian era. The large estates that supplied grain to the cities of the empire were owned by absentee landowners and were cultivated by slave labor under the supervision of hired overseers. As slaves, usually war captives, decreased in number, tenants replaced them. The late Roman villa of the Christian era approached the medieval manor in organization; slaves and dependent tenants were forced to work on a fixed schedule, and tenants paid a predetermined share to the estate owner. By the 4th century AD, *serfdom* was well established, and the former tenant was attached to the land.

Feudal Agriculture

The feudal period in Europe began soon after the fall of the Roman Empire, reaching its height about AD 1100. This period was also that of the development of the Byzantine Empire and of the power of the Saracens in the Middle East and southern Europe. Spain, Italy, and southern France, in particular, were affected by events outside continental Europe.

In the Arab period in Egypt and Spain, irrigation was extended to previously sterile or unproductive land. In Egypt, grain production was sufficient to allow the country to sell wheat in the international market. In Spain, vineyards were planted on sloping land, and irrigation water was brought from the mountains to the plains. In some Islamic areas, oranges, lemons, peaches, and apricots were cultivated.

Rice, sugarcane, cotton, and such vegetables as spinach and artichokes, as well as the characteristic Spanish flavoring saffron, were produced. The silkworm was raised, and its food, the mulberry tree, was grown.

By the 12th century agriculture in the Middle East was static, and Mesopotamia, for example, fell back to subsistence level when its irrigation systems were destroyed by the Mongols. The Crusades increased European contact with Islamic lands and familiarized western Europe with citrus fruits and silk and cotton textiles.

The structure of agriculture was not uniform. In Scandinavia and eastern Germany, the small farms and villages of previous years remained. In mountainous areas and in the marshlands of Slavic Europe, the manorial system could not flourish. Stock raising and olive and grape culture were normally outside the system.

A manor required roughly 350 to 800 ha (about 900 to 2000 acres) of arable land and the same amount of other prescribed lands, such as wetlands, woodlots, and pasture. Typically, the manor was a self-contained community. On it was the large home of the holder of the fief—a military or church vassal of rank, sometimes given the title lord—or of his steward. A parish church was frequently included, and the manor might make up the entire parish. One or more villages might be located on the manor, and village peasants were the actual farmers. Under the direction of an overseer, they produced the crops, raised the meat and draft animals, and paid taxes in services, either forced labor on the lord's lands and other properties or forced military service.

A large manor had a mill for grinding grain, an oven for baking bread, fishponds, orchards, perhaps a winepress or oil press, and herb and vegetable gardens. Bees were kept to produce honey.

Woolen garments were produced from sheep raised on the manor. The wool was spun into yarn, woven into cloth, and then sewn into clothing. Linen textiles could also be produced from flax, which was grown for its oil and fiber.

The food served in a feudal castle or manor house varied—depending on the season and the lord's hunting prowess. Hunting for meat was, indeed, the major nonmilitary work of the lord and his military retainers. The castle residents could also eat domestic ducks, pheasants, pigeons, geese, hens, and partridges; fish, pork, beef, and mutton; and cabbages, turnips, carrots, onions, beans, and peas. Bread, cheese and butter, ale and wine, and apples and pears also appeared on the table. In the south, olives and olive oil might be used, often instead of butter.

Leather was produced from the manor's cattle. Horses and oxen were the beasts of burden; as heavier horses were bred and a new kind of horses was developed, they became more important. A blacksmith, wheelwright, and carpenter made and maintained crude agricultural tools.

The cultivation regime was rigidly prescribed. The arable land was divided into three fields: **one** sown in the autumn in wheat or rye; a **second** sown in the spring in barley, rye, oats, beans, or peas; and the **third** left fallow. The fields were laid out in strips distributed over the three fields, and without hedges or fences to separate one strip from another. Each male peasant head of household was allotted about 30 strips. Helped by his family and a yoke of oxen, he worked under the direction of the lord's officials. When he worked on his own fields, if he had any, he followed village custom that was probably as rigid as the rule of an overseer.

About the 8th century a 4-year cycle of rotation of fallow appeared. The annual plowing routine on 400 ha would be 100 ha plowed in the autumn and 100 in the spring, and 200 ha of fallow plowed in June. These three periods of plowing, over the year, could produce two crops on 200 ha, depending on the weather. Typically, ten or more oxen were hitched to the tongue of the plow, often little more than a forked tree trunk. The oxen were no larger than modern heifers. At harvesttime, all the peasants, including women and children, were expected to work in the fields. After the harvest, the community's animals were let loose on the fields to forage.

Some manors used a strip system. Each strip, with an area of roughly 0.4 ha (about 1 acre), measured about 200 m (about 220 yd) in length and from 1.2 to 5 m

(4 to 16.5 ft) in width. The lord's strips were similar to those of the peasants distributed throughout good and bad field areas. The parish priest might have lands separate from the community fields or strips that he worked himself or that were worked by the peasants.

In all systems, the lord's fields and needs came first, but about three days a week might be left for work on the family strips and garden plots. Wood and peat for fuel were gathered from the commonly held woodlots, and animals were pastured on village meadows. When surpluses of grain, hides, and wool were produced, they were sent to market.

About 1300 a tendency to enclose the common lands and to raise sheep for their wool alone first became apparent. The rise of the textile industry made sheep raising more profitable in England, Flanders, Champagne, Tuscany, Lombardy, and the region of Augsburg in Germany. At the same time, regions about the medieval towns began to specialize in garden produce and dairy products. Independent manorialism was also affected by the wars of 14th- and 15th-century Europe and by the widespread plague outbreaks of the 14th century. Villages were wiped out, and much arable land was abandoned. The remaining peasants were discontented and attempted to improve their conditions.

With the decline in the labor force, only the best land was kept in cultivation, and in southern Italy, for instance, irrigation helped to increase production on the more fertile soils. The emphasis on grain was replaced by diversification, and items requiring more care were produced, such as wine, oil, cheese, butter, and vegetables.

SCIENTIFIC AGRICULTURE

Agriculture: Experimental

By the 16th century, population was increasing in Europe, and agricultural production was again expanding. The nature of agriculture there and in other areas was to change considerably in succeeding centuries. Several reasons can be identified. Europe was cut off from Asia and the Middle East by an extension of Turkish power. New economic theories were being put into practice, directly affecting agriculture. Also, continued wars between England and France, within each of these countries, and in Germany consumed capital and human resources.

A new period of exploration and colonization was undertaken to circumvent Turkey's control of the spice trade, to provide homes for religious refugees, and to provide wealth for European nations convinced that only precious metals constituted wealth.

Colonial agriculture was carried out not only to feed the colonists but also to produce cash crops and to supply food for the home country. This meant cultivation of such crops as sugar, cotton, tobacco, and tea and production of animal products such as wool and hides. From the 15th to the 19th century the slave trade provided needed laborers, replacing natives killed by unaccustomed hard labor in unfavorable climates and substituting for imported Europeans on colonial plantations that required a larger labor force than the colony could provide. Slaves from Africa worked, for instance, in the Caribbean area on sugar plantations and in North America on indigo and cotton plantations. Indians were virtually enslaved in Mexico. Indentured slaves from Europe, and especially from the prisons of England, provided both skilled and unskilled labor to many colonies. Ultimately, however, both slavery and serfdom were substantially wiped out in the 19th century. When encountered by the Spanish conquistadors, the more advanced Indians in the New World had intensive agricultural economies but no draft or riding animals and no wheeled vehicles. Squash, beans, peas, and corn had long since been domesticated. Land was owned by clans and other kinship groups or by ruling tribes that had formed sophisticated governments, but not by individuals or individual families. Several civilizations had risen and fallen in Central and South America by the 16th century. Those met by the Spanish were the Aztec, Inca, and Maya.

The scientific revolution resulting from the Renaissance and the Age of Enlightenment in Europe encouraged experimentation in agriculture as well as in other fields. Trial-and-error efforts in plant breeding produced improved crops, and a few new strains of cattle and sheep were developed. Notable was the Guernsey cow breed, still a heavy milk producer today. Enclosure was greatly speeded up in the 18th century, and individual landowners could determine the disposition of land and of pasture, previously subject to common use. Crop rotation, involving alternation of legumes with grain, was more readily practiced outside the village strip system inherited from the manorial period. In England, where scientific farming was most efficient, enclosure brought about a fundamental reorganization of landownership. From 1660 on, the large landowners had begun to add to their properties, frequently at the expense of small independent fanners. By Victorian times, the agricultural pattern was based on the relationship between the landowner, dependent on rents; the farmer, producer of crops; and the landless laborer, the "hired hand" of American farming lore. Drainage brought more land into cultivation, and, with the Industrial Revolution, farm machinery was introduced.

It is not possible to fix a clear decade or series of events as the start of the agricultural revolution through technology. Among the important advances were the purposeful selective breeding of livestock, begun in the early 1700s, and the spreading of limestone on farm soils in the late 1700s. Mechanical improvements of the traditional wooden plow began in the mid-1600s with small iron points fastened onto the wood with strips of leather. In 1797, Charles Newbold (1764-1835), a blacksmith in Burlington, N.J., introduced the cast-iron moldboard plow. John

Deere, another American blacksmith, further improved the plow in the 1830s and manufactured it in steel. Other notable inventions included the seed drill of the English agriculturist Jethro Tull, developed in the early 1700s and progressively improved for more than a century; the reaper of Cyrus McCormick in 1831; and numerous new horse-drawn threshers, cultivators, grain and grass cutters, rakes, and com shellers. By the late 1800s, steam power was frequently used to replace animal power in drawing plows and in operating threshing machinery.

The demand for food for urban workers and raw materials for industrial plants produced a realignment of world trade. Science and technology developed for industrial purposes were carried over into agriculture, eventually resulting in the agribusinesses of the mid-20th century.

In the 17th and 18th centuries the first systematic attempts were made to study and control pests. Before this time, handpicking and spraying were the usual methods of pest control. In the 19th century, poisons of various types were developed for use in sprays, and biological controls such as predatory insects were also used. Resistant plant varieties were cultivated; this was particularly successful with the European grapevine, in which the grape-bearing stems were grafted onto resistant American rootstocks to defeat the *Phylloxera* aphid.

Improvements in transportation affected agriculture. Roads, canals, and rail lines enabled farmers to obtain needed supplies and to market their produce over a wider area. Food could be protected in transport and shipped more economically than before as a result of rail, ship, and refrigeration developments of the late 19th and early 20th centuries. Efficient use of these developments led to increasing specialization and eventual changes in the location of agricultural suppliers. In the last quarter of the 19th century, for example, Australian and North American suppliers displaced European suppliers of grain in the European market. When grain production proved unprofitable for European farmers, or an area became, more urbanized, specialization in dairying, cheesemaking, and other products was emphasized.

The impetus toward more food production in the era following World War II was a result of a new population explosion. A so-called green revolution, involving selective breeding of traditional crops for high yields, new hybrids, and intensive cultivation methods adapted to the climates and cultural conditions of densely populated countries such as India, temporarily stemmed the pressure for more food. A worldwide shortage of petroleum in the mid-1970s, however, reduced the supplies of nitrogen fertilizer helpful to the success of the new varieties. Simultaneously, erratic weather and natural disasters such as drought and floods reduced crop levels throughout the world. Famine seemed to be imminent in the Indian subcontinent and was common in many parts of Africa south of the Sahara. Economic conditions, particularly uncontrolled inflation, threatened the food supplier and the consumer alike. These problems became the determinants of agricultural change and development.

AGRICULTURAL EXPORTS

In North America, agriculture had progressed further before the coming of the Europeans than is commonly supposed.

Until the 19th century, agriculture in the U.S. shared the history of European and colonial areas and was dependent on European sources for seed, stocks, livestock, and machinery, such as it was. That dependency, especially the difficulty in procuring suitable implements, made American farmers somewhat more innovative. They were aided by the establishment of societies that lobbied for governmental agencies of agriculture; the voluntary cooperation of farmers through associations; and the increasing use of various types of power machinery on the farm. Government policies traditionally encouraged the growth of land settlement. The Homestead Act of 1862 and the resettlement plans of the 1930s were the important legislative acts of the 19th and 20th centuries.

In the 20th century steam, gasoline, diesel, and electric power came into wide use. Chemical fertilizers were manufactured in greatly increased quantities, and soil analysis was widely employed to determine the elements needed by a particular soil to maintain or restore its fertility. The loss of soil by *erosion* was extensively combated by the use of cover crops (quick-growing plants with dense root systems to bind soil), contour plowing (in which the furrow follows the contour of the land and is level, rather than up and down hills that provide channels for runoff water), and strip cropping (sowing strips of dense-rooted plants to serve as water-breaks or windbreaks in fields of plants with loose root systems).

Selective breeding produced improved strains of both farm animals and crop plants. Hybrids of desirable characteristics were developed; especially important for food production was the hybridization of corn in the 1930s. New uses for farm products, by-products, and wastes were discovered. Standards of quality, size, and packing were established for various fruits and vegetables to aid in wholesale marketing. Among the first to be standardized were apples, citrus fruits, celery, berries, and tomatoes. Improvements in storage, processing, and transportation also increased the marketability of farm products. The use of cold-storage warehouses and refrigerated railroad cars was supplemented by the introduction of refrigerated motor trucks, by rapid delivery by airplane, and by the quick-freeze process of preservation, in which farm produce is frozen and packaged the same day that it is picked. Freeze-drying and irradiation have also reached practical application for many perishable foods.

Scientific methods have begun to be applied to pest control, limiting the widespread use of insecticides and fungicides and applying more varied and targeted techniques. New understanding of significant biological control measures and the emphasis on integrated pest management have made possible more effective control of certain kinds of insects.

Chemicals for weed control have become important for a number of crops, in particular cotton and corn. The increasing use of chemicals for the control of insects, diseases, and weeds has brought about additional environmental problems and regulations that make strong demands on the skill of farm operators.

In the 1980s high technology farming, including hybrids for wheat, rice, and other grains, better methods of soil conservation and irrigation, and the growing use of fertilizers has led to the production of more food per capita, not only in the U.S., but in much of the rest of the world. U.S. farmers, however, still have the advantage of superior private and government research facilities to produce and perfect new technologies.

Harvest: Grain

One of the recurring problems of American agriculture in the 20th century has been the tendency of farm income to lag behind increases in the costs of production. The problem began in the 1920s, following a period of exceptional prosperity for U.S. farmers. The period 1910-14 was later taken as a standard for the level of farm prices in relation to the general price level and formed the basis for a concept called parity, aimed at maintaining farming as an essential part of the U.S. economy. After the outbreak of World War I the U.S. became the chief source of food for the warring nations of Europe, with U.S. farmers bringing some 16 million additional ha (about 40 million acres) of land under cultivation and investing heavily in new land and equipment. These measures raised production levels until 1920, when the European demand for U.S. farm products suddenly declined, and prices began a continuing downward spiral.

Although attempts were under way to ease the economic difficulties of the farmer, farm income had not begun to recover when the Great Depression of the 1930s intensified them even more. By 1932 the level of farm prices was only about 65 per cent of the 1910-14 average. Farmers continued to produce almost as much as before, and even increased their production in an attempt to maintain their income. That succeeded only in lowering farm prices further. By comparison, manufacturers could control their production, thereby maintaining price levels to a certain degree. Although prices for industrial goods declined, they did not drop as severely as farm prices, so that by 1932 farmers were receiving only 58 cents from the sale of their products for every dollar they had to pay for nonfarm items.

The federal government, which had done little in the 1920s to help farmers, initiated remedial programs during the administration of President Franklin D. Roosevelt. One approach was to reduce the supply of basic farm commodities. The Agricultural Adjustment Act of 1933 provided payments to farmers in return for agreements to curtail their acreage or their production of wheat, cotton, rice, tobacco, corn, hogs, and dairy products. The act was declared unconstitutional in 1936, but in 1938, after several changes in the membership of the U.S. Supreme Court, a second Agricultural Adjustment Act was passed under which production quotas were set as before. Payments were financed from taxes imposed on processors and were based on the parity concept. The government also lent money to farmers to enable them to withhold crops from the market when prices were low and to store the produce so that it might be available in poor crop years.

A third method to limit production provided payments for shifting acreage of soil-depleting crops such as corn, wheat, cotton, tobacco, and rice to soil-conserving plants such as grasses and legumes and for carrying out soil-building practices. In 1939, an all-risk crop insurance program was initiated for interested fanners to prevent economic distress in case of crop failure for hail, floods, and other natural disasters.

Until World War II the problem of low farm prices was not basically a result of overproduction. Rather, it was a consequence of the cycles of business and weather, and of problems of internal distribution, transportation, and credit. Following World War II, however, overproduction became a serious problem. Both during and immediately after the war, farm prices were generally high. Because production costs also were high, parity payments remained in force. Federal transactions in surplus commodities, principally the sale of such commodities at prices less than those paid to farmers, proved costly for the government. To reduce costs of the federal farm program, the administration of President Dwight D. Eisenhower proposed the substitution of flexible or variable price supports for the rigid 90 per cent of parity that was in force. A bill authorizing a sliding scale of payments at 82.5 per cent to 90 per cent of parity on the basic commodities was enacted by the U.S. Congress in 1954.

The Agricultural Act of 1956, otherwise known as the soil-bank program, authorized federal payments to farmers if they reduced production of certain crops. A subsidy plan was formulated whereby farmers would be paid for converting part of their cropland to soil-conserving uses. In practice, the fanners shared the costs of planting trees or grasses and received annual payments compensating them for the economic loss incurred by the removal of some of their land from production.

The Department of Agriculture in the administrations of Presidents John Kennedy and Lyndon Johnson during the 1960s made control of overproduction a primary goal of farm policy. Farmers were offered what was in effect a rental payment for a part of their land that would be taken out of production during the following year. At the same time, measures were undertaken to expand the export market for agricultural products. During this period the ratio of a farmer's per capita income to that of a nonfarm person increased from about 50 per cent to about 75 per cent.

Direct subsidies for withholding agricultural land from production were phased out in 1973, as a result of a proposal by President Richard M. Nixon. In the same year, net farm income swelled to $33.3 billion.

Poor grain harvests throughout the world, particularly in the Soviet Union, prompted massive sales of U.S. government-owned grain reserves. World climatic conditions also helped keep demand for U.S. produce high through the mid-1970s. Toward the end of the decade, exports lessened, prices dropped, and farm income began to fall without a corresponding decrease in costs of production. U.S. net farm income in 1976 fell to $18.7 billion.

In 1978, a limited, voluntary output restriction was begun by President Jimmy Carter. Called the "farmer-held grain reserve program," the action took grains off the market for up to three years or until market prices reached stated levels. The program was intended also to provide an adequate reserve, lessen food-price gyrations and combat inflation, give livestock producers protection from extremes in feed costs, and contribute to greater continuity in foreign food aid.

On Jan. 4, 1980, President Carter declared a limited suspension of grain sales to the Soviet Union in response to that country's invasion of Afghanistan. Additional restrictions included a prohibition on sales of U.S. phosphate. Despite the grain embargo, the U.S. continued to honor a 5-year agreement already in effect that committed it to sell 8 million tons of grain to the Soviets yearly. The year 1980 was an election year, and despite efforts by President Carter's opposition to void the embargo, it continued. Administration officials argued that the Soviets had never been a major customer or even a reliable buyer. U.S. farmers maintained, however, that the action was at their expense and had made 1980 one of their worst years. In fact, U.S. farm exports in 1980 reached an all-time high of $40 billion, but the continued rise in costs of production and an extremely hot summer with accompanying droughts affected many farmers adversely. A new crop insurance program, passed by Congress in the fall of 1980, offered relief from such conditions rather than having to rely on disaster loans, which amounted to $30 million for feed alone in that year.

Whether the 1980 grain embargo had a strong effect on the USSR was a matter of conjecture. Beef production dropped 16 per cent, pork was off 10 per cent, and milk production fell 4 per cent, but by the end of the year the Soviets had apparently obtained their needed grain from other sources. When President Ronald Reagan took office in 1981, he lifted the embargo and extended the agreement that allowed the USSR to purchase 8 million tons of grain yearly from the U.S. The two nations then signed a new 5-year agreement in 1983 that obliged the Soviet Union to import a minimum of 9 million tons of U.S. grain every year.

Farming Regions.

Hog farming

The U.S. has ten major fanning areas. They vary by soil, slope of land, climate, and distance to market, and in storage and marketing facilities.

The states of the northeast and the Lake states are the country's principal milk-producing areas. Climate and soil there are suited to raising grains and forage for cattle and for pastures. Broiler farming is important to Maine, Delaware, and Maryland. Fruits and vegetables are also important to the region.

The Appalachian region is the major tobacco-producing area of the nation. Peanuts, cattle, and dairy production also are important.

Beef cattle and broilers are the major livestock products farther south in the states of the Southeast; fruit and vegetables and peanuts are also grown. Florida has vast citrus groves and winter vegetable production areas.

In the Delta states, principal cash crops are soybeans and cotton. Rice and sugarcane are grown in the more humid and wet areas. With improved pastures, livestock production has gained importance in recent years. It also is a major broiler-producing region.

The Corn Belt, extending from Ohio through Iowa, has rich soil, good climate, and sufficient rainfall for excellent farming. Corn, beef cattle, hogs, and dairy products are of primary importance. Other feed grains, soybeans, and wheat also are grown.

The northern and southern Plains, extending north and south from Canada to Mexico and from the Corn Belt into the Rocky Mountains, are restricted by low rainfall in the western portion and by cold winters and short growing seasons in the north. But about 60 per cent of the nation's winter and spring wheat grows in the Plains states. Other small grains, grain sorghums, hay, forage crops, and pastures help make cattle important to the region. Cotton is produced in the southern part.

The Mountain states provide yet a different terrain. Vast areas are suited to cattle and sheep. Wheat is important in the north. Irrigation in the valleys provides water for hay, sugar beets, potatoes, fruits, and vegetables.

The Pacific region includes California, Oregon, and Washington plus Alaska and

Hawaii. In the northern mainland, farmers raise wheat, fruit, and potatoes. Dairying, vegetables, and some grain are important to Alaska. Many of the more southerly farmers have large tracts on which they raise vegetables, fruit, and cotton, often under irrigation. Cattle are raised throughout the region. Hawaii grows sugarcane and pineapple as its major crops.

Agricultural Resources

Harvest: Wheat field

The total land area of the U.S. is about 917 million ha (about 2.27 billion acres), of which about 47 per cent is used to produce crops and livestock. The rest is distributed among forestland (29 per cent) and urban, transportation, and other uses (24 per cent).

Approximately 161 million ha (about 399 million acres) make up cropland resources. Almost 83 per cent of cropland is cultivated, including about 23 million ha (about 57 million acres) used for wheat, about 30 million ha (about 74 million acres) used for corn, and about 25 million ha (about 62 million acres) used for hay. More than 50 per cent of croplands are prime farmland, the best land for producing food and fiber.

The nation has another nearly 400 million ha (almost 1 billion acres) of nonfederal rural land currently being used for pastures, range, forest, and other purposes. About 27.5 million ha (about 68 million acres) of this land are suitable for conversion to cropland if needed.

Recent Changes

The history of agriculture in the U.S. since the Great Depression has been one of consolidation and increasing efficiency. From a high of 6.8 million farms in 1935, the total number declined to 2.1 *million* in 1991 on a little less than the same area, about 397 million ha (about 982 million acres). Average farm size in 1935 was about 63 ha (about 155 acres); in 1991 it was about 189 ha (about 467 acres).

About 4.6 million people lived on farms in 1990, based on a new farm definition introduced in 1977 to distinguish between rural residents and people who earned $1000 or more from annual agricultural product sales. The farm population continues to constitute a declining share of the nation's total; about 1 person in every 54, or 1.8 per cent, of the nation's 250 million people were farm residents in 1990.

Total value of land and buildings on U.S. farms in 1990 was $658 billion, substantially less than the value in 1980. Value of products sold was $170 billion per year. Overall net farm income was more than $46 billion in 1989, of which government subsidies accounted for 23 per cent.

Agricultural income in the United States[1]						
Year	1960	1970	1975	1980[2]	1985	1989
Income from livestock and livestock products	18,989	29,563	43,059	68,000	69,800	83,700
Income from crops	15,259	20,976	45,150	71,700	74,200	75,400
Government payments of subsidies and bonuses	702	3,717	807	1,800	7,700	10,900
Nonmoney income: value of housing and products used in farm homes	3,303	3,770	6,689	12,200	11,800	19,200
Gross income[3]	38,894	58,575	100,338	149,300	166,200	189,200
Cost of production	27,376	44,424	75,863	133,100	134,000	142,600
Net income[3]	11,518	14,151	24,475	16,100	32,300	46,700

1. In millions of dollars.
2. Figure since 1980 rounded to nearest hundred million dollars.
3. Includes adjustments to income not otherwise shown.

Not including real estate, major expenditures by farmers in 1989 were for feed ($22.7 billion); fuel, lubricants, and maintenance ($13.1 billion); hired labor ($11.9 billion); fertilizer ($7.6 billion); and seed ($3.7 billion). .

Outstanding farm debt in 1989 was $146 billion, of which about 55 per cent was owed on real estate. Interest payments on the mortgage debt were about $7.6 billion per year.

In 1980 a report based on projections by the U.S. government stated that in the next 20 years world food requirements would increase tremendously, with developed countries requiring most of the increase, and food prices would double. Less than five years later, however, the U.S. farmer was enveloped in a major crisis caused by exceptionally heavy farm debts, mounting farm subsidy costs, and rising surpluses. A number of fanners were forced into foreclosure.

The ailing Farm Credit System, a group of 37 farmer-owned banks under the *Farm Credit Administration* appealed to the government for a $5 to $6 billion fund that would keep the system solvent despite the weak national farm economy. After initial resistance, President Ronald Reagan signed legislation in December 1985 designed to create the Farm Credit System Capital Corp. to take over bad loans from the system's banks and to assume responsibility for foreclosing or restructuring distressed loans.

In December President Reagan also signed the Food Security Act, of 1985, legislation designed to govern the nation's farm policies for the next five years, trim farm subsidies, and stimulate farm exports.

AGRICULTURAL EXPORTS

Harvest: Wheat

The U.S. is the world's principal exporter of agricultural products. In 1989 the value of produce exported was about $39.7 billion, including roughly $1.5 billion in donations and loans to developing nations.

A substantial percentage of the wheat, soybeans, rice, cotton, tobacco, and corn for grain produced in the U.S. is exported. The principal foreign markets for the products are Asia, Western Europe, and Latin America. Japan heads the list of individual countries that import U.S. farm products.

Agriculture

Less than one-sixth of Asia's land is arable. In general the basic unit of production is the village rather than the farm. In South, Southeast, and East Asia agriculture is characterized by small landholdings in alluvial lowlands, too many people on too little land, production largely for subsistence, high rates of tenancy (except in the Communist countries), a heavy dependence on cereals and other food staples, and premodern technologies. Rice is the food-staple crop of South, Southeast, and East Asia. It is usually grown under wet conditions. In South and Southeast Asia, yields are extremely low, controlled irrigation facilities are poorly developed, and double-cropping is seldom practiced. Irrigation schemes in India have helped stabilize annual yields and increase overall production. The example of Japan has shown that small farms and wet-rice agriculture can enormously increase yields and production through the introduction of new high-yielding varieties, careful water management, the application of fertilizers, and the elimination of landlordism.

Despite a wide distribution of new high-yielding varieties of wet rice in many parts of South and Southeast Asia since the late 1960s (the so-called green revolution), production has not risen as hoped. The average rice yields in India, Thailand, and Burma are less than one-half that of Japan. In India, however, high-yielding varieties of wheat, developed in Mexico, have had, in certain areas, an impressive impact on wheat yields, the country's second crop.

Also practiced in the lower latitudes, and in marked contrast with the predominantly subsistence types of agriculture, is large-scale estate agriculture, which produces crops for export, such as rubber, palm oil, coconut products, tea, pineapples, and abaca fiber. Estate production originated during the colonial period in South and Southeast Asia, and many estates remain under foreign ownership and control. Most of the same crops, however, are grown in substantial quantities on small holdings.

In East Asia agriculture is based on wet-paddy cultivation to a latitude of about 35° N in China and about 40° N elsewhere. In contrast to Southeast Asia, yields here are high, double-cropping is common, irrigation is highly controlled, and fertilizer inputs are extremely high, especially in Japan. North of the Huai R. in China, rice gives way to wheat and other dry grains, especially sorghum and corn, all cultivated in a form of intensive horticulture characteristic of Chinese agriculture. Swine, poultry, and fish (in ponds) are raised in both the N and S where possible, but the raising of dairy and beef cattle is common only in Japan and Korea.

In the drier interior regions, some dry-farming of grains is practiced, and the raising of cattle, sheep, and horses is important. Oasis-type agriculture is found in favored locations in W China and in Central Asia. Dry-farming of grains, nomadic herding, and some irrigated oasis-type cultivation are also characteristic of Southwest Asia. For the most part, however, productivity levels are low.

AGRICULTURE, FORESTRY, AND FISHING

Agriculture

About two-thirds of all Indians derive their livelihood directly from agriculture, a proportion that is not much smaller than it was at the beginning of the 20th century. The area cultivated, however, has risen steadily and is now considerably more than half of the nation's total area, a proportion matched by few countries in the world. In the more fertile regions, such as the Indo-Gangetic Plain or the deltas of the eastern coast, the proportion of cultivated to total land often exceeds 90 per cent.

Although the average farm size is only about five acres (two hectares) and is declining, that figure masks the markedly skewed distribution of landholdings. More than half of all farms are less than three acres in size, while much of the remainder is controlled by a small number of relatively affluent peasants and landlords. Most cultivators own farms that provide little more than a bare subsistence for their families. Further, nearly one-third of all agricultural households own no land at all and, along with many submarginal landowners, must work for the larger landholders or must supplement their earnings from some subsidiary occupation, often the one traditionally associated with their caste.

Agricultural technology has been undergoing rapid change in India. Government-sponsored, large-scale canal irrigation projects, begun by the British in the mid-19th century, were greatly extended after independence, but there has been a shift in emphasis toward deep wells (called tube wells in India), often privately owned, from which water is raised either by electric or diesel pumps. Important in several parts of India, especially the southeast, is "tank" irrigation, in which water is drawn from small reservoirs created along the courses of minor streams.

The demand for chemical fertilizers also has been steadily increasing, although since the late 1960s the introduction of new hybrid, high-yielding varieties of seeds (HYVs), mainly for *wheat* and secondarily for *rice,* has brought about the most dramatic increases in production, especially in Punjab (where their adoption is virtually universal), Haryana, western Uttar Pradesh, and Gujarat. So great has been the success of the so-called *Green Revolution* that India was able to build up buffer *grain* stocks sufficient for the country to weather several years of disastrously bad monsoons with virtually no imports or starvation and even managed to become, in some years, a modest net food exporter. During the same period, the production of coarse grains and pulses, which were less in demand than rice and wheat, either did not increase significantly or decreased. Hence, the total per capita grain production has been notably less than that suggested by many protagonists of the Green Revolution, and the threat of major food scarcity has not been eliminated.

Most Indian farms grow little besides food crops, especially cereal grains, and these account for more than three-fifths of the area under cultivation. Foremost among the grains, in terms of both area sown and total yield, is rice, the crop of choice in almost all areas with more than 40 inches (1,000 millimetres) of average annual precipitation, as well as in some irrigated areas. Wheat ranks second in both area sown and total yield, and, because of the use of HYVs, leads all grains in yield per acre. Wheat is grown mainly on the fertile soils of northern and northwestern India in areas with 15 to 40 inches of average annual precipitation, often with supplementary irrigation. Unlike rice, which is mainly grown during the *kharif* (summer) season, wheat is primarily a *rabi* (cool-season) crop. Other important cereals, in descending order of sown acreage, are sorghum (called jowar in India), pearl millet (or bajra), corn (maize), and finger millet (or ragi). All of these typically are grown on relatively infertile soils unsuitable for rice or wheat, while corn cultivation is also favoured in hilly and mountainous regions. After cereals, *pulses* are the most important category of food crop. These ubiquitous, leguminous crops—of which the chick-pea (gram) is the most important variety—are the main source of protein for most Indians for whom the consumption of animal products is an expensive luxury or is proscribed on religious grounds. Nonstaple food crops, eaten in only small amounts by most Indians, include potatoes, onions, various greens, eggplant, okra, squashes, and other vegetables, as well as such fruits as mangoes, bananas, mandarin oranges, papaya, and melons. Sugarcane is widely cultivated, especially in areas near processing mills. Sugar is also obtained by tapping the trunks of toddy palms, which are abundant in southern India, but much of this syrup is fermented, often illegally, to make an alcoholic beverage. A wide variety of crops—mainly peanuts (groundnuts), coconuts, mustard, cottonseed, and rapeseed—are grown as sources of cooking oil. Others, such as the ubiquitous chilies, turmeric, and ginger, are raised to provide condiments or, in the case of pan (betel leaf) and areca (betel nut), digestives. Tea is grown, largely for export, on plantations in Assam, West Bengal, Kerala, and Tamil Nadu, while coffee is grown almost exclusively in southern India, mainly in Karnataka. Tobacco is cultivated chiefly in Gujarat and Andhra Pradesh.

Foremost among the commercial, industrial crops is *cotton.* Maharashtra, Gujarat, and Punjab are the principal cotton-growing states. *Jute,* mainly from West Bengal, Assam, and Bihar, is the second leading natural fibre. Much of it is exported in processed form, largely as burlap. An even coarser fibre is derived from coir, the outer husk of the coconut, the processing of which forms the basis for an important cottage industry in Kerala. Coconuts and oilseeds are also important for the extraction of industrial oils.

Despite the fact that Indians eat little meat, livestock raising plays an important role in the agricultural economy. India has by far the largest bovine population of any country in the world. Cattle and buffalo are used mainly as draft animals but also serve many other purposes—to provide milk, as sources of meat (for those, including Muslims, Christians, and Scheduled Castes, for whom beef eating is not taboo), and as sources of fertilizer, cooking fuel (from dried cow-dung cakes), and leather. Milk yields from Indian *cattle* and *buffaloes* are quite low, although milk from buffaloes is somewhat better and richer on average than from cattle. Because cow slaughter is illegal in many states, scarcely any cattle are raised expressly for providing meat, and most of what little beef is consumed comes from animals that die from natural causes. Cattle that outlive their usefulness, rather than being slaughtered, may be sent to *goshalas* (homes for aged cattle), which are maintained by contributions from devout Hindus or allowed to roam as strays. In either case, they compete with humans for scarce vegetal resources.

While many orthodox Indians are vegetarians, others will eat goat, mutton, poultry, eggs, and fish, all of which are produced in modest quantities. Sheep are raised for both wool and meat. Pork is taboo to members of several faiths, including Muslims and most Hindus, but pigs, which serve as village scavengers, are raised and freely eaten by several Scheduled Castes.

Low-lying, alluvial region in northwestern Uttar Pradesh state, northern India. The Rohilkhand is part of the Upper Ganges Plain and has an area of about 10,000 square miles (25,000 square km). It is bounded by the Ganges River on the south and the west and by the frontiers of China and Nepal on the north. The region is referred to as the Madhyadesh in the Hindu epics the *Mahabharata* and the *Ramayana*. It was later called Hindustan by Muslim historians. The Rohilkhand was settled by the early Aryans and was successively ruled by various Hindu kings before it became part of the territory of the Muslim rulers of Delhi. With the eclipse of Mughal rule, the region was divided by Jat, Rohilla, and Maratha annexations until it passed to the British following the 1857 Indian Mutiny.

Part of the asymmetrical, alluvium-filled Indo-Ganges trough, the plain is broken by the Tarai-Bhabar submontane belt extending eastward from the foot of the Siwalik Hills. The Ramganga, Sukhata, Deona, Sarda, Pilkhara, and Garra rivers follow a northwest-to-southeast course through Rohilkhand. The region was once densely forested, but only a few small patches of forest remain. Agriculture, especially cereals, dominates the regional economy. Large-scale industries (producing textiles, sugar, general-engineering products, and chemicals) are located in urban centres such as Moradabad, Bareilly, Shahjahanpur, and Pilibhit. The region is one of the most densely populated in India and is well served by a network of roads.

Geographic and geologic factors seriously circumscribe the Keralan economy. The amount of arable land is deficient for the needs of the crowded population. The state lacks major deposits of fossil fuels and minerals, except for ilmenite (the principal ore of titanium), rutile (titanium dioxide), and monazite (a mineral consisting of cerium and thorium phosphates), which are found in beach sands. Kerala has great hydroelectric potential, and the Idukki complex is the largest power-generating facility.

The educational system, a developed banking system, and excellent transportion facilities provide optimum conditions for further economic development.

Agriculture is the state's main economic activity. Commercial plantings on less than half of the total land under cultivation earn a sizable amount of foreign exchange but have also necessitated the importation of food for local consumption. Kerala's principal cash yielders are perennial areca nut, cardamom, cashew nut, coconut, coffee, ginger, pepper, rubber, and tea; the major food crops are annual rice, pulses (*e.g.*, peas and beans), sorghum, and tapioca. Commercial poultry farming is well developed. The forests yield valuable timbers such as ebony, rosewood, and teak, as well as industrial raw materials such as bamboo (used in the paper and rayon industries), wood pulp, charcoal, gums, and resins. Foreigners regularly attend the tea and timber auctions held in Cochin. Kerala ranks first among Indian states in fish production.

Most of the population is unaffected by industrialization. Unemployment is acute, and a high level of education among the jobless accentuates the problem.

Traditional low-wage cottage industries, such as the processing of coconut fibre and cashews or weaving, employ most workers. More than one-fourth of Kerala's workers provide services. Food processing is the largest industrial employer. Other products include fertilizers, chemicals, electrical equipment, titanium, aluminum, plywood, ceramics, and synthetic fabrics.

Kerala has well-developed road and railway systems. It is connected with the states of Tamil Nadu and Karnataka by national highways. A railway coming from the east through the Palghat Gap meets with a railway running from north to south through the state and on to Kanniyakumari, the southernmost town of India. There are three major ports—Calicut, Cochin-Ernakulam, and Alleppey—that handle both coastal and foreign traffic. Cochin-Ernakulam also has major shipyard and oil refining facilities and serves as the headquarters for Indian coast guard and navy commands. More than 1,100 miles of inland waterways form main arteries for carrying bulk freight to and from the ports. Trivandrum has an international airport, and Calicut and Cochin have airports for domestic flights.

Union territory of India, comprising two widely separated districts on the country's western coast. Daman, with an area of 28 square miles (72 square kilometres), is an enclave on the state of Gujarat's southern coast, situated 100 miles (160 kilometres) north of Bombay. Diu, an island 15 square miles (40 square kilometres) in area, lies off the southern coast of Gujarat's Kathiawar Peninsula, 40 miles southeast of Veraval.

PHYSICAL AND HUMAN GEOGRAPHY

Daman lies on an alluvial coastal plain, although outcrops of basalt create low plateaus and headlands in the area. The Damanganga River flows through the territory, with Daman town situated where the river enters the Arabian Sea. Mean daily maximum temperatures range from 84 F (29 C) in January to 93 F (34 C) in May. Annual rainfall, received mainly between June and September, averages 81 inches (2,057 millimetres). Diu, whose territory includes a small area of the mainland, is separated from the Kathiawar Peninsula by a narrow swampy creek. Much of the island is covered by sand, silt, and marsh. Temperatures are similar to those of Daman, though rainfall is significantly less (23 inches annually).

The people of Daman and Diu are predominantly Hindu, with small Muslim and Christian minorities. Gujarati is the main language in both districts. One-fifth of Daman's population is tribal. Agriculture and fishing dominate the economies of Daman and Diu. Rice, ragi (also called finger millet), pulses (legumes), and beans are the main crops of Daman; in Diu, however, where only 20 per cent of the land area is cultivated, crops such as bajra (pearl millet) and wheat are more suited to the arid climate. Industrial development remains relatively limited. Daman and Diu, the only towns in their respective districts, are the main commercial and service centres.

Daman and Diu, each organized as an administrative district, together constitute a centrally governed union territory headed by an administrator (the governor of Goa) appointed by the central Indian government.

Island group, Andaman and Nicobar Islands union territory, India, in the southeastern part of the Bay of Bengal. The Andamans constitute one of the two major groups of islands in the union territory. They include more than 300 islands (26 inhabited), with a land area of 2,474 square miles (6,408 square km). The three major islands are North Andaman, Middle Andaman, and South Andaman (closely positioned and collectively known as Great Andaman). Also prominent is Little Andaman, to the south. Of the original inhabitants, only the *Jarawa* and *Onge,* on the islands of South Andaman, Rutland, and Little Andaman, retain a traditional hunting-and-gathering way of life.

The Andamans, situated on the ancient trade route between India and Myanmar (Burma), were visited by Lieutenant Archibald Blair of the Bombay Marine (the East India Company's navy) in 1789. The first European settlement on the islands was at *Port Blair,* situated along the eastern coast of South Andaman. It is now the union territory capital.

The islands are a succession of dome-shaped hill ranges running parallel to each other from north to south. The highest peak is *Saddle,* at 2,418 feet (737 m), on North Andaman. Flat land is scarce and confined to a few valleys, such as the Bitampur and Diglipur. The islands are formed of Tertiary sandstone, limestone, and shale and are highly dissected. Their surface is covered with dense forest, and large mangrove swamps occur in the northern part of North Andaman. Perennial rivers are few, and adequate water supply is a continuing problem.

Agriculture is the principal occupation; crops include cereals, pulses (legumes), coconuts, betel nuts, fruits, cassava, chilies, and turmeric. There is little industry. Only South Andaman has roads; an interisland steamer service connects Port Blair with North, Middle, South, and Little Andaman. Vinayak Damodar (Vir) Savarkar, the great Indian revolutionary, was imprisoned for life (1911-37) in the Cellular Jail (declared a national monument in 1979) in Port Blair. Pop. (1991 prelim.) 240,089.

The wet weather that plagued much of northwestern Europe during the fall and winter of 1998 continued through the spring of 1999 and was followed by a prolonged drought that lasted the entire summer. Despite late planting, many crops matured early, and seed crops, though not as abundant as in some seasons, were of especially high quality, with above-average germination.

Many seed-producing companies continued to expand their operations in China, where favourable climates, inexpensive, well- trained labour, and government support made operations quite profitable for foreign producers. Although the spring was unusually hot in northern China—where most crops were produced—and led to some damage to cool-weather crops like snapdragon, lettuce, and chicory, the season was successful and crops were good.

Seed companies in the U.S. noted significant increases in sales of vegetable seed owing to home gardeners' fears about the impact of the Y2K computer problem on energy- and food-distribution systems. This led to sales increases of more than 20% in some cases and spot shortages of some vegetable seed.

Consolidation also continued in the seed industry. Whereas in previous years ownership changes in the producer and agronomic sectors had been prevalent, 1999 saw retail mail-order nurseries Gurney's and Henry Field's purchased by Foster and Gallagher, which, with holdings that included home garden merchants Breck's of Holland, Spring Hill Nurseries, Michigan Bulb Co., Stark Brothers' Nurseries, and the Vermont Wildflower Farm, claimed to be the largest direct-to-consumer North American distributor of garden products. In a separate transaction, J.W. Jung Seed Co. of Randolph, Wis., purchased R.H. Shumway Seedsman and a host of specialist subsidiaries, including Vermont Bean Seed Co., Seymour's Seeds, Totally Tomatoes, Carolina Seeds, and Horticultural Products and Services.

The European flower-testing organization Fleuroselect did not award a gold medal for outdoor plants for the 2000 season, but it did reward one commercial greenhouse flower with that medal: Delphinium consolida (Annual Larkspur) Sydney Purple, bred by Hamer Bloemzaden of The Netherlands. Fully double, it bloomed in only 8-12 weeks from sowing, at a height of 120 cm (1 cm=0.4 in).

All-America Selections (AAS) made four 2000 awards for vegetables. Hybrid Cabbage Savoy Express was chosen for its earliness, only 55 days from transplant. The half-kilogram (one-pound) heads were suitable for both spring and fall plantings, and at 20 cm high and 15 cm across, they could be spaced only 30 cm apart. Hybrid Pepper Blushing Beauty produced blocky, four-lobed, 10-cm fruits that ripened from ivory to coral and to red and were first ready for harvest about 72 days after transplanting. Plants stood 45 cm, were 40 cm wide, and were resistant to three races of bacterial leaf spot. Resistance to Fusarium wilt and powdery mildew were strong attributes of AAS 2000 winner Mr. Big garden pea. The 1.5-1.8-m (1 m=3.3 ft) plants bore mostly double 11-12-cm pods containing 9-10 peas beginning two months after emergence. The final vegetable award from AAS went to Indian Summer hybrid sweet com. Its 20-cm ears with 16-18 rows of multicoloured kernels matured on 2.1-m plants in about 79 days from sowing. As an sh2 supersweet corn, it had to be isolated from other forms of sweet corn by about 75 m to avoid cross-pollination. The sh2 indicated that the corn had the "shrunken two" gene, which inhibited the normal sweet corn conversion of sugar to starch after harvest, therefore allowing the cobs to remain sweet for 7-14 days after harvest.

Five new flower varieties also won AAS awards for 2000. *Catharanthus roseus* Stardust Orchid was the sole bedding-plant winner. Single orchid blossoms four centimetres in diameter with white centres were held just above glossy green shield-shaped foliage. Mature plants grown in a sunny spot reached 35-40 cm in height and width. *Cosmos sulphureus* Cosmic Orange bore a full covering of golden orange semidouble blooms (5 cm in diameter) on mounding 30-cm plants with moderately divided mid-green foliage. *Hybrid Dianthus* Melody Pink was a new single-flowered annual with serrated petals and stems 55-60 cm long. The first dwarf Mexican sunflower *(Tithonia rotundifolia)* to receive an AAS award was Fiesta del Sol, which grew only 60-90 cm in height and width and had orange daisylike flowers (5-7.6 cm in diameter) with raised yellow centres. Soraya became the first true sunflower to win an AAS award because of its colour—orange petals with a chocolate disc. The 1.5-1.8-m branching plants bore multiple 10-15-cm blooms on long side stems, and those not harvested for bouquets provided seed for garden birds, a trait increasingly rare in new sunflower varieties.

The Perennial Plant Association of the U.S. chose *Scabiosa columbaria* Butterfly Blue as its Plant of the Year. Reliably hardy from U.S.

Department of Agriculture zones 3-9, it grew in full sun to light shade and bore 5-cm lavender flowers on 30-38-cm stems from midspring to early fall if kept deadheaded. The nearly flat basal foliage was gray-green, ovate to lance shaped, and hairy, with its upper foliage more finely divided, forming a mounded rosette 15-20 cm in height and at least 30 cm or more across.

All-America Rose Selections named three winners for the 2000 season. Knock Out was a new hybrid shrub rose developed by William Radler that grew in height and width to 0.9 m. Its 7.6-9-cm cherry red blossoms had five to seven petals that gave off a light tea rose fragrance.

Crimson Bouquet was a hardy disease-resistant grandiflora rose with 10.1-cm-diameter bright red flowers atop 40-45-cm stems on a rounded plant 11 cm in height and 9 cm in width. A coral and cream bicolour, Gemini was a hybrid tea rose with 11-cm flowers, fully double, with a petal count of 25-30 and long stems for cutting. This disease-resistant rose was hybridized by Keith Zary, using Anne Morrow Lindbergh and New Year as the parents.

2

INTRODUCTION TO AGRICULTURE

Introduction

Agriculture, art, science, and industry of managing the growth of plants and animals for human use. In a broad sense agriculture includes cultivation of the soil, growing and harvesting crops, breeding and raising livestock, dairying and forestry (*see* Animal Husbandry; Crop Farming; Dairy Farming; Forestry; Poultry Farming; Soil Management). Regional and national agriculture are covered in more detail in individual continent, country, state, and Canadian provinces articles. Modern agriculture depends heavily on engineering and technology and on the biological and physical sciences. Irrigation, drainge, conservation, and sanitary engineering—each of which is important in successful farming—are some of the fields requiring the specialized knowledge of agricultural engineers. Agricultural chemistry deals with other vital farming concerns, such as the application of fertilizer, insecticides (see Pest Control), and fungicides, soil makeup, analysis of agricultural products, and nutritional needs of farm animals. Plant breeding and genetics contribute immeasurably to farm productivity. Genetics has also made a science of livestock breeding. Hydroponics, a method of soilless gardening in which plants are grown in chemical nutrient solutions, may help meet the need for greater food production as the world's population increases.

The packing, processing, and marketing of agricultural products are closely related activities also influenced by science. Methods of quick-freezing and dehydration have increased the markets for farm products (*see* Food Processing and Preservation; Meat Packing Industry). Mechanization, the outstanding characteristic of late 19th- and 20th- century agriculture, has eased much of the backbreaking toil of the farmer. More significantly, mechanization has enormously increased farm efficiency and productivity (*see* Agricultural Machinery). Animals including horses, oxen, llamas, alpacas, and dogs, however, are still used to cultivate fields, harvest crops, and transport farm products to markets in many parts of the world.

Airplanes and helicopters are employed in agriculture for seeding spraying operations for insect and disease control, transporting perishable products, and fighting forest fires. Radio and television disseminate vital weather reports and other information such as market reports that concern farmers. Computers have become an essential tool for farm management.

II. World Agricultural

Over the 10,000 years since agriculture began to developed, peoples everywhere have discovered the food value of wild plants and animals, and domesticated and bred them. The most important crops and cereals such as wheat, rice, barley, corn, and rye; sugarcane and sugar beets; meat animals such as sheep, cattle, goats, and pigs or swine; poultry such as chickens, ducks, and turkeys; animals products such as milk, cheese, and eggs; and nuts and oils. Fruits, vegetables, and olives are also major foods for people. Feed grains for animals include soybeans, field corn and sorghum. *See also* Grasses; Hay; Gram; Legume; Silage. Agricultural income is also derived from nonfood crops such as rubber, fiber plants, tobacco, and oil seeds used in synthetic chemical compounds, as well as animals raised for pelts. Conditions that determine what is raised in an area include climate, water supply and waterworks, terrain, and ecology.

In the mid-1990s, 48 per cent of the world's labour force was employed in agriculture. The distribution ranged from 61 per cent of the economically active population in Asia to less than 3 per cent in the United States and Canada. In Africa the figure was 60 per cent; in South America, 20 per cent; and in Europe, 9 per cent. Farm size varies widely from region to region. In the 1990s the average for Canadian farms was about 264 hectares (about 654 acres) per farm; for farms in the United States, 190 hectares (469 acres). By contrast, the average size of a single land holding in the Philippines was 2.6 hectares (6.5 acres). Size also depends on the purpose of the farm. Commercial farming, or production for cash, usually takes place on large holdings. The *latifundia* of Latin America are large, privately owned estates worked by tenant labour. Single-crop plantations produce tea, rubber and cocoa. Wheat farms are most efficient when they comprise thousands of hectares and can be worked by teams of people and machines. Australian sheep rations and other livestock farms must be large to provide grazing for thousands of animals. The agricultural plots of Chinese communes and the Co-operative farms held by Peruvian communities are other necessarily large agricultural units as were the collective farms that were owned and operated by state employees in the former Union of Soviet Socialist Republics (USSR).

Individual subsistence farms or small-family mixed-farm operations are decreasing in number in developed countries but are still numerous in the developing countries of Africa and Asia. A back-to-the-land movement in the United States partially reversed the decline of small farms in the 1970s. Nomadic herders range over large areas in sub-Saharan Africa, Afghanistan, and Lapland; and herding is a major part of agriculture in such areas as Mongolia. Much of the foreign exchange earned by a country may be derived from a single agricultural commodity; for example, Sri Lanka depends on tea, Denmark specializes in dairy products, Australia in wool, and New Zealand and Argentina in meat products. In the United States, wheat, corn, and soybeans have become major foreign exchange commodities in recent decades.

The importance of an individual country as an exporter of agricultural products depends on many variables. Among them is the possibility that the country is too little developed industrially to produce manufactured goods in sufficient quantity or technical sophistication. Such agricultural exporters include Ghana, with cocoa, and Myanmar (formerly Burma), with rice. However, a developed country may produce surpluses that are not needed by its own population; this is the case with the United States, Canada, and some western European countries. Because nations depend on agricultural not.only for food but for national income and raw materials for industry as well, trade in agriculture is a constant international concern. It is regulated by international agreements such as the General Agreement on Tariffs and Trade (GATT), the North American Free Trade Agreement (NAFTA), and by trading regions such as the European Community.

The Food and Agricultural Organization (FAO) of the United Nations directs much attention to agricultural trade and policies. According to the FAO, world agricultural production, stimulated by improving technology, reached a record high in the mid-1990s. Further, agricultural output in developing nations increased 54 per cent during the period from 1976 to 1996, as compared to a rise of 13 per cent in developed countries. On a per capita basis, however, food production rose by 38 per cent in developing nations, and actually dropped slightly in developed countries. *See also* Food.

III. History

The history of agriculture may be divided into four broad periods of unequal length, differing widely in date according to region: prehistoric; historic through the Roman period; feudal and scientific.

A. Prehistoric Agriculture

Early farmers were, archaeologists agree, Largely of Neolithic culture. Sites occupied by such people are located in southwestern Asia in what are now Iran, Iraq, Israel, Jordan, Syria and Turkey; in sourtheastern Asia, in what is now Thailand; in Africa, along the Nile River in Egypt; and in Europe, along the Danube River and in Macedonia. Thrace, and Thessaly (historic regions of southeastern Europe). Early centers of agriculture have also been identified in the Huang He (Yellow River) area of China, the Indus River Vally of India and Pakistan; and the Tehuacan Valley of

Mexico, northwest of the Isthmus of Tehuantepec. The dates of domesticated plants and animals very with the regions, but most predate the 6th millennium BC, and the earliest may date from 10,000 BC. Scientists have carried out carbon-14 testing of animal and plant remains and have dated finds of domesticated sheep at 9000 BC in northern Irq; cattle in the 6th millennium BC in northeastern Iran; goats at 8000 BC in central Iran; pigs at 8000 BC in Thailand and 7000 BC in Thessaly; onagers, or asses, at 7000 BC in Iraq; and horses around 4000 BC in central Asia. The llama and alpaca were domesticated in the Andean regions of South America by the middle of the 3rd millennium BC.

According to carbon dating, wheat and barley were domesticated in the Middle East in the 8th millennium BC; millet and rice in China and Southeast Asia by 5500 BC; and squash in Maxico about 8000 BC. Legumes found in Thessaly and Macedonia are dated as easly as 6000 BC. Flax was grown and apparently woven into textiles early in the Neolithic Period. The transition from hunting and food gathering to dependence on food production was gradual, and in a few isolated parts of the world this transition has not yet been accomplished. Crops and domestic meat supplies were augmented by fish and wildfowl as well as by the meat of wild animals. The farmer began, most probably, by noting which of the wild plants were edible or otherwise useful and learned to save the seed and to replant it in cleared land. Lengthy cultivation of the most prolific and hardiest plants yielded stable strains. Herbs of goats and sheep were assembled from captured young wild animals, and those with the most useful traits—such as small horns and high milk production—were bred. The wild auroches was the ancestor of European cattle, and an Asian wild ox of the zebu, was the ancestor of the humped cattle of Asia. Cats, dogs, and chickens were also domesticated very early.

Neolithic farmers, lived in simple dwellings—caves and small houses of sunbacked mud brick or reed and wood. These homes were grouped into small villages or existed as single farmsteads surrounded by fields, sheltering animals and humans in adjacent or joined buildings. In the Neolithic Period, the growth of cities such as Jericho (founded about 9000 BC) was stimulated by the production of surplus crops.

Pastoralism (individual country living) may have been a later development. Evidence indicates that mixed farming, combining, cultivation of crops and stock raising, was the most common Neolithic pattern. Nomadic herders, however, roamed the steppes of Europe and Asia, where the horse and camel were domesticated. The earliest tools of the farmer were made of wood and stone. They included the stone adz, an axlike tool with blades at right angles to the handle, used for woodworking; the sickle or reaping knife with sharpened stone blades, used to gather grain; the digging stick, used to plant seeds and, with later adaptations, as a spade or hoe; and a rudimentary plow, a modified tree branch used to scratch the surface of the soil and prepare it for planting. The plow was later adapted for pulling by oxen. The hilly areas of southwestern Asia and the forests of Europe had enough rain to sustain agriculture, but Egypt depended on the annual floods of the Nile River to replenish soil moisture and fertility. The inhabitants of the Fertile Crescent around the Tigres and Euphrates rivers in the Middle East also depended on annual floods to supply irrigation water. Drainage was necessary to prevent the erosion of land from the hillsides through which the rivers flowed. The fanners who lived in the area near the Huang He developed a system of irrigation and drainage to control the damage caused to their fields in the flood plain of the meandering river. Although Neolithic settlements were more permanent than the camps of hunting peoples, villages had to be moved periodically in some areas when the fields lost their fertility from continuous cropping. This was most necessary in northern Europe, where fields were produced by the slash-and-burn method of clearing. Settlements along the Nile River, however, were more permanent, because the river deposited fertile silt annually.

Organic Farming

Organic Farming, system of agriculture that uses environmentally sound techniques for raising crops and livestock that are free from most synthetic pesticides, growth hormones, and antibiotics. Organic farmers typically rely on pesticides and fertilizers derived from plants, animal wastes and minerals. They incorporate biological methods, such as the use of one organism to suppress another, to help control pests. The methods used in organic farming increase soil fertility, balance insect populations, and reduce air, soil, and water pollution. Organic farming is a small but rapidly growing sector of agriculture in the United States. Sales of organic foods increased from $1 billion in 1990 to $3.5 billion in 1996. Organic food sales now made up 2 to 5 per cent of retail food sales but are projected to increase to 10 per cent by the year 2000. Exports of organic products to Japan and some European countries are also growing, where organic farming typically is more widespread. In Austria, for example, 11 per cent of land is farmed using organic methods.

Organic Farming Techniques

Organic farming combines a variety of methods to maintain the health of soil, prevent soil erosion, and control pests with minimal or no use of synthetic pesticides. Conventional farmers also use some of these methods, but to a lesser degree.

A.Soil Preservation

Fertilizers are used to provide the minerals lacking in some soils, and to replace the minerals removed from the soil by crops as they grow. Many conventional farmers rely on concentrated chemical fertilizers that are rapidly absorbed by plants. These fertilizers produce quick growth, but at the same time may kill important soil organisms, such as earthworms and bacteria. Organic farmers use manure, *compost* (a mixture of decaying organic matter that is rich in beneficial soil microorganisms), and other natural materials for fertilizers that nourish soil organisms, which in turn slowly and steadily make minerals available to plants. Organic farmers are more likely than conventional farmers to rotate crops, a technique that replenishes soil nutrients without the use of synthetic fertilizers. In crop rotation, a field is used for one to several years to grow one type of crop, such as corn or wheat, followed by a season in which a legume such as alfalfa or soybeans is planted. Legume roots harboor beneficial bacteria that incorporate nitrogen from the into the soil (*see* Nitrogen Fixation), enriching the soil and reducing the need for nitrogen-containing fertilizers. Crop rotation also conserves nutrients since the roots of the first crop may be near the surface and the second crop's roots may be deeper, so that nutrients are drawn from different depths in the soil. Soil held in place by plant roots is less likely to blow or wash away, or erode, than bare soil. Organic farmers minimize soil erosion with cover crops—short—lived plants, often grasses or legumes— that protect the soil between the harvesting of one crop and the planting of the next. Many organic farmers also conserve soil by practicing no-till or low-till farming, avoiding the use of plows to turn the soil, or using implements that only slice or slightly turn the soil. They may also leave the unharvested portion of a crop in the field to cover the soil, preventing soil erosion from wind or rain.

B. Pest Management

Conventional farms rely on an array of synthetic pesticides to kill weeds, disease-causing fungi and harmful insects. These pesticides are manufactured by chemically processing petroleum, natural gas, ammonia, and a number of other raw materials. They include active and inactive ingredients, both of which can be highly toxic and long lasting. Organic farmers typically use pesticides primarily derived from chemically unaltered plant, animal, or mineral substances in which the toxic active ingredient breaks down rapidly to become nontoxic after being applied to the crop. Pyrethrum, a substance extracted from chrysanthemums, a variety of soaps, and oil from the neem tree are among the insecticides used by organic farmers. Bordeaux mix, a combination of calcium carbonate and copper, is used by organic farmers to control disease-causing fungi. In addition to using natural pesticides, organic farmers use a variety of methods to control insects and disease-causing fungi. In a technique called intercropping farmers plant different crops in wide alternating bands. This interrupts the movement of disease-causing organisms through a field, since many insects and fungi feed on just one type of crop. Organic farmers also reduce insect damage by spraying crops with bacteria that kill *larvae* (immature insects) and planting crops that attract ladybugs, lacewings, and other beneficial insects that prey on unwanted insects. Organic farmers use many methods to control weeds. Mulching involves covering the soil around crops with straw or other materials that smother weeds.

Cover crops can be planted in the fall and turned under in a few months; they help control weeds by competing with them—an oat crop, for example, grows faster than weeds and deprives them of the nutrients they need to produce seeds. Other types of cover crops, such as cereal rye, release substances from their roots that inhibit weed seed germination.

Organic farmers sometimes use a variety of tractor-drawn equipment to uproot weeds that emerge with crops *(see* Weed Control).

Organic farming is sometimes referred to as sustainable agriculture, although the two concepts have subtle significant differences. Sustainable agriculture seeks to improve the entire food and agricultural system by balancing production and consumption. For example, a farmer practicing sustainable agriculture may use of manure from the animals to fertilize the fields of grain that are grown to feed the animals. Eliminating the purchase of fertilizer reduces the cost of growing grain, and growing grain for animal feed rather than buying it reduces the cost of raising livestock.

Sustainable agriculture also addresses the environmental, economic, and social issues related to agricultural systems. It attempts to ensure that arable land is protected so that current and future generations will be able to farm from it successfully; many involved in sustainable agriculture also seek to preserve the vitality of family-owned farms and rural communities. A sustainable farm may not be organic, and an organic farm may not be sustainable, although they may use similar techniques.

Benefits of Organic Farming

For consumers, the most obvious benefit of organic farming is health-related—the food produced has little or no pesticide residue. Some advocates of organic farming believe that organic food is more nutritious than food produced by conventional farming, although no valid studies support this claim. Organic farming, however, has less obvious, longer-term benefits. Because it preserves and enhances topsoil, it increases the chances that future generations can continue growing food. It helps preserve aquatic life by minimizing the flow of toxic pesticides into streams, rivers, and lakes. And it encourages healthy populations of beneficial insects that keep destructive insects under control. Opponents of organic farming argue that organic farming is less profitable, requiring more labour and management skill than a conventional farm. Savings on pesticides, fertilizers, and fuels, however, usually offset the cost of the extra labour. And the environmental benefits of organic farming represent long-term savings, not just for the organic farmer, but also for future generations.

Organic Farming Standards

In the past, the practices organic farmers had to follow to label their food organic were regulated haphazadly by a wide variety of state and private certification programmes. The standards set by these certification programmes were far from uniform. To correct this problem, the United States Department of Agriculture (USDA) established the 1990 Federal Organic Foods Production Act scheduled for implementation in 1998. This act mandates that all food labeled organic be produced according to the act's guidelines which dictate, among other issues, what is required for a farm to be certified as organic and what chemicals can be used to organic products. These guidelines were still under negotiation in early 1998. One of the first proponents of organic fanning was the English agriculturalist Sir Albert Howard, who, in his 1940 *book An Agricultural Testament,* advocated fanning without synthetic fertilizers and pesticides. Lady Eve Balfour was also involved in the 20th- century organic farming movement. Her 30-year research farm, the Haughley Experiment, was the site of numerous experiments comparing organic and conventional farming. Balfour's book, *The Living Soil* (1943), corroborated Howard's studies and documented the importance of·healthy soil for farming. The work of Howard and Balfour inspired J. I. Rodale to found *Organic Farming and Gardening* magazine in 1942 (now called *Organic Gardening),* which educated the public about organic techniques. Rodale also also established the nonprofit and Health Foundation research center (now called the Rodale Institute). Rachel Carson, a marine biologist with the United States Fish and Wildlife Service, added momentum to the organic farming movement with her book *Silent Spring* (1962), which chronicles the harmful effects of pesticides on wildlife. Also in the United States, Helen and Scott Nearing pioneered in organic farming. Their book *Living the Good Life* and their numerous other publications promoted organic farming and helped inspire the back-to-the-land movement of the 1960s and 1970s.

3

ECONOMIC DEVELOPMENT AND STRUCTURAL CHANGE IN AGRICULTURE

When economics of underdeveloped countries begin to get geared to the process of economic development, they pass through a period of stresses and strains. Chief among these is the stress on agriculture which is fully extended and is called upon to increase the quantity and improve the quality of its output. Over and above the normal requirements of food crops for direct consumption by all and non-foods crops for the use of such industries as already exist, agricultural now has to meet many additional needs. Economic development of underdeveloped economies means rapid industrialization and this implies larger employment for the urban-industrial population and larger incomes and demands in the urban-industrial sector. To meet these enhanced urban-industrial demands, a larger production and marketing of food and raw materials from the agricultural sector becomes necessary. It is true that as people's incomes rise their food consumption does not rise in the same proportion, as the well known Engel's law points out. But this assumes that there are no empty stomachs and no hunger or starvation. Until that happy stage is reached, demand for food could rise very rapidly. If people further try to improve the *quality* of their food—as, no doubt, they would—it is possible that money demand might rise nearly as fast is money incomes. The income-elasticity of demand for food, thus, could be approaching unity. Under such circumstances agriculture is called upon the perform a heroic feat and produce and market much more than before. Failure to do this will create inflationary pressures and the ramifications of this phenomenon are manifold, not to mention that fact that the process of industrialization and development itself cold peter out into stagnation.

Economic growth and industrialization also involve a shift of workers from the rural to the urban-industrial sector. But it is well known that there is no automatic mechanism for transferring to urban-industrial areas the food previously consumed in rural areas by those who have now shifted. The market mechanism is no doubt there and one might be tempted to think that enhanced incomes and demands in urban areas, by raising food prices, might attract more supplies and relieve the strain substantially. But this assumes a well-behaved normal supply curve of agriculture (where by supply, we mean marketed supply). There is, however, some reason to believe on the contrary that the supply curve of agriculture, at least among important categories of farmers, is of a backward-sloping nature. Since in inflationary times the prices of agricultural commodities generally faster than those of manufactures, the terms of trade turn in favour of agriculture, and in order to obtain the same supplies of industrial goods—cloth, kerosene, iron and steel, etc.—the peasant needs to market less of his produce than before. Marketability of agricultural produce thus falls off precisely at the time when the demand has risen in the urban industrial sector and this makes matters worse and the inflationary impact greater. In view of such backward-sloping supply curves, it follows that if inflation is to be contained and industrization not to be thwarted, it is necessary to start with an even higher rate of expansion of agricultural output so that the terms of trade may not turn against the industrial-urban sector and agricultural supplies to this sector may become available in sufficient quantities.

In inflationary situations, expectations also play their part and once the trend of rising agricultural prices gets established, the hoarding propensities of farmers, wholesale traders and retail traders come into free play and result in further withholding of produce from the final consumer market. This can cause further inflation, at least in the short run, and hamper development.

Inflation in the prices of food and raw-material, in a poor-country, could bring more troubles than most of us think. Starting with the factors mentioned above it can create political and social unrest among important groups in society. It can raise the cost of living, stir up the demands of trade unions, cause industrial unrest and a further wage-induced inflation. It certainly makes planning more expensive and requires an expenditure of larger sums of money to achieve the same physical targets. By turning the terms of trade against the industrial-urban sector, it tends to dry up the sources of savings in this sector and makes it more difficult to achieve the targets of the development plans. If an attempt is made to create the required resources through deficit, financing, this leads to a further budget-induced inflation. If the country tries to meet the food and raw-material needs through imports from abroad this causes trouble on the balance of payments front. Undoubtedly such a situation is charged with a good deal of destructive potential.

India's Economic Development and Agricultural Productivity

The factors mentioned above appear to have exerted themselves considerably during the current Indian experience of planned economic development. Investment activity both in the public and the private sectors of the economy has enhanced employment and money incomes. The demand for food has risen considerably both quantitatively and qualitatively. Population has also been rising at a rate faster than was thought probable in 1951. Substantial shifts in population from the rural to the urban sector have occurred. The price mechanism normally might have effected an increase in supplies to match enhanced demands in urban areas, but the relative inelasticity of supply of agricultural produce seems to have exerted itself. The increases in the output of foodgrains as well as of non-food-grains have been precarious and dubious during the decade, 1949-50 and 1958-59, and failed by a large margin to match the demands. Production apart, in some cases, though not all, the backward-sloping curves of marked supply seem to have harmpered increased marketability, and the result has been a significant inflation of food prices. In the process, hoarding activity, too, seems to have proceeded merrily on, making matters worse.

Thus it is becoming abundantly clear day by day that agriculture in this country so far has been unable to cope up with the demands of a planned programme of industrial development. There are little signs yet in the Indian agricultural sector of the establishment of a clear and definite upward trend in production. What is called the record output of 73.5 million tons of foodgrains during 1958-59 represents, after all, a less than 8 per cent increase over the output recorded as early as 1953-54; but meanwhile, population appears to have increased by about 9 to 10 per cent since 1953-54. We thus have no reason to be happy over the behaviour of agricultural output, as our *per capita* production of foodgrains might in fact have declined somewhat. In the absence of such a trend of improvement, we may succeed, partly and for some time, by resorting to expedients like fair-price shops, zoning schemes, control of wholesale trade, procurement, state trading in foodgrain etc., all of which aim at reducing hoarding and other antisocial practices. But none of these expedients, even when operated efficiently, can take us very far it output of agricultural itself not rising as fast as demand.

In the absence of output increases, a policy, say, of state trading in foodgrains is doomed to failure. Even if everything else goes on smoothly, low production must mean low procurement whether it be directly from the farmer's yards of out of arrivals at the market. Then the government's stocks cannot be built up and all else follows. Price stabilization techniques fail, balance between surplus and deficit zones or states cannot be established, terms of trade continue to become unfavourable to the urban-industrial sector, resources continue to dwindle and all the difficulties mentioned earlier (and more) continue to play havoc with the objectives of planned economic development. All that remains then in the armoury of economic weapons is a rigid system of levy, price control, rationing and allocation. But that ends up, as the experience of the war years has shown, in blockmarketing, profiterring, corruption, bureaucracy, and a serious misdirection of resources and leads the economy to the brink of authoritarianism and regimentation.

Even that does not help as the basic shortage in production remains and keeps on exerting pressures on prices, honesty and morale. "The best thing to do to a shortage is to abolish it" and no agricultural policy or expedient which does not care for increased production, in the first place, can hope to achieve anything worthwhile.

CASH AND KIND INCOMES

(Jagir Areas of Hyderabad)

Occupation	*Percentage of cash income to total income*		*Percentage kind income to total income*	
	1948-49	*1953-54*	*1948-49*	*1953-54*
1	2	3	4	5
Cultivation	37.8	39.3	62.2	60.7
Labour	42.4	52.6	57.6	48.0
Profession	63.0	63.4	37.0	36.6
Trade	100.0	100.0	0.0	0.0
Property	46.6	44.9	53.4	55.1
All Occupations	43.9	47.6	56.1	52.4

The Agricultural Economics Research Section of the Delhi School of Economics has some pieces of evidence *(enquiry into the State of Wheat Arrivals in Some Selected Markets of UP and Punjab)* to the effect that at a time when agricultural prices were rising faster than the non-agricultural prices, the marketability of agricultural produce declined. It, is however possible that this decline has been the consequence, not of relative prices (terms of trade) but of expectation of future price rises by farmers. Piecing together some of these evidences, it seems reasonable to say that the secular trend in the economy has been towards marketing (monetization) of a larger share of agricultural produce, but that this secular trend is accentuated further when Agriculture has adverse terms of trade and is retarded when it has favourable terms of trade.

The real clue men, is obviously in increased and continually increasing production, since we have a programme of continually increasing industrialization for years and for decades to come. To be sure, we aim quite high. We fix our targets in agriculture for a 30 or 35 per cent increase in output over a period of five years. We talk of the possibility of having a production of 110 million tons of foodgrains by 1966 and of a doubling of output within a brief span of time. But it seems idle to think that such achievements will come about with a minor tinkering of our techniques of production and without any fundamental change in the organizational pattern and an overhaul of the agrarian institutional frame-work. Agriculture in most countries and at most times has been known to be an occupation much given to conservatism, much dominated by inertia and much averse to change. And yet it is this sector that has to rise and be in the vanguard of industrial development. One of the most competent and experienced economists of our times put the matter sharply when be said, "I define an industrial country as one with a highly developed agriculture", and it does seem that if Indian agriculture is to fulfil the hopes that we attach to it, the shake-up and the enlivenment in this sector will have to be bigger than what most of us think. It will have to be an institutional and structural as well as a technical change, a change so vast and fundamental that one can almost call it a revolutionary change.

Tenure and Size Disincentives

When the problem of raising agricultural productivity and production by very substantial margins is posed squarely, a section of enlightened opinion invites our attention to better seeds and more manures and fertilizers and to irrigation and to new techniques of cultivation. Now, the methods of raising agricultural productivity are no secret to this country, for one major advantage of coming late in the field of economic development is that this country can borrow a large part of the already known techniques of efficient agricultural operations and is only left with the problem of adaptation of these techniques in its own local circumstances and of organizational changes that follow from such adoption. Thus we know the techniques and can arrange through a concerted effort, in local production and imports, for the provision of some of the requirements of agriculture— seeds, fertilizer, water, implements, etc. But how can the planning or the administrative agency provide *all* that is required in the agriculture of a country of huge dimensions and with a vast multitude of farmers arranged in some 40 million farming units? And what is the guarantee that what is in fact provided will be utilized by the farmers? We have had in this country the sad spectacle of 6½ million acres of land potentially brought under irrigation during the era of planning but only about 4 million acres actually utilized by peasants for cultivation with the use of the new source of water. The Community Development Projects have been insisting for eight years now, and the State Departments of Agriculture for much longer, on the use of better seeds and fertilizers. But the targets for fertilizer use fixed in the Second Five Year Plan have remained grossly under-fulfilled in practically every State of the country without exception. The various agricultural research stations working under the auspices of the State Departments of Agriculture have been experimenting with an publishing the most desirable strains of seeds in given conditions, the tools and implements which are likely to be most effective and the techniques which can lift up productivity very substantially—ranging from the Japanese method of paddy cultivation to crop protection devices, from the use of mowers and threshers to the most judicious combination of manufactured and local fertilizers. But judging both by what the authorities tell us about their popularity and demand and by the effect on the final product, the use of improved methods and techniques has been only marginal and insignificant. The rate of capital formation in agriculture has been extremely slow and despite the availability of surplus labour in most places the long list of agricultural processes— fencing, well sinking, manuring, funding, etc.—looks more like a catalogue of what has not been done and has yet to be undertaken by Indian farmers.

Why do Indian farmers, when they know many or all these processes, not apply them in their productive work? Why do targets of irrigation, better seeds, fertilizers and manures and almost everything else remain vastly underfulfilled? The reason cannot be ignorance, for these methods are widely known or are increasingly being made known through various agencies. Nor can it be mainly the lack of credit facilities, for apart from what facilities there do exist, many of these processes do not require credit from outside: they can be undertaken by family labour or with the use of but a little hired labour. The most important clue to the phenomenon of inaction and non-achievement is to be found in the absence or smallness of incentive. Disincentive is the basis of this inertia. A detailed study of the structure of the agrarian economy a we hope to show in subsequent chapters—reveals that this structure in general; and that of certain sub-sectors within the agricultural economy in particular, has become disincentive-ridden to an alarming degree and is not capable of supporting in the present form a progressive improvement. To start with, the sheer pressure of human numbers on land has grown rapidly and the land-man ratio has become so unfavourable that it now stands at less than one acre per person when all agricultural land is spread over the total population and at about 1.5 acres per person when it is spread over the cultivating population. An average family holding sustaining a family of five persons has dwindled down to a size of about 7½ acres. Moreover, the size distribution of holdings shows serious lopsidedness and nearly two-fifths the cultivating families appear to be living on a holding of less than 2½ acres each (i.e. less than ½ acre per person). The last straw is the fragmentary nature of even these pocket-sized holdings—a holding being scattered into three, four, five or more plots. The under-utilization of labour force and of bullock power and implements is consequently enormous and a great deal of wastage arises, making costs of cultivation higher than what they would otherwise be. A high-cost production structure (when the value of family labour is included) relative to the small output which is largely or wholly consumed away, leave little room for saving and capital formation in the conventional sense. Therefore, insofar as technical change itself is closely related to capital formation, techniques remain primitive and unproductive.

Yet another serious structural defect and a serious source of disincentive is the tenurial system. Large chunks of land had historically passed into the hands of absentee administrators (the jagirdars who collected land revenue) and absentee owners (the Zamindars who collected rents from the cultivators). Surpluses accrued in their hands and as this class of landlords, like its counterpart in many other countries, has been given to conspicuous consumption, these surpluses never had a chance of being saved and ploughed-back into agriculture. Land, in other words, got neglected from the side of owners and administrators. On the other hand, the cultivators, who operated but did not own the land, suffered from serious insecurity of tenure. They could be evicted by the land-owners at any time the latter desired and this uncertainty prevented them from taking a long-term view of the situation and investing in land even if they had some surplus. They neglected most of the durable operations—levelling, bunding, drainage, irrigation, fencing etc.—and lived from hand to mouth, getting what they could in the very short run. There was little investment, then, from this source either and the net result has been declining productivity. Insofar as pressure of population and the excessive demand for land in relation to supply has created a high rent structure, the surplus in the hands of tenant-cultivators dwindled down further. Incentive and capital formation is less for this reason as well.

Fragments Per Holding and Per Acre

.Uttar Pradesh			West Bengal		
Size group (ares)	Numbers of Fragments		Size group (acres)	Number of fragments	
	Per farm	Per acre		Per farm	Per acre
under 2.5	3.60	2.02	0.01 — 1.25	3.6	5.3
2.5—5.0	6.26	1.67	1.26 — 2.50	7.1	3.9
5.0—7.5	8.94	1.44	2.51 — 3.75	9.4	3.0
7.5—10.0	1.90	1.34	3.76 — 5.00	12.1	2.8
10.0—15.0	11.76	1.02	5.01 — 7.50	14.3	2.4
15.0—20.0	16.10	0.97	7.51 — 10.00	22.3	2.6
20.0—25.0	22.89	1.02	10.01 — 15.00	17.3	1.4
above 25	24.85	0.60	above 15.00	35.7	1.4

Source: Farm Management Studies.

The second set of disincentives which relates to the tenurial system is, of course, not of universal application; they relate only to the zamindari and Jagirdari tenures and to such ryotwari lands on which tenancy had arisen and cultivation had got separated from ownership. But all this is a relatively small part of the agricultural area. The country remains by and large a country of owner-cultivators. But even these suffer serious disincentives owing to the first set of causes, that is, those associated with an unfavourable land-man ratio, a lopsided size-distribution of holdings and unbridled subdivision and fragmentation.

Now, when the structure in these vast sub-sectors of the agricultural economy is so leden with disincentive and so inimical to capital formation, it seems naive and purposeless to talk merely of seeds, manures, irrigation and implements—at best land management— to the neglect of structural alteration and creation of incentive in place of disincentive. Seeds, fertilizers, etc. the wherewithal of agriculture, are no doubt absolutely essential for increased productivity: without these there can be little advance; but few farmers would be willing to use the improved seeds and fertilizers and new tools and techniques so long as the structure remains what it is— degenerate and disincentive—ridden. We now see why it is that irrigation potential has not been utilized, fertilizer use has lagged so much behind targets, better seeds, new implements and more productive techniques have not been accepted by any considerable section of farmers and the efforts of the various Departments and Ministries— Food and Agriculture, Community Development, Irrigation and Power etc.— have not come to much of a fruition. The truth is that even like the proverbial horse, you can bring the farmer to the pool of water, fertilizers and seed, etc. but you cannot make him drink. You cannot dump from above the numerous inputs of agriculture on 40 million farmers; you have to wait till they themselves desire to use these inputs. And they will desire to do so and in fact become go-getters of these inputs if only the structure were altered and incentives created.

The programme of rapid advances in agricultural production, then, appears to consist of two major planks: structural change for the sake of incentive and provision of the various inputs of agricultural in sufficient quantities. Production cannot rise significantly if any one of these two planks is neglected.

The structural defect in Indian agriculture was shown above to be due to *(a)* the decay of the tenurial system and its concomitants, and *(b)* the decline in the size and the distribution of the cultivation units and the fragmentation thereof. The first is obviously to be reminded through a serious programme of land reforms and it is not our intention to discuss it elaborately here. The second is our primary concern in the present work and our discussion of Co-operative farming must have this consideration throughout as its background. Our references to co-operative farming, as the reader will no doubt presently see, and our advocacy of this system of farming have nothing to do with ideological predilections. It is the economic argument that interests us and we seek an answer to the problem of nearly two-thirds of the cultivation units in the country sinking to a size of 2½ acres and below, with all the resultant disincentives, diseconomies, low consumption and low supluses. No talk of the 'Socialist Pattern of Society', or for that matter any ideological programme seems worth the name until one can show what would be the solution to this problem of uneconomic and non-surplus-generating operation units and how the lower strata of the peasantry would be rehabilitated and made capable of producing an output sufficient for their own legitimate requirements, and marketable surplus for national economic development.

Some Basic Questions and Assumptions

When we think of a suitable programme of agricultural reorganization in the context of the urgent needs of industrialization and economic development, the fundamental questions we have to ask are as follows:

(a) How can we plan and re-organize our agriculture in order to achieve a vastly greater productivity and production?

(b) Which mode of organization would enable us to expand rapidly our gross agricultural area? This involves bringing under the plough all the cultivable land hitherto not cultivated and converting into double cropping all the land that *could be* converted but is now under a single crop.

(c) How best is it possible to utilize effectively the existing labour force in agriculture which is known to be seriously underemployed and marginally quite unproductive?

(d) How is it possible—and this is crucial—to bring about an upsurge of capital construction in the agricultural sector and give impetus in that sector to the numerous investment projects which are possible but have not so far been undertaken? If this could be done it will have the multiple effect of increasing the yield of land, and the productivity of labour, at the same time providing increasing and effective employment.

How can the institutional set-up be so changed as to minimize and eliminate the hoarding of agricultural produce so that a larger marketed surplus becomes available?

Any organization or structural change that answers all these questions satisfactorily and does not quarrel with the fundamentals of our political set-up, should be acceptable to us. We thus have as our constant background throughout the present work, the assumption of a parliamentary democracy, that is, a democracy which starts by asserting the political of all citizens through universal adult suffrage and guarantee of fundamental human rights, but is anxious to give itself quickly a solid economic foundation by asserting on improved levels and standards of living and an equitable distribution of income, wealth and economic opportunities.

We are thus advocating a structural change within a democratic framework and we hope to show in the following pages that the co-operative mode of farm organization answers our requirements better than any other alternative in the poverty-ridden sub-sector of our agricultural economy.

One or two major misconceptions and misunderstandings which crop up in any discussion on co-operative farming may be got out of the way at the very outset. The farm co-operation which is discussed here is an entirely voluntary organization. Force is to be kept out and not totalitarian devices but incentives are to be used in a rapid expansion of the co-operative sector. The decision to join a co-operative farm should be a voluntary decision and should come entirely from the co-operating farmers. In the subsequent working of co-operatives too the Government should have little to do. The role of the Government should merely be to provide the external economies, so to speak, and the right conditions for an appropriate farm organization. But there should be no interference in any matter internal to a farm — no imposition of managers or of farm decisions from a bureaucracy. Our co-operatives then are to be local products, indigenous to the Indian soil and in their voluntariness and freedoms, they are to be contrasted sharply from the collectives and cooperatives of the Soviet Union, Eastern Europe, China and some other countries. (The difference will become clear as the argument proceeds.) The following pages, then, are an attempt to define and describe the mode of operation of Co-operative farming, to compare it with alternative systems and then to discuss the theoretical feasibility as well as the organizational possibilities of this system and finally making what seems to us to be positive suggestions with strict regard to Indian conditions.

In the modern world a number of alternative patterns of agricultural organization have been in vogue at different times in different countries with varying degrees of success. Subjecting ourselves to some simplification, we can categorize these patterns under the following five heads: (a) small peasant farming, *(b)* estate or capitalist farming, *(b)* estate or capitalist farming, (c) state farming, *(d)* collective farming, and *(e)* co-operative farming. Any elaborate treatment of these patterns is neither possible in this study, nor is necessary in view of the fact that these have been dealt with in great detail in various publications on the subjects. A mere reference to their essentials should hare be sufficient to serve our purpose.

Peasant Farming

Peasant farming stands for small-scale private enterprise as applied to agriculture. Here the tiller is also the proprietor who possesses permanent, inheritable and transferable rights in land. Assisted by the members of his family, he manages the farm according to his own free will, plans and produces the crop and retains or sells a parlor whole of this crop without any outside interference. The degree of mechanization generally noticed on peasant farms is negligible or small as the unit of cultivation is small and does not warrant the use of machines. A considerable proportion of the produce is generally for the peasant's own consumption although in some cases, where peasants take to growing cash crops, the whole produce may be sent to the market and, with the cash obtained, food and other requirements might be purchased. In the earlier stages the ryotwari system of India was a near approximation to peasant farming but later, as intermediaries multiplied, this system, in large tracts, degenerated into tenant farming and the difference between it and the landlord-tenant farming, called by various names in the country like Zamindari, Jagirdari, etc. remained one of degree only and not of kind.

Capitalist or Eastate Farming

In what is known as capitalist or estate or corporate farming, land is held in large areas by private capitalists, corporations, or syndicates and is managed like a modern enterprise. Capital is supplied by one or a few persons or by many, in which case the farm runs like a joint-stock company. The unit of organization is large and work in carried on with hired labour and with extensive use of mechanization. Barring a small proportion for seeds and feed of such animals as are retained, almost the entire crop is generally sent to the market. Examples of this system of farming are frequent in the United States of America and in Western Europe; and in India, too, plantations of coffee, tea, rubber and sugarcane are found organized on this basis.

State Farming

State farming, as is evident from its very title, is a case of application of state enterprise to agriculture. Here land is owned and managed by the State which also performs the function of decision-making and risk-bearing, while cultivation is carried on under the direction of State-appointed managers by hired servants in accordance with the plans drawn up by the State. As in estate farming, here too, cultivators have the status of mere wage-earners. Though State-managed farms are to be found in several countries—and there are some in India as well—yet this system of farming typically exists in the USSR, where the experiment has been attempted on a large scale.

Collective Farming

As a pattern of agricultural organization, collective farming has come into much prominence in recent times, and has been adopted by some countries notably by the USSR, Palestine and Mexico. Though the collective farms of different countries (Kolkhoz of the USSR, Ejhido of Mexico and Kibbutz of Israel) reveal considerable differences in their formation, organization, working, etc. yet collective farming clearly has certain distinguishing features. We may take note of these under the three broad aspects of ownership, production and distribution.

In a system of collective farming, the ownership of the land, stock and capital vests in the community as a whole, and generally there are no individual holdings and no private properly in land. A partial exception is, however, made in the case of some collective farms like Kolkhoz and Ejido where individual members are allowed to own some limited livestock and use small strips of land for production on their own account. Apart from this small part of its activity, for purposes of production and management, the whole farm is treated as one unit and naturally the scale of operations is large and the degree of mechanization often high. Work is carried on jointly by the members themselves with little use of outside labour which is resorted to only in special circumstances. Appointment of managers from outside the farm by a government or a bureaucracy is quite common but generally members elect from among themselves a board of management which plans and administers all farm operations. This board or the appointed managers may lay down the pattern of cropping, decide programmes of daily work, direct and supervise the performance of all agricultural operations, organize credit, marketing, supply, etc. and also make arrangements for social services such as education, medical aid and recreation. The members of the collective farms are usually divided into groups or 'work-brigades' which are allotted specific jobs by the management. Each brigade is directed by a foreman who like his counterpart in a capitalist farm, checks up the work done by the members of his brigade, with the only difference that whereas he is elected by the workers from among themselves, the supervisor in a capitalist farm possesses external authority. Methods of distribution as adopted by different collective farms vary considerably. But generally speaking, the share of a member is determined on the basis of the number of 'labour days' contributed by him. These are reckoned not according to the number of actual days for which a member has worked, but according to the amount of work put in by him. This measure is both quantitative and qualitative. Thus an unskilled labourer will require more hours than a skilled one to earn one 'labour-day'. In the case of certain collective farms, to stimulate interest, provision is made for special rewards or bonuses for extra efficiency shown over and above the standards fixed for the particular work. It will thus be seen that, although in collective farming means of production are collectively owned and exploited, the motive of an economic regard in proportion to the work done by individual members is sought to be retained. In other words, while production is collectively undertaken, distribution and consumption are on an individual basis.

Kibbutz is, however, a more complete form of Communistic organization. Here both production and consumption are co-communal. Every member is trusted to work according to the best of his ability and to claim from the common pool of goods available a share in accord with his needs by virtue of being a member and not in proportion to the quantity or quality of work done by him.

Co-operative Farming

While, both in the realm of theory and practice, co-operation has behind it quite a long history, co-operative farming is a relatively recent innovation. Lately a number of countries in Eastern Europe, Mexico, China and others have experimented with co-operative farming, but the nature and the degree of co-operation has varied considerably. As a result, co-operative farming has emerged as a system with a high elasticity of meaning and connotes different things to different persons. While some mean by it the joint use of some essential services without the pooling and joint management of land, that is to say, in a sense not materially different from that conveyed by peasant farming, others identify if with collective farming in which there is joint ownership and joint cultivation. These widely divergent interpretations are a source of confusion and there is little doubt that the present controversy in India about co-operative farming is to some extent traceable to this factor. It is, therefore, necessary, notwithstanding its elastic nature, to note the essential features of cooperative farming, so that we know exactly what is implied when this system is discussed.

Co-operative farming refers to a system of agricultural organization wherein cultivators of an area voluntarily associate together, pool their individual landholdings for purposes of cultivation and manage the whole farm as one unit under as elected management. The individual ownership of land is, however, retained. The private property of the individual member is recognized insofar as he is entitled to a periodic ownership dividend in proportion to the land (and sometimes other assets) contributed by him. Each member working on the farm receives wages at stipulated rates for the work done, and here the quantity as well as the quality of individual work is taken into account. After deducting the wage bill and other costs and contractual charges like interest on capital from the sale proceeds, gross profit is arrived at. A part of gross profit may be withheld for the building up of reserves, and the rest distributed among members generally in proportion to the lands and other assets submitted. Thus co-operative farming essentially is a system wherein there is individual ownership and joint management of land. Its main objective is to combine the incentive of ownership with the "size-economies" possible in agriculture. It seeks to enable the small farmers, without disturbing the institution of private ownership of land, to share the economies of large-scale enterprise which otherwise would not be available to them. It is obvious that this type of organization can serve a useful, and essential, function when the farmers of any area are placed at an economic disadvantage because of petty nature of their individual enterprises.

Sometimes the term co-operative farming is also, and rather loosely, used in another sense. It is used to indicate a system wherein farmers, while owning and managing their lands individually, adopt the principle of cooperation for non-farm operations like marketing, processing, obtaining services like seeds, fertilizers and tools. The is really a case of partial co-operation. Such an application of cooperation to agriculture has come to be called by various names such as "better farming", "co-operative servicing" and "individual farming on co-operative lines" and it should not be confused with co-operative farming as described in the previous paragraph.

Criteria for Choice

In view of the many alternatives thus available, the question does arises as to what criteria agricultural policy should use in India in deciding to support and encourage one alternative or another. Undoubtedly the foremost criterion should be that of making agriculture economically efficient and progressive so that yields per acre increase and costs per unit of output decrease. Maximization of yields per acre is a good enough criterion in a country with a relative shortage of land and whether we go by the consideration of the poverty of Indian farmers, or the shortage of food and the inflation of agricultural prices or the pressure on foreign exchange resources, this criterion would appear to hold the key to the situation. In fact, in the present state of development, the successful implementation of the Five Year Plans would seem to depend, more than anything else, on a rapid rise of agricultural productivity and production. But if efficiency in production is to be the main criterion, the organization of agriculture must be incentive-oriented—incentive for the mass of producers, the large majority of whom are small presents.

Allied to this is the consideration of bringing about an increase in marketed agricultural surplus. This consideration is no less important as marketed surplus is one of main agents in capital formation in the industrial sector and the tempo of development is largely controlled, if not determined by it. In view of this, other things remaining the same, that pattern of organization should be preferred which is likely to prove more effective in increasing and mobilizing agricultural surplus.

Another importance criterion which cannot be neglected is that of employment. Since both from the economic and the social viewpoint, unemployment and under-employment represent a great waste of resources, a strong and very much justifiable bias exists in favour of that pattern of agricultural organisation whose employment-generating potential is high. And the greater the degree of underemployment, the more justifiable is the bias. When we further consider the fact that in the near and the not so distant future the rate of industrialization, on fairly optimistic assumptions, is unlikely to be high enough to absorb (given the natural increase and the large pool of urban unemployed and under-employed) any significant proportion of the present or the growing agricultural population, it follows that increased employment in the agricultural sector itself becomes a prime necessity.

Again, agriculture is not only a source of food and raw materials but also a means of social organization and one of the most important determinants of social and cultural values. Hence, the system of agricultural organization, besides being conducive to greater economic efficiency, delivering larger surpluses and providing greater employment, ought to be helpful in the realization of wider social values accepted as desirable by the community. Together with the programme of conscious economic development, if the goal to be pursued in one of establishing the socialist pattern of society, a social and economic order based upon the values of equality, freedom and democracy, then the pattern of agriculture that we finally choose ought not to be out of tune with these values. It ought, on the contrary, to be helpful in checking exploitation of man by man, reducing disparities of income, wealth, power status, and opportunities, promoting work and initiative on the one hand and freedom on the other to the extent such freedom is consistent with community interest. It should enable the small man to put in his best in the interests of a higher standard of life for himself and increased prosperity for the country. We thus have the criterion of social desirability.

Finally, the practical feasibility or acceptability of a given pattern of farming must be considered. For unless a particular pattern has the capacity of being implemented in the context of conditions in a given country, it would hardly deserve any serious consideration despite its technical or other superiority. Realism and practicability must never by lost sight of, and only that system would have a chance which has not been thought of in complete abstraction and in sheer imitation of foreign ideas, unrelated to the spirit and the tradition of the country.

Suitability of Alternative Systems

The case of small peasant fanning is known to be formidable in very vast tracts all the world over, with some notable exceptions. And in this country too, it has undoubtedly a very great appeal and backing. For one thing, it is a country of small farmers whose only asset is the tiny land holding to which they must cling. For another, there is no stigma attached to the ownership of property and it seems but right that small peasants should be encouraged to pursue their independent activity on their own plot of land. It is felt that the magic of private property, which turns sand into gold, the peasant's feeling that he is his own master, subject to no outside control, and that his reward would directly depend upon his own labour, would drive him to greater and greater efforts. It would induce him to invest, take maximum care of the land and work it to the best of his ability. The result would be maximum soil-fertility and high yields per acre. The criterion of efficiency and productivity would be satisfied, and in the process, it is hoped, some surpluses would also emerge for the market. Again, when farming is undertaken on a small scale, with little use of *machinery*, it would be trusted to provide a high degree of employment to labour; that would be the fulfillment of another criterion. The provision of independent means of support to the vast majority of producers in the country, which peasant farming guarantees, would mean decentralization of economic and social power, promotion of individual initiative and freedom, and the establishment of real democracy in rural society. It is also argued that in addition to these obvious benefits there is nothing in small peasant farming which is out of tune with the social values accepted as desirable by the community and that, bearing in mind the great attachment to land among the peasantry, there can be no question about its feasibility. The system is thus thought to fulfil all the important criteria.

While accepting many of the suggestions in this line of argument, it is important not to lose sight of some serious limitations in the working of this system. The economic benefits claimed for peasant farming are true of farming on holdings of an economic size, that is. on holdings which do not deviate too much from the optimum size. They are not true of peasant farming on tiny, scattered and uneconomic holdings where the small size of land itself becomes a bottleneck to a full utilization of other indivisible resources such as family man-power or a pair bullocks and managerial efficiency is thwarted. This means serious diseconomies of small scale and results in high average costs of production and low yields per acre per man. Under such conditions incomes are obviously low and uncertain and neither high consumption nor substantial investment nor, indeed, the marketing of large surpluses becomes possible. From the point of view of the society, then, industrialisation cannot get going, while for the peasanry itself indebtedness arises. Debt-slavery then makes the ownership of land insecure and dampness incentive to better farming. Besides, under such conditions farmers draw too heavily on the limited natural resources of their farms and this results in soil depletion and poor conservation of resources and is bound to affect very adversely the interests of both the individual and the society as a whole. Thus peasant farming in conditions of an extremely adverse land-man ratio would tend to perpetuate the vicious circle of poverty and many of its normal economic benefits would turn out to be illusory when the scale of operation is far away from the optimum range.

Furthermore, many of the intricate of underdeveloped agriculture cannot be effectively tackled on an individualist basis, as farmers working on separate tiny plots do not have adequate resources in capital and organization to tackle them. Consider for example, the question of scarcity of manure and the use of cattle dung as fuel. Cattle dung is one of the richest types of manures, but to a large extent it is used as fuel. As is well-known, the primary reason for this is the economic necessity and the shortage of any other fuel and not the ignorance or conservatism of the peasantry. The remedy, therefore, lies in making alternative fuel available to peasants at cheap rates. But the small cultivator has neither the capital nor the organization to undertake afforestation schemes. Or, take the question of restoration of soil-fertility through proper rotation of crops. The cultivator knows full well the value of crop-rotation but on a small patch of land, which yields insufficient surplus, he is unable to manage any scientific crop-rotation or mixed farming. Similarly, there can be no solution to the livestock problem on an individualist basis. The small individual farmer cannot afford to set apart any considerable portion of his small holding for growing fodder crops. As has been well said, when the choice is between starvation of himself or of his cattle, it needs very little imagination to perceive which would suffer. Nor can a petty peasant buy machinery in order to replace drought animals. Thus, the inadequacy of small peasant farming arises largely from the extremely small size of the unit of operation. Where the unit is large and viable but not too large and unwieldy for the managerial ability of the peasant, better results can be expected. However, in the Indian context of today, it is difficult to see how the size of the operational unit can be enlarged, within the institutional framework of peasant farming, in a large enough number of cases to make any impact on aggregate production. Reclamation of waste lands may be of limited help in increasing the size of holdings, but neither a programme of reclamation nor of any major improvement can get going as the surplus labour and money resources of individual farmers are too small to start with. Apart from reclamation, the allotment of surplus lands, after imposing ceilings on large farms, to boost up the size of the uneconomic holdings could perhaps be advocated; but while this surplus can be used as an incentive to co-operative and pool existing holdings it can hardly be sufficient to bring about any worth-while enlargement in the average size of individual unit of operation. On its own, it cannot provide much material help. And as far consolidation of holdings, it can only bring together scattered fragments of a farm but cannot enlarge the size of the farm.

Thus, all is not well and is not likely to be well with peasant farming un petty farms, and it seems that this mode of fanning despite its very great appeal in the country fails to fulfil the important criteria of productivity and surplus and is not likely to provide adequate answer to the problems of industrialization and economic development.

The system of large-scale capitalist or estate farming would then seem to provide an alternative to dwarf peasant farming. This mode of organization would enclose substantial areas, establish large units of cultivation, invest capital, mechanize farm operations, increase productivity, market large surpluses and plough back the profits for further expansion. It appears to meet well the requirements of efficiency and surplus generation, and has a reasonably good record of success on these counts, wherever it is practised. But it is on the other criteria of employment, social desirability and feasibility within a given context that this system must be more closely examined.

The main attraction in capitalist farming is to enclose large areas, mechanize farm operations in order to minimize the labour cost which is generally high on large farms and make profits. Now apart from the consequences of farm mechanization, what are the chances of such mechanization in India in the short and the medium run? but it can be stated here that mechanization of farms in general has to wait until a certain degree of industrialization has been achieved; for machines, tractors and their energy requirements are the products of the industrial process and are not domestically available in considerable quantities just yet. The proposal for any large-scale import of these items is bound to meet in the early phase of development with a balance of payments difficulty. Then, on such machines as are available there are, and ought to be, other priorities. Their use in reclamation of such waste land as cannot be reclaimed without machines and tractors is one such priority; diversion of machines which are labour-saving in character to certain specific regions with a relative shortage of labour and high wage rates, is another. If, then, machines are available only in limited numbers and these are directed through policy measures where they are needed most, the chances of any nation-wide move towards capitalist farm organization cannot be bright. For, if machines generally cannot be had and mechanization of farms is not possible, it is difficult to see what attraction is left in capitalist farming with labour-intensive methods, particularly in the context of a generally rising wage rate through social legislation, minimum wage fixing etc. Taking into account the competing uses of capital, the chances of attracting private capital into labour-intensive ventures do not seem to be bright at all.

Even if machines were to be available for capitalist farming, the effect of farm mechanization on employment cannot be neglected. It is true that a kind of partial mechanization is possible which displaces only drought animal power but does not materially affect employment of men; but such a fine and balanced use of machines can be brought about only in a co-operative organization where employment is, among other things, an important consideration. A farm working with only-profit motive, once it has bought a machine or tractor, will obviously displace all the labour that can be technically displaced.

This consideration of employment has more in it than meets the eye. When countries of the western world took to large-scale, enclosed or mechanized farming there has been alongside this enclosure or mechanization an exodus of population from farms in order to man the newly growing industries. As a matter of fact the migration of labour from farms and the rise in agricultural wage rates formed the basic impetus to farm enclosures and mechanization. In the United Kingdom the growth of towns and the shift of labour from agriculture was instrumental in raising agricultural wages and it seemed necessary to have enclosures and change the techniques of production from a greater to a smaller use of labour. In the United States the shortage of workers along a growing agricultural frontier was the primary cause of inventions and labour-saving devices on farms. The Russian and the East European mechanization of farming also co-existed with rapid industrialization and absorption of agricultural population in industry. In other words, either the existing shortage of agricultural workers (and the high wage rate) justified farm mechanization and, in fact, gave a fillip to it, or such mechanization co-existed with a transfer of population from the agricultural sector so that the problem of massive underemployment in agriculture did not arise as a result of mechanization.

But in the context of Indian politico-economic set-up, it is clear now that in the short and the medium run, the most optimistic rate of capital formation in the non-agricultural sector will not absorb any significant proportion of agricultural population. In fact, it is doubtful if in the near future there will be any net absorption at all of agricultural workers in non-agricultural occupations—so large is the existing pool of unemployed and under-employed workers in the non-agricultural sector and so rapid the natural growth of population within that sector. In such a context, to introduce a mode of farming which generally displaces man-power in order to maximize profits, seems patently out of place and harmful.

The argument against capitalist farming does not stop with the charge of unemployment. For very soon, it develops into an argument about social acceptability. A system which, given the context, creates unemployment cannot be generally acceptable except in limited pockets. Moreover, though the surplus generating capacity of capitalist farming is never questioned, it is the generation of surpluses continually and at the time they are most needed that is in doubt. The reference is to the well-known hoarding propensities of large private farms. And from the angle of political feasibility a system which deprives large numbers of agriculturists of their right in land and reduces them to a position of wage-labourers, has little to commend itself on a national scale. Thus, it is that on grounds of employment, social desirability and feasibility, the mode of capitalist or estate farming fails to command itself.

Nor indeed can two other modes of organization, namely collective farming and state farming, be supported in the context of Indian reality, except for experimental and demonstration purposes. Insofar as the two types involve mechanization, they are open to more or less the same objections which we have discussed with respect to capitalist farming. But more importantly, the economic efficiency of state farming is highly doubtful and the bureaucratic management, administrative delays and low incentives which it involves, seem to make it unsuited for emulation. Moreover, state farming as well as collective farming, requires such major upheavals— chief among them being the complete elimination of private property— that the social acceptability of these systems on a nationwide scale is almost negligible. They are not in keeping with the socio-political structure of this country today, nor indeed with the structure which is envisaged in the foreseeable future. They are, then, not feasible.

Our conclusion on this question of the most suitable mode of agricultural organization, then, is that under the assumptions of a severe under-employment of labour and an extremely unfavourable land-man ratio, a political democracy desirous of achieving a high rate of economic growth cannot organize its agricultural on lines of capitalist, collective or state farming. Nor can it benefit at all from dwarf peasant farming in which the mass of individual holdings is of a below-subsistence nature. Only where holdings are of an economic size, there is something to be said for this order. We are then left with the mode of organization called co-operative farming. As the bulk of this eassy is concerned with the examination of the theoretical merits and the practical feasibility of this mode of farming, we propose to restrict our remarks to two general propositions, each of which we shall substantiate in the pages that follow.

The first of these propositions is that co-operative farming, if properly understood and explained to farmers and if propagated within the framework of an integrated land policy, can fulfil ail the criteria we have enumerated above. By joining a co-operative voluntarily, in the face of incentives, the peasants can enlarge the size of the operational unit, get all those things done which a large farm can do and a small one cannot—deep ploughing, soil conservation, rotation of crops—undertake investment through the utilization of surplus labour and thus increase employment on the one hand and productivity and production on the other. Once production increases, both consumption and marketed surplus can increase. Moreover, if the goal of a 'socialist pattern of society' stands for "a social and economic order based upon the values of freedom and democracy, without caste, class and privilege, in which there will be a substantial rise in employment and production and the largest measure of social justice", then there is nothing in co-operative farming, which does not support and strengthen the ideals of a 'socialist pattern of society'.

Our second proposition is that some other brands of cooperations, such as credit cooperation or service cooperation, while they can support and supplement co-operative farming, are no substitutes for it. This is so because service co-operatives tackle only the money resources of farmers which are in short supply any way, and do nothing to utilize the more abundant resources of underutilized labour for constructing capital. Moreover, service co-operatives do little to enlarge the size of the operational unit which, we noticed earlier, is a bottleneck on efficiency and full utilization of resources. But these propositions must be taken up and examined later in much greater detail.

Our discussion of co-operative farming has so far been confined to the possibility of increased employment and the emergence of economies of scale through an enlargement of land, labour and capital. It is important to show at this stage that in addition to these major possibilities, co-operative farming clearly scores over individual peasant farming in its potentiality for generating larger agricultural surpluses.

It is now well-known that the process of industrial development requires the transfer of a large suplus of agricultural production from the agricultural to the non-agricultural sector to meet the increasing tempo of industrial production and the needs of a growing industrial labour force. In a peasant economy this surplus comes in millions of small trickles through the market mechanism. But the availability of this surplus is not an easy matter. Many a planning authority in this and other countries has faced serious difficulty in mopping up agricultural surpluses. Inflationary pressures and adverse terms of trade for the industrial sector, arising out of this shortage of agricultural surplus, have often led to the pruning of developmental plans and downward revisions of programmes of industrialization.

Relation Between Size of Farm Income and Marketed Surplus
(685 farms in 43 ex-jagir villages of 'Hyderabad—1953-54)

Income group	*Average income per form (Rs.)*	*Ratio of marketed surplus to total produce*	*Log. of average Income per farm*	*Log. of the ratio of marketed surplus to total produce*
1	2	3	4	5
1	160	17.3	2.2041	1.2380
2	263	19.9	2.4200	1.2989
3	338	26.4	2.5289	1.4216
4	486	18.9	2.6866	1.2765
5	578	23.2	2.7619	1.3674
6	686	29.0	2.8363	1.4624
7	684	35.8	2.8351	1.5539
8	856	40.2	2.9325	1.6042
9	1004	37.9	3.0017	1.5786
10	1009	41.2	3.0039	1.6149
11	1161	37.3	3.0649	1.5717
12	1258	32.4	3.0997	1.5105
13	1343	42.9	3.1281	1.6325
14	1442	38.8	3.1590	1.5888
15	1730	41.9	3.2380	1.6222
16	2197	49.0	3.3418	1.6902
17	2959	53.9	3.4711	1.7316

There is somewhat similar evidence available in a study entitled *Distribution of Marketed Surplus of Agricultural Produce by Size Level of Holdings in India* : 1950-51 (unpublished so far) by Mr. Dharm Narain of the Institute of Economic Growth, University of Delhi.

What are the hurdles to be overcome before agricultural surpluses become finally available on the market? To start with, agricultural production itself may be too low to admit of any substantial surplus. Or, what little production there is, may be from small peasants who produce so little that they can hardly have their own requirements of feed, seed and wage payment satisfied and therefore cannot send much to the market. It is well-known that there is a good positive relationship between the size of farms or farm incomes and the marketed surplus, whether this surplus is defined in absolute quantities or as a proportion of total output. Even if there were many farms large enough to deliver the surplus, there is the hoarding propensity of the farmers to contend with. Farmers, in any case, would retain a part of their produce for their own consumption needs, for seeds for the following crop, for wage payments in kind and perhaps for some direct barter transactions within the agricultural sector. We may, borrowing the phraseology from the theory of money, call these retentions or stocks of agricultural commodities, transactions stocks. But peasants have stocks other than transactions stocks. In so far as it is technically possible and there are storage facilities available to them, they may retain a part of their production because they think they will get a better price if they sell later, say next year. These we may call speculative stocks. The final availability of surplus on the markets is diminished to the extent these speculative stocks are held peasants. Yet another trouble at a later stage is the stockpiling by wholesale traders and finally by retail traders. This also diminishes surpluses on the final market. These hurdles are often formidable and some of these are the despair of planning authorities. The Foodgrain Enquiry Committee set up by the Government of India in 1957 has made it clear that retentions and speculative hoardings by peasants and traders have played a significant role in causing the relative shortage of food and the high prices of foodgrains which, among other things, have seriously upset the calculations of the Five Year Plans. An urgent need for the control of these hoardings and for short-circuiting the anti-social practices of large farmers, wholesalers and retailers has become too obvious, and an elaborate set of devices ranging from fairprice shops to zoning of production areas and from socialization of wholesale trade to procurement of foodgrain directly from the peasant have been recommended. Procurement from the market and State-trading in foodgrain, in some places supported by price fixation and in others within a free-market, has been experimented with. But probably owing to an unimaginative and haphazard mode of implementation—the mode itself different from State to State—the mopping up of the surplus has been well below expectations and neither speculative hoarding nor agricultural prices could be brought under control. In fact, evidence of a major decline in arrivals in the market; and increased hoardings by farmers has mounted up irrespective of the relative efficiency of these measures and their suitability within given political and other assumptions. The situation points out to the dire need for enhanced agricultural surpluses in a developing and industrializing economy.

Now governments in many countries harassed by these 'leakages' in agricultural surpluses, have from time to time resorted to heavy or procurement of foodgrains directly from farmers to feed industrial populations. But there are numerous loopholes and opportunities for evasion in this system, as any food administrator knows. In any case it is no easy matter to procure grain from so many millions of small farmers. The production of each farm has to be assessed and the farms categorized. Exemption limits have to be fixed for each type of farm; if injustice is really to be avoided, the size of family has to be taken into account as well as the size and nature of farm. The proportion of produce that each category of farmer is to submit has to be carefully determined, so that neither the farm is starved of its basic necessities nor too much grain is left with it for hoarding. Where, for some valid reason, output has been low, discretionary powers have to be used. In other words, the administration of a nation-wide procurement system in a nation of several million farmers is a very difficult proposition. The project is known to be very costly. There are too many evasions and injustices and too many dissatisfactions and grumbles. Moreover, the system is apt to strain severely the integrity of the civil service, the officials and the administrators who run it.

Co-operative farming provides an alternative which is administratively far more feasible. These farms will be much fewer in number, to begin with. They will be submitting their accounts and their records regularly. They will not change hands so often as a private farm does owing to inheritance, mortgage, sale, break up of the joint and so on. Their records will be consistent, continuous and amenable to proper checks and scrutiny. Moreover, once they have run for some time it would be quite simple to estimate their annual needs for retention of their produce for their own consumption and production requirements. Their hoarding over and above the genuine retentions can be reduced to a minimum and can in fact be eliminated. The grain procured from them can go straight into the hands of the government, if so desired, so that a socialization of wholesale trade can be made automatic without causing frictions. Thus a greater availability of agricultural surplus appears to be a very clear possibility in the context of co-operative farming.

Some critics are extremely fearful or sceptical whenever co-operativization as a means of facilitating the mopping up of surpluses is mentioned. They at once imagine the Russian collectives which at one time had been instruments of squeezing the peasantry into delivering "surpluses" of such magnitudes that peasants were left with deplorably low stocks, insufficient even to meet the basic requirements. This caused great discontent. But if the propelling force in our scheme of co-operativization is to be incentive and not force— as we stated earlier and hope to show in greater detail in—it is not necessary at all to allow oneself to be shocked by the thought of similar things happening in this country. For one thing, our whole scheme of co-operativization is motivated by court desire to increase agricultural output through incentive; for another, we have the example of many a country in this respect and know how not to co-operativize through force.

We must also take note of the criticism that if the surplus mopped up via the co-operative farming mechanism is to be of the same size, as can be collected with the existing tax mechanism, there is no need for co-operative farming. If, however, it is intended to collect a larger surplus, then the co-operative becomes an instrument of concealed (or even open) taxation of substantial magnitude. Considering the prevailing state of very low real consumption in the agricultural sector any proposal for additional taxation, it is said, is undesirable and vicious because it squeezes consumption standards further.

The real point, however, is that the process of economic development demands sacrifices from all sections of the community including the peasantry. To feed this growth process, agricultural and other surpluses are a necessity and in fact the required magnitude of these surpluses to bring about growth at the desired rate is larger than what the present rates of taxes on agriculture are able to command. There are difficulties. Governments are often reluctant to raise the required resources through enhanced taxation because of political reasons; moreover taxation carries with it a big risk of evasion. The method of collecting surpluses from a few points and from farms which are more amenable to control is administratively much more feasible-indeed, so feasible it is that the Soviet and some of Eastern European governments were unduly tempted to carry it much too far, as though the collective farms were nothing but instruments of collection. That was so in the peculiar setting of those countries, changed with force and revolutionary ideas. But one need not mix up administrative feasibility with exploitation. Abuse is no argument against use. It the distinction made earlier between transactions stocks and speculative stocks of farms is remembered and if procurement aims at mopping up the latter, there need be no heart-burning on the part of the members of the co-operatives.

As for reduced consumption standards it is absurd to think that the procurement will be at the cost of consumption. Co-operative farming is useless if it does not lead to increased output. Indeed if the co-operative farm has not done any better and not produced significantly more than the several family farms which constituted it it has no justification for survival. If output does increase in the manner shown earlier then it is out of the *increased* output, without reducing consumption any further, that the extra suplus will be obtained. If output increases as it should, then both consumption and surplus can rise. We need not therefore attach to co-operative farming the stigma that it results in tightened belts; and if we did not do that we shall find that its administrative merits are over-whelmingly superior to small family farms when it comes to generation of surpluses. Co-operative farming then becomes as important instrument in activizing the process of economic growth.

Some Other Objections

There are some other much less formidable objections to cooperative farming which can be answered and disposed of rather easily. It is sometimes said that co-operative farms in India have been a failure. Where are the co-operative farms, anyway? One only has to scratch the surface of a so-called co-operative farm in this country and one gets the true picture. Co-operation is only nominal: it is often another name for evasion. Absentee landlords, in order to get over the stigma of absenteeism or to evade existing or immanent ceilings or to gam favours like subsidies, have set up farms in the name of co-operatives. Few of the partners work; absenteeism continues and the whole spirit of cooperation is negatived. If we applied some criterion of genuineness and ask, say: in how many of these farms all or more than half the co-operating families are actually participating in manual or managerial work? We shall get only a small proportion of genuine co-operative farms. In some cases the so-called co-operative is merely a large family farm where the lands have first been split and assigned to different names, perhaps to evade the ceilings provision, and than registered with the registrar of co-operative societies to give it a respectable look. It is true that in Punjab, UP, Bombay and sorne other States there do exist some genuine co-operative farms and in some cases the results are good too. But often these farms, along with private farms, are equally starved of credit, fertilizers, seeds, implements and trained managerial abilities; and the last factor obviously is more important for a relatively larger co-operative than for a small private farm. A judgement on co-operative farming in progressive Indian conditions has to be reserved until we put the conditions rights. The past provides little guidance to the future in this case.

Another objection to co-operative farming brands it as undemocratic. Take a vote about it, it is contended, and you will get a clear rebuttal: the peasants are against it. What has to be remembered here is that change always has a large opposition. Human inertia and conservatism are barriers to all reforms. To take an example, we know for certain—and the peasants know it too—that sub-division and fragmentation of holdings have played havoc with our agricultural system and have raised costs and reduced yields considerably. Too obviously the thing to do is to consolidate the fragments. But the movement for consolidation, when it starts, has its most bitter opponents and but few peasants agree to being removed from some plots and being allotted other plots. If a vote were taken at an early stage the popular voice would be so clearly against consolidation that it would seem awfully undemocratic and authoritarian to go ahead with the programme. But once the programme is pushed ahead, people get adjusted to the new set-up. The over whelming benefits of consolidation begin to catch the eye and the same people begin to proclaim the benefits of the movement and advice others in its favour. The same is the attitude towards the construction of a new multipurpose projects, towards the suggestion for increased taxation for economic development, towards simplification of weights and measures and so on. Each such scheme is initially opposed and agitated partisans invariably use the bogey and ask: Is it compatible with democracy? It has been said that when proposals for the abolition of slavery were afoot in the United States in the last century there were many who raised the question: Is the abolition of slavery compatible with democracy? The truth is that science, knowledge, experience and technical considerations have a major role to play in a democracy; otherwise governments of under-developed countries reconciled to poverty, would not be going ahead with developmental plans which invariably call for major sacrifices from all sections of the population. Often in technical matters democratic governments do not consult the country, through ultimately the people have every right to sit in judgement over the results of a policy and to decide on its continuation or reversal. When the bill for the decimalization of currency was being discussed in the Indian Parliament, some people, to jeopardize it, or to delay it, advocated that the bill be sent to the country for public opinion. The Prime Minister's apt retort came in time: "We do not consult the country about the Theory of Relativity!" And so, if co-operative farming, voluntarily conceived and run with incentives, proves itself worthy, popular support would be automatic. If it fails after a good trial, the people would call for a reversal of policy and they have every right to do so. But to provide the necessary facilities and external economies in order to initiate and to run the experiment is not undemocratic, especially when (as shown in subsequent chapters) coercion is ruled out, incentives both in the formation and the running of co-operative farms are provided and voluntariness is to be the essence of the association.

It has been shown in previous chapters that it is possible, through co-operative effort, to construct capital, enlarge land, provide more employment, obtain economies of scale and larger agricultural supluses. While none of the objections to co-operative farming discussed so far seems to have a real substance, there is one line of reasoning which probably has some merit and deserves to be considered more seriously. It is contended that human beings, being what they are, would any day work and manage their property better on their own than manage it jointly. And the corollary of this is that if they are made to manage things jointly, their contribution would be of a lower order, and that this would reveal itself in smaller production, enhanced costs and general inefficiency of failure.

Even this argument has two sides to it. One is a cynical side. Human beings are essentially selfish, it is contended. They are actuated by individual urges rather than by community urges. They always do things better for themselves than for others. Human nature does not change and so it is no use having new-fangled ideas like co-operative farming which assume, and depend for their success on, a change in human nature. Now this idea that "you can't change human nature" can and has indeed been misused a greated deal. The dominant urge in human beings is, no doubt, the individual urge and it is unusual for people, confronted with a choice between the dictates of the individual urge and those of the community urge to choose the latter. When an army is marching from victory to victory, the community urge among soldiers to advance forward and bring glory to their unit and their country goes hand in hand with their personal urge to secure name, fame and riches for themselves. The two work in the same direction and there is no conflict.

The development of agriculture, so basic for economic development requires two broad sets of measures. One relates, to the provision of essential technical facilities in regard to conservation of soil resources, seeds, manures, irrigation, equipment, crop protection, finance, marketing, etc. and these are necessary for meeting the requirements of scientific agriculture. The second relates to the creation and maintenance of a suitable institutional frame-work. As a matter of fact it is the latter which constitutes the first basic step in the direction of re-vitalization of agriculture and must precede scientific agriculture. To strive for scientific agriculture without first replacing the out-moded and inhibitive institutional frame-work by a just and efficient one would be fruitless and would not lead to any progress worth the name. It is in recognition of the truth that for rapid agricultural development, scientific agriculture should follow and not precede the required structural and institutional chance that many a planned and developing economy has been obliged to resort to a thorough and effective overhauling of the existing institutional frame-work before undertaking the various technical measures of development. Once the defective framework has been replaced by an appropriate one, measures for scientific agriculture will do the rest; but not until then.

India's past experience, too points to the same conclusion. In the pre-independence period, in spite of some efforts to develop agriculture, hardly any worthwhile results were obtained because, apart from the inadequacy and haphazard nature of the efforts, the cart of technical facilities was put before the horse of structural change. The things characteristic of the second stage were taken up before completing the organization of the first stage, viz. the setting up of an institutional pattern favourable to growth. That the need for structural change arises partly because of the decay of the tenurial system and partly owing to an extremely adverse land-man ratio, a crazy size-distribution of land holdings and a serious fragmentation of farms. Concentrating for a while on the first of these factors, we may notice that the pattern of tenure in existence before Independence was the landlord-tenant pattern. Defective from many points of view, this pattern led to increasing concentration of ownership of land into the hands of a small number of absentee landlords and gave rise to all the well-known economic and other evils of a degenerate tenancy system—an endless sub-letting of land, emergence of many intermediaries, rack-renting, vast evictions of tenant-farmers, charging of high rents which left the tenants with an income too insufficient for any plough-back and hardly adequate to maintain themselves and their facilities, and finally the lowering of the dignity of the cultivator. Thus under the prevalent pattern, the real cultivators had neither the means nor the incentive to make improvements in agriculture. The surplus from the land, itself not very large because of low productivity, flowed into the hands of landlords whose habits were not conducive to economic growth inasmuch as most of them were not savers but conspicuous consumers. Typically, they did not invest in the land and took little initiative or interest in the development of agriculture. Thus land was virtually nobody's concern. The institutional framework was such that it cramped both men and crops. And it also disrupted social life by creating rigid classes in the rural society—classes whose interests diverged and brought them into conflict. Within such an economically and socially defective frame-work, extension of credit facilities, irrigation and the like was largely a waste and could hardly be expected to bring about any major outburst of productivity.

While this was so, no serious efforts were made during the days of the British rule to replace this out-moded, semi-feudal pattern to prepare the ground for rapid agricultural development. The only thing that was attempted was the enactment of tenancy legislations in different parts of the country with a view to the reconciliation of the conflicting interests of landlords and tenants. The interests of landlords were sought to be safeguarded by giving them powers of ejectment and recovery of arrears, while tenants were sought to be protected by obtaining for them security of tenure and fair rents. Such a fine balance or compromise was too difficult to be reached and maintained. What resulted from this line of action was a precarious set-up wherein the landlord was reduced to the position of rent-receiver and yet the tenant was not the owner.

The task of scrapping the tenant-landlord pattern and eliminating the intermediaries was taken up with some enthusing immediately after the attainment of Independence. Intermediaries like the Jagirdars and the Zamindars were, no doubt, compensated and abolished throughout the country. But the process of abolition of Zamindari itself is known to have led to the eviction of many a tenant-cultivator as many large absentee landlords have been able to resume their lands from their tenants and re-establish themselves as 'cultivators' or 'bhumidars' owing to much too porous a definition of personal cultivation and an absence of ceilings on land. Then, tenancy laws have been enacted almost everywhere in the country with the object of giving security of tenure, reducing rents and enabling the tenants to purchase the lands and become owners. But this reform too has been at best a partial success. Owing to an extremely sluggish progress of legislation and implementation, it has been possible for land owners to evict tenants before their permanent and protected status could be recognized in law. Numerous evictions have taken place causing destitution and misery and the establishment of the actual cultivator as the owner of the soil has not been achieved in any substantial measure. In but few cases has the right to purchase land, at very much less than the market price, been utilized. Owing to the shortage of land in relation to the demand for it, and the prevailing structure of social relations between land-owners and cultivators, little use has been made of the legislative provisions for rent reduction. There has then been either a serious evasion or a non-implementation of important aspects of land reform and the desired structural change leading to a creation of incentive does not seem to have occurred. Enthusiasm about agricultural re-organization in fact seems to have slowed down in recent years.

Tenurial decay, apart, the other set of factors responsible for the present disincentive was stated to be related to an adverse land-man ratio—too tiny a unit of cultivation, divided into too many fragments. The enlargement of the unit of cultivation with a view to removing the bottleneck which land at presents is on other factors of production, in particular labour, could be achieved through land reclamation, imposition of ceilings on very large (and often less productive) holdings, land re-distribution, consolidation and cooperative farming. But there is very little evidence of any integrated thinking, let alone action in these directions. No serious programme of land reclamation is in sight even though this source is known to be capable of providing about 100 million acres. Apart from Punjab and UP and to a lessor extent in a few other states, very little has been done to consolidate into one unit the scattered bits and pices of farms and stop sub-division and fragmentation of holdings from proceeding further. Vast tracts are in dire need of a programme of consolidation of holdings. The question of redistribution of land and increasing the size of submarginal farms has received a very lukewarm treatment. Ceilings on future acquisition of land have been fixed in many states but those on present holding have either not been enacted or not implemented except in one or two tracts such as a few districts of Andhra Pradesh. Delay on this front has been leading to large land owners partitioning their lands and bringing about bogus transfers of chunks of their lands in the names of near-relatives and others. A fruitful source of land distribution is being frittered away.

As for co-operative farming, in recent years it has attracted considerable attention and has in fact been practised with varying degrees of success in different parts of the country. But in public as well as in academic discussions the non-genuine co-operative farms of which there are very many, have got badly mixed up with the genuine ones. This has provoked misunderstanding and controversy. In its pure and natural form, co-operative farming has yet to be explained to farmers, administrators, political workers and others. The present work does not intend to deal with those aspects of agrarian change which are concerned with tenure, i.e. the set of rights and obligations between the cultivator, the owner and the Government. But it does attempt to tackle the problem of the enlargement of the unit of cultivation and capital formation in agriculture through cooperative farming and all that goes with it.

Increasing Support for Co-operative Farming

Until recently it had looked as though the view-point that cooperative farming is suitable and necessary for the development of agriculture had gained considerable strength in the country. Many an expert committee in recent years had lent its support to this mode of farm organization. The policy Committee on Agriculture, Forestry and Fisheries (1945) recommended the undertaking of "pilot schemes on different types of farm organization befitting each type of soil, climate, tenure, etc., so that the comparative advantages of the different organizations may be determined and a suitable one evolved for wider application". A year later a similar suggestion was made by the Cooperative Planning Committee (1946) that experimental ventures on types of co-operative farming should be undertaken in two selected villages in each district. This was further endorsed by the first Indian Co-operative Congress in 1952 which adopted a resolution to the effect that "active and vigorous steps should be taken by the State and Co-operative movement to organize co-operative farming societies whenever their formation is feasible and desirable". More recently the Indian Delegation to China on Agrarian co-operatives has expressed an emphatic view that co-operative farming is both desirable and necessary for the successful reorganization of Indian agriculture. It has advised the country to take to this mode of farming and with this end has outlined a number of suggestions which relate to the nature of land reform measures to be adopted, the role of leadership— particularly of the party in power—in creating the proper climate, training of necessary personnel, organization of about 10,000 planned experiments in co-operatives farming so as to evolve suitable techniques of co-operative management, and scientific demonstrations of the programme.

The Indian Planning Commission has also expressed a similar view. Apart from other reasons, it regards co-operative farming as essential for bringing into existence sizable farm units to facilitate the application of scientific knowledge on a wider scale, increase in capital investment and rise in the productivity of land. This view-point is clearly reflected in the Five Year Plans. It was suggested in the First Five Year Plan that small and middle farmers should be encouraged and assisted to combine on a voluntary basis into co-operative farming societies and a number of suggestions were made in this connection. A sum of Rs. 40 lakhs was provided for promoting planned experiments in co-operative farming so as to evolve suitable methods and techniques of co-operative organization and management under Indian conditions. State Governments were later requested to draw up phased programmes for co-operative farming. But, on the whole, little action was taken during the First Plan period on these lines. A sustained effort in carrying out planned experiments in co-operative farming for the purpose of evolving suitable techniques of management was lacking, and obviously, therefore, the bulk of the provision remained unutilized.

The position in this regard reviewed in the Second Five Year Plan. The Planning Commission reaffirmed the basic policy of encouraging co-operative farming which was now defined in its essential as pooling of land and joint management. They emphasized the need for an attitude of experiment in all matters of detail and suggested that such steps should be taken during the period "as will provide sound foundations for the development of co-operative farming so that over a period of 10 years or so a substantial proportion of agricultural lands are cultivated on co-operative lines." Co-operative farming societies were to be given special concessions and subsides such as credit on liberal terms, priority in consolidation proceedings, preference in supply of seeds, fertilizers and constructional materials, and special financial and technical assistance for developing ancillary occupations. But it has been emphasized that care should be taken to ensure that concessions are given to genuine societies only.

Regarding targets of co-operative farming to be achieved during the Second Five Year Plan, it had been stated that these would be determined in the first year of the Plan after consulting individual States. The matter was taken up for discussion at the Conference of State Ministers for Co-operation held at Mussoorie in July 1956, and if was decided that the objective should be to bring about the formation of at least one co-operative farming society in every National Extension Block during the Second Plan period and that about 500 experiments should be undertaken in 1956-57.

This question was further examined by the Standing Committee of the National Development Council in September 1957. That Committee recommended that conduct of 3,000 co-operative farming experiments during the remaining period of the Second Plan and a beginning with 600 such farms in 1958-59. The work was to be undertaken in consultation with the State Governments on whom the main burden of the experiment falls. In actual fact, however, many State Governments have been lukewarm in promoting this programme and attitudes have ranged from acceptance to indifference and from indifference to positive hostility. Needless to say, the targets remained grossly underfulfilled.

The point of real importance is that if we want the progress of co-operative fanning to be voluntary and interpret voluntariness to mean a spontaneous desire on the part of the peasantry to rise and organize itself into farm co-operatives without anybody, say a government, telling them to do so and placing incentives in their way, then the peasants will not get organized into such co-operatives. They may not even save themselves from proletarianization. Voluntary action on the part of peasants must not be interpreted in the *laissez-faire* sense of the government doing nothing to provoke such a voluntary action. If small peasants are to form co-operatives voluntarily, not only has great publicity and persuasion got to be used to bring home them *(a)* the absurdity of their present uneconomic units of cultivation, and *(b)* the possible benefits of co-operations, but the other benefits of land reform, land distribution, tax exemptions, subsidies and provision of facilities have to be brought to bear on the situation. Thus, it is in the nature of things that the approach to co-operative farming must intergrade the progress of this mode of organization with the progress of land reforms and other incentives. Otherwise, a voluntary progress of co-operative farming, where voluntariness is defined in the queer sense of *laissez-faire* non-interference, is inconsistent with the fixing of targets of achievement. Targets can be fixed only where voluntariness means voluntary action in the face of incentives and these incentives can come through an integration of co-operative farming with land reforms and other benefits. Fulfillment of targets then becomes a test of the achievement of land reforms, etc. in creating sufficient incentives for co-operativization.

Before the beginning of 1959 there was no evidence of such an integration of co-operation and other aspects of agricultural policy. The situation for a while seemed almost hopeless. However, the All India Congress Committee at its session at Hyderabad towards the end of 1958 appointed a subcommittee to go into the question of agricultural production. The report of this Committee became the basis for the well-known Nagpur Resolution at the 64th Session of the Indian National Congress. This Resolution, for the first time, brought into juxtaposition several institutions and courses of action which were by their nature allied but had not been thought of in an integrated manner. Attention was drawn to the allied role of village panchayats and co-operatives as the spearheads of all developmental activity and the need for strengthening the powers and the resources of these organizations as well as for the diversification of their activity, especially intensive farming, was emphasized. The future agrarian pattern was declared to be of co-operative joint farming and this mode of organization was defined. Service co-operatives were to precede co-operative farming and a target of achievement for the former was fixed. Ceilings on existing and future holdings were to be imposed in all States by the end of 1959 and a virtual integration of ceilings and co-operative farming was for the first time arrived at by declaring that surplus land obtained from ceilings should vest in the panchayats and should be managed through co-operatives consisting of landless labourers and small peasants. Nor were matters like fixing of floor prices for each crop, state trading in wholesale trade in foodgrains and utilization of waste land lost sight of.

Although action on most of the aspects of the Nagpur Resolution is still a matter for the future and it is difficult to predict how seriously implementation will be taken up in the years to come, the Resolution itself seems to us to be of such for reaching importance that we prefer to quote the whole of it below:

The Congress, having considered the Report of the Agricultural Production Sub-Committee, records its general approval of the Report and endorses the recommendations made therein. In particular, the Congress is of opinion that—

(1) The organization of the village should be based on village panchayats and village co-operatives, both of which should have adequate powers and resources to discharge the function allotted to them. A number of village co-operatives may form themselves into a union. All permanent residents of the village, whether owning land or not, should be eligible for membership of the village co-operative which should promote the welfare of its members by introducing progressive farming methods and improved techniques of cultivation, developing animal husbandry and fishery and encouraging cottage industries. In addition to providing credit and discharging other servicing functions, it will arrange for pooling and marketing the agricultural produce of the farmers and for storage and godown facilities for them. Both the panchayat and the co-operative should be the spearheads of all developmental activities in the village and, more especially, should encourage intensive farming with a view to raising the per acre yield of agricultural produce.

(2) The future agrarian pattern should be that of co-operative joint farming, in which the land will be pooled for joint cultivation, the farmers continuing go retain their property rights and getting a share from the net produce in proportion to their land. Further, those who actually work on the land, whether they own the land or not, will get a share in proportion to the work put in by them on the joint farm.

As a first step, prior to the institution of joint farming, service co-operatives should be organized throughout the country. This stage should be completed within a period of three years. Even within this period, however, wherever possible and generally agreed to by the farmers, joint cultivation may be started.

(3) In order to remove uncertainty regarding land reforms and give stability to the farmer, ceilings should be fixed on existing and future holdings and legislation to this effect, as well as for the abolition of intermediaries, should be completed in all States by the end of 1959. This does not mean any ceiling on income, as it is expected that, by intensive cultivation as well as by additional occupations, rural incomes will rise. Such surplus land should vest in the panchayats and should be managed through cooperatives consisting of landless labourers and small peasants.

(4) With a view to assuring a fair return to the tiller, a minimum or floor price should be fixed reasonably in advance of the sowing season with respect

to each crop, and arrangements should be made to purchase directly, whenever necessary, the crops produced.

(5) The introduction of State trading in wholesale trade in foodgrains is welcomed and should be fully given effect to.

(6) Every effort should be made to bring the uncultivated and waste land into cultivation. The Central Government should appoint a Committee to devise suitable measures for the utilization of this land.

The Congress trusts that the steps indicated above will be taken by the Central and State Governments at an early stage; and that the Congress organization, in co-operation with others, will help in every way to mobilize mass enthusiasm and arouse initiative and a spirit of self-help in the millions of farmers in the-country.

While this has been the situation on the political and administrative front, there has been no shortage of academic discussions on the subject; but the majority of works that have been published recently have tended to oppose co-operative farming. In the following chapters we deal in detail with the rationale of these works and state our own reasons for accepting the superiority of the co-operative mode of farming especially for the small cultivators of India. We also suggest a detailed plan of organization and effort which integrates a voluntary progress of co-operative farming with the schemes of land reform. But before we do that, it is necessary to take stock of the existing situation in the country regarding co-operative institutions in general and co-operative farms in particular.

The So-called "Co-operative Farms"

Official sources state the number of co-operative farming societies to be 2,020 on 30 June, 1958.

Credit & Multi-Purpose Co-operative Societies on 30th June 1958

1. Number: 1,65,712	
2. Membership: 11.5 million	
3. Share capital	
(a) Government	Rs. 2,21,07,000
(b) Member's contribution	Rs. 25,98,79,000
(c) Total	Rs. 28,19,86,000
4. Deposits	Rs. 8,62,96,000
5. Total amount of credit made available through Primary credit Societies	

	during 1957-58	
	(i) Short and medium term	Rs. 96 crores
	(ii) Long term	Rs. 10 crores
6.	No. of marketing societies organized during 1956-57 and 1957-58	750

The state-wise distribution of these societies (given hereunder) shows that Punjab accounted for the largest number of societies, i.e. 478 and Bombay and Uttar Pradesh followed with 402 and 255 societies respectively.

Andhra	31	Punjab	478
Assam	170	Rajasthan	105
Bihar	27	Uttar Pradesh	225
Bombay	402	West Bengal	148
Jammu & Kashmir	7	*Union Territories*	
Kerala	55	Delhi	22
Madhya Pradesh	140	Manipur	3
Madras	37	Tripura	12
Mysore	100	—	
Orissa	28	Total	2,020

Not all of these, however, were societies in which land is pooled for joint management. The actual position was:

Tenant Farming and Better Societies	663	
Collective Farming Societies	 381	1,357
Joint Farming Societies	... 966	

Thus about one-third of these or 663 societies were tenant farming and better farming societies where for the sake of economy, efficiency and avoiding exploitation, some services are jointly secured by farmers who work on their separate plots quite independently. Societies in which land was actually pooled and managed jointly were 1,357 in number. Of these, 391 well collective farming societies where land is worked jointly but where workers get only a reward for their labour and no reward for land. This happens generally where people do not submit any of their own lands to the pool but colonize new lands or are settled on lands belonging to the government or some other agency. The remaining 966 societies were what are called joint farming societies, where land is pooled and operated jointly and property rights are preserved in the sense that co-operators receive an annual or other rent or dividend proportional to the land submitted. Those who work, of course, get another payment—a wage or a work bonus. Thus, a first sight, these 966 so-called joint farming societies would seem to approximate to the definition of cooperation farming which we are using in the present work. This, however, is far from the truth. A closer scrutiny of the motives which actuated the formation of these societies and the application of some criteria of genuineness, reveal most of these societies to fall short of genuine co-operative farm organizations.

Some of these societies have been initiated, perhaps without much desire for genuine and permanent co-operation, in order to obtain timely benefits such as allotment of government lands and concessions available under the Land Utilization Act or the Waste Land Reclamation Act. In many cases, displaced persons and ex-servicemen have been settled on lands and quite often they happen to be near-relatives. The society resembles more a large family farm where a family is to be defined in a broad sense including cousins and uncles and nephews. Occasionally some unrelated persons are also included but it is difficult to know if this is a spontaneous.or engineered development. Quite often in order to evade the actual or anticipated ceilings legislation, a landowner splits his large lands and transfers these in the names of near-relatives, and then has the various bits registered together as a co-operative farm. Somctimes to camouflage the family nature of the farm, small pieces of land are transferred in the name of non-relatives or even agricultural labourers in whom the land-owner has confidence (this is a way of attaching them permanently to the farm), and then the whole is registered as a cooperative farm.

These glaring weaknesses and departments from genuine cooperation show themselves in at least three different ways. The management is invariably in the hands of the biggest partner, who is all in all and but few of the other co-operating partners actually undertake manual work. Many live away from the farm and they neither do manual nor managerial work. Secondly, the proportion of relatives to total memberships exceeds considerably the proportion of non-relatives. And many of the relatives are in fact near-relatives. This creates a presumption in favour of the farm being in fact a family farm and the sprinkling of non-relatives is often quite insufficient to detract from the family nature of such an organization. And thirdly, there are vast disparities between the lands held by the largest and the smallest partner and there are obvious signs of a strong correlation between the size of each partner's "land contribution" and his relationship with the main partners or manager who are entrusted with the administration of the co-operative.

Most of these co-operative farms are thus really large capitalist farms or estate farms and have been constituted with motives other than genuine co-operation. This is not to deny their productive role. As farms which are investing a substantial amount in agriculture, using a new techniques, machinery and tractors, reclaiming and improving lands and raising productivity and surpluses, there is merit in them.

However, they are open to the same objections which we raised in the previous chapter against capitalist farming. If by co-operative motive we mean a spontaneous desire on the part of small and equal partners who are need of such co-operation, to come together and work jointly in order to reap proper rewards and save themselves from exploitation, then, most of the so-called co-operatives are but apologies for, if now mockeries of, co-operation. In view of these facts we suggest in a definition of a guanine and standard co-operative farm which we hope will enable us to classify in a more meaningful manner the organizations which pass under the name of co-operatives.

Yield Per Acre (in lb.)

	1952-53	*1953-54*	*1954-55*	*1955-56*
Cotton	339	432	565	341
Paddy	126	400	506	467
Jowar	529	376	592	637
Groundnut	44	69	191	119
Gur	623	592	970	1440

It does not follow from this, however, that there are no cooperative farms in India which will not stand the criteria for genuine co-operation. There are, but they are few and far between. Some of those, though actuated by a proper co-operative motive, do not have competent managers. Others are working in conditions of relative uncertainty on lands obtained from the government on lease and renewable every year. Even so there are co-operative farms whose record is praiseworthy. Their output per acre is higher than that of farms in the neighbouring areas; some of them have been recipients of zonal prizes for highest output per acre; some provide to their members and to the community outside useful services such as schools, books are children, dispensaries, transport to the local market and daily postal service, etc. free of cost. But it is very necessary before looking at the performance of any of the so-called co-operative farms to examine the nature of co-operation. To accept the results of all the societies officially classified as "co-operative joint farming societies" is to mix the results of hoax "co-operatives" with those of the genuine ones. This prevents unbiased judgement and brings the co-operative movement unnecessarily into disrepute. In order to have a fair appraisal we proceed to suggest in a set of criteria which could be used in determining the nature and the genuiness of co-operation. But we must now turn to the theoretical arguments relating to genuine co-operative fanning.

Similarly at Jugroonagar in Bareilly district of UP a joint farming society has shown the following trend of yield:

Average Yield Per Acre (in mds.)

	1952-53	*1953-54*	*1954-55*	*1955-56*	*1955-57*
Wheat	10.0	10.4	14.9	11.3	17.2
Gram	1.7	17.1	18.2	6.0	14.6
Sugarcane	314.0	195.0	360.0	306.0	329.0
Paddy	10.9	13.1	17.0	13.9	14.9

4

DIMENSIONS OF AGRICULTURAL DEVELOPMENT

Concepts and Objectives

Agriculture in the populous countries like India provides the mainstay of livelihood and contributes significantly in the economic development of the nation. It is undoubtedly a well-known fact that the food production in India is not keeping pace with the population increase, where as much as 66 per cent of the working population is engaged in agriculture. The history of agricultural development in India started with the advent of green revolution in the early 1960s when the country underwent a chronic food shortage. The Government of India has introduced many programmes at the district and the block level to increase agricultural production by intensive techniques.

The first stage of new era in Indian agriculture started with the launching of the Intensive Agricultural District Programme (IADP), in selected districts, in 1960-61 (Government of India, 1963). IADP was introduced with the objective of increasing productivity and optimising the use of inputs. Duly recognising the importance of complementarily among various new inputs and practices, the concept of a 'package of practices' was developed and propagated. In 1964-65 and in subsequent years a modified version of the same approach, the Intensive Agricultural Area Programme (IAAP), was extended to several parts of the country.

The second stage of the new era came with effect from 1966 with the introduction of High Yielding Varieties Programme (HYVP) in which special attention was focused on propagating, on a large scale, new high yielding varieties of foodgrains in areas of having assured irrigation facilities. Besides higher yields, greater intensity of cropping was also emphasized at this stage (Raju, 1982).

Needless to mention that the 20 years old Green Revolution has brought about substantial changes in the land utilisation and the cropping patterns and increase in agricultural productivity. But the gains of such changes are said to have been shared unevenly by various factors by production and administrative regions. The studies done by Jha (1974), Shah (1976) and others relating to late sixties and early seventies showed degree of growing inequality in factor shares in agricultural production. The inequalities can be evidently observed specially among the states of food-crop production. In the States like Haryana and Punjab, the causes and consequences of the introduction and spread of new technology are apparently different from those in Rajasthan, Bihar, Madhya Pradesh and Uttar Pradesh depending upon a number of socio-economic and physical factors.

The state of Uttar Pradesh is such where about 75 per cent of working population is engaged in agricultural activities. It will be seen that a balanced regional agricultural development is retorted due to wide regional disparity in agricultural resources and growth potential in the state. It will also be seen that the propagation of new technology in agricultural development (HYV, fertilizers, irrigation and farm machinery) has augmented the increased regional disparities. One of the basic objectives of agricultural development should be a balanced regional growth.

Keeping in view the above statements as the objectives of the present paper, a comprehensive approach to agricultural development has been propounded, wherein the land potentiality and technological development in addition to natural factors have been made responsible. It is realised that the agricultural development should be measured not only by trends in the agricultural production and productivity which reflects the essence of various agricultural developmental measures but also on the basis of the various physical inputs like labour, irrigation, fertilizers, improved seeds, extent of cultivated area, etc. Assessment of the level of development of each of these physical inputs can reflect broadly the potential level of agricultural development which may or may not be reflected in the average productivity levels. The discrepancy in the agricultural sector between the potential production and actual production is often due to the weather conditions (Sharma, 1973). In the present study an attempt has been made to cover the following objectives:

(i) To analyse the levels of inputs utilisation and spatial behaviour in the agricultural pursuit;

(ii) To measure and explain the extent of agricultural land potentiality in the development of agriculture;

(iii) To measure the levels of agricultural development with respect to varying nature of its attributes;

(iv) To explore the possibilities for minimising the inequalities and raising the level of agricultural development; and

(v) To suggest remedies for congenial structure of agricultural development attributes so that agriculture starts playing a significant role in the development of our overall economy.

Methodology and Data Base

Keeping in view the objectives mentioned as above, the method of Principal Component Analysis has been adopted in order to measure the levels of agricultural development and also to drive the components responsible for the existing levels of development. These together would provide the base for remedial measures.

For the purpose, a set of 11 variables has been grouped into two segments: (a) input utilisation and (b) related to agricultural land potentiality. First segment of the selected variables includes agricultural inputs such as (i) tractors, (ii) oil-engines, (iii) electric-pumps, (iv) chemical fertilizers, (v) High-yielding varieties, (vi) agricultural labour force, and (vii) draught force (in terms of ploughs). The second segment consist of the variables like (viii) net sown area, (ix) intensity of cropping, (x) distribution of gross cropped area. Normal annual rainfall has been taken in addition to both the segments but while aggregating the segments this has been treated only once.

Now, the readymade data matrix for fourty-nine observations (districts) eleven variables has been run thrice into the computer on the basis of programming schemes of principal component analysis. Since, the mean value of factor scores of each principle component comes to zero (0) and standard deviation to $\sqrt{\text{eigen}}$ value, the factor scores have been divided into four categories of frequency as (i) $(X + \text{S. D.})$ and above, (ii) X to $(X + \text{S. D.})$, X to $(X - \text{S. D.})$ and (iv) below $(X - \text{S. D.})$ in order to read very high, high, low and very low levels of input utilisation, agricultural land potentiality and agricultural development respectively. The results have been obtained in three principal components.

The district level data have been collected and compiled from the year 1980-81 as average of three years (1979-80, 1980-81 and 1981-82) in order to avoid the possible anomalies. Wherever, there are restrictions for averaging the data, it has been adjusted for the year 1980-81 at the decision making level. Various secondary sources published by the Government of India have been consulted. The main sources are the Bulletin of Agricultural Statistics, Season and Crop Report, Bulletins of Fertilizer Association of India, Rainfall Records from IMD, Livestock Census of India and General Population Tables etc. The hill districts of Uttar Pradesh have not been included in the study because of non-availability of data accordingly.

The analysis and the results of the study have been described in many parts *i.e.*, the analysis of correlates, the analysis of factor loads and the analysis of factor combination and agricultural development.

Analysis of the Correlates

In this part, the degree of relationships between the components of development has been analysed with special reference to high values of coefficients of correlation (+ .500 and above). The characteristic remarks of the relationship between each of the factors have been given below as per the performance of Table 4.1.

(i) Positive Correlates

High positive coefficient of correlation (0.708) has been found between electric pumps (per 1000 ha. of G. C. A.) and fertilizers (Kgs. per ha. of G. C. A.). It is clear that the consumption of fertilizers is high in those areas where intensity of irrigation through electric pumps is high and *vice-versa.*

Oil engines (per 1000 ha. of G. C. A.) and spread of HYV are positively correlated (0. 698). It is worth mentioning that the concentration of HYV area will be high in those areas where higher number of oil engines are available to gross cropped area.

TABLE 4.1

Correlation Matrix

	X_1	X_2	X_3	X_4	X_5	X_6	X_7	X_8	X_9	X_{10}	X_{11}
X_1	1,000										
X_2	0.462	1.000									
X_3	0.373	0.272	1.000								
X_4	0.574	0.708	0.499	1.000							
X_5	0.283	0.474	0.698	0.685	1.000						
X_6	- 0.084	0.207	-0.275	0.184	-0.001	1.000					
X_7	- 0.397	- 0.065	-0.120	0.001	- 0.052	0.350	1.000				
X_8	0.169	0.363	0.267	0.294	0.417	0.010	- 0.083	1.000			
X_9	0.336	0.429	0.296	5.655	0.612	- 0.028	- 0.029	0.157	1.000		
X_{10}	0.075	- 0.413	- 0.129	- 0.469	-0.309	-0.533	- 0.589	- 0.087	-0.267	1.000	
X_{11}	- 0.187	- 0.299	0.264	- 0.033	-0.183	0.313	0.588	- 0.588	0.054	-0.199	1.000

Where,

X_1 = Tractors per 1000 ha. of gross cropped area,
X_2 = Electric pumps per 1000 ha. of gross cropped area,
X_3 = Oil-engines per 1000 ha. of gross cropped area,
X_4 = Fertilizers (in Kgs) per ha. of gross cropped area,
X_5 = Percentage of area under HYV to gross cropped area,
X_6 = Agricultural labourers per 1000 ha. of gross cropped area,
X_7 = Ploughs per 1000 ha. of gross cropped area,
X_8 = Percentage of net sown area to total reporting area,
X_9 = Intensity of cropping,
X_{10} = Gross cropped area (in hectare) per rural capita,
X_{11} = Normal annual rainfall in mm.

Fertilizers and HYV have been found highly correlated (0.685). It implies that the HYV area increases at the cost of fertilizers consumption.

Fourth ranking coefficient of correlation is 0.612 and has been found between HYV and intensity of cropping. The high value of intensity of cropping is explained by the high spread of HYV area.

Tractors (per 1000 ha. of G. C. A.) and fertilizers are ginning a high degree of relationship with each other (0. 574).

High positive correlation have been found between intensity of cropping and HYV area. It is very much apparent that it is HYV, that to a great extent, governs the level of intensity of cropping.

The above patterns of relationship are the indicative of developed agricultural economy in which the emphasis is given to the production of high value crops. Due importance is given to the surplus produce to be marketed and generating rural infrastructure and employment.

(ii) Negative Correlates

The highest negative coefficient of correlation has been between plughs (per 1000 ha. of G. C. A.) and G. C. A. (per rural capita) which shows that wherever the ploughs would be the common tools there would be lesser availability of G. C. A. to rural people. It shows a type of economy where land is scarce and labour is abundant.

The second negative coefficient of correlation is 0.588 and found between net sown area and rainfall. The percentage of net sown area tends to increase towards those area where rainfall volume decreases and *vice-versa.*

Another important remark is also met with the coefficient of correlation of 0.533 between agricultural labourers and G. C. A. availability. It shows that the availability of G. C. A. is less to the areas where human labour is intensively used in agricultural activities. In this type of economy, the agricultural operations are labour intensive and the economy is of subsistence nature.

Input Development

The development of physical inputs is the requisite qualification to make the agriculture dynamic and the rural economy stabilised. The inference drawn on the factor loads into three principal components of input development have been presented in Table 4.2 and the respective factor scores have been depicted. The three principal components accounted for 81.77 per cent in the total variance. New as well as traditional technological variants in addition to rainfall have been discussed here with their merits and demerits in the way of agricultural development.

TABLE 4.2

Factor Matrix of Input Development

	Variables	*1st component*	*2nd component*	*3rd component*
1.	Tractor	0.7005	—0.1451	0.0730
2.	Electric Pumps	0.7513	0.2870	—0.2818
3.	Oil Engines	0.7349	—0.1620	0.3041
4.	Chemical Fertilizers	0.8467	0.3843	0.1505
5.	HYV	0.7884	0.1601	0.1734
6.	Agricultural Labourers	-0.0588	0.8178	—0.2907
7.	Ploughs	—0.2839	0.6850	—0.0977
8.	Rainfall	—0.3933	0.5363	0.4772

The first principal component has been identified as the new technological variables explaining 47.48 per cent of the total variation. It can be apparently seen from Table 4.2 that it is the combination of tractor, electric pumps, oil-engines, chemical fertilizers and the spread of HYV which have posed a positive impact on the agricultural development. Among these, chemical fertilizers, HYV and electric pumps have emerged as highly sensitive inputs with the factor loads of 0.8467, 07884 and 0.7513 respectively. On the other hand, the traditional inputs like rainfall, ploughs and agricultural labourers have shown their negative contributions in the development of agriculture.

The pattern of factor scores of the first principal component discovers that fact that 'very high' concentration of modern technological inputs has taken place in the western part of western Uttar Pradesh, leaving aside the eastern half of region in the vagaries of traditional inputs and monsoons. In terms of developed inputs, the districts of Saharanpur, Buland Shahar, Meerut, Ghaziabad, Aligarh, Agra, Mainpuri, Rampur, Moradabad, Muzaffarnagar and Bijnor have attained an outstanding position in the state. In these districts, per 1000 hectares of G. C. A., tractor numbered about 5, the electric pumps about 10, oil engines about 15, chemical fertilizers about 44 kgs. per ha. and HYV area about 47 per cent of the gross cropped area. The highest score of 7.315 has been registered in Muzaffarnagar district. Besides, these, the districts of Gorakhpur, Deoria, Varanasi, Jaunpur, Faizabad and Barabanki in eastern Uttar Pradesh have also shown some new trends in agricultural development. But these were limited to mere adoption of chemical fertilizers and spread of HYV. Installation of private tubewells has become a prime mover of agricultural development in these developing districts. Low and very low level of modern inputs in agricultural can be seen towards the *Tarai* region in the northern parts of the state. The other area of deprived modern inputs includes the districts of Jhansi, Lalitpur, Banda, Hamirpur, Allahabad and Mirzapur where may be the northern parts of Vindhyan Uplands a great barrier in the way of adopting the developed inputs.

The concentration of factor scores of the second principal component covers the whole of eastern Uttar Pradesh where agricultural development is governed by labour intensive inputs like agricultural labour, plough and rainfall followed by chemical fertilizers and electric pumps. However, the former three inputs with the factor loading of 0.8178, 0.6850 and 0.5363 provide the base for various agricultural operations in eastern Uttar Pradesh. The third principal component is the combination of rainfall, oil engines, HYV and chemical fertilizers and which concentrates along northern districts of the state. The agricultural development in the districts of Basti, Gonda, Baharaich, Barabanki and Kheri is mostly an outcome of the interaction between rainfall and HYV and, chemical fertilizers, which is a highly lacking phenomenon in the southern districts, *i. e.*, Mirzapur, Hamirpur and Banda.

Potential Land Development

The extent of extensive and intensive utilisation of land already put under cultivation and its availability to the people who survive on it is the other aspect of measuring agricultural development. Its business is largely governed by the existing modern input development. Hence, the potential land development and modern input development will hold a positive relationship. The components upto the third principal component have explained as much as 78.06 per cent of the total variance.

Table 4.3. Factor Matrix Potential Development

Variables	*1st component*	*2nd component*	*3rd component*
Net Sown Area	0.3759	0.0908	-0.1921
Intensity of Cropping	0.6803	0.2622	0.1850
Cropped Area Availability	0.6320	0.3186	0.2668
Rainfall	-0.3231	0.1480	0.0204

It is apparent from Table 4.2 that the most powerful variable of potential land is the intensity of cropping followed by cropped area/rural person and net sown area in first principal component which alone accounted for 55.95 per cent of the total variance. Very high level of development of this component lies in the districts of Muzaffarnagar, Aligarh, Mathura, Moradabad and Mainpuri. In these districts, the intensity of cropping has been recorded more than 150, the gross cropped area per rural person equals to 0.32 hectare and net sown area more than 75 per cent. This is to remember that the level of modern input development is also high in these districts. The potential land is highly scarce towards northern parts in the district of

Basti, Gonda, Bahraich, Kheri and Dehradun as well as southern uplands in the districts of Jhansi, Lalitpur and Mirzapur where cropping intensity accounted for about 125 and the net sown area for less than 50 per cent. The factor scores of first principal component very from - 8.090 in Dehradun (the lowest potential district) to + 3.428 in Bulandshahar (the highest potential district).

The constituents of the second principal component are cropped area availability, cropping intensity and rainfall concentration of which lies again in the western districts of Uttar Pradesh. It is apparent from the districts of Dehradun, Saharanpur, Meerut, Ghaziabad, Aligarh, Mathura, Mainpuri and Etawah have attained an outstanding position. This component is not significantly different from the first component. It is further evident from Table 4.3 that these are the cropped area availability and intensity of cropping which constituted the third principal component. The factor loads of this component are less powerful as compared to those in 1st and 2nd components.

Aggregate Level of Development

The total variables considered for the present study accounted for as much as 83.73 per cent of the total variance at the third principal component. The first principal component alone registered 62.13 per cent of total variance and consists of modern technological inputs like chemical fertilizers. HYV, tractor and electric pumps having factor load 0.8196, 0.7433, 0.7251 and 0.7018 respectively (Table 4.4).

It is interesting to note that the districts of Saharanpur, Muzaffarnagar, Meerut, Bulandsahar, Ghaziabad, Aligarh, Mathura, Agra, Mainpuri, Etawah, Rampur and Moradabad avail very high level agricultural development by securing the factor scores more than + 1.83. The district of Muzaffarnager gained the highest position of agricultural developing in the state as a whole by securing the factor score + 5.712. Among 11 components taken for the purpose, the natural and traditional forces like rainfall, agricultural labourers and ploughs indicate no more positive impact on agricultural development. Ploughs and agricultural labourers maintaining factors of—0.76315 and—0.65706 respectively stand responsible for weaker status of agricultural economy.

The pattern of first principal component explains that the agricultural development is endowed with modern technological inputs in the western Uttar Pradesh. In both the parts of U.P. *i.e.*, southern and eastern U.P., very low level of agricultural development has been experienced. The districts of Jaunpur, Gorakpur and Deoria register the higher level of development as compared to its surrounding districts, although it falls below the normal regional level. On the other hand, the district of Azamgarh is highly deprived by securing the highest negative factor score (5.171) and having acknowledged as one of the most under developed district with very low level of agricultural development. Besides these, other districts which need immediate transformation and that having very low level of agricultural development are Dehraduru Ballia, Ghazipur, Jaunpur, Gonda, Faizabad, Sultanpur, Pratapgarh, Varansi, Allahabad and Banda.

TABLE 4.4

Factor Matrix of Overall Development

	Variables	*1st component*	*2nd component*	*3rd component*
1.	Tractor	0.7251	–0.0906	–0.0662
2.	Electric Pumps	0.7018	0.0952	0.1212
3.	Oil Engines	0.2054	0.5131	0.0954
4.	Chemical Fertilizers	0.8196	0.2250	0.0503
5.	HYV Area	0.7433	0.0657	0.0781
6.	Agricultural Labourers	–0.0621	0.0258	0.3721
7.	Ploughs	–0.2109	0.0808	0.2856
8.	Net Sown Area	0.1253	–0.1165	0.0958
9.	Intensity of Cropping	0.3697	0.1059	0.1028
10.	Cropped Area Availability	0.5481	0.1802	—0.0503
11.	Rainfall	–0.3345	0.4584	0.3571

The second principal component is a combination of rainfall, oil engines, land availability and chemical fertilizers. Its major concentration is towards northern parts. The districts of Gorakhpur, Baharaich, Kheri, Saharanpur and Dehradun are having highly increased amount of oil engines as these areas fall in the high amount of rainfall zone. Due to low density of population the per capita cropped area is comparatively high in these districts.

The third principal component of agricultural development comprises the high amount of rainfall, intensive use of agricultural labourers and ploughs. These factors are highly pronounced in the central part of the state, *i.e.*, the districts of Sitapur, Hardoi, Unnao and Baranbanki. The similar characteristics prevail in the districts of Ballia and Mainpuri also.

A comprehensive combination of the factors with relative significance to the existing agricultural development pattern has been depicted in. It can be seen that (a) the districts of eástern U. P. having human labour intensive-cum-traditional technology are characterised by low to very low level of agricultural development and (b) whereas the districts of western Uttar Pradesh particularly Meerut Division having machine-mixed modern technological inputs and larger combinations of factors are representative of high level of agricultural development.

As the analysis is made on the basis of eleven variables selected for the purpose, it appears to be more relevant and decisive to analyse in combination than that of the factors separately. The governing components responsible for development in the districts of eastern Uttar Pradesh and northern part of Vindhyan Uplands may be distinctly grouped into three in order of priority only: (a) ploughs, (b) rainfall and (c) agricultural labourers. The district of Ballia having accounted for the highest negative factor score of 9.645 is deemed to be the most deprived district for all components and their responses to agricultural development. Besides this, the districts of Azamgarh (—8.506), Ghazipur (—6.782), Varanasi (—6.318), Sultanpur (—5.380) and Pratapgarh (—5.308) are also indexed the same. Hence, such districts may be called as agriculturally backward districts where the planning for overall development is highly needed.

On the other side, higher level of agricultural development in western part of the State is accompanied by the higher number of components in combination showing modernisation of agricultural operations. A combination of 5 to 8 factors is a common phenomenon with the areas of high level of agricultural development. Components tend to increase in number with hierarchy of the level of development. For instance, the districts of Saharanpur, Muzaffarnagar, Meerut, Ghaziabad, Bulandshahr, Aligarh, Mathura, Etah, Badaun, Bijnor, Moradabad and Rampur (factor scores above + 7.757) comprising high level of agricultural development are privileged generally with 8 components, all being adequately suitable for modernishing the agriculture. These components are placed in order of their relative significance. The combined components may be mentioned as: (1) gross cropped area, (2) tractors, (3) oil engines, (4) net sown area, (5) HYV, (6) electric pumps, (7) fertilizers and (8) intensity of cropping. The district of Bulandshahr has high level of agricultural development by securing the factor score + 4.198 followed by the districts of Muzaffarnagar (+ 3.363), Moradabad (+ 3.345), Meerut (+ 3.102), Aligarh (+ 3.063), Etawah (+ 3.041) and Rampur (+ 3.025).

In order to minimise the disparities in the levels of agricultural development special attention is needed to pay for minimizing the disparities in man-land ratio. Let us put some comparative instances taking from western and eastern U.P. In general, in eastern U.P., ploughs, rainfall and agricultural labourers are maintaining first, second and third positions respectively in the combination whereas in western U.P. gross cropped area, tractors and oil engines are holding the respective positions.

Concluding Remarks

Through the above study, few strategic points and concluding remarks regarding agricultural development may be submitted as such:

1. The present analysis leads to the conclusion, that low level of agricultural development is mainly due to low level of modern agricultural devices.

Hence, the human labour and primitive tool-intensive agricultural inputs of Eastern Uttar Pradesh need be immediately substituted by scientific and modern technology. Preferably, it should be adopted in the form of bullock and tractor-mixed technology.

2. The land potential factors in aggregate are dependent on input utilisation factors. As for example, the high level of agricultural development in the districts of western Uttar Pradesh is highly correlated with the high degree of concentration of both the aspects, whereas the situation is quite reverse in the districts of eastern Uttar Pradesh.
3. Agro-industrialisation in western Uttar Pradesh is an outcome of its agricultural development. It is highly desirable to point out that the surplus agricultural raw materials available must be retained and processed in the areas of production. Its export of flow to other areas *i.e..* distantly away from the respective area of production, will yield to economic leakage resulting into low level of agro-economic development and level of employment opportunity.

5

AGRICULTURAL PRICE POLICY

The price policy plays an important role in the development of agriculture. It has been recognised both as an incentive to greater production and as an insurance against the vagaries of the market. An important aspect of the agricultural price policy is the assurance to the farmer that his efforts to augment production through adoption of improved technology would to become unremunerative through the price factor. Such a policy is the *sine qua non* of building a modern scientific agriculture.

The inadequacy of marketing facilities in the country-side has resulted in wide fluctuations in agricultural prices in the past. The biological character of agricultural production and its dependence upon weather conditions leads to variations in the volume of agricultural output from year to year. It has been observed that due to the inadequacy of marketing facilities in years of increased agricultural output, prices tend to fall sharply; sometimes to such low levels that they act as disincentive to increased agricultural production. In subsequent years, the farmer is tempted to slacken production efforts or to divert his land to other crops. The agricultural price policy should therefore aim at reducing the unduly large fluctuations in the prices of agricultural products, so that the risk of loss to the producer should be minimised even in the event of an exceedingly good harvest.

The policy of fixing the minimum guaranteed support prices can go a long way in creating a favourable climate for inducing the farmer to augments his investment in agriculture and to adopt improved technology. At the minimum support prices, government undertakes to purchase all quantities offered to it so that the prices of agricultural produce do not fall below a specified level. The minimum support prices policy becomes essential in two ways; firstly it acts as a psychological incentive to the cultivator that his efforts in increasing production will not go in vain. The level at which minimum prices are fixed also plays an important part in encouraging the farmer to undertake investment in improved technology.

In determining the level of minimum prices, a number of factors have to be taken into consideration. These include prices of agricultural produce in the open market, returns from competing crops and also trends in the expenses to cultivation. Since the 1964-65 season, the minimum support price policy has been extended to cover wheat, paddy, jowar, bajra, maize and gram among the foodgrains and sugarcane, cotton and jute among commercial crops. These prices have been gradually revised upwards since that year, keeping in view the changing conditions in the economy.

It may be mentioned that minimum support policy is one of the instruments for promoting agricultural growth; the main instruments being technological and institutional changes. Price support policy contributes to growth by inducing the farmer to adopt improved technology without fear of an excessive price fall.

Increase in input prices would influence the capacity of the farmer to undertake technological changes in agriculture. Thus, the prices of important inputs such as fertilizers, pesticides and improved seeds. Would play an important role in encouraging the farmer to step up their utilization. Inadequacy of supplies of inputs especially of fertilizers an pesticides, as well as improper coordination in the matter of suppliers would hamper improvements in agriculture to a great extent. Price policy could to that extent play a secondary role if the other factors of cultivation are inadequate. Together with the inadequacy of inputs, thee is also the problem of increased costs of input. A sharp rise in inputs prices may result in a corresponding increase in costs of cultivation leading in turn to the vicious circle of price increases of agricultural products, industrial goods, wages and agricultural inputs. It would be necessary to keep a close watch on the costs of inputs needed for increased production. This would involve collection and maintenance of data, on a regular basis, on input index and costs of cultivation.

In order that the benefits of the minimum price policy should percolate to the farmer, it is necessary that machinery should exist at the primary level to guarantee such a price. It should be possible for the government to honour its commitments that the minimum support price will be offered to the producer in all areas and at all times in case prices fall below the minimum. There is need therefore for stepping up the machinery for purchase operation at the support prices. Such a machinery should be established on a permanent footing; its role should be oriented towards price stability. The mere presence of such machinery would ensure that the trade would not exploit the producer and that the cultivator gets an adequate price. The machinery, for purpose of assuring a guaranteed support prices should be set up not only in the surplus states but also in surplus pockets and surplus areas in the deficit states, so that it can step in and buy the requisite quantities of foodgrains and other agricultural commodities offered to it at the minimum prices. This would be essential if the farmer has to be assured of a certain level of process for his produce.

The inadequacy of marketing facilities leading to sharp fluctuations in prices has been mentioned elsewhere. Linked up with the question of assuring the producer of a guaranteed price for his produce, is also that of improvement in the structure of marketing. At present the bulk of the marketable surplus of the producer is sold into villages or, in some areas, in assembling markets. The price received by the producer for his produce is depending on various factors such as the prices obtaining in key consuming and distributing markets, the distance between the production and consuming centres, the channels of intermediaries through which the produce moves, the arrangements for storing of the commodities, credit facilities as well as the agency which the farmer approaches, unwarranted recoveries, multifarious charges and deductions, inhibit the producer from getting a fair return form his produce. In a large number of states, the regulated markets have not made much headway. The setting up of such markets helps in defining and specifying market charges and regulating the transactions by licensing of market functionaries such as commission agents, brokers and weighmen. It ensures introduction of standard weights and measures, assists in the setting up of suitable committees for settlement of disputes as well as the management of the market, arranges for the display of market information with regard to prices arrivals, stocks, etc. makes arrangements for storage, sales etc. All of thee factors assist the farmer in obtaining higher prices for his produce. The efficiency of marketing is therefore one more important factor to improve the earning capacity of the farmer.

The main role of the minimum support price policy is to create conditions favourable to the achievements of the long-term objective of raising agricultural output. But so long as the gap between supply and demand persists, price policy has also to aim at stabilising prices at the consumer level. This is especially important in the case of foodgrains, which constitute an essential item of consumption and whose prices set the pace an pattern of price behaviour in other sectors of the economy. In a situation of shortage, available supplies of foodgrains in the free market would generally be at very high prices. The poor and vulnerable sections of the community would be put to extreme hardship in such a situation. Both from the welfare angle i.e., for protecting the interests of the vulnerable sections of the community, and also from the narrow economic standpoint, i.e., of keeping the impact of the supply-demand imbalance of prices within reasonable limits, a sound price policy is essential. For achieving the objective of stabilising prices at consumer levels, the government may adopt systems of public distribution to meet at least a part of the consumer demand at fixed reasonable prices. The public distribution system can be in the form of rationing or fair price shops or both. Whatever be the policy of public distribution followed by government, it cannot operate unless the necessary stocks are acquired by it. One way of acquiring the stocks would be through the open market. But in a system of shortages, it is a moot point whether government would be able to acquire the needed stocks without raising the prices in that process. Other means of procurement are by a levy on producers or on traders or millers depending upon the system of procurement followed in the different states. In determining the procurement prices, due regard would have to be given to the trends in the level of market prices during a particular season, the surplus or deficit nature of the state, the crop conditions in a particular year, the system of procurement followed as well as the varietal differences of the crop. Determination of prices would also have to take into account the distribution costs an margins as well as the marketing and processing costs. However, the efficiency of the systems of marketing prevailing in the different states and regions will determine very largely the share that the producer will get in the prices ultimately paid by the consumer. Obviously procurement prices should be such that they assist and not hamper production in the long run.

Position Prior to Fourth Five-Year Plan

Agricultural pricing policy has an important role in development planning, particularly in a predominantly agricultural economy like India's. With regard to foodgrains a policy of fixing support prices for important cereals and gram was adopted throughout the country in 1964.

In the subsequent year the Agricultural Prices Commission was set up to advise the government from time to time on appropriate price policies for agricultural commodities. In the same year was also established the Food Corporation of India which, for the first time, sought to provide an all-India machinery for purchase of foodgrains. Purchases were undertaken in different contexts, namely procurement through levy and other means on times to shortage and price support in times of plenty. Procurement prices for major foodgrains were also being fixed for the purpose. During the drought years of 1965-66 and 1966-67 the minimum support prices remained mostly inoperative on account of high market prices. In 1967-68, following a bumper wheat crop, the government adopted a policy of providing support at procurement prices with a view to providing incentives to the cultivators for increasing production and consequently procurement. This policy was continued during 1968-69. Besides minimum support and procurement prices, issue prices were also being fixed for the supply of grain through the public distribution systems mainly for the purpose of protecting the interests of the vulnerable sections of the population.

As regards commercial crops, minimum support prices were being fixed for jute for several years but these were mostly ineffective on account of the inadequacy of the purchase organisation in the field. In the case of cotton, ceiling and floor prices fixed on a statutory basis till 1967-68. In that year the statutory control on cotton was lifted and minimum support prices were fixed in 1967-68 and 1968-69 at a level higher than the previous floor prices. In the case of sugarcane, statutory minimum prices payable by the sugar factories were being fixed for a long time.

Position during the Fourth Five Year Plan

Foodgrains. The Fourth Five Year Plan laid considerable emphasis on streamlining the arrangements for providing effective support as a means of facilitating the adoption of modern technology and increasing agricultural production. It fact the assurance of incentive prices to the producers has been an essential ingredient of the New Strategy of agricultural development adopted since 1966-67. In the case of foodgrains, the price policy adopted comprises fixation of minimum support prices for all important cereals. Although minimum support prices have been announced for kharif cereals during the first four years of the Fourth Five Year Plan, support to the market has been provided at the level of procurement prices which are generally higher than minimum prices and include considerable incentives for the producers of food grains. The policy of providing support at procurement prices has aided the adoption of new technology particularly in the case of wheat which has brought about a substantial increase on foodgrains production. Production of wheat has almost doubled since the adoption of the New Strategy; its procurement has also shown considerable increase having reached 5.1 million tonnes in 1971-72 and 1972-73 marketing seasons as compared to neglectable quantity procured before 1967-68. In the case of rice, though some increase has taken place in its production in recent years, there is no evidence yet of a real breakthrough. Among coarse grains, bajra has recorded better performance but the overall production of coal cereals has been fluctuating considerably from year to year on account of weather. In the case of pulses minimum support prices and procurement prices, have not been fixed during the Fourth Plan period. However, despite high market prices production of pulses has remained by and large stagnant or even decline dover the last few years.

The government has been maintaining a large network of ration/ fair-price shops for supply of grains to the vulnerable sections of the population at reasonable prices. Issue prices are accordingly fixed for this purpose. Government had also undertaken to build up a buffer stock of five million tonnes in foodgrains during the Fourth Plan with a view to even out fluctuations in availability over the years and to ensure stability in prices.

Along with the policy of buffer stocks, the government took over the wholesale trade in wheat in 1973 with the primary objective of providing fair prices both to the producer and the consumer.

Commercial Crops

The policy of fixing minimum support prices for jute and cotton has been continued during the Fourth Five-Year Plan. In order that the minimum support prices fixed for jute are really effective, they were fixed on a statutory basis for the 1972-73 season. During 1971-72, minimum support prices for cotton were not announced. However, following a bumper crop during the year, support was provided to the cotton market at prices higher than the previous minimum support prices. For this purpose a special purchases price was fixed, corporations were set up to handle, among other things purchase and sale operations in respect of jute and cotton in order to make the support price policy more effective. Purchase prices were also fixed for soyabean. In the case of sugarcane, the policy of fixing statutory minimum prices payable by the sugar factories was continued. A system of public distribution of sugar at fixed issue price was maintained to meet the minimal requirements of the people. To regulate production, distribution and prices of sugar, the government adopted a policy of compete control/partial decontrol of informal arrangement involving procurement of a part of the production of sugar at fixed prices/complete decontrol at different periods depending on the overall availability and prices position of sugar in the country.

The Future Policy

The significant progress in the production of foodgrains, particularly wheat in recent years is mainly attributable to the adoption of high yielding varieties and the accompanying technology aided by the policy of providing support at procurement prices which include considerable incentive for producers. In the case of wheat, this policy of providing support at incentive prices, has however, resulted in the diversion of area of wheat from less profitable crops, like gram and barely. With the elimination of imports from January 1972 which indicently was made possible by the remarkable progress in which production; release of wheat through the public distribution system at the present procurement and issue prices (Rs 76 and 78 per quintal) involves a huge subsidy. It is estimated that distribution of wheat from the public distribution system would involve a subsidy of nearly Rs 150 crores during 1972-73. Further, in years of good production of other foodgrains, this policy is also likely to result in the accumulation of stocks with the Governments as was the experience in 1970-71. although practically all the stocks are likely to exhaust by March end, 1973, wheat procurement of the current year will again leave us with a stock of about 7 million tonnes by July end, 1973. While it would be necessary to continue to ensure incentive prices to the producers of wheat, all avenues for reducing the burden of subsidy involved in the distribution of wheat including adjustment in procurement prices and maximum economy in distribution cost should be the main objective to the future price policy.

In the case of rice the slow rate of progress in production can be attributed mainly to the non-availability of suitable varieties and other technological factors. Water management presented a major problem during the kharif season for achieving breakthrough in rice production. The extension of area under summer rice offered greater possibilities of stepping up production. While the policy of providing.

1. Imports of the order of 2 million tonnes of wheat have been made for 1972-73. Since these imports are on commercial basis, they will not a make any change in the subsidy, particularly because prices of wheat in the international market have also increased.
2. The original procurement target of 8.4 million tonnes was reduced to 6.0 million tonnes and actual achievement is hardly expected to be 5.0 million tonnes, incentive prices to the producers during the Fifth Five Year Plan would be a step in the right direction, main effort at increasing the production of rice will have to be on the technological revolution.

In the case of coarse cereals (jowar, bajra and maize), production continues to fluctuate widely despite the policy of providing support at incentive prices. The main causes of instability in their production have to sought in weather conditions, proneness of hybrid varieties to pests and diseases, etc. While it would be necessary to continue the policy of providing support at incentive prices to these grains during the Fifth Five Year Plan; the major effort in securing a sustained increase and imparting a measure of stability to their production should lie in the field of (i) research for evolving drought and disease resistant varieties, and (ii) developing and spreading dry farming techniques.

Price policy can hardly achieve anything in so far as pulses are concerned. The main cause of stagnation in production was the absence of high yielding varieties. The wide intra-seasonal fluctuations in prices also pointed towards the need for improving the, marketing structure to protect the interests of the producers. Whatever the government might do to bring about stability in the prices of pulses, nothing will possibly work unless the production is increased to meet the ever-increasing demand of the consumers. Since there is hardly any promise to do anything in regard to pulses, efforts will have to be directed towards augmenting the available supplies of protein from alternative sources like groundnut and soyabean meal. This would call for more emphasis on new crops like soyabean.

Commercial Crops

The avoid any hardship to the producers in a situation of glut in the market, minimum support prices should continue to be fixed for cotton, jute, important oilseeds and other important crops. Since the production of major commercial crops is subject to considerable fluctuation and is generally short of requirements, a policy of undertaking limited purchase and sale operations for encouraging the production of these crops and for ensuring stability in prices will have to be followed.

In the case of sugarcane, the present policy of fixing statutory minimum prices for cane delivered to the sugar factories is right. Efforts will also have to be made to impart greater stability to sugar production through a long-term policy. A policy of building up a moderate buffer stock supplemented with export/import adjustments should be adopted during the Fifth Five Year Plan. With regard to distribution, the present policy of partial decontrol under which the vulnerable section of the population is provided their minimum requirements at controlled prices, will be that best policy for at least a few years till there is sufficient improvement in the sugar situation.

As for cotton, the position is rather precarious. It has been observed during the last two years that with the coming into picture of the Cotton Corporation of India, the cultivator is being put into a difficult situation in the immediate post-harvesting period. There was practically no buyer for the Punjab as well as Gujarat cotton during 1971-72 season and even during the early part of this year the cultivator faced the same difficulties. Now that certain high-yielding varieties like H-4 in Gujarat, MCU-5 in Karnatka an Tamil Nadu have ben evolved and aerial spraying in Punjab gives promise of increasing the yield of cotton, there is an urgent need to provide remunerative price to the cultivator failing which he will be tempted to divert areas from cotton to other suitable crops.

The position of jute as well as groundnut is a story of complete failure on the production front. Although an intensive jute cultivation programme was sanctioned by the Planning Commission at a cost of Rs 60 millions aimed at giving an additional production of 8.4 lakh bales by 1973-74, there is no success visible in this respect so far. The Planning Commission's target of 7.5 million bales by the end of the Fourth Plan may flop like the earlier schemes. More or less same is the position with regard to groundnut the other oilseeds where no breakthrough in production seems to be possible in the near future. The only silver-lining in this case is the introduction of sun-flower and soyabean. While the progress of soyabean is not very encouraging on the basis of the last few years' experiments, sun-flower seeds appear to be quite a success. With the prevailing market prices of oils and oilseeds there is perhaps little that the government can do to curb them. In the interest of the consumer it may become necessary to augment the supplies through imports of cheaper oils such as palm oil and soyabean oil. In case of jute, the present policy of announcing statutory minimum prices at reasonable levels will perhaps be enough to provide at least a psychological support to the producer. Commercial purchases by the Jute Corporation at the prevailing market prices in the immediate post-harvesting period, particularly in the distant areas away from the main consuming centres—Calcutta—is sound. To avoid transport bottlenecks, it should be decided as a matter of policy that all future expansion of the jute industry should be outside West Bengal, in such States as Assam, Orissa, Bihar and Uttar Pradesh where there are possibilities of expanding the area under jute.

Since the demand for higher levels of support rests on the plea that the guaranteed prices should be incentive prices, the question of price incentives to increase production deserves careful consideration. While there is a great deal of evidence to show that the production of many individual commodities is highly responsive to price, in an economy where shortage is not confined to a few agricultural commodities but runs across the board, what is involved is the response to aggregate agricultural output, rather than of its individual commodity constituents, to change in price. Although firm empirical evidence for India is wanting, a *priori* reasoning as well as available evidence for other countries indicate that the response of aggregate agricultural output to price is feeble. However, unlike in a period of stable technology, such response would be expected to be greater in a period of technological change. But since the incentive for larger production flows from enhanced profitability, it can derive as much from higher prices as from lower unit costs of production. So viewed, improvement in technology become an alternative to higher prices to induce larger output. This competitive relationship between technology and price, however, does not imply that there is no complementary relationship between them; in fact, the existence of favourable price climate does help accelerate the absorption of improved technology.

Beyond the range of complementary relationship, however, the choice between the two instruments for achieving output goals hinges upon other considerations including the objectives of income policy. Here it needs to be emphasized that higher prices of agricultural products imply a transfer payment from the consumers to the producers of these commodities, whereas improvements in technology benefit the producers as well as the society as a whole by making it possible to obtain a larger output from the same productive resources. Further, higher prices benefit more the bigger producers who have more to sell than the smaller ones, and hurt those who, on balance, are net buyers of these commodities. Higher prices thus contribute to an accentuation of income inequalities. The benefits of improved technology on the other hand, are independent of the extent to which one produces for self-consumption or for the market: thus, given the necessary organisational infrastructure, the benefit of seed-fertilizer based technology, for example, can accrue as much to the small producers as to the bigger ones. The goal of income policy being to reduce income inequalities, the choice between the two instruments for increasing production therefore favours emphasis on improvements in technology rather than on higher prices.

Even part from the above considerations, the emphasis on technological improvements is required on yet another consideration. When technological breakthrough in one segment of agricultural production casts its impacts on other segments, there are limits on the extent to which the price instruments can be wielded even for correcting the consequent imbalances in the production basket. The provision for irrigation, for example, has the effect of shifting the area from pulses to cereals like wheat. The enhanced profitability of wheat cultivation consequent upon the introduction of high-yielding varieties of seeds has accentuated this process. The result has been a certain stagnation or even a decline in the production of pulses and the consequent increase in their prices. The fact that the rising prices of pulses relative to those of cereals have not been able to stem the shift of the area, and the further fact that pullses are important sources of protein for poorer people, bring out the limitation of the price mechanism in providing the necessary corrective to the situation. It is obvious that the only satisfactory solution of the problem lies in matching the technological advances in respect of cereals with like advances in other crops, be they pulses or other cash crops.

In many crops like pulses and groundnut, the problem is not one of uneconomic prices but of highly fluctuating prices. For reducing the uncertainty flowing from wide fluctuations in the price of these crops which are grown largely under unirrigated conditions, what is required is not support to prices at artificially high levels but other stabilisation measures like open market operations and appropriate credit an import or export policies.

The danger of providing support at high levels is strikingly brought out by the case of wheat. Administered prices, once pitched high, even if for good reasons, become difficult to lower subsequently when production increases. High support prices then require as in wheat, large subsidies. Even apart from its desirability, the feasibility itself of a policy of large-scale subsidies to agriculture for a country like India is open to question. In a country like the United States of America, for example, where agriculture accounts for about 3 per cent of the Gross National Product, non-agricultural sector with a weight of 97 per cent would be feel the sting of subsidising the much smaller agricultural sector. But in India, where agriculture accounts for a little less than half of the national income, its subsidy by the non-agricultural sector on any significant scale cannot obviously be stingless. The income levels in agriculture here have therefore to be lifted not through a policy of high prices involving subsidies but through a continuing drive at improving its efficiency.

Agricultural Prices Commission

The Government of India appointed the Foodgrains Prices Committee in 1964 under the chairmanship of Shri. L. K. Jha, to advise on the determination of producers' prices for foodgrains on an all-India basis, the reasonable margins for the wholesalers and retailers, and the consumer prices for the 1964-65 crop. The Government also asked the Committee to suggest suitable terms of reference for an agency which would provide such advice on a continuing basis.

After a study of the trends in wholesale prices in the post-harvest periods, the Jha

Committee recommended minimum support prices which were substantially higher than the average of the post-harvest prices actually realised by the farmers in the immediate past. For instance, while the averages wholesale prices of coarse paddy during trie three post-harvest seasons, 1961-62 to 1963-64, ranged in different states between Rs 28.56 and Rs 38.36 per quintal, the minimum prices suggested by the Committee ranged between Rs 33.50 and Rs 39.00 per quintal. In addition to the producers' prices, certain maximum wholesale and retail prices were also suggested by the Jha Committee for paddy and rice for the 1964-65 season. These were enforced after appropriate revision in the light of the views of the state government. In view, however, of the difficulty in enforcing the price ceilings, the maximum prices were subsequently withdrawn by most of the states.

The Jha Committee also suggested the setting up an Agricultural Prices Commission to advise, on a continuing basis, on price policy for the agricultural crops, with a view to evolving a balanced and an integrated price structure and resolving the claims of competing crops on limited resources in the perspective of the overall needs of the economy. The Agricultural Prices Commission was established early in 1965 to advise the government on the price policy of agricultural commodities with due regard to the interest of the producer and the consumer. While recommending the price policy and the relative price structure, the commission is required to keep in view:

(i) The need to provide incentive to the producer for adopting improved technology and for maximising production;

(ii) The need to ensure rational utilisation of land and other production resources;

(iii) The likely effect of the price policy on the rest of the economy, particularly on the cost of living, level of wages, industrial cost structure, etc.

since its inception, the Commission has been in constant examination of the trends in prices, production, etc. of the various agricultural commodities and has submitted to the government a number of reports on price policy for the major crops. The Commission generally recommends the minimum support prices which are in the nature of an insurance against an excessive fall in prices in the wake of an increase in the output of agricultural commodities an also, in respect of foodgrains, the procurement prices at which the government should procure the grains, keeping in mind the trends of market prices, the size of the crop and the need for public distribution. With a view to making price policy more effective, the Commission has also been making recommendations on related matters, such as agencies for procurement, storage and movement of the procured grains and development of facilities for quality assessment and orderly marketing of the agricultural produce.

The actual policy announcements are made by the Government of India in the light of the recommendations of the Commission and after consultations with the state governments. A statement showing the recommendations made by the Commission from time to time and those announced by the government are shown in Table 5.1 to 5.3.

TABLE 5.1

Sugarcane: Statutory Minimum Price

Season	*Minimum price at 9.4 per cent recovery (Rs per quintal)*	*Premium for every 0.1 percentage point increase in recovery (Paise per quintal)*
1966-67	5.68	4.00
1967-68	737	5.36
1968-69	737	5.36
1969-70	737	5.36
1970-71	737	6.60
1971-72	737	6.60
1972-73*	8.00	9.40

* The price linked to recovery of 8.5 per cent with a premium of 9.4 paise per quintal for every 0.1 per cent increase in recovery above 8.5 per cent. *Source:* Directorate of Sugar and Vanaspati, Ministry of Agriculture.

Food Corporation of India

The Food Corporation of India which was established in January 1965 to undertake purchase, storage, movement, transport, distribution and sale of foodgrains has been gradually extending its scope of operations. It has set up a number of regional and district offices. Besides setting up modern rice mills and mechanical paddy dryers, the Corporation has been responsible for the supply of imported wheat to the roller floor mills and for the distribution of wheat products in various states. It played a major role for the first time in the pricurement operations of wheat in Punjab and Haryana when production in these two states touched an all-time record.

The organisation of the Corporation was strengthened during 1969 with the transfer to it of a major part of the Food Department. With the completion and the transfer of the executive be functions of that Department, the Food Corporation of India has become thé sole agency of the Central Government for state trading in foodgrains. In a short span of five years it expanded its activities both in terms of area of operations and the volume of purchased and sales. The internal purchases of

TABLE 5.2

Wheat: Procurement Prices as Fixed by the Government

State	*Variety*	*1965-66*	*1966-67*	*1967-68*	*1968-69*	*1969-70*	*1970-71*	*1971-72*	*1972-73*	*1973-74*
Bihar	Indigenous Red				71.00	71.00	71.00	71.00	71.00	71.00
	Mexican				76.00	76.00	76.00	76.00	76.00	76.00
	Common White	52.00	54.00	95.00	76.00	76.00	76.00	76.00	76.00	76.00
	Superior				81.00	76.00				
Gujarat	Indigenous Red	55.00			73.00			72.00	72.00	72.00
	Mexican							76.00	76.00	76.00
	Common White	60.00			78.00			76.00	76.00	76.00
	Superior	65.00			83.00			76.00	76.00	76.00
Haryana	Indigenous Red					66.00	74.00	74.00	74.00	74.00
	Mexican			88.60*	76.00	76.00	76.00	76.00	76.00	76.00
	Common White	52.00	59.50*	94.45*	76.00	76.00	76.00	76.00	76.00	76.00
	Superior	56.00	63.50*	96.85*	81.00	76.00				
M.P.	Indigenous Red	52.00	55.00	65.00	74.00	74.00	74.00	74.00	74.00	74.00
	Mexican			75.00	76.00	76.00	76.00	76.00	76.00	76.00
	Common White	57.00	59.00	70.00	76.00	76.00	76.00	76.00	76.00	76.00
	Superior	61.00	63.00	75.00	81.00	76.00				
Mharashtra	Indigenous Red					70.00	70.00			
	Mexican									
	Common White									
	Superior									

State	Variety	1965-66	1966-67	1967-68	1968-69	1969-70	1970-71	1971-72	1972-73	1973-74
Punjab	Indigenous Red									
	Mexican			74.50*	76.00	76.00	76.00	76.00	76.00	76.00
	Common White	52.00	59.00*	76.00	76.00	76.00	76.00	76.00	76.00	76.00
	Superior	56.00	63.50*	81.00	81.00	76.00				
Rajasthan	Indigenous Red	47.00		77.00	74.00	74.00	74.00	74.00	74.00	74.00
	Mexican			79.50	76.00	76.00	76.00	76.00	76.00	76.00
	Common white	51.25*		81.00	76.00	76.00	76.00	76.00	76.00	76.00
	Superior	55.00*		85.00	81.00	76.00				
Uttar Pradesh	Indigenous Red	60.00	80.00	72.00	74.00	74.00	74.00	74.00	74.00	74.00
	Mexican			85.00@	76.00	76.00	76.00	76.00	76.00	76.00
	Common White	65.00		85.00	76.00	76.00	76.00	76.00	76.00	76.00
	Superior				81.00	76.00				

* Average of prices at which purchases were made.

@ Later reduced to Rs. 80.00.

TABLE 5.3

Rice: Procurement Prices as Fixed By the State Governments

State	*Standard Variety*	*1964-1965*	*1965-66*	*1966-67*	*1967-68*	*1968-69*	*1969-70*	*1970-71*	*1971-72*	*1972-73*
Andhra Pradesh	Akkullu	63.00	63.45	65.82 (26.12.66) 66.50 (15.4.67) + 6.00 (11.8.67)	72.69	72.69	72.69 74.22 (10.1.70)	80.32	80.32	83.37
Assam	Winter Sali	61.23	63.34	80.90	107.77	97.65				
Bihar	Coarse	59.67	68.00 69.66 (1.3.66)	69.66 90.82 (1.1.67) 1.25.00 (17.6.67)	95.25 +5.75	95.25 +5.75	95.25	95.25	95.25	95.25
Gujarat	Sathi	64.10	68.10	72.20						
Haryana	Begmi	60.00	60.00	69.50	86.00	86.00	87.00	88.80	88.80	91.80
Kerala	Palghat Matta	71.76 76.18 (3.9.65)	76.18	76.18	90.14	90.14	90.14	85.88	85.88	
Madhya Pradesh	Gurmatta	58.23	64.80	73.43	85.00	85.00	85.52	86.30	56.30	
Maharashtra	Coarse	65.50	69.40	68.50	76.15	76.15	78.50	81.00	83.00	83.00
Mysore	Coarse (Raw)	60.70	67.00	70.00	70.00	74.30				
Orisa	Common	59.50	64.00	74.15						
Punjab	Begmi	60.00	68.00	69.50	85.00	87.00	87.49 87.45 (5.8.70)	88.80	88.80	91.80
Tamil Nadu	Kattai Samba	64.45	70.01	72.80	72.80					
Uttar Pradesh	Grade III	60.00	66.50 68.00 (16.12.65)	69.50	90.00	90.00	89.00	89.00	89.00	92.00
West Bengal	Common	61.82 64.30 (1.1.65)	64.30 66.00 (16.1.66)	72.00 93.75 (7.4.67)	93.75 + 8.00	94.00 + 8.00	91.20 + 7.50	92.90 + 7.50	92.90 + 7.50	92.90 + 7.50

\+ Bonus

Note: Prices are inclusive of cost of gunny in Gujarat, Haryana, Kerala, Madhya Pradesh, Mysore, Orissa, Punjab and Tamil Nadu in all the years, in Andhra Pradesh, Bihar and Uttar Pradesh for 1964-65 and 1965-66.

the Corporations out of the 1968-69 crop increased to about 4.7 million tonnes. The purchase and sales turnover has expanded form about Rs 290 crores in 1965-66 to about Rs 1,500 crores during 1973. The Corporation, which was set up with the object of securing a commanding position in the foodgrains trade, is now playing a dominant role in purchase, storage and distribution of foodgrains in the country.

The Corporation procured about 47 lakh tonnes of foodgrains from the 1968-69 crop. This represents an increase of nearly six lakh tonnes over the previous year. The performance of the Corporation during the year 1972 in respect of the purchase, movement and storage of the bumper wheat harvest in Punjab and Haryana was particularly noteworthy. The procurement operations are increasing in magnitude every year. Today when the 1972-73 wheat crop is expected to reach of level of twenty six million tonnes and the government has decided to takeover the wholesale trade in wheat, the Corporation is expected to play a major role to implement the price policy of the government in respect of foodgrains.

Jute Corporation of India

In order to assure the jute grower a remunerative and fair price for his produce and thereby sustain his interest in jute growing, the Government has adopted a price support policy. Under this policy, minimum support prices are fixed in consultation with the Agricultural Prices Commission for delivery at Calcutta as well as for upcountry markets. Grade and variety differentials are also laid down to determine prices with reference to the standard, *viz*.. Bottom grade of Assam variety of white jute. Even before the Agricultural Prices Commission was set up, such an arrangement existed. It has thus been accepted that the primary requirement for attaining maximum production of fibre is the assurance of a fair price to the cultivator. While dealing with the price support policy and the minimum support price for the 1968-69 season, the Agricultural Prices Commission recommended *inter-alia* that the government might consider the advisability of charging some agencies with the standing responsibility for price support operations in raw jute and making it obligatory for such agencies to buy whatever quantities are offered for sale at the support prices at specified assembling points. Such a centralised purchase was necessary to ensure that market prices did not sag below the support price at any time. It was in this context that government decided that a Jute Corporation should be set up for conducting price support as well as buffer stock operations which would impart stability to prices and contribute to the growth of jute production in the country.

The functions of the Jute Corporation which came into existence early in 1971 broadly are:

(i) To conduct purchase operations to ensure the maintenance of support prices;

(ii) To procure and maintain a buffer stock of fibre (including imported fibre, if necessary);

(iii) To conduct commercial operations in a judicious manner;

(iv) To import fibre, as and when necessary to supplement indigenous availability;

(v) To export fibre (to the extend authorised by government);

(vi) To process and export jute goods;

(vii) To undertake export promotion measures for jute goods.

A centralised agency with its judicious purchasing policy should be able to ensure for the grower a better price than he is getting today by a process of elimination of the many tiers of intermediaries now operating in the jute market. This is the long term objective of the Corporation which should, in course of time, save the grower from the clutches of the middleman. In this endeavour, the Corporation is making arrangements to buy jute directly from the jute growers and build up its direct relationship with them. In short, the Corporation is expected to assume a commanding position in the jute market and be in a position to act as the most powerful stabilising factor in the market.

In the field of exports of jute goods—function which may be taken up by the Corporation in due course—particularly sackings, the Corporation might play a very useful role. As regards export promotion, considerable work has to be done. The Indian Jute Mills Association at present maintains three foreign offices (New York, London and Brussels); the Association has been pressing the government to take over these offices. This may be one of the first functions of the Jute Corporation in the field of export promotion.

In brief, the Jute Corporation will combine in itself all the functions of (i) Buffer Stock Agency, (ii) Jute Board, and (iii) Jute Marketing Board envisaged in the report of the survey made of India's jute products by the Administrative Staff college on behalf of U. S. Aid.

With the setting up of the Jute Corporation, the State Trading Corporation of India has ceased to be responsible for jute purchases with effect from the 1971 -72 season onwards. The superstructure of the S.T.C and expertise gained by it over a period of years in the purchase operations in the jute market is now available to the Jute Corporation.

The Jute Corporation has completed just two years of its existence and it is difficult to assess its work during this short period. The major problem for it is to establish the agency through which it can make purchases.

The cooperative structure in most of the jute growing states is weak. Hence, the Corporation decided that, while giving preference to growers' cooperatives wherever they function effectively, Departmental Purchase Centres should be set up at other places. During the 1972-73 season, 14 such Departmental Purchase Centres have been set up to buy jute directly from the growers. Buying jute at the growers' level presents certain difficulties. Jute is sold by growers in an unassorted form and the buyer has to be technically and commercially skilled enough to correctly estimate the grade composition of the ungraded jute and its moisture content and to evaluate it in money terms. Recruitment of experienced personnel trained in these skills was found to be much more difficult than anticipated.

Neither the approaches made to the state governments nor the efforts at direct recruitment form the market have been entirely successful. Besides, direct purchases by DPC's would mean that in most cases, payment has to be made on the spot and in cash, and as the Departmental Purchase Centres have to cover a number of weekly hats and feeder markets, this involves difficult problems of financial control. For the each DPC, a baling press and a godown have to be hired with the required complement of labour.

The Cotton Corporation of India

In its report on the cotton price policy for the 1969-70 season, the Agricultural Prices Commission recommended, *inter-alia,* the setting up of an agency in the public sector for the purchase, sale and equitable distribution of domestic cotton and also to serve as a vehicle for the canalisation of import of cotton. The government appreciated the recommendation of the Commission but, since the cotton season was than about to commence, it was decide to make every effort to evolve a new framework ready for introduction in the cotton season 1970-71.

Accordingly, a new framework has been evolved and the Cotton Corporation of India has been established as a public sector agency. The Corporation has been over the entire import trade in cotton. The Corporation consults the user mills in regard to their requirements and concludes contracts in consultation with them. In effecting purchases and concluding contracts, the Corporation draws upon the experience of those who have been engaged in this trade in the past. In order to be able to provide such services, the Corporation enrols the private indenting houses as its associates and works out with them the terms and conditions on which their services are utilised. The Corporation will gradually develop its technical competence so that its dependence on associated is reduced over time.

In the fields of domestic trade, the Corporation made a beginning with undertaking purchase for certain specific objectives. The growers of cotton are assured of an agency which gives them the necessary price support. The more enterprising cultivator growing the new varieties has now an agency to buy the extra long staple cotton at fair prices. The necessary expertise is being developed for procuring the cotton for the mills under the purview of the National Textile Corporation. The Corporation is expected to assist private sector mills in their programmed of cotton purchases. The Cotton Corporation of India Ltd, came into being on 31st July 1970, with an authorised capital of Rs 5 crores with a paid-up capital of Rs one crore and headquarters at Bombay with the following main objectives.

1. To purchase, sell and otherwise dispose of and deal in raw cotton imported from outside India.
2. To purchase cotton grown in India and to sell and otherwise dispose of an deal in such cotton.
3. To purchase sell and otherwise dispose of and deal in 'kapas' or unginned cotton as the Central Government may direct from time to time.
4. To generally implement such special arrangement for imports or internal trade of distribution of raw cotton as the Central Government may specify from time to time in public interest.
5. To organise and undertake purchases, sale and transport of cotton (imported into or grown in India) and/or kapas including such allied duties as may be entrusted to the company for such purpose by the Central Government from time to time in India or anywhere else in the world.
6. To organise and undertake purchase, sale and transport of natural and/or man-made fibre or fabrics of any king whatsoever (imported into or produced in India) including such allied duties as may be entrusted to the Corporation for such purpose by the Central Government from time to time in India or anywhere else in the world.
7. To generally carry on business as exporters and importers of cotton.

The Corporation started functioning from 1st September 1970.

6

EVOLUTION OF AGRICULTURAL POLICY

Farming methods and practices in India have a long history, and have been evolved out of the experience of ages. They have well stood the test of time. The technological advances in the world. e.g., the use of fertilizers, insecticides, fungicides,irrigation pumps, machines, etc., have mostly taken place in the present century and are hardly sixty years old. It should cause no surprise if a country of India's size with millions of small farmers has not yet been able to adopt them as fully as has been done by some countries of the West and Japan in the East.

Agricultural Department

Government took little interest in the problems of agriculture before the mid-seventies of the 19th century. They were pre-occupied almost exclusively with problems of revenue police and judicial administration. It was the great Bengal and Assam Famine of 1866 and the experienceof organizing relief that forced the attention of the Government to the need for the development of agricultural as a means of averting famine and mitigating the hardships of the people. The first proposal to constitute a special Department of Agriculture at the Centre was put forward by the commission appointed in 1866, following the severe famine of that year but was turned down as premature.

Guided by the needs of Lancashire textiles mills, the British Government undertook measures in 1869 to improve the production of long staple cotton in India. This could perhaps be cited as the initial measure towards some action in developing Indian agriculture. In the same year the proposal for a Special Department was repeated by the Manchester Cotton Organization which led to a serious consideration of the subject by the Government. However, this attempt also did not achieve much success.

It was only in 1871 that the Department of Revenue, Agriculture and Commerce was created. It was recognized that provincial Departments of Agriculture must form an essential part of any scheme of agricultural development. The Department had a very brief existence of about seven years and was merged in the Home Department in 1879 for reasons of financial stringency. Its only achievement was to evolve a system for the collection of agricultural statistics and other data.

Very useful suggestions were made by the successive Famine and Irrigation Commissions for the development of agriculture. But, as observed by the Royal Commission on Agriculture, "The Government took no immediate action and the next ten years were spent mainly in conferences." It was the result of the recommendations of the report of the Famine Commission, 1880-81 that a separate Department of Revenue and Agriculture was revived in 1881. It was fully realised even in the initial stages that without a provincial agency, no programme of agricultural improvement, which emanated from the Headquarters, could be productive of tangible results, that the main responsibility for agricultural research and experiment must fall on provincial governments and all that the Centre could do was to give a general lead. The new department was directed to take up immediately the question of the form to be given to provincial department.

Provincial Departments of Agriculture were established in U. P. (then known as the NWP), Bombay, Bengal, Madras and C. P. between 1875 and 1905. Their functions were clearly defined by the Government of India and consisted largely of collection of statistical and economic data. There was some research notably in Bombay, and a few sporadic experiments in the cultivation of American cotton, groundnut, potatoes and fruit in some other provinces. But not much progress could be made by such efforts in the absence of a sound policy and an efficient organization.

Veolcker Report

It was realized very early by the Secretariat that without technical advice no advance was possible. Dr. J. A. Voelcker, distinguished agricultural chemist, was brought over to India in 1889 to conduct an enquiry into the character of the soils and agricultural conditions. This is regarded the "first serious step to frame a policy of agricultural research suited to the conditions of India." This was followed by the appointment, in 1891, of an Inspector-General of Agriculture with advisory functions. His duties were specified as "the systematic study of Indian agriculture, its conditions and remediable defects; the supervision and development, of provincial agricultural departments; the establishment of improved agricultural methods and new staples and generally the direction of the agricultural policy of the government." An imperial mycologist and an entomologist were added to the department in 1901-3. Among the developments of this period mention should be made of the establishment of the Imperial Institute of Veterinary Research in 1889 and the Botanical and Zoological Surveys in 1890.

The Famine Commission Report, 1901

The report of the Famine Commission of 1901 inaugurated the next significant stage in the growth of agricultural departments. The commission recommended the strengthening of the scientific staff of agricultural departments and legislation to set up mutual credit societies. During the period of Lord Curzon these recommendations were implemented and agricultural departments were thoroughly reorganized. There was an expansion of the scientific staff of the departments and of their activities. With the setting up of the Imperial Agricultural Research Institute at Pusa and bringing together with it the scientific staff of the department, under a single roof, organized agricultural research may be said to have begun in India. Directors of Agriculture were appointed in all the provinces. Circles were marked out and placed under Deputy Directors who were responsible for experimental seed and demonstration farms within their jurisdiction. Agricultural colleges were reorganized and research stations and experimental farms were also set up. The Imperial Agricultural Service was constituted in 1906. Altogether the years from 1905 to 1914were years of the steady growth in agricultural research and education and other services. The substantial advances made during these years by provincial departments were largely due to the stimulus and guidance of the central government and grants made by it.

The 1919 Reforms

After the reforms of 1919, agriculture became a transferred subject. Under the new arrangements, the Central Government was to look after the central agencies and institutions for research. The provincial governments could not make much headway since the resources at their disposal were limited. During this period some irrigation works were completed. Under 1935 Reforms, the sphere of central government's responsibilities was further restricted. It was to look after the subjects of all-India importance only.

Royal Commission on Agriculture

A Royal Commission on Agriculture was appointed in 1926. It was required to examine and report on the prevailing conditions of agriculture. The Commission was asked to investigate, particularly, (1) the measures being taken for the promotion of agricultural and veterinary research, experiment, demonstration and education, the compilation of agricultural statistics, the introduction of new and better crops and improvement in agricultural practices, dairy farming, and the breeding of stock; (2) the existing methods of transport and marketing of agricultural produce and stock; (3) the methods by which agricultural operations were financed and credit afforded to agriculture; and (4) main factors affecting the rural prosperity and welfare of the agricultural population.

The recommendations of the Commission were made with a view to bringing about all round efficiency in the field of agricultural production. The aim was to make agriculture a profitable industry rather than a mode of living. The Commission covered such subjects as fragmentation and subdivision of holdings, irrigation, improvement of livestock, rural education, cooperation, marketing, etc. One of the basic recommendations of the Commission was to change the outlook of the cultivator himself so that he may take an active interest in bringing about improvements. It suggested that "rural problem should be tackled as a hole in all its various aspects simultaneously." It assigned and charged the government with specific responsibility for bringing improvements in rural life. The Commission's Report is still considered a reference publication on agricultural policy.

Russet Report

In accordance with the recommendations of the Royal Commission to review periodically the working of the Empirical Council of Agricultural Research which had been established on its recommendation, Sir John Russell, Director, Rothamsted Experimental Station, England, visited India in 1936.

According to him, the Council had admirably discharged its primary duty of co-ordination and promotion of agricultural research in the field of crop production. He stressed the need for further use of existing knowledge, rather than the accumulation of more knowledge and stated that knowledge gained in the laboratory should be applied to cultivator's fields. He suggested that the burden of the Council should be lessened by asking universities to undertake all investigations of a scientific or non-technical nature.

He recommended that the results of research in agriculture should be made widely public. It had not been done in the past. He suggested that the Council should take the responsibility of putting the result of research into practice and act as a Development Commission. He further remarked that the Council's investigations should be directed to increasing the output per care of food crops with a view to ensuring food supplies and liberating land for the growth of the supplementary and fodder crops for the production of milk.

Position before the War

The achievements up to 1937 may be described as the creation of the Central and State Departments of Agriculture for agricultural research, education and extension and provision and intensification of facilities for research and education. In spite of considerable expansion in the staff and activities of agricultural departments, research and education facilities were very inadequate. There were few men for agricultural extension work. Stationed at the revenue divisions, they could hardly contact farmers and much less offer any technical advice to them. The demonstration farms maintained by government failed to carry much conviction to the farmers as the facilities they had were considered something very different from their own farming situations. Agricultural education was not very effective in turning out scientists and practical farmers.

The village community spirit had been considerably weakened under British rule. A void had been created in the village adminitration in spheres in which the community exercised control before. A powerful class of new vested interests in the form of moneylenders and landlords had been created. The landlords its a class lacked the traditions which their counterpart in Britain had developed and many of they were interested only in rent. The moneylender was armed with traditions of English Law. All restrictions on accumulation of interest or increase in the burden of farmer's debt imposed by the community under the old village panchayat system hail been thrown to the wind. The farming community groaned under the exploitative economic system in which the landlord and the moneylender figured prominently.

Agriculture could hardly prosper under such an exploitation. The discouragement to village and cottage industries destroyed the economic balance which characterized the older system. Persons displaced from these occupations added to the pressure on land which had already become disquieting.

Partition

The years that followed were those of upheaval in the Indian history. The Congress ministries took office in the States. World War II started. The country subsequently became free but was partitioned. The balance between agricultural resources and the industries based on them was upset. There was unprecedented movement of population resulting from partition of the country. The enterprising farmers of Punjab moved to the unsettled areas in U.P. and other States. One good effect of such a shift in their habitation has been the improvement in the standard of crop and animal husbandry in the areas where they settled. The war, partition, and the growing population aggravate the problem of food.

The Grow More Food Campaign began in 1942. Sines then food production has received top priority. In respect of land reforms most States initiated action for the abolition of landlordism and intermediaries. Another notable development has been the setting up of a wide network of National Agricultural Extension Service. There has been considerable expansion in research and training programme as well. It was decided by the government in 1951 to channel the development of the total economy of the country through a regular planned course.

Planning

An era of planning thus started in 1951, when the First Five-Year Plan was launched. This was followed by the Second Five- Year Plan (1955-56) and the Third

Five-Year Plan (1960-61). All these developments, from the Grow-More Food Campaign in 1942 to the planned economy, are discussed separately. On the whole, the country lacked any well-defined policy of agriculture up to the end of Second Five Year Plan. Many improvements have taken since then.

Co-operative farming, declared at one stage as the ultimate form of organization for agricultural production, did not receive wide acceptance and that accounts for lack of any serious effort or enthusiastic response to its implementation. To achieve social justice, ceiling on the size of holdings is being imposed but there is considerable confusion in matters of detail and execution. Adequate investigations are lacking to justify the measure economically. There is good cultivable land which is lying waste in some States not because the ownership rights are vested in persons who are not able to farm it themselves but also because they do not lease it out for fear losing their title to it. To the extent such land is brought under cultivation, the measure may prove beneficial. But where the holdings are pushed down below a certain minimum size, it is not easy to predict the effect on agricultural efficiency and production.

Quite often the emphasis on increasing the production of foodgrains has led to a fall in the production of commercial crops, leading to shortage of such products. In turn a cry is raised to increase the production of cash crops. Such an action can hardly be called a policy. Not much thought was given to preparing the estimates of cost-benefit ratio in undertaking programmes of large-scale irrigation and land reclamation. Therefore, emphasis has been changing from major to minor irrigation projects and vice versa; similarly from land reclamation to raising the standard and intensity of cultivation on land already under the plough.

These instances make it clear that hardly any well-defined policy existed and much less was pursued. A sound policy calls for precise and detailed information on cost-benefit relationships of various developmental projects, efficiency standards, and farms in various size groups in different farming type regions. First, it is necessary to have a census of agricultural resources, surveys on land use, investigations on factors inhibiting agricultural efficiency regionwise and the demand projections for the current and future needs in respect of food, cash crops and export. Then taking the country and regions as units, sound plans need to be drawn up for land use, agricultural production and needed supplies. To supports the fulfillment of these targets, ancillary measures need be taken, namely, land reclamation, assistance in raising the intensity of cultivation, credit support, educational programmes, irrigation, etc. All such measures, for successful implementation, need a watchdog in the form of adequate farm management and agricultural economics research to guide and evaluate simultaneously the execution of the programmes in order to make them efficient and effective and to stop those measures which, on experience, have a retrograde effect.

No doubt, steps have been taken towards the direction suggested above but major lacunae in them may be listed as follows:

(1) The carrying out of developmental activities on a piecemeal basis with the result that full benefits expected from such programmes have not been obtained. In U. P., tube-wells, for example, were dug but could not be commissioned for lack of energization.

(2) The budgetary provision for efforts in various directions were made on the basis of administrative divisions and not on a properly estimated requirement of expenditure with the consequence that the efforts were too thinly spread and proved almost futile. The early Grow More Food Campaign is an appropriate example of this.

(3) The approach where an investment could be most effective from the point of view of natural factors like land, climate, or economic factors like demand, particularly in case of livestock products, has been conspicuous by its absence.

(4) Duplication and overlapping has often led to futility of efforts which could never be enough according to needs and urgency.

If we examine any one of the policy measures on prices of food-grains, production, distribution, land tenure, irrigation, land reclamation and settlement, we will see that each one has some woeful tale to tell. In the functions of the Ministry of Community Development and Co-operation and the Ministry of Agriculture, there was a considerable overlapping till they were recently merged. Between them and even within the same ministry there is duplication of positions. This leads not only to wastage of scarce funds but also to confusion in working.

Some Important Reports

A number of committees have from time to time to gone into the problem of agricultural administration and machinery. All of them have helped to expand agricultural organization. Mention may be made of the Patel Committee (1945-46) as a result of whose recommendations the Directorate of Economics and Statistics became a separate unit of the Ministry; the enquiry by N. Gopalaswami Ayyanger (1949), which recommended the creation of the Bureau of Natural Resources and Agricultural with a coordinating minister for the Ministries of Agriculture, Food and Works; the W. R. Natu Enquiry 1951; and more recently, the enquiry by Mr. Keifer (1965). Fullscale general enquiries have also been made of the machinery of the Government of India leading to proposals for its reorganization by Richard Tottenham in 1945, by Shri R. A. Gopalaswami in 1952 and Dr. Appleby in 1954. It may also be mentioned here that the Food Grains Policy Committee 1947, the Grow More Food Committee 1952 and the Balwantrai Mehta Committee 1947 also have made recommendations for the reorganization of the machinery of agricultural administration at different levels. The desire for administrative improvement has been continuous, the suggestions for reform numerous, but the practical results of the enquiries have been rather disappointing. Piecemeal changes and adjustments have been made from time to time. There has been some bureau shuffling off and on but no thorough reorganization that could assure the efficient functioning of the machinery for accelerating development.

Agricultural Production Board

In order to streamline agricultural production in the country, an Agricultural Board was constituted in 1963. The main function of the Board was to bring about expeditious implementation of the programme for the development of agriculture including animal husbandry, forestry and fisheries. The working of the board was recently reviewed and the Minister of Food, Agriculture, Community Development and Cooperation is its Chairman and the Minister of Irrigation and Power, the Minister of Planning, the Minister of State in the Ministry of Finance, and the Member Agriculture, Planning Commission, are its members. The function and powers of the board are: (i) to secure effective and continuous co-ordination between the concerned ministries and departments at the Centre and between the Centre and the States, in matters relating to the formulation and execution of development programmes; (ii) to review the progress of development programmes, ensure efficient arrangements for supplies and services, locate difficulties, if any, and endeavour to remove them; and (iii) to take such other measures as would ensure speedy and effective implementation of the development programmes at the Centre and in the states. The decisions of the Board will be treated as the decisions of the Government of India as long as they are within the plan provisions and within the framework of the approved budget provisions of the ministries concerned.

Reoganization of the Ministry

The Ministry of Agriculture was reorganized in 1964 on the initiative of the then Minister of Food and Agriculture who had found it a poor instrument for carrying out a dynamic and forward-looking policy aimed at making the country self-sufficient and even surplus in the foodgrains and assuring the prosperity of the farming population. The reoganization was aimed at making the Ministry "function more effectively and to enable it more actively to associate itself with the implementation of development schemes by the States." It attempted (i) a redistribution of subjects and responsibilities among the various units of the Ministry so as to make the assignments of the units compact and specific; (ii) to assign to scientific and technical officers of the Ministry executive functions and operational responsibilities for planning, implementation and supervision of programmes, in addition to their advisory functions, to secure through "better administrative machinery"; and (iii) closer and more continuous contact between the Central and State Governments in the Implementation of agricultural plan schemes and to improve Centre-State relations.

The apex of the structure of administration consists of the Minister of Food and Agriculture, who is a senior member of the Cabinet, and Ministers of State and Deputy Ministers, who constitute the political executive. The Secretary of the Ministry is the official head of the Ministry and the principal adviser to the Minister. The Department of Agriculture which is one of the three departments included in the Ministry is organized in eight wings, and has besides five specialised offices, a number of supervisory and operational agencies and some cells and units for certain specific purposes.

The eight wings of the department are: (1) Production; (2) Forestry; (3) Fisheries; (4) Inputs (F) (Fertilizers); (5) Inputs (M) (Machinery); (6) Land problems; (7) Special development programmes; and (8) Co-ordination. The specialised offices are: (I) The Indian Council of Agricultural Research; (2) The Directorate General of State Farms; (3) Agricultural Prices Commission; (4) The Directorate of Extension; and (5) The Directorate of Economics and Statistics.

Lastly, reference may be made to the numerous subordinate and attached offices of the Department which operate various programmes of the Ministry, expert committees and other agencies. The more important attached offices are the Directorate of Plant Protection, the Office of the Agricultural Marketing Adviser, the National Seeds Corporation, the Warehousing Corporation, etc.

Committees, both official and unofficial, expert of representative of interests, have been established to advise the Ministry on many important matters. Fifteen of these have been listed as functioning at present. These are panel of Economists, Panel of Experts on Agricultural Administration, Panel of Agricultural Scientists, and the Panel on Nationwide Agricultural Demonstration and finally Panel of Farmers to advice the Agricultural Prices Commission.

The other agencies falling in this category are: Committee for the Improvement of Agricultural Statistics; National Committee on the Preparation of Farm Management; Manual for the use of extension workers; Study Group on Wild Life and Wild Life Products; Indian Board for Wild Life; Central Forestry Commission; Board for Agricultural Machinery and Implements; Standing Committee of Experts on manures and Fertilizers; National F.A.O. Liaison Committee; Central Council of Gosamvardhana; Central Variety Release Committee; Central Board of Fisheries; and Central Fisheries Research Committee.

The areas covered by these boards and committees are of vital importance to the development of agriculture and it is necessary, therefore, that they should be active bodies manned by competent personnel.

The Planning Commission's role in agricultural administration needs to be examined. It is directly and intimately connected with agricultural development at all stages from policy formulation and planning to plan implementation and evaluation. The Planning Commission's concurrence is required for the Ministry's policies and programmes in detail. In fixing the targets of production, in making financial allocations in the five year plan and in the annual plans, in negotiations with the States, the Planing Commission plays a leading, if not dominant, role. The working relations between the Planning Commission and the Ministry have not been always smooth and are not infrequently marked by friction.

Fourth Plan

Encouraged by the results of the new techniques and recovery on the economic front, "Approach to the Fourth Five Year Plan" described "growth with stability" as its main aim. During the previous few years, instability on the economic front had primarily been associated with wide fluctuations in the agricultural production and consequent rapid rise in prices. The "Approach Document," therefore recommended the building of buffer stocks of main agricultural commodities to meet such fluctuations besides evening our physical supplies between years. Such stocks were also to be utilized for stabilizing the levels of agricultural prices. All this could, however, be possible only in the context of a steady rise in agricultural production. The other major object of the Fourth Plan was to move towards self-reliance as speedily as possible because it was felt that the process of development sustained by continuous foreign aid (inclusive of food aid) was not healthy. This could be possible both by reducing imports and increasing exports. Here again agriculture was required to play a major role in both of these objectives.

If the country could achieve self-sufficiency in foodgrains and remove constraints on supplies of most of the traditional export commodities based on agricultural raw materials, this objective could be fulfilled. The general approach for stabilization of agricultural prices and supplies from buffer stocks and open sale arrangements could in part look after this. There was also need to investigate new types or varieties of agricultural products when international demands and inducements come to increase their supply. Where domestic consumption of these products was to be restrained in order to make physical availability for export possible, the means used would have to be the imposition of excise or direct purchase by a state agency for export. Programmes for research and market surveys in order to find new uses and new markets for traditional export commodities and new markets for traditional export commodities and adequate measures for augmenting their production and availability for export were also recommended.

In the agricultural sector, the Fourth Plan had two main objectives. The first objective was to provide conditions necessary for a sustained increase of about five per cent annum over the next decade. The second was to enable as large a section of the rural population as possible, including the small cultivator and the farmer in dry areas, to participate in development and share its benefits. Accordingly, the priority programmes of development in agriculture fell broadly in two categories, namely, those which aim at maximizing production and those which aim at remedying imbalances.

The creation of conditions necessary for a five per cent agricultural growth was basic to the approach of the Fourth Plan. The crucial position of agriculture and the extreme urgency of achieving rapid growth in production were brought home, more than ever before, by the experience of previous years. The pace of development in the agricultural sector sets a limit to the growth of industry, of exports, and of the economy as a whole and constitutes a major condition for achieving economic and social stability and improving the standard of living for the mass of the people.

Role of the Central Government

The activities of the Central Government in relation to agriculture at present may be summed up as follows:

(1) The planning of agricultural development jointly with the Planning Commission;

(2) Coordinating state agricultural plans, watching their implementation and evaluating performance;

(3) Providing financial assistance to States for their agricultural schemes by way of grants-in-aid and loans;

(4) Ensuring the supply of agricultural inputs such as improved seeds, implements, fertilizers, etc., in sufficient quantities and in time;

(5) Providing credits and facilities for marketing, storage, and transport of produce;

(6) Providing economic incentives to farmers in the form of remunerative prices and enforcing the minimum and maximum prices;

(7) Extension work among farmers including extension education to make farmers receptive to scientific farming practices;

(8) Providing technical advice and assistance to State in implementing programmes;

(9) Development of livestock, dairying and fisheries;

(10) Development of forests;

(11) Land reclamation, soil conservation, utilisation of water resources;
(12) Fundamental and applied research in agriculture, veterinary science, forestry, agricultural economics, fisheries etc.; maintaining the standards of higher education;
(13) Home science and nutritional needs of the population; and
(14) The administration of external assistance.

National Commission on Agriculture

The National Commission on Agriculture was constituted during 1970. The Commission is expected to look into almost all the problems that were gone into 48 years ago by the Linlithgow Commission. Though much has been done to develop agriculture since then, and especially after independence, the basic problems remain almost the same.

For instance, the Linlighgow Commission (1925) was to examine and report on the "present conditions of agriculture and rural economy in India and to make recommendations for the improvement of agriculture and promotion of the welfare and prosperity of the rural population." Another important term of reference of the 1925 Commission was to investigate "the existing methods of transport and marketing of agricultural produce and stock."

The new Commission too will "examine and report on the present conditions of agriculture and rural economy and make recommendations for the improvement of agriculture and the promotion of the welfare and prosperity of the people". It will also report on the "development of infrastructures including facilities for transport, marketing, together with the lines of future development."

In view of the vital changes that have occurred since the last Commission reported, the terms of reference of the new Commission have been widened in their scope and made as broad-based as possible.

One of the important terms of reference is to investigate the structure and organisation of the agricultural services, spelling out the relative responsibilities of the Centre and the States. An equally reimportant term of reference is to establish is clear terms the interrelationship between industry and agriculture.

The Commission will also investigate and report on:

(1) The pattern of land utilisation and the nature of the cropping thereon and the methods of land improvement.
(2) The principles of development of water resources.
(3) Utilisation of the science of soil conservation and land shaping.

Land Reforms

(4) Implementation of the land reforms policy and the changes that may be necessary in the pattern of crop sharing

(5) The new strategy of agricultural development based on the package of practices, and modifications required for better results.

(6) Forecast for the decade 1970-71 to 1980-81 on the magnitude of the requirements of fertilizers, good seeds, pesticides and agricultural implements and machinery; the organisation necessary for their production and distribution.

(7) Forecast for the decade 1970-71 to 1980-81, the requirements of credit for the agricultural programme.

(8) The strategy of agricultural research, education and extension, and

(9) Integrated development of agriculture, animal husbandry farm forestry and inland fisheries, to avail of the economies of mixed farming.

The Fifth Five Year Plan

With good agricultural season during 1967-68 and the adoption of the new strategy over the years, the agricultural development proved a great success and the country reaped the richest crop of 108.4 million tones of foodgrains during 1970-71. The subsequent two years have however again proved that nature can play havoc and reduce the effect of the 'new strategy' as well. Agriculture in rainfed areas presents a more formidable problem. There it is not only low yielding, but also precarious, and poverty is acute and widespread.

In the past the focus of the overall agricultural development effort was, to a large extent, on areas with the maximum promise of agricultural growth. In the Fifth Plan, while intensive agricultural programmes such as multiple cropping and high yielding varieties programme generally based on irrigation will continue to receive attention, it will be necessary simultaneously to place stress on agricultural programmes in rainfed and other less promising areas. The Fourth Plan provided for (i) intensification of research into improved dry-farming technology, and (ii) pilot projects for testing the efficacy of the new technology and for providing demonstration-cum-training in this technology. The progress so far has been rather slow on account of, among other things, a late shart, personnel problems and absence of effective tie up between the pilot projects and the research centres. It is proposed to intensify efforts towards evolution and propagation of appropriate dry farming technologies. Many of these areas are suitable for the development of animal husbandary which will receive added emphasis. Intensive efforts to develop animal husbandary in the rainfed areas could make an uppreciable addition to the incomes of the rural poor. Finally, public rural works will have to form an important component of the employment generating activities in dry farming and other rural areas where normal programmes of agricultural development are not likely to make a significant impact.

In recognition of the considerations mentioned above, certain special programmes such as the Small Farmers' Development Agency (SFDA) and Marginal Farmers projects. Crash Scheme for Rural Employment and Drought-Prone Areas programme were initiated in the Fourth Plan. By and large, these programmes were conceived in isolation and their operations were locationally dispersed. In the Fifth Plan, it will be necessary not only to accelerate the pace of implementation of these programmes but also to make distinct operational improvements. The experience with these programmes has shown that, if an impact is to be made, the development programmes in general and the special programmes in particular will have to be integrated. In these areas, an effort towards integrated are development will be a necessary condition for improving the economy of small and marginal fanners and agricultural labourers. Such integrated rural development programmes will vary in terms of different land and water resource situations and will have to be tailored to suit local conditions.

New Package Approach

According to Dr. Minhas, Member, Planning Commission, a new approach to agricultural development is being throught of. What is now proposed is legislation to make consolidation compulsory. He has suggested topographical surveys and grading of land, and the construction of irrigation channels and drains before the consolidated plots are redistributed. Such a programme would given inpetus to moisture conservation and the drilling of tubewells.

Finance for this large scale land development programme could be made available under the existing allocations for agriculture, irrigation, soil conservation and flood control as well as the new rural works programme, indeed a land development and consolidation programme could generate large scale productive employment and increase the receptivity to and gains from the new agricultural technology. Rural electrification and the development of a network of rural roads would also greatly strengthen the village infrastructure and create new outlets and opportunities for marketing, agricultural processing and other rural industries. None of these programmes is entirely new and have been taken up randomly in one part of another of the country. What is proposed, however, is a concentrated and combined effort to implement a mutually reinforcing package of programmes on the ground and to finance these through an aggregation of allotments separately made for a series of related schemes. There is much promise is such an approach in terms of employment generation, the improvement of agricultural prospects and the development of rural services.

Rural electrification, which would form part of the package is being pushed forward by the Rural Electrification Corporation which in the first 20 months of its existence has made a fair start. It has sanctioned loans aggregating nearly Rs 90 crores for 137 projects that have energised over 200,000 pumps-sets and provided power to some 32,000 rural industries in 108 districts, nearly half of them backward areas. The Corporation has attached great importance to standardisation of materials and methods as a means of achieving improvement in rural economy through rural electrification. It is trying to promote rural electric co-operatives and has decided to apply less rigorous viability tests for projects in backward areas including certain under-developed hill districts.

Programmes of this nature being small and widely dispersed lack the grandeur of size and patronage associated with giant schemes. But it is those projects that will truly transform the countryside.

7

LAND REFORMS AND AGRICULTURAL DEVELOPMENT

Land reforms, abolition of Jagirdari and zamindari, the community development and other and rural development programmes, the cooperatives and the panchayats have revolutionised rural life in Rajasthan in less than 15 years. During the time of the princely states, Jagirdari was the principal system of land tenure in the State—59 per cent of all cultivated land in Rajasthan was under the jagirdars. In some states, the proportion was even higher. In Jodhpur 82 per cent of the land, and in Jaipur 65 per cent of the land was under jagirdars. The Jagirdari was an old deudal system under which the jagirdar became the *de facto* owner of the land. He paid a small 'tribute' to the prince in lieu of enjoyment of the right to the jagir, but had full rights to give out the land for cultivation on payment of rent by tenants, in cash or kind. The tributes, many of which had been fixed centuries ago, were nominal, but the jagirdars could increase the rents payable by the tenants. In some states, such as Jaipur and Jodhpur, legislation had been passed to regulate the terms of tenancy, in order to protect the cultivators from rack renting by the jagirdars and to give them security of tenure. But in most other states, such protective legislation did not exist. Moreover, since the jagirdars were locally very powerful—the larger ones were regarded as virtual rulers of their areas—the small cultivators could hardly challenge their authority. Finally, since rents were in many areas fixed as a proportion of the produce, they rose automatically with increase in production. The share of produce to be given as rent range from one-fourth to one-half in different areas, or for lands of different productivity. Under this system, the cultivators had little security of tenure or incentive to increase production. A share of increased production would go to the jagirdar and the cultivator might be evicted from the land at any time at the jagirdar's pleasure. Accordingly, there was considerable dissatisfaction with the system, and with political awakening among the masses, the demand for its abolition had begun to be made even in the 1930s. Dissatisfaction with the system increased in the 1940s, when prices of agricultural commodities rose and the tenants felt that a disproportionate part of their income was being taken away by the jagirdars.

In addition to Jagirdari, there were also the zamindari and 'biswedari' systems which obtained in about, 5,000 villages spread over eight districts of northern and eastern Rajasthan. Under these systems, the Zamindars and the biswedars paid a fixed amount of revenue to the prince; but the cash rents payable to them by the tenants were not fixed. The tenants could also be ejected by the Zamindars or the biswedars without much difficulty.

The feudal land systems were an anathema to the cultivators, and the new democratic government established after the formation of Rajasthan decided to abolish them as soon as possible. The Government had also to institute a uniform system of maintenance of land records and of land revenue administration. Conditions in this respect also varied tremendously in the areas of different states. While some of the more advanced states such as Jaipur and Jodhpur had well-developed systems of land records and land revenue administration, patterned after the system of the British provinces of the Punjab or U.P., conditions in the smaller states were much less satisfactory. In some of the latter, as also in many Jagirdari areas of even the advanced states, there were virtually no land records and the tenants were at the mercy of the prince or the jagirdar and their functionaries.

The first major land reform law, the Rajasthan Resumption of Jagirs Act was passed in 1952. It gave the State the right to resume the jagirs on payment of compensation. However, the process of reform had started much earlier, in fact with the formation of Rajasthan. As the popular government which came to power in the new State was committed to abolishing Jagirdari and zamindari a large number of landlords started evicting their tenants, so that they could retain the lands as their own personal cultivation *(kihud kasht)* lands. This led to considerable unrest among the tenants and, if allowed to continue, would have made land reform meaningless. Accordingly, the state passed in 1949, an ordinance called the Protection of Tenants Ordinance, which reinstated the ejected and dispossessed tenants. At the same time, another ordinance was passed dividing the state into divisions, districts, sub-divisions and tehsils for purposes of programme of replacement of fucntionaries (Patwaris, Tehsildars) of the jagirdars with revenue stage of the state, were started throughout the state. Survey and settlement were particularly necessary in order to determine the rights and fix the cash rents in respect of each plot of land. This step and the institution of a system of land records maintained by the functionaries of the State were essential to give effect to the land reforms which were envisaged.

The Resumption of Jagirs Act was challenged by the jagirdars in courts of law, on ground of being unconstitutional. Simultaneously, a State-wide agitation against it was started by an organisation of jagirdars called the Bhooswami Sangh. The most aggrieved groups were the very small jagirdars of Marwar who had subsisted on small rent incomes from their lands and would now become destitute. The courts upheld the law, but since this decision was not acceptable to the jagirdars, it was agreed by both the jagirdars and the State Government that the Prime Minister of India, Shri Jawaharlal Nehru, be requested to act as an arbitrator and give an award. Shri Nehru gave his award in 1955 which was accepted by both the parties and its terms were incorporated in the amended Jagir Resumption Act. The process of jagir resumption then started in full swing and by 1958 all but the smallest jagirs had been resumed. In 1960, the State Government enacted a law to abolish zamindari and 'biswedari'.

The Nehru Award is much more generous to the jagirdars than the original Jagir Resumption Act had been, in payment of compensation for jagirs. In addition to compensation, which had been fixed at seven times the annual income of a jagir, it awarded payment of rehabilitation grants to the small jagirdars. The grants, fixed on a sliding scale, decline with rise in the income of the jagir.

The Resumption of Jagirs Act had started the land reform process. It was completed by the Rajasthan Tenancy Act of 1955 and the Rajasthan Land Revenue Act of 1956. The Tenancy Act confers the right of *khatedari* tenant on the great majority of the cultivators. The right includes the right of mortgage, sale and bequest and makes the cultivators virtual owners of land. The jagirdars are allowed to retain lands under their personal cultivation as *khudkasht* lands and have the same rights and obligations as the *khatedars.* Sub-tenants are recognised, reflected in a number of amendments, has been to confer *khatedari* rights on this class of cultivators also so that the ideal of 'land to the tiller' becomes a reality. The Rajasthan tenancy law is among the most progressive pieces of land reform legislation in the country. The Rajasthan Land Revenue Act of 1956 established by an Act of the State Legislatue the structure of revenue administrations which has been established earlier by an ordinance. With the enforcement of this Act and completion of settlement and survey operations for most of the State, the land revenue system has been established on a firm and stable basis.

Since 1954, village panchayats have been given certain important powers in land management and revenue administrations. Pasture lands, *abadi* lands and certain other classes of wastelands have been vested in the panchayats for management. The Sarpanch of the panchayat is associated with allotment of culturable wastelands in the panchayat area. He and other members of the panchayat have the right to supervise *gasht-girdawari* (crop inspection) conducted by the Patwari. Mutations of most classes are sanctioned by the panchayats. These measures give the people's representatives important function and also protect the cultivators from malpractices of petty renvenue officials.

Land reforms, by giving the cultivators security of tenure and a new dignity, have laid the foundations of progressive agriculture. A programme of agricultural development, aimed at making agriculture remunerative and progressive, was started by the new Sate from its inception. The size of the programme and the activities included within it were greatly increased after the inception of the community development programme in 1952. The principal ingredients of the programme are the following: (i) Extension—advice in improved agricultural and animal husbandry practices: (ii) Supplies—of improved seed, chemical fertilizers, pesticides, improved implements and other essentials of modem agriculture; (iii) Services—of stores, co-operative marketing societies, dispensaries for treatment of animals, agricultural workshops, soil testing laboratories, etc. and (iv) Financial assistance—short-term and medium-term loans through the co-operatives, *taccavi* loans for purchase of bullocks, for construction and repairs of wells and tanks to increase irrigation facilities, and for land reclamation, soil conservation and other improvements.

The extension organisation is built round community development blocks. Each block is a unit of about 100 villages having a population between 40,000 and 80,000. The blocks in the desert area have lower populations. In each block, there are 10 Village Level Workers, each having responsibility for 10 villages. The VLWs are matriculates with two years' training in agricultural extension and other subjects. At the block headquarters, there is a team of specialists, in various fields—agriculture, animal husbandry, co-operation, panchayats, etc. to look after different aspects of rural development. The VLW's and Extension Officers work under the direction of the Block Development Officer. The entire extension organisation now functions under the panchayat somites, popular institutions composed of Sarpanchs and other village leaders, which have been give full responsibility for rural development. Each block has a panchayat Samiti. Development at the district level is coordinated by the zila parishad. The panchayat Samiti and zila parishads were constituted in October 1959. It is a remarkable fact that Rajasthan, which was the citadel of the feudal order, was the first State in the country to introduce the democratic structure of panchayat Samitis and zila parishads.

Alongside the establishment of the extension organisation, efforts were made to build up agricultural research and training. The older research stations were expanded and a number of new ones were established. Similarly, facilities for training of agricultural graduates were rapidly increased. In 1961 an Agricultural University was established at Udaipur.

Rajasthan has a large co-operative movement. In 1962 over a million cultivators, about 40 per cent of all cultivators in the State, were members of co-operative societies. Loans advanced by the cooperative societies during that year totalled Rs. 18.78 crores. Besides, a network of co-operative marketing societies has been established. These societies market the cultivators' produce, thus affording them protection from the malpratices of money-lenders and traders. They have also the major responsibility for distribution of improved seed, chemical fertilizers, pesticides and other requisites of improved agriculture. The progress of the co-operative movement is truly phenomenal when one considers the fact that in most of the princely sates, which now form Rajasthan, the movement hardly existed.

The structure of the agricultural extension organisation has been strengthened further and agricultural credit and supplies are given much more liberally in the Intensive Agricultural Production Districts and Blocks. In these, the number of village level workers per block is larger and teams of specialists in various branches of agriculture— agronomy, soil science, horticulture, plant protection, etc. have been established at district and block levels. In the IADP districts, plans for increasing production are prepared by the VLWs and co-operative extension officers for each farm family. Specialised service institutions, like soil testing laboratories, have also have been established.

The agricultural improvement programme incudes animal husbandry. This aspect is especially important in Rajasthan, with its enormous cattle wealth. Apart from programmes for improvement of cattle, which provide facilities for systematic breeding, treatment of cattle disease and improvement of fodder supplies, a special sheep and wool improvement programme has been undertaken in recent years. Under this programme, wool-grading and shearing centres have been established and various other measures are being taken to increase production and improve the quality of wool. The programme is extremely important for the State and for the country, because Rajasthan is the principal wool-producing State in the country. India is deficit in wool and has to import large quantities of it at enormous foreign exchange cost.

The result of these improvement measures is reflected in increase in agricultural production and rise in the income of the farmers. Agricultural production has been increasing steadily, and, as explained earlier, it has risen by 40 per cent in less than a decade. However, it is impossibly to express in statistical terms the effect which land reforms such as abolition of Jagirdari and zamindari, the grant of rights in land and the initiative gained in managing their own affairs including the development programmes, have produced among the rural people. These are seen in the emergence of a new generation of farmers—literate an freer from superstition, increasingly aware of the modem world, self-reliant and progressive.

The farmers of Gujarat tend to give more emphasis to commercial crops than to food crops. The area under non-food crops is as high a 46.70 per cent. The tangible consequences of this pattern in food deficit for the State on the one hand, and an appreciably more prosperous peasantry, relatively speaking, on the other. The value of the cash crops, moreover, covers more than the cost of foodgrains that have to be imported. The attitude towards production and cost-mindedness is, perhaps, the distinguishing characteristic of Gujarat farmers. It makes them takes more readily to a mixed farming pattern. Almost every farmer keeps a mulch animal or two and feeds them on fodder grown along fences and farm border ridges. The development of co-operative dairy industry in several districts has given a definite financial value of milk production and consequent adoption of improved animal husbandry practices.

The important food crops of the States are bajra, jowar, paddy and wheat. The important non-food crops are groundnut, cotton and tobacco. The Sate deficit foodgrains is variously assessed at between 20 to 40 per cent of the total requirements. The main causes of the deficit in foodgrains are, low yield per hectare of foodgrains and comparatively larger proportion of total cultivated area under commercial crops. But the dependence on other States for its requirements of foodgrains is more than compensated by exports of the crops of higher value. Gujarat ranks first among all the States of the Indian Union in so far as the production of cotton and groundnut is concerned. Gujarat produces about 29.9 per cent of the cotton and 17 per cent of the groundnut produced in India. In the production of tobacco it ranks second only to Andhra Pradesh. Cotton and groundnut crops sustain a large part of the trade and processing industries in the State. Cotton in particular has also given a marked fillip to the cooperative processing and marketing activities.

Agricultural practices are in some respects different from those in vogue in other parts of the country. Some of these in respect of the more important food crops of the State are described below.

Bajra

In terms of area and outturn bajra is the most important food crop, which is grown in all the districts of the State. In 1958-59 it was grown in 16.61 lakh hectares and produced some 6.09 lakh metric tonnes. The main important bajra-producing districts in order of importance are the Banaskantha, Mehasana, Junagadh, Surendranagar, Kutch, Kaira and Jamnagar. In addition to the local strains, the improved strains of bajra popularised among the farmers are bajra-207 and Babajipura bajra which give about 10 per cent higher production than the local strains. The average yield per hectare of bajra was 168 kilograms in 1958-59.

This crop requires rainfall ranging from 45 to 63 cm. and is grown in medium to black *goradu* soils. The sowing season is normally June-July. But in certain parts of Kaira pre-monsoon sowing is also prevalent. The harvesting season commences in September-October. In certain places where irrigation facilities are available, bajri is taken as a summer crop and is sown in February and March.

Generally, the ground is prepared for the crop by two ploughings followed by one to two harrowings to destroy the weeds and to keep the soil in fine tallith. In some cases plighting is dispensed with and the field is harrowed lengthwise and corrosive with a heavy blade harrow *(karab).*

In Kaira, Panchmahals, Ahmedabad and Baroda 10 to 15 cartloads of farmayard manure is applied per hectare of land in the month of May. Groundnut cake is also used for manuring.

Sowing is done with a three-cultured drill *(tarfen)* the spacing between the lines being 22 to 30.38 to 45 cm, and in some places 50 to 60 cm. The seed rate is 9 kg to 11 kg per hectare if sown alone. It is, however, generally mixed with pulses such a *mung, guvar* and in a few cases, *tur.* After sowing the seed is covered by a *samar* (wooden plank). The crop needs thinning which is done by *dental* (tooth harrow) by drawn it cross-wise in the field when the crop is 10 cm. tall. When the crop to check its growth and thereby encourage tillering. This operation is done after interculturing. Two hand weedings (with a simple tool called *kharapadi)* and two interculturings are given to the crop. The usual rotations are followed which vary according to soil types and extent of rainfall, (i) bajra-paddy or tobacco (ii) bajra-wheat, jowar or cotton (iii) *tal* and bajra-cotton.

Jowar

The second important staple food crop of the State, jowar was grown in 1958-59 on 14.50 lakh hectares with an outturn of 4.36 lakh metric tonnes. The crop is grown in all the districts of the State but important among them are Rajkot, Bhavnagar, Surendranagar, Banaskantha, Mehsana, Kutch, Ahmedabad and Dangs. The improved strains of this crop which are being popularised are B.P. 53, Jowar no.8 C-10-2 which have proved to be superior in yield by 8 to 19 cent more than the local strain. These strains are being developed at research stations located at Surat, Broach and Viramgam. The yield per hectare for jowar for Gujarat State was 286 kg. in 1958-59.

Jowar is essentially a crop of deep and heavier soils and good results are obtained in places with rainfall ranging 63.5 to 101 cm. The crop is principally grown in kharif and rabi seasons. Where irrigation facilities are available, fodder jowars are grown during summer. Regions receiving 56 to 76 cm. rainfull grow early kharif varieties of jowars in lighter types of soils and late kharif varieties in heavier types of soils.

Kharif or early sown jowar varieties are important in the Baroda and Mehsana districts. The land is made ready for sowing by the end of June. Usually two the three harrowings are given in the month of May. Generally manure is not applied to this crop. Nitrogenous fertilizers are given as a top dressing at sowing time. In some districts farmyard manure is given once in three years as it is believed that farmyard manure is effective up to 3 years.

Gavar, mung, chora, did, tal are grown as mixture with jowar. In the black soil of

Surat, *udid* and *mung* are the pulses sown with jowar. The ordinary seed rate is about 11 kg. per hectare when grown alone and 7 to 8 kg. jawar seed and 1 to 3 kg. oilseeds or pulses in the case of mixed crop.

After levelling the surface of the soil, the seed is drilled and covered with a *samar*. As a rule, in well prepared soil, jowar requires only one hand weeding. The crop shades the ground quickly and weeds are more or less suppressed. Bullock interculturing commences when the crop is 15 cm. high and generally three interculturings at intervals of 15 day are given to keep down the weeds and loosen the soil. Kharif jowar is ready for harvest by January-February.

Rabi or late sown varieties of jowar are confined to the Broach district. For rabi crop one or two good plughings and 3 to 4 harrowings are given during the rainy season. The seed is drilled during September-October, about 5 cm. or more deep to secure satisfactory germination. The area is then covered and pressed by a plank. Two interculturings are given with bullock hoes. The harvesting season for this crop commences in February March.

For the summer crop which is primarily grown for fodder where irrigation facilities are available, the seed is sown thick and as a result fine thin *kadab* (stalks) of excellent quality is produced. The seed rate for fodder crop is 67 kg. per hectare. The seed is drilled at 30 cm. distance between the two rows. Jowar is usually rotated with cotton, wheat and groundnut.

Paddy

The third important staple food crop of Gujarat is paddy. It was grown on 4.87 hectares, with an outrun of 5.36 lakh metric tonnes in 1958-59, the average yield per hectare being 947 kg. The important varieties of paddy found in Gujarat are *sathi-34-36, kada* 176-12, *sukhwel*-20, E.K. 70, early *sutarsal* 39, Z-31, K.42, jirasal 280, *kamod*-118, *pankhali-203, sarin* 15-14 etc. The principal rice producing districts are Panchmahals, Surat, Baroda and Kaira.

Paddy thrives in warm and moist climate. The crop is capable of withstanding higher temperature as compared to other cereal crops. It requires about 100 cm. rain, though a few strains suitable for areas having 37 cm. to 75 cm. rainfall have been evolved have been evolved. The crop is grown in *kyari* lands, *gorat, besar* and medium black soils.

There are three methods of raising paddy crop, namely drilling, transplanting and broadcasting. The method followed in any area largely depends upon the amount of rainfall, irrigation facilities and soil conditions of the area. During method is most prevalent in rained areas, while transplanting is generally confined to *kyari* and irrigated lands. The third method is the least important and rarely followed.

Sowing and transplanting take place in June-July, while the harvesting of the crop falls in October-November. If the crop is to be transplanted, after the first rain in June the seed is sown broadcast in the seed beds. Twenty-seven kilograms of seed is considered sufficient to raise seedlings for one hectare of land. In July when the seedlings are grown 12 or 15 cm. they are transplanted. The distance between the plants and that between the rows vary from 30 x 30 cm. 22 X 22 cm., 10 X 10 cm., etc. Where irrigation facilities are available seedlings are raised in early June, and fields for transplantation are carefully embanked so as to allow the water to spread evenly in the fields and be retained at a desirable depth.

The crop needs to be well manured and in addition to farmyard manure, groundnut cake and ammonium sulphate are applied. In the *kyari* lands of Surat district green manuring is widely practised. Where paddy is grown by the drill method, after one or two showers or rain in June the seed is drilled at 40 cm. distance by a three-cultered drill. Pre-monsoon sowing is also done in some places, and the seed rate for this method is 33 to 40 kg. per hectare. In drilled paddies two hand weedings and two to three interculturings are given.

Tur is sometimes taken in drilled paddy as a subordirnate crop.

Wheat

What is the fourth important cereal crop of Gujarat and had 4.14 lakh hectares with an outturn of 2.93 lakh tonne in 1958-59. The average yield per hectare was 785.12 kg. The crop is grown is all parts of the State but the most important wheat-producing districts are Ahmedabad and Mehsana. The crop grows best on deep black and retentive soils such as are found in Ahmedabad, Surat and certain parts of Baroda district. For irrigated crop well drained, light alluvial are suitable. The improved strains grown in Gujarat are A.206, A.624, N.P. 710 and N.P. 718.

It is taken both as dry and irrigated crop, though the largest area in the State is under dry crop. The crop is grown by October-November and harvested by February-March. The seed rate per hectare varies from 45 to 56 kg. for dry crop and from 67 to 90 kg. for irrigated crop. Planking is done after sowing as it brings our moisture from the lower layer of the soil which helps satisfactory germination of the seed.

Soil is ploughed 15 to 18 cm. deep at the commencement of rains followed by frequent harrowing which suppresses the growth of weeds and allows thorough absorption of rain water and creates mulch to prevent the loss of water from the soil due to evaporation. This is very important for the dry crop as it has entirely to depend for its growth on the rain water absorbed by the soil. Where kharif crops are taken prior to rabi crops, for example bajra in the Kaira district, two to three harrowings are given after the harvest of the kharif crop and the soil is prepared for sowing wheat. In districts like Mehsana ploughing is clone in September-October or irrigantion crops. When the land is kept fallow for wheat two to four ploughings are given as in the Panchmahals, Surat, Mehsana and Baroda districts.

Dry wheat generally receives no manure, though sometimes well decayed farmyard manure is applied. An irrigated crop is necessarily manured. Chemical fartillizers, groundnut cake, and manure mixture are also applied.

Irrigated wheat is grown alone or rotated with cotton, groundnut, jowar. Dry wheat is grown year after year in many places or rotated with cotton, jowar, paddy, gram. The dry crop is sometimes mixed with linseed and gram.

Maize

Maize is also grown fairly extensively. In 1958-59 the area under it was 2 lakh hectares with an outturn of 1,093.21 lakh metric tonnes, the average yield per hectare being about 950 kg. The important maize producing districts in Gujarat are Panchmahals and Sabarkantha and to a certain extent, Baroda.

With about 50 to 75 cm. rainfall the crop thrives well on rich *goradu* to medium black soils of Panchmahals. It is usually grown as a kharif crop.

The land is first well ploughed in the beginning of the rains and 5 to 10 cartloads of farmyard manure are added to the soil. The seed is sown early in June, the usual seed rate being 11.2 to 22.4 kg. per hectare when grown as fodder crop, the seed rate varies fro 56 to 67 kg. per hectare. The crop is usually sown alone as it grows quickly and does not make it a good companion for subordinate mixtures. But recently the multipurpose project in the Sabarkantha district experimented successfully by introducing cotton in alternate rows.

The crop is once hand-weeded, and intercultural two to three times with a bullock hoe. It ripens within three to four months and is ready for harvest from the middle of September to the middle of October.

The kharif crop of maize is usually followed by a rabi crop of what or gram. In Panchmahals the crop is rotated with *kodra* and in this region white *sameri, dudh mogri,* local yellow and reddish yellow varieties of maize are grown.

Bavto is grown in parts of the Panchmahals, Ahmedabad, Surat and Baroda districts on lands which are too light for rice, or too steep to be converted into terraced rice fields. *Bavto* is said to be very nutritious and the grain can be preserved for more than fifty years.

Kodra is grown in parts of Karia, Panchmahals, Surat, Broach and Ahmedabad. It is particularly well adapted for the poorest soils under favourable conditions of climate and rainfall. The usual mixture is *kodra, tur* and Sesamum with a sprinkling of *ambadi* (Bombay hemp). In low-lying damp fields in Gujarat rice and *kodra* are grown mixed.

Kang is grown in the Panchmahals district on medium light soils and does better than other millets under low rainfall conditions.

Pulses

About 11 per cent of the total food crop area is devoted to the cultivation of pulses. The total area under pulses in Gujarat in 1958-59 was 6 lakh hectares and production about 1.28 lakh metric tonnes.

The more important types of pulses are *tur* and gram. Other pulses of lesser importance are *urid, mung* and *moth.* The major pulse-producing districts of Gujarat are Kutch, Baroda, Surat, Banashantha and Mehsana, though Kutch accounts for the highest area. The important tur-producing districts are Surat, Baroda, Kaira and Mehsana, while the main gramproducing districts are Panchmahals, Mehsena, Junagarh and Jamnagar.

The soils on which *tur* is grown vary with the requirements of the crop with which it is sown. It thrives best on a medium moist which allows its roots to penetrate deep down. *Tur* is generally sown in June-July and is ready of harvest by February-March. It is seldom irrigated because its roots penetrate deep and tap moisture from lower levels of soil, although the upper layers of the soil are dry.

Gram is grown is all types of soils from the heaviest call to the highest loam. It gives high yield in heavy soil. In light soils, it is grown as a mixed crop with wheat or linseed. When grown as a sole crop of the year it is rotated with wheat or jowar. It also follows rice in *Kyari* lands as a second crop.

Oilseeds

The total area under oilseeds in Gujarat is about 19 per cent of the total cropped area (18.75 lakh hectares). Important among the oilseeds grown are groundnut, Sesamum and castor seed and, to a certain extent, rape, mustard and linseed. Rajkot division is the major producer of oilseeds.

Groundnut. It is grown in all the districts of the State but it constitutes an important crop in the Rajkot. Jamnagar, Junagadh, Bhavnagar, Sabarkantha. Amreli and Panchmahals districts.

Groundnut is grown both as a dry and irrigated crop. It can stand a rainfall up to 125-137 cm. Light sandy soil, well-drained sandy loans and good alluvial loams are particularly suited to it. It is also grown on ordinary red loam, black cotton soil and heavy soil of rice fields. Ploughing is done to a depth of about 15 cm. after the first rains. The crop is sown in rows of about 15 cm. for the spreading varieties and of 22 cm. for the erect varieties. The seed rate for the spreading varieties is about 45 to 56 kg. per hectare and that for the erect varieties is about 90 kg. per hectare.

Groundnut is grown generally as the only dry crop of the year. Usually it takes about six months to mature. But there are early varieties which take about 90 to 100 days. It is also grown as a mixed crop with jowar, cotton, *tur,* bajri, etc. Irrigated groundnut is rotated with paddy but other crops like potatoes and vegetables are also grown in rotation with it. The yield varies between 1608 to 1905 kg. per hectare.

Sesamum. This oilseed crop is losing its importance in view of the rapid spread of groundnut. It is grown in almost all districts, more particularly in Junagadh, Banaskantha, Bahavnagar, Surendranagar and Mehsana.

Sesamum is grown as a rain-fed crop in areas where the annual rainfall is slightly above 50 cm. It is a rabi as well as kharif crop. Heavy rains ruin the crop when the flower are being fertilised. It is grown in a variety of soils. The kharif sesmum requires a sandy or light soil while the rabi crop grows best in black soil or in soils retaining moisture. As the seeds are small, they are mixed with ash or manure before being sown in order to secure even distribution. The seed rate is 1. 12 to 3.36 kg. per hectare according as it is drilled or broadcast. It takes about 3½ to 4½ months to mature. When grown as a rabi crop, it follows fodder jowar, *udid,* etc. If sown as a sole crop in June-July it is rotated with jowar or other dry crops. Sesamum is believed to exhaust the soil; the other dry crops are, therefore, well manured in order that the fields are maintained in good condition. It is also grown either as a mixed crop along with *tur,* castor, etc. or as a subordinate crop to bajra. The average yield of Sesamum, when pure, runs to 336 to 448 kg. per hectare, though yields up to 673 kg. per hectare are also obtained.

Castor. This is both a kharif and a rabi crop and can grow at any altitude in dry arid climates and can stand heavy rains and foods. It is grown generally in deep red loams both sandy and clayey, in good light alluvial loams, even in rough gravelly uplands. A coarse open texture is favoured, while black cotton soils are avoided. It can be grown in rice fields but the soil has to be sandy and well drained. In such soils castor is grown as an irrigated crop, as a shade crop or on the margins of the fields. The kharif crop is usually sown in June-July and it ripens in Decembers.

Cotton

With 5.19 lakh hectares, cotton occupies about 18 per cent of the total cropped area and is the most important cash crop of the State. The leading cotton-growing districts are Broad, Broach, Ahmedabad, Surat, Bhavnagar, Sabarkantha and Mehsana.

The State is divided into three important cotton zones in which different varieties of cotton are grown. The Kalyan zone comprises parts of the Ahmedabad district, Mehsana and Banaskantha, all to the west of the Sabarmati. East of the Sabarmati is the Vijay zone, comprising a part of Ahmedabad, Sabarkantha, Kaira, Panchmahals, Baroda and the portion of Broach district which lies to the north of the Narmada river. *Vijay* and *digvijay* varieties are long staple and fetch good prices. South of the Narmada river comprising parts of the Broach district and Surat grow 2087 *(vijalpa)* variety of cotton. It is superior both in staple length and yield as compared to *vijay*. In addition, long staple American cotton varieties, viz. 170 x CO2 and 134 x CO2 are found suitable for this region.

Cotton grows best in places where the rainfall ranges between 50 to 75 cm. The rainfall has to be well distributed; there should be soaking rains in July, followed by moderate showers alternating with sunshine in August, heavy rains in the second fortnight of September. Rains after October are not desirable are may even prove detrimental to the crop.

Cotton cultivation is generally restricted to the deep black soils. Medium and lighter types of soils which are well drained also give satisfactory yields.

The preparatory tillage consists of harrowing the land with the local blade harrow two to three times during the period, after the harvesting of the previous crop and before the setting in of the monsoon. In North Gujarat, the crop is not manured directly but feeds itself on the manure given to the previous crop. In other cotton zones 5 to 7 cartloads of farmyard manure are given; some progressive farmers apply about 224 kg. per hectare of groundnut cake and chemical fertilizers. Once the monsoon sets in, the seed is drilled at the rate of 8.96 to 17.92 kg. per hectare. The operation of sowing is generally done during the last week of June or at the beginning of July. The spacing given generally is 45 to 90 cm. between two rows and about 15 to 22 cm. between the plants in the row. As the plant grows, interculturing is done three to four times with small blade hoes and hand weeding is done twice.

The crop usually begins to flower at the end of November or the beginning of December. Picking of cotton is done during February and March. Women and children are employed in the operation. The crop rotations followed in different parts of the States are: (i) North Gujarat: first year cotton second year jowar, bajri or wheat (ii) Central Gujarat: first year cotton, second year rabi jowar or wheat, and (iii) South Gujarat: first year cotton, second year jowar mixed with *tur*.

Tobacco

Cultivation of tabacco, mainly concentrated in the Baroda and Kaira districts, accounts for 0.71 lakh hectares with an output 0.447 lakh metric tonnes. The average yield per hectare in 1959-59 was 715 kg.

The soils best suited to tobacco are alluvial type, found in river beds and red and sandy loams. Black cotton soils are also suitable. The soils should be deep, retentive and well drained since the tobacco crop has deep roots and they are sensitive to moisture.

Tobacco crop is transplanted and hence seedling nurseries are required. In July, after the rains have started, the nursery is prepared. The seed beds are generally heavily manured by farmyard manure at the rate of 80 cart loads per hectare. The land is ploughed and harrowed twice. The seed is broadcast at the rate of 10 gm. per bed and is well mixed with soil. The seeds sprout after a period of about five to seven day. In order to obtain healthy and strong seeding castor and groundnut cake is added at the rate of 224 to 448 kg. per hectare.

If there is continuous rain during the growth of the seedlings, ash mixed with castor oilcake is spread which prevents the damping off and enhances the growth of the seedlings. Seedlings are transplanted after about a month and half, i.e., in the second fortnight of August; 360 square feet of seed-bed area is sufficient for transplanting into an acre of tobacco crop and about 4 oz, of seed is required to be sown on the seed-bed for this purpose. At the time of transplanting the seedlings have about four to five leaves. The operation is best carried out in drizzling rain.

Before the seedlings are transplanted, the effect is levelled with the plank and is marked out in crosswise directions at a distance of 30 by 30 cm., 45 by 45 cm. or 40 by 90 cm. and seedlings are transplanted with these spacings. If at this time there are no rains, the field is irrigated and for this purpose it may be necessary to make and furrows at appropriate distances.

After the seedlings are well established hand weeding and hoeing is done twice. Light earthling up is also done. The crop is then intercultured four to five times with a blade hoe at an interval of a week. Irrigation is not given until plants show signs of drying and is delayed as long as possible. Usually three to four watergins are given once every fifteen days. Once to 1½ months after transplantation, flowering shoots appear; they are nipped off and 8 to 14 leaves are left on the plant. While nipping the tops, the lower leaves which are yellow and small are also removed. Suckers which begin to per.after this operation are removed each week till the crop ripens.

The crop is considered to be ripe for harvest when the colour of the leaves turns from green to golden yellow with brown spots and the tips show signs of drying. The plants are ready for harvest in 100 to 160 days between January to March depending upon the variety. The K-49 variety takes 100 to 120 days to ripen, G-6 a late ripening variety takes 150-160 days, two other popular verities *saijpuri* and *piliu* take 120 to 135 days and 100 to 110 days, respectively.

Tobacco is not rotated with any other crop because there is general belief that the planting of tabaccos every year gives best yield and the material fetches the highest prices in the market. Sometimes *sundhia* (foddr) jowar is sown in the hot season in tobacco fields. In some places the rotation followed is: first year tobacco, second year bajri mixed with *guvar,* and third year *kodra* paddy mixed with *tur.* The yield of tobacco varies according to the variety. Varieties grown under dry conditions yield about 673 kg. to 897 kg. per hectare and those under irrigated conditions about, 1, 121 kg. to 1.345 kg. per hectare.

During the last decade agricultural production has undergone certain basic changes. Foodgrains production has almost doubled itself 1951-52 without any marked increase in the acreage. Cash crops like groundnut, cotton and tobacco, as also areas under fruits and vegetables on the other hand, have registered considerable increase both in terms of area and outrun.

As between various food crops, there has occurred an interesting shift. The output of rice has gone up from 59,944 metric tonnes in 1951 to a maximum of 4,43,890.4 metric tonnes in 1958-59, with only about 8 per cent increase in area under the crop. Similarly, the outrun of wheat has increased substantially, largely through additional areas brought under the crop. As against this, both jowar and bajri have recorded increased production on a declining acreage. The maximum increase in acreage has taken place in regard to groundnut. The next highest increases have occurred in tobacco, followed by wheat and cotton in that order. Pulses show the highest decline.

Both the farmer and his wife work together in ploughing, harrowing, manuring, sowing, irrigation, protecting the crop and in harvesting. Irrigation of the fields occupies the farmer throughout the year and both men and women spend considerable time in this operation. Men attend to the *mhot* while women irrigate the fields. Much of the work of watching the crop-in-grain is done by women. In harvesting also women spend four times the time men. do. Except for the months of July, August and September, women spend considerable time in threshing and winnowing.

Ploughing

Depending on the nature of the soil ploughting is done every year. Usually iron ploughs are used now, with 2,3 or 4 pairs of bullocks, depending on whether the land is being ploughed for the first time or subsequently in the season. Lighter soils are ploughed every year to a depth of 15 to 22 cm. Ploughing is done soon after the first monsoon showers in June which soften the soil and immediately after the harvest of a crop. If the land is free of weeds fewer ploughings are necessary. The tendency of the farmer is to go in for too many ploughings even when the land is clear of weeds. Another undesirable feature is that not infrequently he ploughs the land in the same direction as the slope, thereby increasing erosion. Extensive propaganda is being conducted by the State Agricultural Department to curb this tendency. In certain soils known as *chapan* ploughing of sub-soil is done.

Harrwoing

Next to ploughing is important as it breaks the clods. Sometimes a wooden plank, known as *maind,* is used to break up the clods. Harrowing is done with a blade-harrow known as *kulav* or *vakhar.* The use of the blade-harrow also helps in compacting the soil at a depth of a few centimetres and leaving a firm and uniform seed bed. Harrowing is also done before and during monsoon to keep the land of weeds.

Manuring

The fanner in Maharashtra is well aware of the necessity of manuring the land and as far as possible he prepares an conserves farm-yard manure. About 40 or 50 year ago it was the general practice not the allow farmyard manure to be taken away or sold from a village. The whole village would obstruct any attempt to remove manure from a village for use in another village. Gradually, however, the village *bania,* who had been acquiring lands from the farmers in various villages surrounding his own residential village, managed to purchase farmyard manure either by increasing the price per cart-load or by other means such as creating dissensions among the cultivators.

In spite of the fact that constant propaganda is being conducted for the last 40 or more year to make the fanners prepare compost in a proper way one hardly finds heaps of properly prepared farmyard manure. In majority of villages refuse heaps are found in the oustskirts, each family claiming a certain spot for throwing the refuse from the cattle yard and the kitchen.

A recent survey conducted by the Agricultural Development of the State in the twelve districts of western Maharashtra shows that 54 to 91 per cent of the holdings are not manured. In the Kolhapur district, 46 per cent of the holdings were manured, while in the Sholpapur district only 9 per cent holdings were manured

Whenever possible farmers apply 10 to 20 cart-loads of farmyard manure per hectare when unirrigated crops are taken. The amount of dressings is considerably more in the case of irrigated crops and may range from 40 to 80 cart-loads per hectare. Manure is brought to the fields in bullock carts and distributed in small heaps in straight lines. Later on it is spread over the field.

The use of chemical fertilizers has become common for irrigated crops such as sugarcane, paddy and vegetables. Dressing of inorganic fertilizers commonly recommended are 22 4 kg. of nitrogen per hectare and about the same quantity of phosphate for ordinary irrigated crops and paddy. However, larger applications of both can be probability if applied according to the phases of the growth of the crops.

Sowing

The common practice in Maharashtra is to saw crops in lines by drilling by means of wooden drills known as *pabhar*. These usually are 2,3 or 4-tyned. The distance between the two tynes varies from 22 to to 30 cm. For crops which are sown dry under limited moisture conditions, such as rabi jowar, 45-cm. drills are becoming popular. In certain areas such as Kolhapur and Sangli dibbling of seed is quite common. After drilling or dibbling of seed a light harrow or a light plank is drawn over the sown field to cover the seed and to press it lightly

With the exception of a few farmers the majority do not realize the importance of good seed. For supply of seed they tend to depend on the village *bania* or some well-to-do cultivators who store grain and sell it as seed at high prices at the sowing time.

Seed rates for the different crops vary. There is a conspicuous tendency on the part of the farmer to use too much seed for sowing. This is due to habit as well as to the feeling that he must obtain of feeding material for his animals. Another reason is the indifferent seed he uses for sowing. The larger seed rates make allowance for all kinds of mortality in the seed beds.

Thinning

Within 3 to 4 weeks after germination it is necessary to thin out crops like jowar and *bajri* to obtain more even stand of plants. The practice is being advocated in jowar-growing areas

Interculturing and Weeding

The farmers practise very commonly interculturing of crops For this purpose they use bullock-drawn hoes. Usually two hoes can be drawn by a pair of bullocks with two men manipulating the hoes. The soil between the rows is scraped and stirred and weeds and removed. Crevices in the soil are broken up and filled up with loosened soil.

Earthing up

This operation consists of digging the soil around the plants and building it in a heap around the plants. The practice is common with irrigated crops such as sugarcane, vegetables like potatoes and ginger and fruit trees like bananas. It is doubtful if 'earthing up' is as beneficial as claimed. In sugarcane, experimental evidence goes against the practice. Earthing up disturbs the root system and gives a setback to the plant However, more investigation is necessary to justify the farmers' general faith in this operation.

Irrigation

The common practice is to irrigate crops in rows, like sugarcane, or in rectangular beds. Under canal irrigation the practice is to soak up type land as much as possible, as the interval between two waterings is of longer duration than is required. This often results in over-irrigation of the crop. In the case of well irrigation of fields, for obvious reasons, it is not over done, but irrigations are too frequent, and can be reduced to some extent without loss of yield.

The necessity of levelling the land for irrigation is not fully realised. Frequently one finds water running down from one end to the other of the row or the bed. In certain irrigated areas such as Kopergaon, where sugarcane has become the dominant crop, contour furrow irrigation is being taken up by the farmers.

Protection of Crops

The grain crops are watched by a woman standing on a *machn* built in the middle of the field. She ascares the birds away by hurling clods or small pebbles from a sling. Farmers are slowly taking up modem methods of plant protection and the use of insecticides and fungicides is becoming popular. The knowledge regarding how and when to use the protective chemicals is not yet as universal and well understood as one would wish with the result that a farmer is often too late in resorting to control measures. The cost of modem pesticides is also too high for the average farmer. Use of sulphur for control of jowar smut has become very common, especially in western Maharashtra. The Deccan wingless grasshopper has been controlled by aerial spraying.

Harvesting

Crops are harvested in a variety of ways. Cereal crops are usually cut with a sickle close to the ground. The harvested crop is laid out in rows in the field and after 2 or 3 days tied in bundles for removal to the threshing-yard. Pulse crops are removed in a similar way or uprooted.

For threshing, ears are removed from stalks in the case of *bajri* and jowar. They are spread in a circle with a certain thickness over which bullocks go on trampling in a circle to loosen the grain from the ears, Sometimes a bullock cart is used for this purpose. In certain areas rolling stone-threshers, especially for jowar, are being used. Wheat is also threshed by spreading the whole mass on the threshing floor. Paddy is threshed by striking the bundles against a wooden stand.

The most common form is to winnow grain in the prevailing wind currents. Threshed mass is pourced from a height by a man standing over a tall three-legged stool. While the man is pouring the threshed material, a woman removes the heavier chaff from the fallen heap of grain and also supplies the man with baskerfuls of threshed mass for winnowing.

Storing

The cleaned grain in filled in gunny bags if it is to be sold soon. In case the grain is to be stored for some time, it is deposited in circular bins made of bamboos or stalks of cotton or *tur*. These are plastered with a mixture of mud and cow-dung. The surface of the filled grain is dressed with leaves *of neem* to prevent insect attack.

Agricultural Implements

The chief agricultural implements of the farmer are *mangar* (plough), *vakhar* (harrow), *pabhar* (seed-drill), *Kolapi* (hoe) *mhot* (leather water bucket). In addition to these the farmer has the necessary number of hand tools such as sickles, *phavada* (shovel) and *kudal* (flat spade) for digging the soil. Bund-formers are also becoming popular. With the exception of the iron plough, hardly any other 'improved' implement has gained universal popularity and has replaced the corresponding indigenous implement.

Plough

In spite of the popularity of the iron rough the wooden one is still widespread. It is manufactured from wood of *babul* trees which are grown by the farmers on the bunds of their fields. A wooden plough last from 10 to 15 years.

Vakhar

Next to the plough, *vakhar* is an important and versatile implement. The blade harrow is used for pulverising the soil after ploughing. It can also be used for the removal of stables. It is also used for mulching and covering seed-beds.

Pabhar (seed drill)

The country-made seed drills are most common. So for no improved seed drill has made headway. Those who have gone for such new machines have had to give them up for one reason or another. The simplicity of the indigenoues seed drill is still unmatched. The wooden seed drills in use have 2,3, or 4 tynes. In certain areas 6-tyned drills are also in use. Two-tyned ones are usually used for sowing kharif crops.

Kolapi (hoe)

Hoes are of two kinds, one with the whole blade which works between the rows and another with a slit so that the blades work on the two sides of a row. Two hoes can be drawn by a pair of bullocks. The shorter-bladed hoes are while longer-bladed ones are used in interculturing cotton and rabi jowar.

Mhot

The use of the leather bucket is the most common method of lifting water from wells in the up-ghat regions. Two buckets are used at a time in drawing water, and each is pulled by a pair of bullocks. The platform on which the bullocks operate has an incline away from the well. After the buckets empty water in the throught the animals are pulled backward. In certain areas like Sangli, Kolhapur and Satara, iron *mhots* are also in use.

In Konkan, *rahat* (Persian wheel) is more common than *mhot*. In this area *okti* is also used. This is a primitive device which is worked annually. An iron bucket is tied to a vertical pole which is hinged to a horizontal beam fixed across the well.

New Trends

Certain trends towards better agriculture are noticeable. There is a considerable reduction in the number of wooden ploughs during the period from 1956 to 1961. The reduction is very heavy in the case of Vidarbha (actually 3,83,408 wooden ploughs less in 1961). There is some decrease (actually 27,501) in Marathwada, but in the other two areas Konkan and western Maharashtra, there is increase of 26,557 and 19,835 wooden ploughs, respectively. There is a general increase in the number of iron ploughs in all the regions.

As regards tractors there is a slight upward trend in western Maharashtra and no increase in Konkan. But the reduction is very heavy in the other two areas. Vidarbha shows a decrease in tractors from 1,486 in 1956 to 512 in 1961, and in Marathwada the reduction was to extent of 920 tractors. The 1961 figures do not include tractors owned by the Government. It is obvious that a large number of machines have gone into disuse for lack of repair facilities and spare parts.

As regards major-lifting devices for irrigation purposes no figures are available for *mhots*, but other means reflect an interesting trend. Persian wheels are used extensively in Konkan and Marathwada. The two districts Bhir and Nanded have 2,234 and 2,934 Persian wheels, respectively. In the case of both oil-engines and electric pumps there is increase in all the areas.

8

AGRICULTURAL EDUCATION

The Famine Commission, 1880, was the first to draw the attention of the Government to the need of agricultural schools. But no action was taken until 1890, when Dr Voilcker's proposals in regard to agricultural education were examined by the representatives of the Provincial Governments. The decisions taken were that:

(1) Agricultural degrees, diplomas, or certificates should be placed upon the same footing as corresponding science degrees, etc., in qualifying for admission to Government appointments and more particularly those connected with land-revenue administration.

(2) There should not be more than four institutions giving a high class diploma, viz. at Madras, Calcutta, Bombay and some place in the North-Western Provinces, and that these should be utilised by other provinces.

(3) The diploma should eventually by compulsory in the case of certain appointments, e.g. agricultural teachers in training schools, assistants to the Director of Agriculture etc.

(4) The practical instruction of candidates for certain subordinate appointments at a school, or an experimental farm, should be further considered.

(5) A special school course leading up to the agricultural diploma degree, or certificate, is required.

(6) The practice of allowing school teachers, either before or after appointment, to pass through a course for a few months on a Government farm is one which deserves consideration.

At the time the Government of India Resolution on Education Policy (1904) laid down a new policy regarding agricultural education at the collegiate level, only five institutions in British India were imparting agricultural education and there were 219 students on roll in them. They were the agricultural branches of the colleges at Poona and Sibpur (Bengal), the College of Agriculture at Saidapet (Madras) and the agricultural schools at Nagpur and Kanpur. Only at Poona, the course extended over three years and L. Ag. degrees of the Bombay University were awarded. Admission to this course was open to students who had passed first year examination in Arts.

The college at Saidapet gave instruction in a three-year course at the end of which an examination was held by the Commissioner of Government Examinations. The agricultural branch of the Poona College was opened in 1897 and that of Sibpur College in 1899.

The Agricultural Schools at Kanpur and Nagpur offered three types of courses, viz. one for training revenue subordinates, second for teachers, and the third for land-holders.

The Resolution in reviewing the agricultural education facilities observed: "While the necessity for developing agricultural education is being recognised, India possesses no institution capable of imparting a complete agricultural education. The existing schools and colleges have not wholly succeeded either in theory or in practice. They have neither produced scientific experts, nor succeeded in attracting members of the land-holding classes to qualify themselves as practical agriculturists." It was suggested to:

(1) Create a central institute for training experts capable of study ing the problems of Indian agriculture and of applying principles of Western science to bring about an improvement in the methods usually employed by Indian agriculturists;

(2) Create an agricultural college in each important province. It was further point out in the Resolution that research would ordinarily be more active and better sustained if it was associated with lecturing, which would check any tendency to the investigations of problems unlikely to lead to practical results; and

(3) Remove the defects of the agricultural colleges observed in the Resolution, i.e. their failure in producing scientific experts and in attracting members of the land-holding classes.

The Central Research Institute at Pusa (in Bihar) was created and started its work in 1908. It was transferred to Delhi after the earthquake of 1934.

Organised instructions in agriculture in India thus began as early as the beginning of this century when five agricultural colleges were established during 1907 offering diploma courses. The first Agriculture College at Poona had however been opened in 1903. These institutions were later upgraded to degree level. The earliest post-graduate courses were organised towards the early twenties and it was only about 1930 that the post-graduate colleges and institutes secured recognition for imparting training at M. Sc. and Ph. D. levels in agriculture. Before independence in 1947, there were 17 agricultural institutions offering degree courses. Although higher agricultural education was costly and the resources available were rather limited, a large number of new agricultural colleges sprang up during post-independence decade under pressure of regional interests and in response to accelerating demand for agricultural graduates during the First Plan period.

Venterinary Education

Organised training of veterinary personnel in India may be said to nave started with the establishment of an Army Veterinary School at Poona in 1862; thus taking a lead over agriculture education. The first civil veterinary school was established at Bobugarh (Uttar Pradesh) in 1877. The training imparted in these two schools was, however of a low standard, more or less comperable to the present day courses for stockmen and compounders.

The first Veterinary College to be set up in India was the Punjab Veterinary College, Lahore (now in Pakistan), started in 1882. By 1903, four institutions were functioning, the three new colleges were at Bombay (1886), Belgatchia near Calcutta (1893) and Madras (1903). Central institute like the Indian Veterinary Research Institute which was primarily developed as a research institute and which over the years developed certain educational functions continued to offer specialised post-graduate courses in veterinary sciences leading to awards of diplomas or certificates.

Development of Agricultural Education During Post-Independence Period

When India attained Independence in 1947, there were 17 agricultural institutions offering degrees. Although higher agricultural education was costly and the resources available were rather limited, a large number of new agricultural colleges sprang up during post-Independence decade under pressure of regional interests and in response to accelerated demand for agricultural graduates during the First Plan period. The number of agricultural colleges went up by about three times and that of veterinary colleges almost doubled, during the period 1951 to 1961. Intake capacity and output of degree colleges reached a peak in 1965-66. In that year, 90 colleges (70 agriculture plus 20 veterinary) admitted 11,562 students and turned out 6,250 graduates. Today, there are 93 degree colleges in all but the enrolments are declining due to unemployment of agricultural graduates with the result that a number of State Governments had to reduce admission by upto 50 per cent.

Although most of these institutions came up as Government colleges under the administrative control of Departments of Agriculture or Animal Husbandary, in some of the states like Uttar Pradesh, a number of affiliated private colleges sprang up without much planning or preparation. Many of these colleges did not have adequate farm, laboratory and library facilities. Uttar Pradesh alone accounts for about 40 per cent of the total intake capacity.

The number of post-graduate agricultural colleges increased from five in 1951 (unrolment 167) to 26 in 1962 (enrolment 654) and to 40 in 1971. Only one veterinary college offered post-graduate courses in 1956 (enrolment 9) but in 1961 the number went up to five (enrolment 123) and in 1972 there were 20 veterinary colleges admitting about 300 post-graduate students. This expansion of post-graduation, unlike under-graduate education, has continued until recently although it now appears to be levelling off. There has been, however, a sharp increase in the number of students preparing for doctorate. Currently, about 25 per cent of agricultural and veterinary graduates go on to post-graduate studies.

While enrolments at some of the affiliated sub-standard colleges have shown abrupt decline, agricultural universities continue to attract larger numbers and better students. Obviously the element of competition between institutions has come into play and performance of the sub-standard institutions has started recoiling on them.

According to statistics published by UGC the total enrolment in agriculture and veterinary has gone up by about 16 times since 1960. The numbers of Bachelor's and Master's degree holders have been increasing at the rate of 14 per cent per annum since 1950. The postgraduate enrolments have registered an increase of 5.8 times over those in 1950, but those in research have gone up by about 14 times over 1960 figures.

Most of the colleges referred to above were primarily concerned with the training of agricultural graduates. The State Departments of Agriculture and Community Development executed the programmes of research and extension respectively. The relationship between agricultural colleges and research departments, therefore, was not close enough to ensure maximum transfer of information in practice. The production technology developed at these educational institutions did not keep pace with the fast changing requirements of our agriculture.

As a result of this it became hard to convince State Governments that investments in agricultural education and research would pay lasting dividends. The rate of investments in agricultural education, therefore, remained deplorably low, virtually stagnant. Moreover this expansion took place along traditional pattern. It was not oriented to meet the needs of economic progress and change that was overtaking the country. It was soon realised that both the system of education as well as the set up of the institutions in the field of agriculture and animal sciences needed to be reorganised, if it was to serve as effective vehicle of agricultural progress and development.

Reorganisation of Agricultural Education and Development of Agriculture Universities

The need to bring about a rapid increase in food production in the years immediately following Independence necessitated re-examination of the existing system of agricultural education. It was soon realised that the goals of increased production could be achieved only through application of science and technology to agriculture and that trained personnel was vital for developing science-based culture. Accordingly, Government of India appointed in 1948-49 the University Education Commission headed by Dr. S. Radhakrishana our former President, which examined the entire pattern of higher education in India and made recommendations for improving it.

The Commission suggested three definite objectives as follows, corresponding to the three main aspects of agricultural education, viz. teaching, extension and research.

1. Training of farmers' sons who will go back to their farms and work on them more efficiently.
2. Training of persons for carrying the results of modern agricultural research to the peasant through extension work and demonstration.
3. Training of persons for carrying on the work of research, developmental and fundamental, relating to problem of agriculture and animal husbandry.

The Commission recommended that the institutions that would take care of these objectives would be (1) basic and post-basic schools, (2) farm institutes situated on model farms and associated with residential high schools, and (3) rural universities for higher education, respectively.

The Joint Indo-American Team appointed by the Government of India to make recommendations for improving the facilities for agricultural education and research in India, suggested in 1955 that the Indian Council of Agricultural Education should play an increasingly important role in the development of higher education in agriculture, veterinary science and other related fields, both at the, under-graduate and post-graduate levels. Having regard to the recommendations of the Team, the Government of India in consultation with the Indian Council of Agricultural Education, decided to make certain modifications in the constitution of the Council. The revised functions of the Council as announced on 8th January, 1957 were as follows:

(a) The Council will be purely an advisory body.

(b) It will co-ordinate agricultural education (including education in Animal Husbandry and Dairying) programmes so that a uniform standard of agricultural education might be achieved throughout the country.

(c) It will concern itself with only the broad principle of agricultural education and will advise State Governments on such problems as may be referred to it.

(d) It may prepare model syllabi for adoption in the various teaching institutions concerning agriculture, animal husbandry and dairying.

Second Indo-American Team on Agricultural Research and Education

As a result of the project for assistance to agricultural research education and extension under the Technical Cooperation Programme between the Government of India and the Government of United States of America, a Joint Team consisting of five representatives of the Indian Government and three American specialists of Agricultural Research and Education was selected to make a comparative study of agricultural educational institutions. It reviewed the existing system of agricultural education and made certain recommendations. Some of these, which are of far-reaching importance are given below:

1. The Team recognises the importance of the recommendation of the University Education Commission that the system of University examinations requires rethinking and reorientation. Much study for agriculturally related higher education should be undertaken by the Indian Council of Agricultural Education, and suitable recommendations developed.
2. The Team endorses the view of the University Education Commission that a determination of a student's final standing should be based not only upon the final examination but likewise upon the quality and chapter of his work in the college. The Team accordingly urged that the teacher should evaluate the quality of the work of the student at least upon the completion of each course and that for the purpose of the final assessment certain credit may be given to the work done by the student during the college year.
3. In teaching institutions greatest care should be taken to ensure that the syllabus does not tend to become static, but keeps pace with the advance of science and the progress of pertinent research. The successful teacher is one who sees beyond the mere outlines of the syllabus and covers a current findings of research which may be even mere significant to the student and his immediate future than the subjects formally included in the syllabus.
4. The Team believes that India has facilities and staff that are under-utilised in service to agricultural research and post-graduate education and recommends that the facilities of the Central Agricultural Institutes as well as these administered by agencies other than the Ministry of Food

and Agriculture should be carefully considered by the Indian Council of Agricultural Education and co-oriented in overall schemes for post-graduate instruction.

5. The Team endorsed the recommendation of the University Education Commission that a Rural University would include "a ring of small resident, under-graduate colleges, with specialised and University facilities in the centre." For the complete development of a Rural University the Team would envisage in the beginning the location on the same campus and in the juxtaposition a college of Agriculture and a college of Veterinary Science to which should be added in due course, colleges of 'applied' liberal Arts and Sciences, a college of Technology using this term in the broad sense of engineering and industries, and with a group of villages to be used as a laboratory for the students. While one would hope that the students entering a Rural University would come largely from the villages, other students would not be excluded. One Rural University should have exactly the same autonomy as is accorded to other universities, and the degrees which it may confer should have the same standing and prestige as these conferred by other universities.

The Team took the view that rural institutions were a halfway house between the rural universities recommended by the University Education Commission and the manjri type of vocational agricultural schools. The Team was of the opinion that each state should have a university and each district a manjri type school.

The question of agricultural education was again examined by the Committee on Rural Education which submitted its report in 1958. They reviewed the position of all the four types of existing institutions, i.e. rural institutes, manjri type, basic-agricultural schools and Janta colleges. This Committee did not favour the view to combine agricultural schools and Extension Training Centres. It thus recommended that the establishment and continuance of agricultural schools should not be based on the number of gram-sevaks required. The Committee was in favour of integration of the agricultural course in multipurpose Higher Secondary Schools and the Agricultural Schools. Janta Colleges according to the Committee had failed to fulfil their avowed purpose of providing mass adult education.

Agriculture Universities

With a view to providing a model legal base to Agricultural Universities, Government of India appointed an Agricultural Universities Committee in 1960. The report of the Committee submitted in 1962, provided the guidelines for developing the Model Act by the ICAR. This Model Act given out in 1966 provides in sufficient details, the legal framework for the development of an Agricultural University in India.

By now the concept of agricultural universities forms an integral part of the National Policy on Education. It has been accepted that at least one Agricultural University may be established in each state. Seventeen Agricultural Universities have already come into existence, the first one being the Uttar Pradesh Agricultural University in Pantnagar in 1960. Out of the total 93 agriculture and veterinary colleges in the country today, about 50 are constituent colleges of agricultural universities. The Indian Agricultural Research Institute, where the new system of post-graduate education was first adopted in 1958, has the deemed university status. The Indian Veterinary Research Institute and the National Dairy Research Institutes continue to offer M. Sc. and Ph. D. degrees of affiliating universities like the Agra and Punjab Universities.

The transfer of Statewise agricultural research, education and extension to the Agricultural Universities is the first requisite if the objectives of integration are to be achieved and these institutions are to be make their contribution to the agricultural production of a State.

The concept of such as integration has already proved its worth through the remarkable progress made in the field of agricultural education, research and extension by the new Agricultural Universities.

There is perceptible improvement in the quality of education. There are more competent teachers, better equipped libraries, laboratories and farms. The internal examination system is geared for continuous preparation on the part of both the students and the teachers.

The reform in agricultural education introduced in agricultural universities besides improving in quality has brought down to a large extent the wastage in higher education in agriculture. This not only saves costs but provides training opportunities for more students. These institutions today are serving as fountainheads of new knowledge earned through purposeful, problem-solving research and have become main centres of dissemination of useful knowledge to the farming community. Some of the best training for farmers is offered by the agricultural universities. A large number of functional specialists who have confidence born out personal experience in successfully applying scientific knowledge to the solution of practical problems as well as those who have the competence to earn new knowledge that helps to enrich educational programmes are coming out of our agricultural universities.

With the good working conditions and incentives that they offer both to the faculty and the students, agricultural universities are providing opportunities for productive and prestigious work, are filtering unprecedented team spirit and a healthy change in the outlook of all the teachers, the researchers, the farmers and the administration. They are winning the confidence of both the farmers and the

Governments. They have already assumed leadership in sciences, education and extension in a number of cases and are producing future leaders for Indian agriculture. Their direct contribution to programmes like pedigree and production, fertilizer-use and National Crop Demonstration is highly impressive. Above all, they are aware that there is no substitute for quality, and competence and for this they are striving ceaselessly. In this connection the inter-institutional collaboration needs special mention.

Affiliated Colleges of Agriculture and Veterinary Sciences

Although agricultural universities have been established in most of the states, about one-third of the agricultural and veterinary colleges still remain outside this orbit and are affiliated to traditional universities. They account for about 40 per cent of the total output of agricultural graduates. The ICAR therefore, continues to give assistance for the development of such of these colleges which have the necessary potential for development. As in the case of agricultural universities, assistance was formerly given as per the UGC pattern. According to the revised pattern of assistance recently adopted by the ICAR affiliated colleges will also receive assistance on 100 per cent basis on 11 selected items of development with an overall ceiling for the Plan period as well as for each item. With a view to bringing about uniformity, introduction of the three-year degree course after 12 years of schooling has received special attention in the case of states universities where it had not been adopted so far. It is intended to extend, through ICAR assistance, the concept of integration of teaching, research and extension to these colleges also.

Agricultural Education at the School Level

At the school level in the existing system to education, agriculture forms part of the curriculum of studies in most general schools. It is one of the crafts studied in basic schools. It is an optional subject of study for the matriculation in the ordinary high schools and is one of the streams in four hundred out of the two thousand and odd multipurpose high schools in the country. Agriculture is taught as a vocation in a variety of special schools—such as Agricultural Schools, Manjri Schools or Rural Institutes and Gram Sawek Training Centres which have been established since 1952 in the wake of Community Development programme. Agricultural education up to the technician certificate or diploma level is imparted in the former institutions while the Gram Sewak Training Centres offer a two-year integrated course for village level workers in agriculture and extension.

The large number of schools and the variety of educational programmes have not assured the country a standard of agricultural education at the primary and secondary levels adequate to meet the needs of development. A reform of the school system of education lies along the lines indicated by the Education Commission, 1964-66, on the place of agriculture in general education. The Commission has recommended that in India some orientation in agriculture should form part of the general education at the school stage and at the university as well as a teacher's training. In its view of the period at school should be utilised to impart a sound general education with particular emphasis on mathematics and science to serve as a preparation for coping with the rapid scientific and technological changes that may be anticipated in Indian agriculture in the near future. The Education Commission has further recommended the establishment of specialised institutions which will provide vocational education in agriculture at the post-matriculation level. These institutions to be called Agricultural Polytechnics, are to be responsible for training the non-professional specialists required as farm-mechanics, farm managers, craftsmen and technicians in agro-industries, assistants in agricultural credit and insurance organisations. The Commission has suggested that the course at these polytechnics should be practical in nature including a significant element of experience. They should be terminal in character and lead to specific and immediate employment.

Training for Farmers and Extension Workers

As progress in agriculture depends ultimately upon the adoption by farmers of improved agricultural technology, their education is of fundamental importance. This is primarily adult education and is in the nature of inservice training. The Central Government in collaboration with state governments has initiated a training programme for farmers and farm women recently. This includes specialised training courses for a selected number of progressive farmers in agriculture and allied as well as course of training for farm women in agricultural production. The courses are given by the Farm Sevak Training Centres, Agricultural Colleges and Government Farms where adequate facilities for practical training exist.

This type of training programme has been neglected so far. There is need to enthuse the state governments to take up this type of training more vigorously. But for Maharashtra and a few other states nothing tangible has been done so far in this respect, although the Fourth Plan proposal is to cover at least 10 per cent of the 60 million farming families in the country.

The training of village level workers is carried out in 144 Gram Sevak Training Centres spread throughout the country. These centres had trained 77,000 Gram Sevaks and 9013 Seviks by the end of the Third Plan. The major emphasis in the curriculum is on agriculture and allied subjects and extension education. The pre-service training phase is now almost over and the main subject during the Fourth Plan will, therefore, be the qualitative improvement of the training given. To increase the technical competence of the present village level worker following steps may be considered as very useful.

(i) Encouraging village level workers, who are eligible for admission to agricultural colleges, to study for their degrees.

(ii) Instituting an advanced Diploma Course of a year in agriculture for old village level workers and upgrading all the Gram Sevak Training Centres by the end of the Fourth Plan.

(iii) Increasing the number of refresher training programmes at all the centres so that all village level workers are able to attend them.

Growth of Dairy Education in India

The establishment of Imperial Institute of Animal Husbandry and Dairying in 1925 at Bangalore (later known as the Indian Dairy Research Institute) was the first landmark in the history of Dairy Education in India. The first course in dairy farming was started in 1924. Later the Indian Dairy Diploma (IDD) course of two years duration was started. In 1945, the Associateship course was initiated. The Institute has also been admitting honorary research workers who could earn M.Sc. and Ph. D. degrees of affiliated Indian universities. In 1955, the headquarters of the Indian Dairy Research Institute was shifted to Karnal and the Institute was renamed as the National Dairy Research Institute. The Dairy Science College was organised as part of NDRI in 1957. The degree course in dairying was first introduced in that year itself under the auspices of the Punjab University. This was a four-year degree course. This course was later divided into two branches, viz. B.Sc. (Dairy Husbandry) and B.Sc. (Dairy Technology) in 1961. This was also the year for starting M. Sc. Dairy course. From the same year the IDD training was bifurcated into Indian Dairy Diploma (Dairy Husbandry and Dairy Technology). The education and research functions of the eight divisions of this Institute are fully integrated although the Institute has a separate Division of Dairy Education and Training largely with coordinating functions. During the Fourth Plan period dairy education at NDRI has further been strengthened adding three more divisions viz, (i) Dairy Cattle Nutrition and Physiology, (ii) Dairy Cattle Genetics and Breeding, and (iii) Nutrition and Dietetics. So far about 70 IDD, 200 B. Sc. (Dairy), 150 M. Sc. (Dairy) and 20 Ph. D. students have passed out from National Dairy Research Institute. The Institute continues to remain the strongest single centre of education and has provided most of the trained manpower available in this field in the country.

The Dairy Science College at Anand, started in the year 1961-62, is the only other Dairy Science College in the country. The college prepares students for IDD course and B. Sc. (Dairy Science) degree of the Ballabh Vidyapeeth, Anand. At present IDD course is offered at five centres including NDRI and Anand.

Growth of Fisheries Education in India

Until the fifties, fisheries departments recruited graduates of zoology or biology. Some of them organised inservice training courses and also started polytechnics to award diploma courses in fishery technology and navigation. The Central Government also started the Inland Fisheries Training Course at Barrackpore and Marine Fisheries Training Courses at Mandapan in Tamil Nadu. By and large these ad-hoc arrangements proved to be unsuccessful and had to be wound up.

It was only in 1958 that the Government of India decided to take steps to organise fisheries education on more permanent basis. The Central Institute of Fisheries Education at Bombay was set up in 1961 and the Central Institute of Fisheries Operatives, at Cochin in 1963. In the same year a post-graduate diploma course in Fish Processing and Technology was started at Mangalore by the Government of Mysore with Japanese collaboration. Some of the universities have since introduced Fish and Fisheries as a special subject in the M. Sc. course. Some have started full-fledged M. Sc. courses in Marine Biology and Oceanography. In 1969 the Calicut University started a B. Tech. Fisheries course and in the same year the University of Agricultural Sciences has started a full-fledged Fisheries College at Mangalore.

The Central Institute of Inland and Marine Fisheries training referred to above offers a number of Diploma and Certificate courses of duraton varying from 6 to 15 months. Short courses of about one month's duration in Fist and Technology are offered by the Central Food and Technological Research Institute, Mysore, as well as under the Indo-Norwegian Project. There is, no doubt that fisheries education needs to be strengthened considerably to meet the increasing demand for trained manpower in this field.

Agricultural Engineering Education in India

The Allahabad Agricultural Institute was the first institution in India to offer in 1942 the courses leading to the Bachelor's degree in Agricultural Engineering of the University of Allahabad. The Indian Institute of Technology, Kharagpur, started a similar degree programme m 1932. Six more constituent colleges of agricultural universities—U.P. Agricultural University (Pantnagar), University of Udaipur (Udaipur, 1964), Punjab Agricultural University (Ludhiana 1965), OUTAT (Bhubaneswar, 1966), JNKVV (Jabalpur, 1967) and MPKV (Poona, 1969) have started an agricultural engineering college each. These colleges have an admission capacity of about 300 students per year and a total strength of about 1,200 students. About 1,000 students have already graduated from these institutions in the past. It is evident that all the engineering colleges are at present located in North India. There is room for opening such institutions in the south. This regional imbalance in the growth of agricultural engineering education is responsible, at least partly, for the unemployment problem of agricultural engineering graduates witnessed during past few years and needs to be corrected as early as possible.

Home Science Education in India

Home Science education is not new and a large number of traditional universities have been offering education in Home Science for quite some time. However, in response to the need for orientation of Home Science education towards agriculture, three of the Agricultural Universities have organised constituent Home Science colleges after 1965. These are Andhra Pradesh Agricultural University, Hyderabad, Punjab Agricultural University. Ludhiana, and Udaipur University. At least five more agricultural universities have developed plans to start Home Science colleges in the near future. These colleges are organised around four main disciplines—Food and Nutrition, Child Care and Development, Home Management and Clothing and Textiles.

One of the main hurdles coming in the way of the development and growth of Home Science education for farm woman is the paucity of trained personnel as well as excessive mobility of staff.

Diversification and Specialisation in Higher Agricultural Education

With the advancement of science and technology and the need for greater degree of specialisation to tackle the new and complex problems that are coming up in the field of agriculture, there has been considerable diversification and specialisation in higher agricultural education in India. The subject of Agronomy which was organised only in late forties, as a distinct subject of Agriculture, has since developed into further fields of specilisation such as Crop Husbandry, Water Technology, Weed Control, Plant Tissue Test, and Soil Fertility-Likewise, the subject of Agricultural Botany has since grown into distinct fields like, Genetics, Plant Breeding, Plant Physiology, Plant Protection, Seed Technology and Horticulture. Horticulture has developed further into Pomology, Olericulture, Floriculture and Fruit Preservation. A number of new subjects like Food Technology, Human Nutrition, Business Management, Agricultural Credit and Nematology have come up in recent years. In the field of Veterinary and Animal Husbandry also new subjects like Poultry Science, Public Health, Animal Products Technology and Bio-chemistry have been added. With a view to reducing India's dependence for foreign training, the programme for the establishment of "Centres of Excellence" in selected but specialised areas has been taken up lately. Thus there has been considerable spurt in the diversification and specialisation of agricultural education in recent years.

ICAR and Coordination of Agricultural Education

ICAR was established in the year 1929 in pursuance of the recommendation of the Royal Commission on Agriculture (1926). The main functions of the ICAR inter-alia are to coordinate, aid and promote agricultural research and education and to serve as a cleaning house of information relating to these functions. Although the statutory, authority for coordination, accreditation etc., of higher education rests with the University Grants Commission (UGC) in practice the coordinating functions in the field of technical, medical and agricultural education have been entrusted to the Central Ministry concerned. Accordingly, ICAR under the aegis of the Ministry of Agriculture, performs for all practical purposes, the same role in relation to agricultural education as the UGC does in the case of Basic Sciences are Humanities. The Council is responsible for coordination and promotion of higher agricultural education in the country. It provides incentives and leadership for the development of new and need-based agricultural education programmes, introduction of educational reforms, and institutionlisation of educational innovations. One of the important functions of the Council is to achieve uniformity of standards of agricultural education between institutions. The ultimate goal is to provide well trained manpower needed to give a sound scientific and technological base to India's agriculture with a view to accelerating the pace of agricultural development and social transformation. The main emphasis in the programme of agricultural education is, therefore on achieving "excellence" and "relevance". For this, the Council gives assistance to agricultural universities and colleges for their development programmes, such as construction of the laboratories, libraries, hostels and purchase of books and equipment. It also provides technical guidance in the development of curricular materials, revision of syllabi and production of books. Organisation of Summer Institutes and in-service training programmes, award of scholarships and fellowships to the students, organisation of workshops, seminars and symposia and development of their-institutional training programmes also fall under the purview of the activities of the Council. Provision of foreign training of faculty members and acquiring the services of foreign experts are some of its other important activities.

As mentioned earlier, the early post-Independence era witnessed rapid growth of higher education in agriculture and it become necessary for the Indian Council of Agricultural Research, in its capacity as national coordinating agency for agricultural education, to play an increasingly important role in the maintenance and improvement of standards of agricultural education. Accordingly, the Indian Council of Agricultural Research set up under its aegis, the Indian Council of Agricultural Education in the year 1951. The main functions of ICAE were *(a)* to co-ordinate agricultural education programmes (including education in animal husbandry and dairying) so that uniform standards might be achieved throughout the country, *(b)* to prepare modern syllabi for adoption by various colleges, and *(c)* to advise the State Governments on problems of agricultural education and develop a consensus with regard to reforms that need to be introduced in agricultural education. The first meeting of the ICAE held at Hyderabad setup a Standing Committee with Dr B. P. Pal as Chairman to ensure the follow-up action and continuity of work between the Council meetings. Later, as per recommendations of the first Joint Indo-American Team, the composition and functions of the ICAE were further modified in the year 1957. Inter alia a whole-time Chief of Agricultural Education to serve as an ex-officio Secretary to the ICAE was appointed. A quarterly journal of the ICAE was also published from 1958 to 1964.

With the reorganisation of ICAR in 1966, in pursuance of the recommendations of the Agricultural Research Review Team (1963) and the formation of the Division of Agricultural Education in the ICAR, the ICAE, has been replaced by the Standing Committee on Agricultural Education. This Standing Committee is assisted by the two Scientific Panels on Agricultural Education and Animal Sciences Education. According to the recommendations of the Education Commission (1964-66) it is envisaged that the ICAR will eventually have within it a body on the pattern of the UGC to advice on the development of Agricultural Education.

Financial Support for Agricultural Education

Until the later part of the Third Plan, Agricultural Education received little financial assistance from either the UGC or the Ministry of Agriculture. It was only towards the mid-sixties that a definite provision was made for giving financial assistance to agricultural universities and colleges on an ad hoc basis. After the reorganisation of ICAR and the establishment of Agricultural Education Division in the year 1966, concerted efforts have been made to develop agricultural education systematically and for this purpose financial assistance has been provided in a planned manner.

Until 1969-70, assistance was given as per the UGC pattern. Accordingly, institutions which could not find the matching grant could not avail themselves of ICAR assistance and this came in the way of educational development. A revised pattern has, therefore, been adopted since 1970-71. Assistance is given under this pattern on 100 per cent basis for sixteen selected items of development of agricultural universities and eleven items for agricultural veterinary-colleges with a ceiling of Rs. 2. crores for an agricultural university, Rs. 8 lakhs for a post-graduate college and Rs. 4 lakhs for the undergraduate college during the Fourth Plan period. The total outlay for schemes of Agricultural Education is Rs. 28 crores during the Fourth Plan period. This support is in addition to the central support that these institutions receive for participation on various research and development schemes.

Current Schemes of Agricultural Education

The following schemes of Agricultural Education have been in operation during the Fourth Plan period:

1. Establishment and development of agricultural universities.
2. Development of Agricultural, Veterinary and Dairy Science colleges.
3. Improvement of Agricultural colleges of central universities.
4. Award of scholarships and fellowships to undergraduates and post-graduate students of agriculture.
5. Establishment of centres of excellence.
6. Organisation of Summar Institutes.
7. Improvement of the competence of teaching facilities.
8. Higher education in Food and Nutrition at Agricultural universities/ colleges.

9

AGRICULTURAL RESEARCH

Several unsuccessful attempts to introduce western farming techniques to improve Indian agriculture over a long period led the Government of India to the realisation that only the scientific study of the conditions of farming and experiments in agricultural practices in the Indian context would lead to any real improvement. Dr Voelcker, an agricultural science who was invited to India in 1889 to advise on the steps to be taken to improve agriculture, stressed the need for research and this was later emphasised by the Famine Commission of 1901 and the Irrigation Commission of 1903. Subsequently, funds were provided for the development of agricultural research and education in the provinces and a separate Department of Agriculture was constituted in most of the provinces. Agricultural colleges were also founded in Poona, Kanpur, Nagpur, Lyalipur, Coimbatore and Sabour. Till the turn of the last century, however, the provincial departments of agricultural were small and were engaged largely in the collection of statistics of the area and production of crops and were concerned with research.

The setting up of the Veterinary Research Institute at Poona in 1889 which was subsequently transferred to Mukteswar in 1893 was the first concrete step taken towards technical research in agriculture in India. The close of the nineteenth century witnessed a revolution in agriculture in the Western world. The inventions of inorganic fertilizers by Liebig in 1861, study of agricultural chemistry and its scientific application to cropping pattern, application of guano, bone-meal and phosphates, improved breeding of livestock, etc., can be cited as a few examples. The beginning of the twentieth century marked a further awakening about technical improvements in agriculture in the West. With the munificance of Mr Henry Phips— an American— and a generous donation of thirty thousands pounds by him in 1903, agricultural research began on a more regular basis with the setting up of the Imperial (now Indian) Agriculture Research Institute at Pusa (Bihar) in 1905. The Institute was transferred to New Delhi in 1936.

There was not much progress after this for some time. Even after the constitutional changes of 1919, under which agriculture was placed under the direct control of the Provincial Governments, the position did not take any positive turn and the stalemate continued until the famous Report of the Royal Commission on Agriculture, 1928, high-lighted the backwardness of Indian agriculture and recommended the setting up of the Central Council of Agriculture Research with the following aims and objects:

1. To undertake aid, promote and co-ordinate agricultural and animal husbandry research and education.
2. To act as a clearing house of information not only in regard to research but also to agricultural and veterinary matters generally.
3. To establish and maintain a research and reference library in pursuance of the objects of the Council with reading and writing rooms and to furnish the same with books, reviews, magazines, newspapers and other publications.
4. To do all other things which the Council may consider necessary, incidental or conducive to the attainment of its objectives.

The Imperial (now Indian) Council of Agricultural Research came into being in 1929 as a statutory body by an Act of the Central Legislature. Further duties like those of functioning as a link between agricultural institutions in India and those in foreign countries and education as well as development work were brought within the scope of the activities of the Council. It has by now grown into a premier organisation which guides, finances, and co-ordinates research problems connected with agricultural and animal husbandry in the widest sense of the term.

The Council today covers practically every technical aspect of agriculture as well as animal husbandry and the statistical problems pertaining to both under the Board of Agriculture and Animal Husbandry which has two wings—the Crops and Soils. Wing and the Animal Husbandry Wing. Annual Meetings of the two wings are held in alternate years and all the scientific problems are discussed there.

For education, the Council has set up two more Councils under it—The Council of Agricultural and Animal Husbandry Education. For purposes of statistics, it had originally a Statistical Wing, which is now grown into a full-fledged institute known as the Indian Agricultural Statistical Institute. Besides undertaking research in statistical techniques and training in statistics, the Institute has completed a number of project works like estimation of crop yields, survey of fertilizer practices and agronomic research, study of milk yields and cost of production.

Production Programme Started with the Help of the Ford Foundation

The organisational functioning of the Council was arranged in such a manner that the research and technical side was being handled by technical experts and the administrative side by the Secretariat staff; the latter being part and parcel of the Central Ministry of Food and Agriculture. This separation of administrative and scientific duties was intended to help the technical persons to devote their undivided attention to scientific pursuits.

Central Research Institutes

The dream of Lord Curzon was fulfilled in the later years when a number of Central Research Institutes were set up in various parts of the country to deal with different aspects of the problem. Given below is the list of all the Central Research Institutes functioning in the country along with their location and year of starting.

Commodity Committees

Research work and co-ordination of the same with regard to particular crops are dealt with by the Central Commodity Committees which are semi-autonmous bodies and are generally located near the main production areas of the crop concerned. The commodity committees have in most cases also the Central Research Stations financed by them and working under their control. The first commodity committee to be organised was the Indian Central Cotton Committee which was established in 1921. There were eight such committees as shown in Table 9.1.

TABLE 9.1

Commodity Committees in India

Name of the Committee	*Location*	*Year of Starting*
1. Indian Central Cotton Committee	Bombay (Maharashtra)	1921
2. Indian Lac Cess Committee	Ranchi (Bihar)	1931
3. Indian Central Jute Committee	Calcutta (West Bengal)	1936
4. Indian Central Sugarcane Committee	New Delhi (Delhi)	1941
5. Indian Central Tobacco Committee	Madras (Tamil Nadu)	1945
6. Indian Central Coconut Committee	Ernakulam (Kerala)	1945
7. Indian Central Oilseeds Committee	Hyderabad (Andhra)	1947
8. Indian Central Arecanut Committee	Kozhikode (Kerala)	1949

As research on crops included in the commodity committees is looked after by them, the ICAR pays special station to problems of foodcrops like wheat, rice, barley, pulses, millets, maize, tuber crops, grasses and fodder crops, spices, horticulture and other matters common to various commodities and such as control of plant disease and pests, manurial trials, improved implements and dry farming in the field of agriculture, and investigation and control of animal diseases, animal nutrition, animal breeding and dairying in the field of animal husbandry.

In order to avoid a lack of co-ordination between the work of various committees, a Board of Research and Extension has been constituted under the ICAR. All the Directors of Research Institutes and Secretaries of these commodity committees are represented on this Board. These committees have since been dissolved and their research functions including the Administrative control of various research institutes have been assumed by the Council. Agricultural Research on cross-commodity basis is now being conducted under 17 Composite Regional Co-ordinated Research Stations for Cotton, Oilseeds and Millets. Since oilseeds and millets are grown in rotation with cotton, an integrated approach to investigations in respect of these crops in quite useful.

Agricultural research is also being conducted under the Department of Agriculture of the Ministry of Food and Agriculture, Government of India, at some centres other than its Central Research Institutes. Soil conservation research, for instance, is conducted under the soil conservation board of the Ministry. There are also certain other organisations for specific problems like the Central Arid Research Station ad Jodhpur (Rajasthan) to undertake problems connected with desert and dry areas.

Each State Department of Agricultural conducts research on important crops grown in the State and also deals with the other problems of farming, in the field of agronomy, soil sciences, plant breeding, plant diseases, pests, fruit crops, etc. For important crops they have one or more experimental stations in the regions important for the crop. Usually for each subject there is a subject matter specialist at the State level but in the case of important crops there is more than one specialist in many states.

Research in agriculture or allied sciences is also receiving some attention in certain universities and laboratories or agencies endowed by private benefaction. Examples of this type are the Indian Institute of Science, Bangalore, and the Anand Agriculture Institute (Bombay). Again tea, coffee, rubber and silk boards are functioning under the Ministry of Commerce and Industry, although the problems tackled by then pertain mainly to agriculture.

Similarly, the Council of Scientific and Industrial Research is looking after certain research centres like the Central Food Technological Research Institute at Mysore and the National Botanical Garden, Lucknow. The Autonomic Energy Commission, through its Biological Wing, conducts research on irradiation and induction of mutants and related fundamental genetics research.

The Indian Constitution

Under the Constitution of India which came into effect on January 26, 1950, agriculture including agricultural research and education remained a state subject whereas promotion of special studies or research as well as co-ordination and determination of standards in institutions for higher education and research came within the purview of the Union Government. Both the Union and the states could thus assume a share or responsibility for ensuring in agricultural research. The weightage will appear to be in favour of the states. The fact, however, remains that so far as agricultural research is concerned, the Central Government is taking more initiative than the state governments. Though certain state governments have set up very good agricultural research stations and educational institutions research is badly neglected in many other states. Support for research was provided by the Government of India through a number of central research institutes, commodity committees and directorates.

The working of the Indian Council of Agricultural Research as well as the organisation of agricultural research and education in the country as a whole has come for detailed study and critical review by expert teams from time to time since Independence. Two joint Indo-American Teams had studies the position in the years 1954 and 1959 respectively and made various recommendations for improving the efficiency of the research and educational institutions in the field of agriculture and for better co-ordination of research. A really comprehensive study was, however, made by the Agricultural Research Review Team in 1963 which comprised eminant scientists farm three different countries—India, United Kingdom and the United States of America. The Team submitted its report on March 19, 1964, and the Government agreed to implement its recommendation within a year. A complete reorganisation of not the Indian Council of Agricultural Research but of agricultural research and education in the country as a whole was initiated after getting the report of this Team and is still in progress.

New Developments

Three important developments which were to radically change our approach to agricultural research and education in the country took place during the last decade. These were:

(a) The establishment of Agricultural Universities in different states starting with Uttar Pradesh Agricultural University at Pantnagar in the year 1960;

(b) The reorganisation of the Indian Council of Agricultural Research as a central body for co-ordinating, directing and promoting agricultural research and education in the whole country; and

(c) The formulation and implementation of a number of all-India co-ordinated research projects in the fields of agriculture and animal sciences.

Agricultural Universities

The Agricultural University in India is not, as many people have wrongly concluded, an Indian version of the Land Grant College system of the United States of America nor a purely alien innovation. In fact, the seed for this was contained in the Report of the University Education Commission which was headed by the eminent Indian scholar and educationist Dr S. Radhakrishan.

The First Joint Indo-American Team also endorsed the recommendation of the University Education Commission that a Rural University should include a ring of small resident, under-graduate colleges with specialised and university facilities in the centre. For the complete development of a rural university the Team envisaged in the beginning the location on the same campus and in close juxtaposition a college of Agriculture and a college of Veterinary Science to which should be added in due course a college of Home Science, a college of "Applied" Liberal Arts and Science, a college of Technology using the term in the broad sense of engineering and industries, and with a group of villages to be used as a laboratory for the students. The Team also recommended that the development of Rural universities should be definitely encouraged by substantial grants-in-aid from the Centre so that they will be able to secure autonomy and efficiency of operations.

The Uttar Pradesh Agricultural University, which was the first to be established in the chronological order, had not attempted to take over the state-wide responsibility of research from the State Department of Agriculture nor had the State Government accepted that position from the very beginning. So.far as Punjab is concerned, the State Government handed over the entire responsibility of research and education in the field of agriculture and animal sciences to the newly established Punjab Agricultural University. The specialists of the University were also given a chance to play a prominent role in the field of extension education. Those who were at the helm of affairs tried to build up a good working relationship between the scientists and the experts of the university and the team of extension workers under the technical directorates dealing with agriculture and animal husbandry.

The universities which came up subsequently generally tried to work on the model of the older and well-developed universities like the U. P. Agricultural University and the Punjab Agricultural University. In certain cases, the implementation was either slow or half-heared and the universities did not always get the full support from their own state governments. This has resulted in the stunted growth of some of these universities, avoidable duplication of efforts and lack of adequate progress in the field of agricultural education and research. The concept of an Agricultural University has, however, been welcomed by all the states and agricultural universities have already been established by almost all the states. The State of Maharashtra has established two Agricultural Universities whereas the State of Himachal Pradesh has established one composite multifaculty university with a separate Faculty of Agriculture with a more or less autonomous status functioning like an agricultural university. So far as the State of Gujarat is concerned, though the enactment for establishing the university was passed in the year 1969 the university is yet to be established.

Many of the agricultural universities have already become strong centres of agricultural research and education and are playing a very important role in the transformation of traditional agriculture into one based on modern science and technology. Under the UNDP Project attempts are now being made to create centres of excellence in respect of certain importance disciplines in some of the agricultural universities which are considered specially strong in those disciplines.

The functions of Agricultural Universities are visualised as follows:

(1) The University should be responsible for statewise agricultural research, teaching and extension education.

(2) The University should have control over all research and extension education programmes in agriculture. It should integrate these with teaching.

(3) The University should include colleges of agriculture, veterinary science and animal husbandry, home science, agricultural engineering and a school of basic sciences and humanities preferably on a single campus.

(4) The University should devote itself to the education problems of the rural people, to the development of leadership among them and the improvement of their standard of living.

Reorganisation of the Indian Council of Agricultural Research

According to the recommendations of the Agricultural Research Review Team, all the Central Research Institute which were under the direct administrative control of the Ministry of Food and Agriculture or under the different commodity committees became the constituent units of the India Council of Agricultural Research. The word "constitutent units" is being deliberately used here since it is the avowed policy of the new Council that institutions which are to do scientific research should not be organisationally subordinated. The institutes functions as the executing arm of the Council and all research programmes that are to be undertaken by the Central Agency itself will be through the agency of the Institutes. At present the Council has got 23 Research Institutes in the fields of agriculture and Animal Sciences including fisheries. This includes two new Research Institutes which have been set up by the Council—the Institute of Horticultural Research, Hessaraghatta (Bangalore) and the Central Soil Salinity Research Institute, Karnal. Immediate steps were taken to improve not only the pay scales of the scientists in agriculture, but also in the recruitment methods as well as working conditions. All the Commodity Committees referred to earlier were not called Directorates of Developments for each commodity and came under the direct control of the Department of Agricultural at the Union Ministry of Agriculture.

All the Research Institutes are also expected to maintain Research Project Files in which the names of the scientists attached to a particular project are to be specified. The Directors and other senior scientists have to ensure that only those who are actively engaged in a particular piece of research work are shown as partners in the project relating to that work. The Council also organises periodical conferences of Directors, senior scientists and administrators of the Council. An open discussion and exchange of ideas in a forum of this nature where the Directors who are actually in charge of the research programmes and the senior scientists who serve as advisers of the Council are present is bound to yield good results. Various administrative bottlenecks resulting from cumbersome procedures, insufficient or faulty delegation of authority, inadequacy of staff etc. come to light during these discussions. Earnest attempts are being made to find solutions to the problems brought to light in such conferences. The Indian Council of Agricultural Research also organised an inter-organisational meeting of different scientific organisations where scientists of different organisations discussed problems of common interest and suggested ways and means of better and more effective inter-organisational cooperation. A small beginning has been made which, if taken to its logical conclusion, is bound to yield high dividends.

The major suggestions made above may be briefly stated as follows:

(1) A close link between research, teaching and extension must be established and suitable organisational steps to achieve this must be taken in order to make teaching as well as research effective in handling the problems of the farmers.

(2) Larger resources of the country should be devoted to research because the present programmes of research are inadequate.

(3) Separation should be effected between supply and service responsibilities on the one hand and extension on the other in order to enable extension workers to devote more time in field advisory work.

(4) Indian Council of Agricultural Research should develop strong regional

research centres for the more fundamental types of research on regional and national agricultural problems.

(5) A memorandum of understanding should be drawn up between ICCR or the Government of India and the state governments defining in clear terms the respective spheres of responsibility in research as in other activities of the Centre and the states.

(6) The service conditions of agricultural research workers are not satisfactory enough to attract persons of superior intelligence and promise to agricultural research as a career. This should be remedied.

(7) Liberal decentralisation of authority is necessary so that control points are not for removed from the levels at which research is carried on.

(8) Several areas of research have not received the attention which they deserve. Among these are study of soil structure and physics, study of organic matter in relation to physico-chemical and micro-biological changes in the soil, designing and testing of agricultural implements and machinery, soil microbiology, seed technology, use of radiation and other mutagenic agencies to bring about useful hereditary changes in crop varieties, studies on bacterial and virus diseases, breeding of vegetable crops, nematology, weed control, and judicious water use. These may developed with the help of coordinated approach between the Centre, the States and agricultural universities.

The salient features of the reorganised Council with effect from 1st April, 1966 are as follows:

(i) The rules have been suitably revised so as to make the Indian Council of Agricultural Research a competent and adequately autonomous research organisation.

(ii) The Governing Body of the Council has been reconstitued making it pre-eminently a body of scientists and those with interest in or knowledge of agriculture.

(iii) A career scientist has been appointed as its first Director-General and Vice-President of the Council, thereby replacing the tradition that such a post is to be occupied only by an administrator.

(iv) The administrative and financial control of all the Research Institutes has been transferred to the reorganised Council.

(v) In view of the deficiencies of making recruitment to scientific and technical posts through the Union Public Service Commission, recruitment to such posts under the Council is now being made on the recommendations of

Selection Committees consisting of outstanding scientists in the particular disciplines, constituted by the Council itself.

(vi) While the Council is the decision-making and executive authority for planning, developing and co-ordinating research policy and programmes, actual research work is being undertaken through its Research Institutes and Stations. The Directors of the Research Institutes have, accordingly, been delegrated adequate administrative and financial powers, so as to enable them to function in an atmosphere of comparative freedom with the requisite amount of flexibility, elasticity and autonomy.

(vii) The reorganised Council has also developed a series of problem and production-oriented research project on all-India basis which are being implemented with the active-co-operation of the agricultural universities and State Governments. These projects are proving to be highly effective not only in making research activity more efficient, expeditious and productive but also in the development of valuable research potential in the country.

(viii) The Indian Council of Agricultural Research is also responsible for co-ordinating and fostering agricultural education and plays more or less the same role in this field as the University Grants Commission does in the field of basic sciences and humanities education. It provides assistance to agricultural universities and colleges for their development programmes, for improving the quality and standards of teaching, for purchase of books and equipment, expansion of library facilities, construction of hostels, etc. Financial assistance amounting to Rs 2.5 crores was given to the Agricultural Universities in the country during the year 1968-69.

All-India Co-ordinated Projects

The present strength of agricultural research in India lies in the partnership that has been formed between the Central and State agencies for the attainment of a common national goal. This is what has attracted the attention of many visiting scientists from other countries. It has already been pointed out that in the federal Constitution of India, agricultural development and agricultural research are important state subjects. The Union Government can also play a vital role in promoting scientific research and education. If, however, the two agencies were to work exclusive of each other it would result in wasteful duplication of efforts. In order to get the best results out of any programme, one has got to take advantage of the facilities available at every place and make use of the best taken the country to the maximum extent irrespective of their located in a Central Government's laboratory or State research organisation.

(i) Research has to be problem-oriented;

(ii) Research efforts should be intensified at selected centre for tackling important problems to support the new strating of agriculture; and

(iii) Research is to be strengthened in a Centre where leadership and the desired facilities were available or could be available.

In formulating the All-India Co-ordinated Research Projects, the problem is visualised for the country as a whole, and research centres and sub-centres are located in different parts of the country to meet the needs of areas covered by distinct agro-climatic zones and regions rather than on the basis of individual states. A selective approach is necessitated due to limitation of funds, paucity of centres of research in the country having reasonably good physical facilities for research and the scarcity of well-trained and experienced scientists for manning various research positions. Thus, instead of spreading too thin, efforts have been made to consolidate and intensify research at a few selected centres representing different soil and climatic conditions. In these co-ordinated projects, the Central Research Institutes and the state agencies (Agricultural Universities or State Departments of Agriculture concerned with research) work as partners. The programme is drawn up in such a manner that all efforts tend to become supplementary and complementary to one another rather than competitive. This, however, does not eliminate competition for good quality of work. In fact such competition is only a healthy sign.

Another important feature of these co-ordinated projects is their multi-disciplinary approach. Thus, for example, in the co-ordinated project for crop improvement where the main emphasis is on breeding of superior varieties and hybrids, research work is simultaneously carried out in related fields like agronomy, plant pathology, entomology and biochemistry so that all the facets of research which are necessary to bring about an overall improvement in crop production, both in terms of quality and quantity are brought to bear on the problem. Good team work amongst the scientists working in different disciplines as well as different centres is the keynote to the success of the All-India Co-ordinated Research Programmes. The co-ordination of the scientific work under the Project is done by the Project Co-ordinator who, one among equals, is also an accepted leader and active research scientist. The technical programme of the project is drawn up by him on an All-India basis at the annual or periodical workships or meetings in which the scientists engaged in the project participate. The progress made from time to time also comes for review and discussion in this workship.

Fourth Five Year Plan Proposals

The Planning Commission allocated Rs. 85 crores for Agricultural and Education Research during the Fourth Five Year Plan. The allocations proposed for each major head are shown below:

	Major Head	Allocation (Rs crores)
1.	Agricultural Education	28.00
2.	Research Institutes :	
	(a) Agricultural Institutes including Nuclear Research Laboratory	11.85
	(b) Animal Sciences Institutes	8.98
3.	Research Schemes:	
	(a) Agricultural schemes relating to crop and soils	22.03
	(b) Agricultural research statistics schemes including strengthening of the institutes.	2.00
	(c) Animal Sciences schemes	11.64
	(d) Research schemes of centrally administrated areas.	00.50
	Total	85.00

Under the present organisational structure of agricultural research in the country we have 23 well-equipped and well-staffed research institutions all over the country directly supported by ICAR. In addition, there are eight Soil Conservation Research Centres. We have 19 agricultural universities with nearly all states having one agricultural university each and Maharashtra having four such universities. Through the All-India Co-ordinated Research Projects of ICAR an excellent machinery has been created for integrated research on all problems of national importance in agriculture, animal husbandry and fisheries. There are 69 such projects currently under operation and a unique feature of these projects is that they provide a mechanism for bringing together scientists working in different disciplines and under different administrative hierachies into a common working family. The strength of our agricultural research system lies in the national grid of co-operative testing programmes and the rigor of the screening programmes of every experimental finding before it is passed on the farmer. As much as Rs 106.58 crores has been spent by the ICAR on agricultural research during the last 25 years. It increased from Rs 6 lakhs in 1947-48 to nearly Rs 46 crores during 1971-72.

APPENDIX

RESEARCH INSTITUTES IN INDIA

1. Indian Agricultural Research Institute, New Delhi
2. Indian Veterinary Institute, Izatnagar (U. P.)
3. National Dairy Research Institute, Karnal (Haryana)
4. Central Potato Research Institute, Simla (Himachal Pradesh)
5. Central Rice Research Institute, Cuttuck (Orissa)
6. Central Arid Zone Research Institute, Jodhpur (Rajasthan)
7. Indian Grassland and Fodder Research Institute, Jhansi (Uttar Pradesh)
8. Central Tuber Crops Research Institute, Trivandrum (Kerala)
9. Central Sheep and Wool Research Institute Avikanagar (Rajasthan)
10. Technological Research Laboratory (Cotton), Bombay
11. Jute Agricultural Research Institute, Barrackpore (West Bengal)
12. Technological Research Laboratory (Jute), Calcutta
13. Central Tobacco Research Institute, Rajahmundry (Andhra Pradesh)
14. Central Coconut Research Station, Kayangulam (Kerala)
15. Central Coconut Research Station Kasaragod (Kerala)
16. Central Arecanut Research Station, Vital (Mysore)
17. Indian Lac Research Institute, Ranchi (Bihar)
18. Central Inland Fisheries Research Institute, Barrackpore (West Bengal)
19. Central Marine Fisheries Research Institute, Mandapam Camp (Tamil Nadu)
20. Central Institute of Fisheries Technology, Ernakulam (Kerala)
21. Sugarcane Breeding Institute, Coimbatore (Tamil Nadu)
22. Indian Institute of Sugarcane Research, Lucknow (U. P.)
23. Institute of Agricultural Research Statistics, New Delhi.

10

MODES OF PRODUCTION, AGRARIAN STRUCTURES AND WOMEN'S WORK

The objective of my talk is to describe the present situation of the female labour force in agriculture in the Socialist Federated Republic of Yugoslavia. The evaluation of working relations between the sexes covers both sectors of agriculture; the socialist sector — which includes only about 14 per cent of a total of some 10 million hectares of arable land — and the private, peasant farming sector — which comprises about 86 per cent of the total arable land, and is cultivated by 2.6 million peasant families, each holding an average of 3.5 hectares of arable land.

The prevalence of small holdings requires different forms of organising production, so as to promote a faster and more effective development of agriculture; for the majority of peasant family farms, production is based on traditional work patterns, even when modern farm technology is provided. Socialist farms supplied with modern means of production— co-operatives, agro-enterprises, etc. have not introduced significant changes in productive relations. In both production modes, women are still relegated to a marginal social position, as substitutes for men working off the farm.

In what follows I will summarise the existing situation of women in the agricultural labour force. The analysis is based on comparative statistical data from the Agricultural Censuses of 1948,1953,1961 and 1969. Several issues could be delineated.

First, women provide a substitute agricultural labour force on both co-operative, i.e. socialist, and private farms. They take over the work left by men who have moved out of agriculture. As a rule this work is associated with the lowest qualifications and is rewarded with the lowest income levels. Further, it is work without prestige. None the less, farm women have no alternative other than the agricultural profession; employment in agriculture is for them a social and economic necessity rather than a result of their own free choice.

Second, women now represent the majority of the agricultural labour force, that is, the work force in agriculture has been feminised. However, this feminisation seems to be a temporary phenomenon, generated by the unequal intersectoral demands for male and female labour. It tends to occur in the early phases of industrialisation and/or in regions undergoing rapid industrial change.

Several indicators show that the future trend in the sex ratio of the agricultural labour force will be towards a reduction of female labour participation. The most obvious one is that the rate of farm girls completing eight years of compulsory education is equalising with that of farm boys. In addition, their enrolment in high school and vocational education is also increasing. This implies that rural girls and women are gaining the formal basis for vocational selection, and that in the future gender may not be the only selective criteria for farming.

It should be stressed here that the recent activity of the extension service centres in rural communities is oriented towards attracting rural women so as to provide them with information on new farm technology and methods. Specialisation of production on farms large enough to turn to market production is also encouraged. Although at present fewer women than men attend these centres—a proportion which does not compare with the sex ratio of the agricultural labour force — those who do will tend to influence their female neighbours by transmitting their new skills and greater knowledge. Social action oriented towards mobilising farm women has been recently initiated, among others through organising clubs for women of co-operatives. Thus, as agriculture gains in status as a socially and politically recognised profession, the rural labour force will tend to become professionally stable. A series of new laws supports this trend.

Third, farm women are bearers of a threefold role. They carry out between 60 to 70 per cent of all the farm work, and are engaged in agricultural production for at least two-thirds of their normal working time. In addition, they do all the household work together with the majority of chores related to raising children. Yet, rural women appear to be an underutilised human resource. If given better technical and social training, they could make a greater contribution both to agricultural and rural development, as well as to the development of society in general. The latter point should be particularly emphasised since in such overworked and limited living conditions, in addition to being socially isolated in the slowly changing patriarchal rural community, rural women are perceived as not being fully conscious of their status. Yet, we might ask, why do they tend to leave the farm after completing elementary education and why are they increasingly marrying non-farming men, which takes them out of farming.

Thus, we could conclude that rural women in general and farm women in particular in Yugoslavia are conscious of their unequal social position when compared with the social position of urban and/or non-farm women, on the one side, and of rural and/or farm men, on the other. Yet they are still not dealing directly with the problem, but instead trying to change their social position in indirect ways.

A tentative conclusion can be drawn: women in the agricultural labour force in a post-revolutionary, early socialist society, continue to be in a marginal social position, thus representing a neglected resource for agricultural and of rural development. It is a negative consequence resulting from their occupational position in farming, and their specific role in society and the peasant family.*

A METHODOLOGICAL APPROACH TO ANALYSING THE EFFECTS OF CAPITALIST AGRICULTURE ON WOMEN'S ROLES AND THEIR POSITION WITHIN THE COMMUNITY

Kate Young

There seems to be almost universal agreement that one of the commonest effects of development is to relegate women to the subsistence sector. Technology, credit and know-how are concentrated in the hands of men while women are assigned the non-modernised, subsistence sector, in other words food production for domestic consumption. I am somewhat concerned by this seemingly common agreement. That is not to say that a common effect of development is not to concentrate modern agricultural machines and implements and new varieties of seed and so on into the hands of men, nor to deny that many of the assumptions that westerners hold about correct gender roles have led to men being exclusively given agricultural training, even in societies in which women are traditionally the farmers. My unease stems from the relegation argument.

It appears to be suggested that the marginalisation of women to food production (sometimes called the feminisation of agriculture) is a world-wide phenomenon.

---

* More extensive information on the subject can be found in:

First-Dilic, Ruza, 1974 : Zena - aktivni poljoprivredni proizvodjac ("Women as active agricultural producers"), zena, Vol. 32, No. 2, pp. 18-26.

First-Diilic, Ruza, 1974 : Polozoj poljoprivrednice u radu i samoupravljanju ("Position of farm women in work and self-management"). *Marksisticke Sveske*, vol. 4, NOs. 1/4, pp. 313-326.

First-Dilic, Ruza, 1974 : Sex roles in rural Yugoslavia. *The International Journal of Sociology of the Family*, Vol. 4, No. 2, pp. 161-169.

First-Dilic, Ruza, 1977 : Zena u socijalistickom razvoju poljoprivrede ("Women in the socialist development of agriculture"). *Socijalizam*, Vol. 20, Nos. 7/8, pp. 1392-1405.

First-Dilic, Ruza, 1978 : The productive roles of farm women in Yugoslavia. *Socialogia Ruralis*, Nos. 2/3. Forthcoming.

However, my own limited knowledge of the effects of development in Latin America would suggest that this is hardly a sufficient analysis even though it is possible that women are participating to a greater extent how in all phases of production of food crops than in the past. Rather I would suggest that a common effect of development policies is to make women central to the production of crops destined for the market. This production may either be on a household basis in which case the women work as unpaid labourers for the man/men of the household, or production may be on a much larger scale in which the women provide very cheap seasonal labour for the owners of the enterprise. In this latter case the women may form part of a familial labour group (payment going to the "head of the family"), or member of an all-female work gang (payment going to the gang leader for redistribution to the workers), or be hired as individual labourers.

Women's role in cash crop production may or may not be associated with an increasing percentage of female labour time also being spent on food production on small household plots. Frequently however, we are looking at labour flows between self-provisioning sectors of the rural economy and the more capitalised sector producing for the market. The insistence in the literature on the pushing of women into subsistence, or their marginalisation as it is often described, may possibly come from an unequal number of studies being carried out in certain parts of the world in which this is in fact an effect of capitalist development in agriculture. On the other hand, it may well stem from our inadequate understanding of rural production systems and the ways in which they have been integrated, and the extent to which they have been integrated, into the market economy. It seems to me that we may well have an inadequate understanding of the gender specific roles assigned to members of productive units within these rural systems, and indeed of the nature of the productive unit itself. As a part of this, there is often a lack of specification of the place and importance of female roles, how they interlock with, or are functionally complementary to, male roles. Lastly, little is written about the representation of these roles in the system of beliefs and practices stemming from them, in other words, how these roles are represented in ideology and reproduced over time.

Having made this criticism, it is clear that I am bound to be asked what I would consider an adequate analysis of a given production system (and the totality of women's roles within it) which would provide a basis for understanding how it has been modified in the process of integration into the market, and how this in turn has affected inter-gender relations.

Firstly, the nature of the productive system prior to integration must be delineated. All elements of production should be covered including the production of the following generation (not merely childbirth, but also child-care and socialisation). Each of these elements would correspond to what I call for convenience sake (although the concept is clearly problematical) a labour process. Taking each labour process individually, a number of questions about all aspects of control over the means of production have to be posed. For example, is one dealing with formal or real ownership? In other words, is a distinction made between possession and ownership; this is a particularly important point in systems where women are "owners" of land, but in fact social ordinances forbid them working it directly. It is also important to note that ownership of the important element of the means of production, land for example, is not a sufficient condition to ensure that a labour process can be set in motion. In other words, access to a variety of other inputs (also means of production) are essential to carry through the labour process: for example, seeds, fertilizers, water, animal traction, labour. It is common enough in the literature to read of women who have land but lack seed or fertilizer, or having all three lack the ability to command labour to bring the three inputs into working relation. This latter point is one which I think is often insufficiently elaborated in the literature: that is that the sexes often have very different access to various categories of labour - female, male, young, old - even in situations where there is a clear-cut gender allocation of tasks and a high degree of non-substitutability of tasks. Women typically can call upon their young children, or their female relatives; in some areas they may also call upon their brother (s) (but they are constrained by the demands made on their own time by their own familial requirements). When what is needed is male labour, and no kinsmen are about, women may be unable to ensure that a labour process they have set in motion can be carried through. This labour command bottleneck may be resolved where labourers (non-kinsmen) are available for payment in kind and women have access to the necessary resources, for example, millet for beer brewing. Men may be similarly affected when there are tasks within a total labour process which absolutely require female labour; it seems to be quite common that female labour cannot be acquired even for payment in kind.

These are the sorts of constraints that allow demographic hazards (too many sons, too many daughters) to play havoc on the ability of a production unit constituted by a single household to survive under certain conditions of labour shortage.

Lastly, we have to ask who controls the proceeds of each labour process. This can be in terms of rights over distribution within the production unit or beyond it. We may be looking at a system in which the head of a large compound with a number of constituent households (either conjugal couples, or men with several wives) dispenses the totality of the compound's product, or a system in which each unit within the compound is allocated a certain quantity of the produce for consumption according to numbers while the seed grain for the following year's sowing is zealously guarded by the compound head; or we may be dealing with a system of peasant households in which the women are charged with "husbanding" their household's food supplies throughout the year and also ensuring an adequate supply of sprouted seed for the next year's planting. We must beware of a number of assumptions about the distribution of the proceeds, in particular the common assumption that the household is a unit within which all members have equal access to and rights over household resources. In fact in many societies social rules as to the correct way in which the proceeds should be spent differentiate considerably between males and females; for example, expenditure of the so-called household budget on cigarettes and alcohol by men is often paired to strong sanctions against such consumption by women. Food consumption within the household varies widely with, in many cases, an age and sex hierarchy operating, elder males getting the "lion's share" with children next and adult women last. Expenditures during life-crisis rituals again may well differ by gender: for example, male circumcision ceremonies may demand heavy outlays of produce (including the typically male produce -livestock) while female circumcision, where it takes place, may require none or a small-scale expenditure of grains (considered a female product).

Having made such an inventory of labour processes, the next step must be to attempt to see how they relate to one another and how they are co-ordinated so as to ensure social reproduction, in other words, continuance of the system over time. There are clearly a number of important questions to be asked but in no way is it possible to produce an exhaustive check list: instead it is probably more suggestive to indicate areas of questions. Firstly, one should ask in what way are the district labours within labour processes and between labour processes linked? Are they related to each other in a hierarchical or non-hierarchical way? In other words, is there a single (or a small number of) key labour processes upon which the others turn? This should give one an insight into relations of domination and subordination. In addition, is there privileged access of one gender to coordinating roles? Or to turn this question on its head, are women excluded from those labour processes which one has identified as playing a key co-ordinating role? Lastly, it is essential to analyse how this hierarchy and the privileged access are represented in other aspects of social life (political or religious institutions, ideology). This may give one an insight into how what could be called the concrete manifestation of an abstract division of labour (and in particular the sexual division of labour) is internalised and then reproduced by the participants in actual behaviour and justification of behaviour. In some cases, this may take the form of an assumption that the sexual division of labour derives from natural attributes of men and women (all of us have heard all too often that women are "naturally" better equipped to do meticulous, time-consuming, repetitive work and so on). In others, women's subordination is maintained by a series of repressive measures rather than internalised conceptualisations.

Since the foregoing discussion has been somewhat abstract, perhaps it would be useful to illustrate the approach from my own field data from Mexico.

In the nineteenth century in the area where I worked, women were renowned for their skill as weavers. They acquired raw and unseeded cotton from local traders (male), which they, with the help of their children (possibly of both sexes at least while they were young) cleaned, spun and wove lengths of tough cotton cloth. These lengths they then made into skirts and tops for themselves and their daughters, into trousers and blouse-like tops for the men, and shawls for both. Lengths surplus to domestic requirements were bartered locally or sold to the traders. From the traders the women were able to acquire inputs for other labour processes such as metal tools (machetes, plough tips) or oxen, as well as ornaments for the church, dynamite and alcohol for religious festivities. The women controlled the labour process involved, most of the inputs into it, the timing of the totality of the tasks necessary, and the proceeds. From such activity they apparently derived a high standing in their households and in the community. None the less women were barred from the administration of the church although they were absolutely essential participants in the most important rituals of the village and without them the health and safety of the village and all its members could not be ensured. But again, the principal actor in the ritual was the man not his wife. These church rituals provided, I would argue, the dynamic to surplus production (which in poor harvest years ensured adequate provision of subsistence). Women were also barred from participation in the local *(i.e.,* village) government and did not thus have a formal position of effective power in the administration of village affairs. One of the primary concerns of the village government was the allocation of rights in the means of production upon which most of the other labour processes in the village were based: land. Women did not work the land (although they provided labour at peak periods, such as harvest) and even as owners of rights in land, did not have effective possession of it. They did not have effective control over the supply of raw cotton but were dependent for this upon the traders who appear to have been "outsiders" to the areas. This relationship is, in fact, not very clear in that the traders who provided raw cotton may also have appropriated the finished product (i.e. giving space for manipulation of prices to the detriment of the producer).

It could be argued that the basis for the reproduction of the system was in some measure dependent upon the women's activity as weavers. None the less, this key role in society was not represented in the political sphere, rather it was denied. In the religious sphere the necessity for the male-female pair in ritual could be interpreted as a representation in part of the complementarity of productive activity through the exchange of the product for inputs needed in other vital labour processes. As far as co-ordinating roles are concerned, they were exclusively given to males who, for example, appear to have been primarily responsible for non-local exchanges of cotton cloths for other necessary inputs. They were also exclusively involved in the public organisation of the ritual life of the community.

Having established the nature of the productive system and its mechanisms for reproduction, the next task is to analyse the nature of the contradictions set up by the intervention of capital.

Clearly, here one has to be able to specify what type of capital is involved: are we looking at mercantile or industrial capital. The importance here is the difference in the ways in which a surplus is extracted from the local system: mercantile capital typically appropriates surplus labour through the product (often by means of unequal exchange), while industrial capital expropriates it through the buying and selling of labour power. Here one would have to make some perhaps rather simplistic assumptions about the nature of capitalist development — the drive to accumulate, to extract a surplus, to free labour, to increase productivity, etc.

At the same time, the notion that capital has a free hand in such transformation is dubious. In other words, I would very much question the analytical usefulness of the approach which assumes that everything is as it is because it serves some purpose (function) to capital. Rather the critical concept of contradiction and constraints should be kept at the forefront of the analysis.

Firstly then, we should ask why the local system is no longer able to reproduce its own conditions of production and reproduction. This question can be phrased another way — how has it been integrated into the wider economy? Has the intervention of capital resulted in the commoditisation of labour and the means of production but not the product of labour, or has the product been commoditised but not the labour that produces it? Or have both been commoditised? Have new forms of production been introduced and with them new ways of organising production? Have new labour processes, or partial processes, been introduced? Have these new forms of organisation of production or the new labour processes led to contradictory effects at the level of access to means of production, access to employment, rates of fertility, and so on?

To illustrate these somewhat programmatic remarks, I will again refer to my own field work.

In the late nineteenth century, domestic manufacture of cloth declined rapidly in part because raw cotton supplies dwindled—the cotton producers were directing their product to England because of the American Civil War. At the same time there was an expansion of cotton production which drew men out of the area into temporary wage labouring; but women apparently were not utilised as held labour. One could note a number of clear disbalances developing between men and women under these circumstances: men's greater familiarity with the wider society including command of Spanish; their ability to shed responsibilities for child-care through absence and the conversion of this responsibility into the gender-specific responsibility of the wife, recognised through handing over some part of the wage; and the reversal of economic roles — from the man now came the essential inputs into other labour processes (including that of the wife who could now only weave if she had manufactured yarn). This out-migration flow ceased quite rapidly after coffee growing was introduced forcibly into the area. Each household was obliged to grow a certain number of coffee bushes by the local political boss (or *cacique)* who combined with the coffee merchants to "encourage" production and yet keep coffee prices low. Land, as will be remembered, had always effectively been controlled by men, both at the household and the community level, even though inheritance was bilateral. With the perennial cash crop (coffee bushes, once planted, continue to produce for some 30 years or more) there was a double pressure to expand land use: both for coffee bushes and for the traditional subsistence crops, for coffee growing was grafted onto the traditional food cropping and did not replace it. This led to individualisation of land holding as well as to the gradual exclusion of women from land ownership. In turn there was a process of dispossession: this affected those families who already had claims to very small amounts of land as well as those who were not politically strong or adept enough to maintain what rights they had.

Coffee growing also implied a changed organisation of production: large—or small-scale reciprocal labour exchanges rapidly diminished, and production both of coffee and of the subsistence crops (but not so markedly) was carried out on the basis of house-hold labour resources. This individuated mode of working was accentuated by competition between households for land and privileged access to the merchant and/or *cacique.* The social conditions were thereby laid for an emphasis on the importance of familial labour, in other words, for large families. At the individual level, larger numbers of sons and daughters ensured sufficient supplies of unpaid familial labour for subsistence and coffee production, but at the community level, since land supplies were restricted, it meant ever increasing pressure on land and ever larger numbers of families without access to land, or access only to portions of land so small that they could not produce enough for necessary consumption. There was in effect a surplus of labour in the area and at this point (around the turn of the century) no alternative jobs.

The effect on women was varied. They did not become more involved in subsistence labouring than before; rather, they became deskilled. They fell form the status of income-generating skilled weavers to unpaid familial labourers controlling none of the proceeds of their labour nor in charge of the totality of the labour process. The nature of the labour process was also such that the women's annual cycle of labour expenditure greatly worsened: by this I mean that as weavers, women could fit their activities into the general routine of housekeeping, food preparation, child-bearing and caring throughout the whole year. There were only a few seasonal peaks requiring extra expenditure of energy: the most important being the maize harvest which came at a time of relatively abundant food supplies (maize is harvested dry). With the change in the nature of production and the organisation of production, as well as the change in the means of access to labour, the women's annual cycle was greatly altered. Two seasonal peaks now had to be accommodated as well as most probably an increase in time and energy spent in pregnancy and loctation.

In this brief excursion into an empirical example, the contradictions brought both to the productive system itself and to women, in particular, are quite clear. The introduction of coffee sets in train both a reorganisation of production, of the relations of production and a different pattern of use of land, this in turn encourages individuation and differentiation and the "freeing" of certain categories of people from ownership of the land at the same time that a higher rate of population growth is encouraged. The two forms of production (embodied in a number of concrete labour processes) are insufficient to reproduce the material conditions for the maintenance of the population. In part, this is a consequence of the nature of the way in which integration into the wider economy has been achieved - through the intervention of mercantile capital which has commoditised the product of a set of labour processes but not labour itself. Women's participation in the total labour process of producing coffee is minimal: they are not in effective possession of the plots of land, nor do they co-ordinate the total labour process; rather, they intervene at only one part of the production process (albeit a critical one, that of harvesting the coffee berries). Lastly, they are not in control of the product of their labour. Through the loss of control of their own speciality their subordinate position vis-a-vis their men is accentuated, and with growing individuation of households, the potential for their social isolation is increased. At the same time, the reorganisation of production and the need for supplies of unremunerated labour lays stress on their child-bearing capacities: women become more and more centred upon child-bearing and rearing, and pronatalist elements in the culture become stressed. High rates of child mortality necessitate high rates of fertility, while privatisation of land accentuates the importance of legitimate paternity. Women's importance as economic agents becomes subordinated to their importance as bearers of children. But pronatalism for the community as a whole spells disaster.

In such a meeting as this one at the ILO, one should perhaps try to outline why one's approach is at all useful for those who are trying to devise policies which will at best forward the aspirations of women, or at least ameliorate their living conditions.

It seems that only when one has a clear understanding of the processes of social change, and in particular an understanding of the working of the capitalist system, even if it is at quite a high level of abstraction, is it possible to begin to see how this will affect rural production systems in a number of district ways. This in turn should give one a basis upon which to suggest interventions which will not merely alleviate women's burdens but potentially lead to a more fundamental restructuring of the particular sets of social relations in which women are enmeshed often to the detriment of their own personal development.

We have tried to show in this brief note how the assumption that development inevitably pushes women into the subsistence sector is not adequate for a total understanding of the process. We would like to argue that women's new role as unremunerated familial labour on systems of small-scale production for the market is as prejudicial to their interests as their marginalisation in the subsistence sector.

11

AGRICULTURAL PRODUCTION

The performance of Indian agriculture, which has shown resilience in recent years, remains dependent on monsoon. The influence of monsoon, however, has been moderated by the initiation of various development programmes. Thus, even during the severe drought of 1987-88, the decline in agricultural production was only 0.8 per cent when compared with 1986-87, although foodgrains output declined by 2.1 per cent. The favourable weather conditions, together with the adoption of thrust programmes during the post-drought year, brought about significant increases in agricultural production. It increased by as much as 21.0 per cent in 1988-89 and by 1.7 per cent in 1989-90. Foodgrains production increased to 169.9 million tonnes in 1988-89 and further to 170.6 million tonnes in 1989-90. In 1990-91, with an exceptionally good monsoon and continuance of various development programmes, fixation of more remunerative farm prices, increased allocation of investible resources and water management, the production of foodgrains is likely to hit another peak and may even surpass the targetted level of 176.5 million tonnes. The production of other crops like oilseeds, sugarcane, jute and mesta has also recorded increases of varying proportions.

The commendable achievement in agricultural production is largely attributed to the policies, strategies and programmes adopted by the Government as also the intensive efforts of the agricultural scientists and the dynamic response of the farmers. The average annual growth in agricultural production during the Seventh Plan period was 4.1 per cent as against the plan target of 4 per cent. However, average annual growth in agricultural production during the Seventh Plan was lower than the average annual growth (with base 1979-80) achieved during the Sixth Plan when it was about 6 per cent.

Monsoon

The year 1990 was the third successive year which experienced a good monsoon. Some of the special features of 1990 monsoon were : (1) over 95 per cent of the Indian landmass received normal to excess rainfall; (2) rainfall was very well distributed over time and space; (3) there were no major floods and there was no meteorological sub-division in the "scanty rain" category; (4) apart from a more or less timely onset, it maintained a sustained tempo throughout the four months spell; and (5) in north-west India, monsoon rains commenced about two weeks earlier than usual and the monsoon started withdrawing about two weeks later than the normal dates, resulting in a welcome prolongation of the rainy season in the traditionally rain-deficient areas. This facilitated the execution of various agricultural operations with enthusiasm. However, the delayed monsoon in the Saurashtra region of Gujarat affected the groundnut sowings. The rainfall situation during the decade, meteorological sub-divisions-wise is given in Table 11.1.

TABLE 11.1

Monsoon Rainfall (June — September)

Year	*Number of Meteorological Sub-divisions*			*Per cent of districts having normal to excess rainfall*
	Excess/ Normal	*Deficient/ Scanty*	*Total*	
1981	28	7	35	69
1982	24	11	35	48
1983	32	3	35	85
1984	25	10	35	64
1985	26	9	35	65
1986	21	14	35	52
1987	14	21	35	43
1988	32	3	35	88
1989	29	6	35	72
1990	32	3	35	84

In totality, the monsoon this year ranks as the third best during the past decade (after 1983 and 1988 monsoons), the precipitation being 106 per cent of the normal. The data set out in Table 11.2 indicates the comparative performance of the total rainfall over the country as a whole in 1990.

The regional distribution of rainfall as reflected in the All-India Cumulative Rainfall Index for the entire season constructed by using the average area (for triennium 1980-81 to 1982-83) sown under kharif rice in each meteorological zone as weights, indicates that rainfall recorded this year was 102.3 per cent of normal. Except East and South zones, the rainfall was much better in other zones when compared with a year ago. Unlike last year when southern zone had recorded the highest, it recorded the lowest rainfall this year. The details are given in Table 11.3.

TABLE 11.2

Performance of Monsoon

Year	*As per cent to normal rainfall*
1981	100
1982	85
1983	113
1984	96
1985	93
1986	87
1987	81
1988	119
1989	101
1990	106

TABLE 11.3

Regional Rainfall Indices (June to September)

Year	*All India*	*Zones*				
		West	*North*	*East*	*South*	*Central*
1979	77.0	85.2	52.1	84.4	94.8	69.0
1982	89.4	83.5	94.9	87.8	88.8	93.1
1986	85.3	78.4	88.9	83.7	89.9	86.5
1987	88.7	75.6	62.5	103.6	75.5	77.4
1988	109.6	121.2	116.5	107.6	123.9	90.3
1989	98.7	96.6	90.9	101.3	114.6	82.2
1990	102.3	113.9	112.0	98.0	93.2	112.4

The areas which received excess, normal and deficient rainfall were:

Rainfall	*Meteorological Sub-divisions*
(A) Excess	(1) Arunachal Pradesh; (2) Haryana, Chandigarh and Delhi; (3) Punjab; (4) West Rajasthan; (5) West Madhya Pradesh; (6) Gujarat; (7) Vidarbha; and (8) Telangana.
(B) Normal	(1) Assam and Meghalaya; (2) Nagaland, Manipur, Mizoram and Tripura; (3) Sub-Himalayan West Bengal and Sikkim; (4) Gangetic West Bengal; (5) Orissa; (6) Bihar Plateau; (7) Bihar Plains; (8) East Uttar Pradesh; (9) Plains of West Uttar Pradesh; (10) Hills of west Uttar Pradesh; (11) Himachal Pradesh; (12) Jammu & Kashmir; (13) East Rajasthan; (14) East Madhya Pradesh; (15) Saurashtra, Kutch and Diu; (16) Konkan & Goa; (17) Madhya Maharashtra; (18) Marathwada; (19) Coastal Andhra Pradesh; (20) Rayalaseema; (21) Tamil Nadu and

		Pondicherry; (22) Coastal Karnataka; (23) North Interior Karnataka; and (24) South Interior Karnataka.
(C)	Deficient	(1) Andaman and Nicobar Islands; (2) Kerala; and (3) Lakshadweep.
(D)	Scanty	NIL.

Reservoir Situation

The excellent behaviour of the monsoon and the widespread showers during the fag-end of the monsoon boosted the water levels in tanks and reservoirs. The total live storage in 47 important reservoirs monitored by Central Water Commission till the end of September, 1990 was higher by 18 per cent at 93.4 TMC against 79.4 TMC during the corresponding period of 1989. The water storage in 1990 year was nearly 88 per cent of the designed utilisable storage as against 75 per cent recorded during 1989. In fact, the water level in 1990 was more than the average of last six years live storage of 74.2 TMC.

Production Performance

In the terminal year of the Seventh Plan, the tempo of higher production built up in the preceding year continued as the production of foodgrains increased to an all time record of 170.6 million tonnes and was marginally higher than the production of 169.9 million tonnes achieved in 1988-89. Thus, the average annual foodgrains production during the Seventh Plan period was about 155 million tonnes as against the average of 138 million tonnes achieved during the Sixth Plan.

The trends in foodgrains output in recent years have exhibited some significant qualitative changes. Following the Special Rice Production Programme together with other programmes, the contribution of rice to the increase in output has been much higher than the contribution of wheat. The share of rabi output in total foodgrains output has steadily increased, although in 1988-89 and 1989-90 its share went down. On the whole, the share of rabi crops in total foodgrains, which averaged 42 per cent in the Sixth Plan, increased to an average of 43.7 per cent in Seventh Plan. The share of coarse cereals and pulses has gone down in the total basket of foodgrains (Table 11.4).

The performance of foodgrains production during the Seventh Plan was constrained by unfavourable weather conditions in the first two years and severe drought that occurred in 1987-88. As a result, the targets of foodgrains production could not be achieved in the first three years of the Seventh Plan. However, with a considerable thrust of various development programmes and the Special Foodgrains Production Programme (SFPP) launched in 1988-89, there was a remarkable recovery in foodgrains production as also in other crops and output reached their respective peak levels. The average annual increase in foodgrains production during the Seventh Plan was 3.6 per cent as against the targetted level of 3.7 per cent. The average annual increase in the production of other crops was 7.1 per cent in oilseeds, 5.6 per cent in sugarcane, 6.3 per cent in jute and mesta and 8.3 per cent in cotton (Table 11.5).

TABLE 11.4

Agricultural Production

(Million Tonnes/Bales)*

Crop	1978-79	1979-80	1980-81	1981-82	1982-83	1983-84	1984-85	1985-86	1986-87	1987-88	1988-89	1989(P)90
Rice	53.8	423	53.6	53.2	47.1	60.1	58.4	63.8	60.6	56.8	70.5	74.1
	(2.1)	(-21.4)	(26.7)	(-0.7)	(-11.5)	(27.6)	(-3.0)	(9.4)	(-5.0)	(-6.1)	(23.9)	(5.0)
Wheat	35.5	31.8	36.3	37.4	42.8	45.5	44.1	47.1	44.3	46.2	54.1	49.7
	(11.6)	(-10.4)	(14.1)	(3.0)	(14.4)	(6.3)	(-3.1)	(6.6)	(-5.7)	(4.3)	(17.1)	(-8.1)
Pulses	12.2	8.6	10.6	11.5	11.9	12.9	12.0	13.4	11.7	11.0	13.8	12.6
	(1.7)	(-29.5)	(23.2)	(8.5)	(3.5)	(8.4)	(-7.0)	(11.7)	(-12.7)	(-6.0)	(25.5)	(-8.7)
Coarsegrains	30.4	27.0	29.0	31.1	27.8	33.9	31.2	26.2	26.8	26.4	31.5	34.3
	(1.3)	(-11.2)	(7.4)	(7.2)	(-10.6)	(21.9)	(-8.0)	(-16.0)	(2-3)	(-1.5)	(19.3)	(8.9)
Kharif Foodgrains	78.1	63.2	77.7	79.4	69.9	89.2	84.5	85.2	80.2	74.6	95.6	100.9
	(0.5)	(-19.1)	(22.8)	(2.3)	(-12.0)	(27.6)	(-5.3)	(0.8)	(-5.9)	(-7.0)	(29.5)	(4.5)
Rabi Foodgrains	53.8	46.5	51.9	53.9	59.6	63.2	61.0	65.2	63.2	65.8	74.3	69.7
	(105)	(-13.8)	(11.9)	(3.9)	(10.6)	(5.9)	(-3.3)	(6.9)	(-3.1)	(4.1)	(12.9)	(-6.2)
All Foodgrains	131.9	109.7	129.6	133.3	129.5	152.4	145.5	150.4	143.4	140.4	169.9	170.6
	(4.4)	(-16.8)	(18.1)	(2.9)	(-2.9)	(17.7)	(-4.5)	(3.4)	(-4.7)	(-2.1)	(21.0)	(0.4)
Groundnut	6.2	5.8	5.0	7.2	5.3	7.1	6.4	5.1	5.9	5.8	9.7	8.1
	(1.6)	(-6.4)	(-13.8)	(44.0)	(-26.4)	(34.0)	(-9.9)	(-20.3)	(15.7)	(-1.7)	(67.2)	(-16.5)
Rapeseed & Mustard	1.9	1.4	2.3	2.4	2.2	2.6	3.1	2.7	2.6	3.5	4.4	4.1
	(18.7)	(-26.3)	(64.3)	(4.3)	(-8.3)	(18.2)	(19.2)	(-12.9)	(-3.7)	(30.8)	(29.4)	(-6.8)
Oilseeds@	10.1	8.7	9.4	12.1	10.0	12.7	13.0	10.8	11.3	12.7	18.0	16.8
	(4.1)	(-13.9)	(8.0)	(28.7)	(-17.4)	(27.0)	(2.4)	(-16.9)	(4.6)	(11.5)	(42.9)	(-6.7)
Sugarcane	151.7	128.8	154.2	186.4	189.5	174.1	170.3	170.7	186.1	196.7	203.0	222.6
	(-14.3)	(-15.1)	(19.7)	(20.9)	(1.7)	(-8.1)	(-2.2)	(0.2)	(9.1)	(5.7)	(3.2)	(9.7)
Cotton (Lint)*	8.0	7.6	7.0	7.9	7.5	6.4	8.5	8.7	6.9	6.4	8.7	11.4
	(11.1)	(-5.0)	(-7.9)	(12.9)	(-5.1)	(-14.7)	(32.8)	(2.4)	(-20.7)	(-7.2)	(35.9)	(31.0)
Jute & Mesta*	8.3	8.0	8.2	8.4	7.2	7.7	7.8	12.6	8.7	6.8	7.9	8.4
	(15.3)	(-3.6)	(2.5)	(2.4)	(-14.3)	(6.9)	(1.3)	(61.5)	(-31.7)	(-20.9)	(16.2)	(6.3)

* 170 kgs. each for cotton and 180 kgs. each for jute and mesta. P:Provisional

`@ Nine major oilseeds including groundnut, castorseed, Sesamum, rapeseed and mustard, linseed, saf-flower, nigerseed, sunflower and soyabean.

Figures in brackets are per cent increase or decrease over the year.

With favourable monsoon conditions and the continuance of the various thrust programmes, foodgrains production in 1990-91 may be slightly higher than the targetted level; the foodgrains production in 1990-91 is expected to be about 177 million tonnes.

TABLE 11.5

Targets and Achievements of Agricultural Production During Seventh Plan

(Million Tonnes/Bales)

	Crop	*1985-86*		*1986-87*		*1987-88*		*1988-89*		*1989-90*		*1990-91*	
		Target	*Achievement*	*Target*	*Achievement*	*Target*	*Achievement*	*Target*	*Achievement*	*Target*	*Achievement*	*Target*	*Likely Achievement*
1.	Rice	63.5	63.8	65.0	60.6	64-65	56.9	68.0	70.5	72.5	74.1	73.7	75.0
2.	Wheat	49.2	47.1	49.0	44.3	50-51	46.2	52.3	54.1	54.0	49.7	54.5	54.6
3.	Coarse Cereals	33.0	26.2	32.0	26.8	32-32.5	26.4	33.0	31.5	33.8	34.3	33.3	33.4
4.	Pulses	13.5	13.4	14.0	11.7	14-14.5	11.0	13.3	13.8	14.7	12.6	15.0	14.2
5.	Total Foodgrains	159.2	150.4	160.0	143.4	160-163	140.4	166.6	169.9	175.0	170.6	176.5	177.2
6.	Sugarcane	191.0	170.7	185-190	186.1	180-185	196.7	195.0	203.0	212.0	222.6	220.0	233.4
7.	Oilseeds	13.6	10.8	14.8	11.3	14-15	12.7	15.7	18.0	16-17	16.8	18.0	19.2
8.	Cotton*	8.5-8.6	8.7	8.8	6.9	8.8	6.4	9.8	8.7	10.0	11.4	11.5	10.2
9.	Jute & Mesta@	8.7	12.6	8.5	8.7	8.6	6.8	9.2	7.9	9.5	8.4	9.0	9.0

*Bale of 170 kgs. @Bale of 180 kgs.

There has been deceleration in the rate of growth in agriculture during the Seventh Plan compared with the rate achieved during the Sixth Plan. The average annual rate of increase in foodgrains production was 3.6 per cent during the Seventh Plan as compared with 6.3 per cent recorded during the Sixth Plan (with base 1979-80). The average annual growth rate of kharif foodgrains declined from 7.1 per cent during the Sixth Plan to 4.3 per cent during the Seventh Plan and that of rabi foodgrains declined from 5.7 per cent to 2.9 per cent over the same period. The slower growth in foodgrains output is mainly attributable to deceleration in the growth of yield per hectare from 5.8 per cent per annum in the Sixth Plan to 3.4 per cent per annum in the Seventh Plan. In the case of coarse cereals, it is worth noting that there has been a negative growth rate in its acreage during both the plan periods. The average annual growth rate of coarse cereals production declined from 3.6 per cent in the Sixth Plan to 2.6 per cent during the Seventh Plan and that of pulses declined significantly from 7.4 per cent to 2.1 per cent over the same period. Among the non-foodgrains crops, oilseeds and sugarcane also recorded a lower growth in the Seventh Plan compared with the growth achieved in the Sixth Plan mainly on account of lower growth in productivity. It was only in the case of cotton, jute and mesta that the growth in output during the Seventh Plan was higher than in the Sixth Plan. The trends described in this paragraph are outlined in Table 11.6.

The State-wise production performance during the Seventh Plan was significant in Bihar, haryana, Orissa, Punjab, Tamil Nadu, Uttar Pradesh and West Bengal where the average foodgrains output increased by more than 15 per cent over the average for the Sixth Plan. In fact, it was as high as 34 per cent in West Bengal, 24 per cent in Haryana, 23 per cent in Punjab, 21 per cent in Bihar and 18 per cent in Uttar Pradesh. However, average production declined in Gujarat, Kerala and Rajasthan.

TABLE 11.6

Annual Average Growth Rates (per cent)

		Sixth Plan			Seventh Plan		
	Crops	*Area*	*Production*	*Yield*	*Area*	*Production*	*Yield*
1.	Foodgrains	0.3	6.3	5.8	0.1	3.6	3.4
2.	Kharif Foodgrains	0.2	7.1	6.4	0.2	4.3	3.8
3.	Rabi Foodgrains	0.5	5.7	5.2	(-)0.1	2.9	3.0
4.	Coarse Cereals	(-)1.0	3.6	4.4	(-)0.7	2.6	3.4
5.	Rice	1.0	7.8	6.4	0.6	5.4	4.6
6.	Wheat	1.3	6.9	5.5	(-)0.1	2.8	2.8
7.	Jowar	(-)0.8	0.2	0.8	(-)1.2	4.5	5.4
8.	Bajra	0.4	12.1	10.3	2.4	15.5	7.3
9.	Maize	0.3	9.2	8.6	0.3	5.3	4.3
10.	Pulses	0.5	7.4	6.8	0.6	2.1	1.0
11.	Gram	0.2	7.2	8.0	(-)0.3	1.8	0.6
12.	Arhar	3.1	8.8	5.3	2.6	1.5	(-)1.2
13.	Oilseeds	2.3	9.6	6.5	4.0	7.1	2.6
14.	Groundnut	0.2	5.8	4.5	4.4	8.5	2.6
15.	Rapeseed & Mustard	3.3	18.6	14.1	5.1	7.6	2.1
16.	Cotton	(-)1.9	3.5	5.6	0.2	8.3	7.5
17.	Jute & Mesta	(-)1.1	(-)0.1	1.4	(-)2.3	6.3	6.8
18.	Sugarcane	2.9	6.4	3.4	3.0	5.6	2.6

Crop production was affected in many states because of adverse weather conditions during the second and third years of the Plan. The year 1988-89 and 1989-90 brought about a sharp recovery in almost all the states. The average annual growth in foodgrains production in different states during the Seventh plan works out to be as high as 6.8 per cent in West Bengal followed by 4.8 per cent in Haryana, 4.6 per cent in Punjab, 4.3 per cent in Tamil Nadu, 4.2 per cent in Bihar, 3.6 per cent in Uttar Pradesh and 3.1 per cent in Orissa. There was, however, a decline in production in the states of Gujarat, Kerala and Rajasthan.

Rice

Rice is the most important cereal crop grown over an area of about 42 million hectares with an irrigation coverage of only 43 per cent. A large area is rainfed and most fluctuations in food production in the country are primarily due to variability in production of rice. The production of rice had declined in 1986-87 and 1987-88 due to drought conditions to 60.6 million tonnes and 56.9 million tonnes respectively. With the good monsoons of 1988-89 and 1989-90, the country has been able to step up its production considerably over the last two years with a peak production of 74 million tonnes achieved in 1989-90. The increase in production in 1989-90 was due to an increase both in acreage (1.1 per cent) and in yield per hectare (4.0 per cent). The increase of 3.6 million tonnes (5 per cent) in 1989-90 over 1988-89 was largely contributed by major rice producing states of Assam (3.6 lakh tonnes), Haryana (2.6 lakh tonnes), Orissa (9.9 lakh tonnes), Punjab (17.7 lakh tonnes), Tamil Nadu (6.4 lakh tonnes) and West Bengal (3.6 lakh tonnes). The increase in rice production and productivity can be attributed, inter alia, to the effective implementation of the special Foodgrains Production Programme for rice which covered 14 major States and the Special Rice Production Programme in Eastern States.

Considering the trend in rice production, its output in 1990-91 has been targetted at 73.7 million tonnes (Kharif Rice 65.7 million tonnes and Rabi Rice 8.0 million tonnes). With the excellent monsoon, over both space and time, recorded in 1990, the overall kharif crop condition in the major rice growing States was reported to be satisfactory. Owing to satisfactory monsoon in rabi rice areas, it is expected enough water will be available in canal commands as well as in tank irrigated areas for rabi rice cultivation. During 1990-91, the SFPP-Rice programme was proposed to be expanded. The programme implementation, inter alia, included the developmental programmes for increasing the basmati rice production for export promotion in 19 identified districts in Uttar Pradesh, Punjab and Haryana. Taking all these factors into account, it is expected that the rice production during 1990-91 may reach the level of 75 million tonnes.

Wheat

Wheat has played a very important role in stabilising the foodgrains production in the country over the past few years. It recorded an all time high level of production of 54.1 million tonnes in 1988-89 which was higher than the target of 52.3 million tonnes. However, in 1989-90 it declined to 49.7 million tonnes due to premature rise in temperature in January-February 1990 and consequent early flowering of the plant. The production, except in Punjab, suffered in most of the major wheat producing states, particularly in unirrigated areas of Uttar Pradesh, Madhya Pradesh, Rajasthan, Gujarat, Bihar and Haryana. Apart from decline in productivity, there was shrinkage of area to the extent of 2.7 per cent over 1988-89, although the acreage has generally been registering a rising trend reaching the level of 24 million hectares in 1988-89.

The target of wheat production for 1990-91 has been fixed at 54.5 million tonnes. Since the soil moisture conditions are favourable and the water storage conditions are satisfactory, the area under the crop in 1990-91 is likely to be about 24 million hectares. The main thrust of the strategy for increasing wheat production has been through improvement in productivity by cultivation of identified high-yielding varieties with a recommended package of practices. These, inter alia, include expansion of the area wherever possible; provision of high quality seeds of new high-yielding varieties at reasonable rates; use of optimum and balanced doses of fertilisers; efficient water management to provide irrigation at critical stages of crop growth; weed control at the proper time; and to ensure adequate supply of power and diesel to run tubewells and pump sets at the time of sowing and irrigation at appropriate timings. Along with these, the SFPP for wheat which has been launched from 1988-89 in 71 identified districts is proposed to be increased to 100 identified districts. As a result of these measures and favourable weather, the production of wheat in 1990-91 may register a substantial increase and is expected to be higher than that of 54.1 million tonnes, the peak production level achieved in 1988-89 and may be around the targetted level of 54.5 million tonnes.

Coarsegrains

The development of coarse cereals has assumed a greater importance since it directly benefits the small and marginal farmers. Coarsegrains are grown under rainfed conditions. The crops included in Coarsegrains are jowar, bajra, maize, ragi, small millets and barley. These were grown in an area of 38 million hectares forming about 30 per cent of the total area of 127 million hectares under foodgrains in 1989-90. There has been a noticeable expansion, for the second year in succession, in the output of coarsegrains, which reflects the spread of benefits in favour of dry and unirrigated areas where small and marginal farmers predominate. The production of coarse cereals which had increased by 19.3 per cent to 31.5 million tonnes in 1988-89, advanced further by about 9 per cent to 34.3 million tonnes in 1989-90, surpassing the peak of 33.9 million tonnes achieved in 1983-84. The increase in production in 1989-90 came through the increase in productivity as area under the crop had declined by 2.7 per cent. The increase in the output of coarse cereals was largely contributed by increase in production of jowar, maize and ragi. However, the increase was partly offset by decline in production of bajra and barely.

The area under coarsegrains has been declining. The average annual reduction in acreage during the Sixth Plan was about one per cent. It fell further by 0.7 per cent during the Seventh Plan period. However, with the help of increase in productivity, the production of coarsegrains recorded an average annual growth of 2.6 per cent during the Seventh Plan as against 3.6 per cent recorded during the Sixth Plan. The main constraints in increasing coarsegrains production are: (i) low coverage of area under high yielding varieties/hybrids because of non-availability of seeds in sufficient quantities, and (ii) reluctance of the farmers to invest on inputs as these crops are sensitive to climatic conditions. Incidentally, it is necessary to note, that for the first time, the production in the terminal year of Seventh Plan exceeded the target. In other years, it remained much below the targetted levels.

In the background of achievement in the past and the need to further increase the production of coarse cereals, the target for production in 1990-91 has been fixed at 33.3 million tonnes (27.9 million tonnes during kharif season and 5.4 million tonnes during rabi season). For achieving the target, SFPP-Maize is being implemented in 50 districts spread over 10 states, SFPP-Jowar in 33 districts of six states, SFPP-Bajra in 20 districts of four states and SFPP-Ragi in 20 districts of six states. The programme components include distribution of inputs like certified seeds, pesticides, plant protection equipments, farm implements etc. at subsidised rates. With development measures and favourable weather, the production target is expected to be achieved and may even be exceeded.

Pulses

The importance of pulses as an important source of protein in a balanced diet has special significance in Indian conditions. It is essential to note that the per capita availability of pulses has declined over the years. Pulses are grown over an area of 23 million hectares with production in the range of 13-14 million tonnes (accounting for 18 per cent of total area and less than 8 per cent of total foodgrains production) and the yield per hectare ranging 500-600 kgs. These are grown mainly under unirrigated conditions and the irrigated area accounts for about 9 per cent of the total area. The major pulses (arhar and gram) continue to be produced in the states of Madhya Pradesh, Maharashtra, Orissa, Rajasthan and Uttar Pradesh. The share of arhar and gram in the total production of pulses in 1989-90 was 21.4 per cent (2.7 million tonnes) and 33.3 per cent (4.2 million tonnes) respectively.

The production of pulses which had declined in the drought years of 1986-87 and 1987-88 to around 11 million tonnes, recorded a substantial increase to reach an all time peak level of 13.8 million tonnes in 1988-89 as against the earlier peak level of 13.4 million tonnes recorded in 1985-86. In fact, this was the only year when it exceeded the target of 13.3 million tonnes. However, in 1989-90, there was again a setback in its production when it fell to 12.6 million tonnes wholly because of decline in yield per hectare since there was a marginal expansion in acreage. The fall was recorded largely in rabi pulses (mainly gram)— the decline was about one million tonnes due both to shrinkage in area and lower productivity. The marginal fall in kharif pulses was on account of decline in yield per hectare. Almost all the major gram producing States recorded a decline in production during 1989-90 due to adverse weather conditions which affected the yield per hectare.

The production of pulses in 1990-91 has been targetted at 15 million tonnes (5.5 million tonnes during kharif and 9.5 million tonnes during rabi season). This has been sought to be achieved through the continuance of pulses development programmes like National Pulses Development Project (NPDP) and Special Foodgrains Production Programme. The main objective of NPDP is to increase the production by adopting location specific technology. NPDP comprises distribution of improved seeds, block demonstrations, adaptive trials of promising varieties, pest control, extension work and plant protection. NPDP is supplemented by a special programme under SFPP launched in 13 states covering schemes (a) plant protection for gram and arhar and (b) expansion of area under summer moong/ urad through distribution of seeds at concessional rates. Besides, pulses have also been brought within the purview of Technology Missions. With these measures and an excellent monsoon, the production of pulses during 1990-91 is expected to reach another record level of about 14.2 million tonnes.

Oilseeds

Oilseeds, which constitute an important group of crops, are grown in an area of about 23 million hectares, mostly limited to rainfed areas (80 per cent) in contrast to cereals like wheat and rice which are grown largely under irrigated conditions. The bulk of the vegetable oil production is derived from nine cultivated oilseeds, namely, groundnut, rapeseed/mustard, sesamum, safflower, nigerseed, soyabean, sun-flower-forming the edible group and linseed and castorseed-forming the non-edible group. Groundnut and rapeseed/ mustard are the most important crops which together account for about 73 per cent of the total oilseeds production of 16.8 million tonnes in 1989-90. Soyabean and sunflower, though introduced relatively recently, have also augmented the domestic availability of edible oils. Among the kharif oilseeds, Gujarat is the largest producing state while Uttar Pradesh is the largest producer of rabi oilseeds. The major oilseeds producing states are Gujarat, Andhra Pradesh, Karnataka, Madhya Pradesh, Uttar Pradesh, Maharashtra, Rajasthan, Orissa and Tamil Nadu.

The production of oilseeds, which had declined in 1985-86 from the peak level of 13.0 million tonnes in 1984-85, increased to a record level of 18 million tonnes in 1988-89 and much above the target of 15.7 million tonnes largely due to favourable weather conditions. However, this level could not be maintained in 1989-90 and production fell to 16.8 million t6nnes. This was on account of a steep fall recorded in groundnut production. The adverse weather conditions in Saurashtra and some other parts of the country resulted in a shortfall of 1.6 million tonnes of groundnut in 1989-90. Both kharif and rabi groundnut production declined. The production of rapeseed and mustard, the second major oilseed crop, also declined from 4.4 million tonnes in 1988-89 to 4.1 million tonnes in 1989-90 mainly on account of sizeable loss of crop in Rajasthan and Madhya Pradesh. However, the fall in oilseeds production was partly offset by increase in the production of soyabean and sunflower during 1989-90. The production of different oilseeds is given in Table 11.7.

TABLE 11.7

Production of Oilseeds

(Lakh Tonnes)

Oilseeds	*1985-86*	*1986-87*	*1987-88*	*1988-89*	*1989-90*
Groundnut					
Kharif	37.6	44.3	41.8	74.9	61.2
Rabi	13.6	14.5	16.7	21.7	19.7
Total	51.2	58.8	58.5	96.6	80.9
Castorseed	3.1	2.3	2.0	4.1	5.1
Sesamum	5.0	4.5	5.8	6.8	7.1
Rapeseed & Mustard	26.8	26.0	34.6	43.8	41.2
Linseed	3.8	3.2	3.9	3.6	3.5
Nigerseed	1.9	1.3	1.8	1.8	1.8
Safflower	3.5	3.5	4.6	4.4	4.9
Sunflower					
Kharif	1.7	2.5	3.8	2.2	3.2
Rabi	1.1	1.7	2.5	1.5	2.7
Total	2.8	4.2	6.3	3.7	5.9
Soyabean	10.2	8.9	9.0	15.5	17.1
Total					
Kharif	59.5	63.8	64.2	105.3	95.5
Rabi	48.8	48.9	62.3	75.0	72.0
Total	108.3	112.7	126.5	180.3	167.5

During 1989-90, out of the total fall of 1.3 million tonnes, Gujarat alone recorded a fall of 1.1 million tonnes. Other major oilseeds producing states which recorded lower production include Andhra Pradesh, Rajasthan and Karnataka. However, the decline in production in these states was partly offset by increase in production in the states of Madhya Pradesh, Maharashtra and Tamil Nadu.

Despite steady progress achieved in oilseeds production, the country has not yet achieved self-sufficiency. The supply of edible oils still falls short of demand. As a result, edible oils continued to be imported in 1989-90, although it was much less than the imports in 1987-88. The imports during the oil year 1989-90 (November-October) aggregated 6.1 lakh tonnes valued at Rs. 328 crores as against 18.2 lakh tonnes valued at Rs. 1,061 crores imported in 1987-88. The increasing dependence on imports has therefore been somewhat halted which was made possible by substantial increase in edible oilseeds production which in turn was the result of the development measures undertaken by the Government. The imported oil is being predominantly distributed through the public distribution system and to some extent used for market intervention operations of National Dairy Development Board (NDDB). Blending of edible oils has also been permitted so as to enlarge the production/ consumption of minor oilseeds/oils.

Considering the need to augment the supply of edible oils and thereby reduce the import dependence, a production target of 18 million tonnes of oilseeds has been fixed for 1990-91:10 million tonnes of kharif oilseeds and 8 million tonnes of rabi oilseeds. In order to achieve the target, the two centrally sponsored schemes, namely, National Oilseeds Development Project and Oilseeds Production Thrust Project, which were operating till 1989-90, have now been merged during 1990-91 into a single scheme namely Oilseeds Production Programmes (OPP). This scheme essentially provides financial assistance to the states for production and distribution of quality seeds, plant protection measures including supply of plant protection chemicals and equipments and organising demonstrations of improved technology.

In addition, the efforts of the Technology Mission on Oilseeds—established in May, 1986—in harnessing the best of production, processing and management technologies to accelerate self-reliance are being enhanced. The Government also announced an integrated oilseeds policy on 5th January, 1989. The policy highlighted commitment to support farmers with technology, inputs and attractive prices and to safeguard the interest of the consumers with reasonable prices. As a sequel to this policy, the Government appointed NDDB as the Market Intervention Agency for procurement of oilseeds and oil for building a buffer stock so as to ensure an incentive price to the farmers and release the same during the lean season at moderate prices to the consumers.

The production of kharif groundnut in 1990-91 has been affected considerably due to absence of rains at the time of sowing in the Saurashtra region—the main groundnut growing area in Gujarat. However, the production of other kharif oilseeds like soyabean, sunflower seed, castorseed and Sesamum is expected to register a sizeable increase in production largely as a result of application of improved technology and timely and adequate rainfall recorded in their respective growing areas. The good monsoon has also helped the production of rabi oilseeds, particularly rapeseed and mustard, which is expected to reach another record production of about 53.2 lakh tonnes during 1990-91 mainly on account of large area expansion in Rajasthan, Haryana, Gujarat, Madhya Pradesh etc. The loss in kharif oilseeds is expected to be offset by increase in the output of rabi oilseeds and the overall level of oilseeds production in 1990-91 may exceed the target of 18 million tonnes.

Cotton

Cotton which is basically a rainfed crop (irrigated area under cotton being only 30 per cent), is grown over an area of 73 lakh hectares. The production is mainly concentrated in Gujarat, Maharashtra, Haryana, Madhya Pradesh, Punjab, Andhra Pradesh, Karnataka, Rajasthan and Tamil Nadu. These states together account for 99.4 per cent of total area and 99.6 per cent of the out-put in 1989-90. While area under cotton declined by 0.2 per cent, the production of cotton registered a leap forward from 87.4 lakh bales in 1988-89 to 114.1 lakh bales in 1989-90 as against the target of 100 lakh bales. This was wholly on account of an increase in per hectare yield of lint; it increased from 202 kgs. in 1988-89 to 265 kgs. per hectare in 1989-90.

The country produces a wide range of varieties of cotton which include short staple, medium staple and superior long staple. The north-western belt comprising Punjab and Haryana and Rajasthan specialises mainly in short and medium staple varieties, while southern and western parts of the country produce basically long and superior long staple varieties. Substantial increases were recorded in all the major cotton growing states.

Despite a significant increase in cotton production, the country is still in short supply of medium and long staple cotton. In order to increase the production of these varieties, a centrally sponsored Intensive Cotton Development Programme, which was started in 1971-72, was revised during the Seventh Plan and emphasis was laid only on increasing the production of long and medium staple cotton. With the country becoming surplus in certain varieties of cotton, exports have been selectively allowed. Keeping in view the progress made, a production target of 11.5 million bales of cotton in 1990-91 has been fixed. The main objectives in cotton production during 1990-91 are: (1) increasing the production of medium and long staple cotton and (2) increasing productivity and improving quality of cotton. To achieve these objectives, the existing Intensive Cotton Development Programme was continued in all the major cotton growing states, namely, Andhra Pradesh, Gujarat, Haryana, Karnataka, Madhya Pradesh, Maharashtra, Punjab, Rajasthan, Uttar Pradesh and Tamil Nadu. The year 1990-91 was, in the beginning, slated for another bumper cotton crop. But subsequently, due to excessive rains in the later part of the season, the crop is reported to have been damaged by pests in major cotton growing districts of Punjab and Haryana. Even then the size of the crop in 1990-91 is expected to be about 102 lakh bales.

Jute and Mesta

In contrast to a continuous fall in acreage, production of jute and mesta increased from 7.9 million bales in 1988-89 to 8.4 million bales in 1989-90 due to a substantial increase in per hectare yield. The productivity increased by 7.5 per cent from 1,540 kgs. per hectare in 1988-89 to 1,656 kgs. per hectare in 1989-90. The increase of 6.2 per cent in the production of jute and mesta was shared by West Bengal and Assam.

The production of raw jute which had declined to as low a level as 57.9 lakh bales in 1987-88 increased by 15.9 per cent to 67.1 lakh bales in 1988-89 and further by about 6 per cent to 71.1 lakh bales in 1989-90 due mainly to increase in productivity. While production in West Bengal increased by 10.4 per cent from 45.3 lakh bales in 1988-89 to 50,0 lakh bales in 1989-90 on account of substantial increase in productivity, in Assam the increase was of the order of 18 per cent where output advanced from 6.7 lakh bales in 1988-89 to 7.9 lakh bales in 1989-90. In Bihar, the jute crop was smaller by about 14.8 per cent in 1989-90 mainly due to fall in acreage. Orissa also recorded a fall in production.

The production of mesta, following the increase in area and yield per hectare, increased from 11.5 lakh bales in 1988-89 to 12.4 lakh bales in 1989-90. The increase came through a substantial increase in productivity which increased from 909 kgs. per hectare in 1988-89 to 972 kgs. per hectare in 1989-90. About 49 per cent of the crop was produced in Andhra Pradesh followed by Orissa 17 per cent and Bihar 15 per cent. Maharashtra, Tripura, Karnataka and West Bengal also produced some quantity of mesta.

Considering the overall performance of jute and mesta production, a target of 9 million bales for 1990-91 has been fixed. With the continuance of Special Jute Development Programme for improving the productivity and quality in potential areas in different states and the reported increase in area following favourable weather conditions, the targetted level of production is likely to be achieved in 1990-91.

Sugarcane

Sugarcane is cultivated over an area of 33-34 lakh hectares both in sub-tropical and tropical regions of the country. In the subtropical belt, main sugarcane growing states are Uttar Pradesh, Bihar, Haryana and Punjab and the tropical belt mainly comprises Maharashtra, Andhra Pradesh, Tamil Nadu and Karnataka. The production of sugarcane during 1989-90 crossed the peak level production of 203 million tonnes recorded in 1988-89 to a new high of 223 million tonnes in 1989-90—an increase of 9.7 per cent as against a target of 212 million tonnes. The favourable weather conditions together with price incentive and other developmental measures were responsible in the expansion of output. Even the area and productivity crossed their previous peak. While acreage increased by 2.4 per cent to 3.4 million hectares, productivity increased by 6.6 per cent to 65 tonnes per hectare during 1989-90. Except for Tamil Nadu and Karnataka, all other major sugarcane growing states recorded increase in varying proportion.

The rising trend in cane production is the outcome of various policy measures undertaken by the Government and developmental efforts initiated by the sugar industry in general and the concerted efforts made by the cane growers in particular. To make the cane cultivation more profitable, the Statutory Minimum Prices (SMP) for sugarcane are being revised upwards in regular intervals. These prices were increased from Rs. 19.50 per quintal in 1988-89 sugar season to Rs. 22 per quintal for 1989-90 sugar season and further to Rs. 23 per quintal for the 1990-91 season. To encourage the farmers to grow more cane, SMF for 1991-92 has been announced in advance at Rs. 24 per quintal. As against the SMP, sugar industry pays a still higher price (as recommended by State Governments) which serves a greater inducement to cane cultivators for increasing both production/productivity and quality of cane.

A target of 220 million tonnes has been fixed for 1990-91. The target is sought to be achieved through (a) production of quality seed cane, (b) increasing irrigation facilities, (c) judicious and timely use of fertiliser application, (d) better management of ratoons, (e) greater and effective participation of sugar factories in cane development programmes, (f) transfer of technology through various extension systems and (g) larger coverage under plant protection measures. Besides, states have been advised to avail of the loan facilities from the Sugar Development Fund. As a result of these measures, production of sugarcane in 1990-91 is expected to be around 233 million tonnes. As per the latest information available, with the continuance of the policy of partial decontrol together with the fiscal and financial incentives, sugar production in 1989-90 touched a peak level at 109.9 lakh tonnes. Considering the need to step-up sugar output in 1990-91, the Government continued to allow, as an additional incentive, higher free-sale quota (75 per cent against normal 55 per cent) of sugar produced during October 1,1990 to November 15,1990. Indications are that sugar output in 1990-91 sugar season may be above 115 lakh tonnes.

Horticultural Crops

The horticulture sector, which started receiving attention from the Fourth Five year Plan, has recorded an impressive growth. The production of fruits and vegetables has increased at the annual growth rate of 2.2 and 11.7 per cent respectively during the Seventh Plan period. Fruits occupy an area of 3.2 million hectares with an estimated production of 261 lakh tonnes and India is the third largest producer of fruits next only to Brazil and U.S.A. In spite of India being the second largest producer of vegetables, next only to China, the production of vegetables has remained almost stagnant. Vegetable crops occupy an area of 5.5 million hectares with a production of 539 lakh tonnes. Various centrally sponsored and central sector schemes have been in operation in the case of fruits development; these include the package programme on pineapple and banana and improved technology for quality apple production and elite progency orchard scheme. Since coconut is valued both as oilseed and food crop, the Coconut Development Board has implemented various programmes for integrated development of coconut industry during the Seventh Five Year Plan. Similarly, integrated programme for development of spices, which include pepper, ginger, turmeric and chillies, was implemented during the Seventh Plan period. Production of cashewnuts, cocoa, arecanut, etc has expanded during the Seventh Five Year Plan through various development schemes.

Tea

Tea grown in an area of over 4 lakh hectares, is one of the most labour intensive industries employing more than a million workers. It plays an important role in the economy of the country. Its production has increased significantly from 544 million kgs. in 1979-80 to a peak of 703 million kgs. in 1989-90. The increase in production during this period has emanated largely from increase in productivity from 1,490 kgs. per hectare to around 1,681 kgs. per hectare. Despite the substantial increase in production, the target fixed for different years, except for 1985-86, of the Seventh Plan could not be achieved.

The North-Eastern parts of the country generally account for a major portion of tea production (75-80 per cent)—Assam and West Bengal are the bulk producers. Among the South Indian states, Tamil Nadu and Kerala are major producers of tea. Of late, the increase in production is largely shared by production in South India. There is a considerable difference between yield rate of North India and South India gardens arising essentially to agro-climatic factors. The average productivity in South India is about 2,000 kgs. per hectare, while in North India the yield rate is 1,600 kgs. per hectare.

Following good agro-climatic conditions, the production of tea in 1990-91 has increased to 719 million kgs. as against 703 million kgs. recorded in 1989-90. The government has been making continuous efforts to increase production of tea. Twenty five top tea estates have been identified for increasing production of tea. Besides, efforts are being made to increase the area under cultivation in the states of Uttar Pradesh and Orissa. The Tea Board is at present operating a number of schemes for stepping up the production. These schemes provide financial support in the form of loans and subsidies. The loan schemes include Tea Plantation Finance Scheme, Tea Machinery and Irrigation Equipment Hire Purchase Scheme, setting up of Tea Nurseries in cooperative sector, New Tea Unit Finance Scheme and Sophisticated Tea Packaging Arrangement Loan Scheme. The subsidy schemes on the other hand include Replanting Subsidy Scheme, New Tea Unit Financing Scheme, Darjeeling Interest Subsidy Scheme, Tea Extension Planting Interest Subsidy Scheme and Irrigation/Drainage Interest Subsidy Scheme.

Alongwith the increase in production, the domestic consumption (about two-third of out-put) of tea has also been increasing. The domestic demand has increased from 438 million kgs. in 1985-86 to 478 million kgs. in 1988-89 and further to 500 million kgs. in 1989-90. Exports of tea on the other hand, in quantitative term, declined from 209 million kgs. in 1988-89 to 203 million kgs. in 1989-90. In 1990-91, exports were of the order of 199 million kgs. as compared with 203 million kgs. during 1989-90. However, with the increase in average per unit value realisation from about Rs. 31 per kg. to about Rs. 43 per kg. export earnings increased from Rs. 644 crores in 1988-89 to Rs. 862 crores in 1989-90. In 1990-91 export earnings were Rs. 1,045 crores as against Rs. 862 crores during 1989-90. In order to boost foreign exchange earnings through tea exports, concerted efforts are being made to increase tea exports through fiscal and financial incentives. The exporters of tea received support from Brand Promotion Fund Scheme in the form of interest-free loan.

Coffee

The production of coffee in the country has been fluctuating as a consequence of fluctuations in weather conditions. The production which had reached to 2 lakh tonnes in 1984-85 came down to 1.2 lakh tonnes in 1989-90, although there has been expansion in acreage. Over 1988-89, coffee output declined by 43 per cent owing to adverse weather conditions in coffee producing states. During 1990-91, the production of coffee is expected to be around 1.7 lakh tonnes. However, the target of 1.8 lakh tonnes set for the terminal year of Seventh Plan had already exceeded during 1986-87 (1.9 lakh tonnes) and 1988-89 (2.1 lakh tonnes).

The Coffee Board, which is statutorily charged with the responsibility of the marketing of coffee produced in India, continued to extend its support in the fields of research, extension and credit to improve both quality and production of coffee. The bulk of coffee production is being exported, although domestic consumption has also increased over years. The country exported 1.3 lakh tonnes valued at Rs. 363 crores in 1989-90 as against about one lakh tonnes valued at Rs. 337 crores in 1988-89. The exports durings 1990-91 were about one lakh tonnes valued at Rs. 276 crores. The decline in exports is due to poor crop during 1989-90 available for marketing in 1990-91 and fall in the international prices due to suspension of quota. The on-going Developmental Loan and Subsidy Scheme continued to remain in operation to assist coffee production.

Rubber

Natural rubber, one of the most valued raw materials, is grown in about 4,4 lakh hectares predominantly by the small growers. The production is estimated to have increased from 2.6 lakh tonnes in 1988-89 to about 3.3 lakh tonnes in 1990-91. The target of production for the terminal year of Seventh Plan was fixed at 2.6 lakh tonnes against a base level of 1.9 lakh tonnes in 1984-85. The growth in rubber plantation industry is characterised by an increase in production, area and productivity.

The remunerative prices and financial and technical support are major factors contributing to appreciable increase in new plantings and replanting.

Natural rubber is mainly cultivated in the States of Kerala and Tamil Nadu. Of the total area, about 85 per cent is accounted by Kerala. As a result of promotional efforts, rubber is now successfully grown in the non-traditional areas of Karnataka, Tripura, Assam, Meghalaya, Mizoram, Manipur, Nagaland, Andaman and Nicobar Islands, Goa, Maharashtra and Orissa.

With the increased production of automobiles, the consumption of rubber has increased over the years. The consumption of natural rubber increased from 3.1 lakh tonnes in 1988-89 to 3.7 lakh tonnes in 1990-91. The gap between the demand for and supply of natural rubber is being met through imports. The import in 1989-90 was 26.5 thousand tonnes as against 51.3 thousand tonnes in 1988-89. During 1990-91, 40 thousand tonnes of natural rubber was imported. In view of the steadily increasing gap between domestic supply and demand for natural rubber, the Rubber Board has continued its efforts in the direction of achieving greater production. Under the Plantation Development Scheme liberal financial and technical assistance is given to growers for new plantings/ replantings. Other important measures implemented by the Rubber Board include the distribution of higher yielding planting materials, supply of relevant inputs at concessional rates and encouragement for group processing and marketing centres by providing financial and technical assistance.

Agricultural Inputs

Agricultural inputs play an important role in accelerating the growth of agricultural production and productivity. Considering the progress made in the production of foodgrains, in particular wheat and rice; through introduction of high yielding varieties of seed together with an input package of assured irrigation facilities, fertilisers, weedicides and pesticides, agricultural credit and extension, greater efforts have been made to enlarge the availability of agricultural inputs and extend its coverage to all the agricultural crops. The developments made in this regard are elaborated in the following paragraphs.

Irrigation

Expansion of irrigation has been the main element of strategy for increasing foodgrains production. Irrigation support has been planned to be provided through major and medium irrigation projects, command area development, tank irrigation and use of ground water. Since minor irrigation schemes are quick maturing and labour intensive by nature, increasing emphasis was laid on the development of minor irrigation structures during the Seventh Plan. The irrigation potential created during the pre-plan period was 22.6 million hectares (9.7 million hectares from major and medium and 12.9 million hectares from minor irrigation projects). By the end of sixth Plan, the cumulative irrigation potential was 67.5 million hectares. The achievement during the Seventh Plan period is anticipated to be about 12.2 million hectares (2.9 million hectares under major and medium projects and 9.3 million hectares under minor irrigation projects) against the target of 12.9 million hectares. Thus the total potential anticipated to have been created by the end of Seventh Plan adds up to 79.7 million hectares.

TABLE 11.8

Development of Irrigation Potential and its Utilisation

(Million Hectares)

	Irrigation (additional area)	Achievement during VII Plan				Cumulative achievement upto VII Plan Plan
		Assumed base level (84-85)	1985-89 (actual Achievement)	1989-90 (Ant. Achi-(evement)	Total VII Plan	
1.	*Major and Medium Irrigation*					
	Potential	30.0	2.0	0.9	2.9	32.9
	Utilisation	25.3	2.0	0.6	2.6	27.9
2.	*Minor Irrigation*					
	Potential	37.5	6.6	2.7	9.3	46.8
	Utilisation	35.2	5.8	2.5	8.3	43.5
3.	*Total*					
	Potential	@67.5	8.6	3.6	12.2	79.7
	Utilisation	@60.5	7.8	3.1	10.9	71.4

@ Cumulative level.

Despite huge investments having been made on creation of irrigation potential, particularly in major and medium irrigation projects, the utilisation of created potential is considerably low. As against an additional potential of 12.2 million hectares anticipated to have been created during the Seventh Plan period, the utilisation is expected to be about 10.9 million hectares. The cumulative gap in utilisation of potential till the end of Seventh Plan works out to about 8.3 million hectares. The gap (5 million hectares) is more in the case of major and medium projects, while it is 3.3 million hectares in the case of minor irrigation projects. The main reasons for this gap are that the farmer takes longer time in adjusting to the revised cropping patterns consequent upon the switchover from dry farming to irrigated farming and delay in the development of on-farm works like construction of field channels, land levelling and adoption of the warabandi system.

A Centrally Sponsored Command Area Development (CAD) Programme was initiated in 1974-75 to bridge the gap between the irrigation potential and its utilisation. The programme broadly covers on-farm development works like construction of field channels, land levelling/shaping, implementation of warabandi for rotational supply of water to ensure equitable and assured supply of irrigation water to each and every farm holding. It also includes arrangements for supply of inputs and credit, agricultural extension, construction of markets and godowns and development of ground water for conjunctive use. The programme at present covers 131 selected irrigation projects with a total culturable command area of around 18.5 million hectares in 20 States and 2 Union Territories, cumulatively, till March, 1990 an area of 11.3 million hectares is expected to have been covered under field channels, 2.0 million hectares under land levelling and 5.2 million hectares under warabandi.

Minor irrigation schemes include ground water and surface water schemes. While ground water schemes include dugwells, shallow tubewells and pumpsets, the surface water schemes include tanks and reservoirs, diversion schemes, lift irrigation from rivers and streams etc. These schemes have been accorded special attention under the Special Foodgrains Production Programme. The Seventh Plan had envisaged the creation of additional irrigation potential of 8.6 million hectares in minor irrigation. As against the target, the likely achievement is 9.3 million hectares.

The progress of irrigation development has not been uniform among the States. Some states such as Bihar, Orissa, Madhya Pradesh, Gujarat and Maharashtra are lagging in the development of major and medium irrigation potential. In respect of ground water development, Punjab, Haryana and Uttar Pradesh have shown excellent progress, while States of Karnataka, Andhra Pradesh, Maharashtra, Bihar, Assam, Orissa, West Bengal, Madhya Pradesh and Kerala lagged behind.

Seeds

In recognition of the importance of quality seeds, efforts are being made to maximise production and distribution of foundation and certified seeds through National Seeds Corporation, State Farms Corporation of India and State Seeds Corporations, the high yielding varieties (HYV) programme introduced in 1966-67 and covering paddy wheat, jowar, bajra and maize, has covered large area under cultivation under different crops and by the end of Sixth Plan the total area covered under the HYV was 54 million hectares. During the Seventh Plan the area covered under HYV increased by 16.7 per cent to about 63 million hectares. The coverage during 1990-91 reached about 67 million hectares. The cropwise coverage of area under HYV is given in a Table 11.9.

TABLE 11.9

Area Under High Yielding Varieties of Seeds

(Million Hectares)

Crop	*1984-85*	*Seventh Plan Target*	*1985-86*	*1986-87*	*1987-88*	*1988-89*	*1989-90*	*1990-91**
Paddy	22.8 (55.3)	32.0	23.5 (57.0)	24.0 (58.4)	22.1 (57.3)	25.4 (60.8)	27.6 (65.4)	29.2 (68.5)
Wheat	19.1 (81.0)	22.0	19.1 (83.0)	19.2 (83.0)	19.7 (85.4)	20.2 (83.8)	20.7 (88.1)	21.9 (90.5)
Jowar	5.1 (32.9)	6.5	6.1 (37.6)	5.5 (34.5)	6.1 (38.7)	6.1 (41.8)	6.8 (45.6)	7.6 (51.7)
Bajra	5.2 (43.7)	6.5	5.0 (46.8)	5.3 (46.8)	4.0 (45.4)	5.9 (49.0)	5.2 (47.7)	5.4 (52.4)
Maize	2.0 (34.8)	3.0	1.8 (31.0)	2.2 (37.0)	2.2 (38.8)	2.5 (42.4)	2.8 (47.4)	2.9 (47.5)
Total	54.1	70.0	55.4	56.2	54.1	60.1	63.1	67.0

*Provisional

Note: Figures in parentheses give the percentage of HYV area to total area under the crop.

The distribution of certified/quality seeds has increased substantially from 25 lakh quintals in 1980-81 to about 57 lakh quintals in 1989-90. In 1990-91 the distribution of certified/quality seeds is expected to be of the order of 60 lakh quintals; the certified seeds will be 45 lakh quintals. The year to year fluctuations in distribution of quality seeds are largely attributed to (a) variation in demand of a specific crop variety, (b) diversification of area from one crop to another and (c) choice of taking up crops which requires less irrigation. As the quality seed (non-certified seed) does not come under the purview of the Seeds Act, 1966, separate monitoring for certified and quality seeds has been started from 1989-90. The distribution of certified/ quality seeds over years can be seen in Table 11.10.

In order to create increased infrastructural facilities commensurate with the projected seed production target and thereby increase the agricultural production, a World Bank aided National Seeds Project Phase-Ill has been launched with effect from March 28, 1990. The thrust of the project is to support efforts to assist the farmers by ensuring timely and adequate availability of certified/quality seeds of suitable varieties at reasonable prices and to provide facilities for the growth of private seed sector through adequate institutional financing. The project will cover initially the States of Andhra Pradesh, Assam, Bihar, Gujarat, Karnataka, Madhya Pradesh, Maharashtra, Orissa, Rajasthan, Uttar Pradesh and West Bengal. The National Seeds Corporation and State Farms Corporation of India will also be covered.

TABLE 11.10

Distribution of Certified/Quality Seeds

Year	*Distribution (lakh quintals)*	*Percentage increase over the preceding year*
1980-81	25.0	—
1981-82	29.8	19.2
1982-83	42.1	41.1
1983-84	45.0	6.9
1984-85	48.5	7.8
1985-86	55.0	13.5
1986-87	55.8	1.5
1987-88	56.3	0.8
1988-89	56.8	0.9
1989-90	57.0	0.4
1990-91	60.0	5.3

A New Policy on Seed Development, introduced with effect from October 1,1988, is aimed at to secure to the farmer high quality seeds available anywhere in the world to maximise his yields, increase productivity and farm income. This is sought to be ensured by permitting access to best available seed by placing planting materials and seeds of vegetables and flowers under O.G.L. and some through regulated imports. As a result of the new policy, there has been significant increase in the import of high quality seeds, particularly those of oilseeds and vegetables.

Fertilizers

Consequent upon the expansion of area under irrigation and greater use of HYV seeds, consumption of fertilizers in the country recorded highest ever quantum jump of 2.3 million tonnes in 1988-89 to reach a level of 11.1 million tonnes of nutrients (NPK). It increased further to 11.7 million tonnes in 1980-90 as against the target of 12.5 million tonnes. The average per hectare consumption of fertilizers increased from 47.4 Kgs. in 1985-86 to an estimated 66.9 Kgs. in 1989-90. In 1990-91, following the third good monsoon year in succession, the consumption is anticipated to increase to 12.7 million tonnes. The details are given in Table 11.11.

In addition to continuance of the existing supporting services and development programmes like ensuring supply of fertilisers at subsidised and uniform prices throughout the country, the other measures taken to increase the consumption include promotion of support services through the District Lead Fertiliser Scheme, extension support by fertiliser industry, opening of additional retail outlets in remote, hilly and inaccessible areas, supply of fertilisers through mobile vans and removal of registration/licensing for small stockists /dealers.

TABLE 11.11

Consumption of Chemical Fertilizers

(Million Tonnes of nutrients)

Year	*Nitrogenous*	*Phosphatic*	*Potassic*	*Total NPK*
1980-81	3.7	1.2	0.6	5.5
1981-82	4.1	1.3	0.7	6.1
1982-83	4.2	1.4	0.7	6.3
1983-84	5.2	1.7	0.8	7.7
1984-85	5.5	1.9	0.8	8.2
1985-86	5.7	2.0	0.8	8.5
1986-87	5.7	2.1	0.9	8.7
1987-88	5.7	2.2	0.9	8.8
1988-89	7.3	2.7	1.1	11.1
1989-90	7.4	3.1	1.2	11.7
1990-91	8.0	3.4	1.3	12.7

Rainfed areas which constitute 70 per cent of the cultivated area consume only about 20 per cent of the total fertiliser. In order to promote fertiliser use and strengthen soil testing facilities in the low fertiliser consumption and rainfed areas, a National Project and Development of Fertiliser Use in Low Consumption Rainfed Areas has been introduced since 1988-89 in certain identified districts of various states.

Consistent with the increased consumption of fertiliser, its production has increased gradually from 3 million tonnes in 1979-80 to about 9 million tonnes in 1988-89. In 1989-90, however, the production recorded was 8.5 million tonnes showing a fall of 4.7 per cent over 1988-89 production. The target for 1989-90 was set are 9.4 million tonnes. The shortfall in production has been primarily due to disruption in supplies of raw materials particularly imported phosphoric acid and ammonia. Other factors responsible were equipment break-down, power and labour problems. The production in 1990-91 is expected at about 9 million tonnes. With the increase in domestic production, the level of imports has declined from 3.6 million tonnes in 1984-85 to about 1.6 million tonnes in 1988-89. However, to meet the shortage, imports increased substantially to 3.1 million tonnes in 1989-90. Following the introduction of fertiliser retention price and subsidy scheme with effect from November 1, 1977, and the increasing trend in production/consumption, subsidy on fertilisers has increased from Rs. 604 crores in 1979-80 to as much as Rs. 4,542 crores in 1989-90. The estimated subsidy for 1990-91 is Rs. 4,388 crores which may increase as a consequence of Gulf crisis and higher petroleum product prices. The details are given in Table 11.12.

Pesticides

Crops including cereals, oilseeds, pulses, cotton and sugarcane are very vulnerable to pest attack. To save the crops from pests, increasing quantity of pesticides are being made available. As against a target of 75 thousand tonnes for the terminal year of the Seventh Plan, the consumption of pesticides was of the order of 52 thousand tonnes in 1985-86, 50 thousand tonnes in 1986-87, 49 thousand tonnes in 1987-88, 55 thousand tonnes in 1988-89 and an anticipated level of about 86 thousand tonnes in 1989-90. In 1990-91, the pesticide demand is placed at about 82 thousand tonnes. However, in the absence of any major pest incidence, the consumption may not be of this magnitude. The overall availability of pesticides in the country is regarded as satisfactory. Besides the increase in domestic production, imports are also being made.

TABLE 11.12

Fertilizers: Production, Import and Subsidies

Year	*Production (N + P) (000 Tonnes)*	*Imports (000 Tonnes)*	*Subsidies (Rs. crores)*		
			On imported fertilisers	*On domestic fertilisers*	*Total*
1979-80	2983	2005	283	321	604
1980-81	3005	2759	335	170	505
1981-82	4093	2041	100	275	375
1982-83	4404	1132	55	550	605
1983-84	4533	1355	142	900	1042
1984-85	5181	3624	727	1200	1927
1985-86	5756	3399	324	1600	1924
1986-87	7070	2310	197	1700	1897
1987-88	7131	984	114	2050	2164
1988-89	8964	1608	201	3000	3201
1989-90	8543	3114	771	3771	4542
1990-91	9044	2758	658*	3730*	4388*

*Provisional

In order to reduce the dependence on the use of chemical pesticides for reducing the crop losses, emphasis has been laid on the adoption of integrated pest management. The integrated pest management is a broad ecological approach that minimises pest population below economic injury level by employing all available pest control techniques such as culture, mechanical, biological, chemical and crop management practices in a compatible manner.

Agricultural Credit

The thrust of agricultural credit policy continues to be on providing adequate and timely credit to farmers, with a focus on small and marginal farmers or the weaker sections, through institutional agencies like co-operatives, commercial banks and regional rural banks in order to support agriculture and allied activities. The major objective of the policy is to enable the farmers specially the small and marginal farmers and other weaker sections to adopt modern technology and improved agricultural practices for increasing production and productivity. The other objective is to provide adequate credit support to areas covered under Dryland Farming, Pulses Development Programme, Special Foodgrains Production Programme (SFPP) and Special Rice Production Programme. In order to achieve the production targets in the selected districts of nineteen States under SFPP, National Bank for Agriculture and Rural Development (NABARD) has taken various steps to provide adequate refinance facilities to the cooperative institutions. The measures, inter alia, include granting of loans to State Governments to enable them (i) to contribute to additional share capital needed by Primary Agricultural Credit Societies (PAGS) and Central Cooperative Banks for managerial assistance to PAGS, (ii) assistance for non-overdues, (iii) blocking of overdues of non-wilful defaulters, (iv) subsidy by the State Governments to small and marginal farmers for purchasing share for becoming members of PACS, and (v) assistance for Agricultural credit Stabilisation Fund of the State Cooperative Banks. Besides, the Government has also provided central assistance to those Central Cooperative Banks which are not eligible to operate credit limits sanctioned by NABARD on account of heavy overdues.

The volume of institutional credit for agriculture has increased from Rs. 2,550 crores in 1979-80 to Rs. 11,225 crores in 1988-89 and further to Rs. 13,022 crores in 1989-90. The target set for 1990-91 was Rs. 13,240 crores. Agency-wise disbursements of agricultural credit are given in Table 11.13.

Agricultural credit is provided at a concessional rate in comparison with commercial rates of interest. The rate of interest till September 21,1990 ranged from 10 per cent per annum for loans upto Rs. 7,500 to 14 to 15.5 per cent for loans over Rs. 25,000. The rate of interest on agricultural loans has been revised with effect from September 22, 1990. The new rates of interest range between 10 per cent on loan upto Rs. 7,500 and 14 per cent on loan exceeding Rs. 50,000. The Produce (Marketing) Loans Scheme has been introduced from December, 1988 on a pilot basis in specified districts in each of the 14 states where SFPP is under implementation. Loans under this scheme are available to those farmers who have availed of crop loans from the credit institutions for raising paddy, wheat, groundnut, rapeseed/mustard, gram and arhar. The quantum of loan under this scheme is however restricted to twice the amount of the crop loan taken by the farmer or 75 per cent of the value of produce at Government announced procurement/support prices subject tomaximum of Rs. 10,000 per individual.

TABLE 11.13

Disbursement of Agricultural Credit

(Rs. crores)

	1985-86	*1986-87*	*1987-88*	*1988-89*	*1989-90*	*1990-91 (Target)*
Cooperatives						
Short-term	2747	2824	3320	3594	4223	4708
Medium-term	393	531	547	602	416	394
Long-term	543	560	690	719	868	957
	3683	3915	4557	4915	5507	6059
Commercial Banks/Regional Rural Banks						
Short-term & Term Loans	3110	3796	3914	6310	7515	7181
Grand Total	6793	7711	8471	11225	13022	13240

Despite phenomenal increase in the overall agricultural credit, the crucial problem of mounting overdues has been inhibiting its expansion to the desired extent. Overdues have been persisting around 40-42 per cent during the last 3.4 years. Increasing overdues have eroded the lending capacity of cooperative institutions for want of recycling of funds and ineligibility of the institutional agencies to borrow additional funds from the higher financing agencies.

In order to enable the farmers to improve the productivity, a scheme, as announced in the Central Budget 1990-91, has been introduced for providing debt relief upto Rs. 10,000 to the non-wilful defaulters of public sector banks and regional rural banks. For banks in the co-operative sector, a scheme on similar pattern was put into operation by State Governments. The Agricultural and Rural Debt Relief Scheme, 1990 which came into force from May 15,1990 covers borrowers of public sector banks and those engaged in agriculture and other allied activities and artisans engaged in any activity of rural development relating to cottage and village industry, handicrafts, weaving, etc. A sum of Rs. 1,500 crores has been released to Reserve Bank of India under the scheme. The disbursement under the commitment of debt relief, in 1990-91 may actually be much higher than the amount provided in the Budget. The debt relief, however, could adversely affect on profitability of banks and erode their lending capacity.

Crop Insurance

In order to provide a measure of financial support to the farmers in the event of crop failure as a result of natural calamity, a Comprehensive Crop Insurance Scheme has been in operation since 1st April, 1985. The scheme is voluntary in nature and covers all the farmers availing crop loans from cooperative, commercial and regional rural banks for raising wheat, paddy, millets, oilseeds and pulses. The premium payable is two per cent of the sum insured for wheat, paddy and millets and one per cent for oilseeds and pulses. Fifty per cent of the premium payable by small and marginal farmers is subsidised which is being shared equally by Central and State Governments.

TABLE 11.14

Progress of Crop Insurance Scheme

Season	*No. of States/ UTs*	*No. of farmers covered (Lakhs)*	*Area covered (Lakh hectares)*	*Sum Insured (Rs. crores)*	*Premium collected (Rs. crores)*	*Claims paid/ payable (Rs. crores)*
Kharif 1985	13	26.4	53.7	542.7	9.4	83.9
Rabi 1985-86	16	12.1	23.2	238.4	4.5	3.1
Kharif 1986	18	39.6	77.4	856.2	15.0	167.9
Rabi 1986-87	17	11.3	21.0	242.4	4.5	4.6
Kharif 1987	21	46.3	84.1	1140.7	19.1	277.2
Rabi 1987-88	19	21.3	32.4	475.4	8.8	12.1
Kharif 1988	13	29.6	52.4	547.9	8.8	29.2
Rabi 1988-89	9	8.7	10.1	164.1	3.1	3.9
Kharif 1989	17	42.3	66.5	873.9	14.5	34.4
Rabi 1989-90	17	6.6	9.6	151.6	2.8	1.8
Kharif 1990	17	19.4	34.1	515.1	7.6	95.4

Coverage under the scheme has been continuously expanding since inception. During the year 1989-90, the scheme was implemented in 17 States and Union Territories against 13 States and Union Territories in 1988-89. In all 42 lakh farmers in kharif 1989 and 7 lakh farmers in rabi 1989-90 were covered as against 30 lakhs and 9 lakhs respectively during the same period of the previous year. However, due to drought conditions and floods in 1986-87 and 1987-88, claims which had shot up from Rs. 84 crores in kharif 1985 to Rs. 277 crores in kharif 1987, came down to Rs. 34 crores in kharif 1989 following the good monsoon. Consequent upon the quantum jump in claims and heavy losses, the scheme was temporarily suspended in January, 1988. But it was re-introduced in September, 1988 with some modifications. The sum insured was, however, reduced from 150 per cent to 100 per cent of crop loans and limited to a maximum of Rs. 10,000 per farmer.

The progress of the scheme can be seen from Table 11.14.

Animal Husbandry

Animal husbandry plays a vital role in the country's rural economy. It is providing gainful employment particularly to small/marginal farmers and agricultural labourers, and other rural poor. It also supplements the nutritional deficiency in human diet. The gross value of output from this sector has increased from Rs. 280 billion in 1987-88 to Rs. 315 billion in 1988-89 which is about 24 per cent of the total agricultural output of Rs. 1335 billion. The animal husbandry programmes are aimed at meeting the requirement of milk, eggs, meat, wool, draught power and rural transport. Efforts are being made to improve the vast population of livestock by providing basic infrastructure and adopting improved technology.

With its large livestock population, the country has vast potential for meeting the growing needs of animal protein and raw material for the industry. The main objectives of the Seventh Plan in this direction were (i) to consolidate the gains achieved under various programmes of animal husbandry, (ii) to provide the infrastructure necessary to achieve accelerated growth in livestock production and (iii) to enable as large a section of rural population as possible, including small and marginal farmers, agricultural labourers and other rural poor to improve their nutritional and economic status by providing them with gainful employment through livestock rearing. As a result of the developmental efforts, production of milk and eggs, has increased significantly. Milk production during the Seventh Plan increased by about 24 per cent from 41.5 million tonnes in 1984-85 to 51.4 million tonnes in 1989-90. Production of eggs on the other hand increased by more than 41 per cent from 14,252 million eggs in 1984-85 to over 20,204 million in 1989-90. The per capita availability of milk has risen from 56 kgs in 1984-85 per annum to 63 kgs. per annum in 1989-90.

Dairy Development

As complementary to agriculture in India dairying provides supplementary income to vast majority of farmers, most of whom belong to the small or marginal category. In order to provide a linkage between the producers and the consumers, a rational approach to dairy development has been evolved. An integrated dairy development programme commonly known as Operation Flood was launched in 1970 to correct the deficiencies of the dairy sector by organising village milk producers in the dairy cooperative societies. The first phase, Operation Flood-I, was aimed at capturing a liquid milk market in the four metropolitan cities by linking 27 milk sheds. During the second phase, Operation Flood-II (1980-85), the programme was extended to almost all the states in the country—about 34,500 dairy cooperative societies had been organised in 136 milk sheds. Operation Flood-III (1985-94) is now under implementation with financial assistance from the World Bank and commodity assistance from EEC in the form of skimmed milk powder and butter oil. In all 62,800 dairy cooperative societies have been organised in 174 milk sheds involving about 7.4 million farmer members to procure 9.7 million kgs and marketing 8.0 million litres of milk a day in over 523 cities and towns of the country as against total milk processing capacity of 14.6 million litres a day and milk powder production capacity of 667 tonnes a day. A technology Mission for Dairy Development has been launched in order to systemetise and coordinate dairy development with various field level development programmes. The Mission will also pursue specific research targets and focus on policy issues that will contribute to sustained dairy development.

Fisheries

Fisheries have been recognised as an important source of food, foreign exchange earner and employment. With various developmental measures, fish production has gradually increased from 7.5 lakh tonnes in 1950-51 to a record of 38.2 lakh tonnes in 1990-91. The increase in 1990-91 came both from the marine sector and inland sector. The export earnings in 1990-91 were of the order of Rs. 875 crores. The main thrust of the fisheries development programme is promote the extensive and intensive fish farming activity in the inland sector and encouraging deep sea fishing in the marine sector so as to optimise fish production. In recognition of the important role of inland fisheries in the overall production of fish, the Government has been implementing two important inland fisheries development programmes since the Sixth Five Year Plan. These are (a) development of aquaculture and (b) National Programme of Fish Seed Development. These programmes have led to a significant increase in inland fish production; its production in 1990-91 was about 14.7 lakh tonnes. Its share in the total production increased from 36 per cent in 1980-81 to 38 per cent in 1990-91. There has been a quantum jump in marine fish production from 18.2 lakh tonnes in 1988-89 to 22.8 lakh tonnes in 1989-90. In 1990-91, marine fish production was 23.5 lakh tonnes.

Agricultural Marketing

Regulation and development of agricultural markets, grading and standardisation of agricultural and allied commodities, assistance for creation of infrastructural facilities in agricultural produce markets and assistance for setting up of rural godowns are the major activities under agricultural marketing. The role of cooperatives in the marketing of agricultural produce has been progressively expanding. The public sector commodity corporations like the Food Corporation of India, Cotton Corporation of India and the Jute Corporation of India whose social objectives converge with those of cooperatives, have been increasingly utilising the services of cooperatives to procure specified agricultural commodities from the farmers. To help the farmers in getting remunerative prices, cooperatives have assumed a major role in price support operations in coarsegrains, oilseeds, copra and pulses and in market intervention for potato, onion, ginger and apples.

The network of marketing cooperatives exist almost in every mandi in the country. At present there are 2,636 general purpose primary cooperative marketing societies and 3,290 special commodity cooperative marketing societies dealing in oilseeds, cotton and fruits and vegetables. In addition, there are 157 District/Central Societies, 29 general purpose State Co-operative Marketing Federations, 16 State level Special Commodity Marketing Federations with the National Agricultural Cooperatives Marketing Federation (NAFED) at the national level. The turnover of the cooperatives in 1989-90 has increased to Rs. 6,282 crores as against Rs. 5,415 crores in 1988-89; these remained much above the targetted levels. The marketing operations of cooperative marketing societies are mainly in foodgrains; the share of other agricultural produce is very small. The NAFED has been appointed as a Central Nodal Agency for under-taking marketing operations of a number of agricultural commodities covered under the price support operations. Such purchases, arising from the price support operations, are being made by NAFED through the State designated agencies i.e. State Marketing Federations. In case of agricultural commodities which do not fall within the purview of price support operations, market intervention scheme is announced on the specific request of the State Governments and losses in these operations are shared on 50:50 basis between the Centre and State Governments. The market intervention scheme and the price support operations enabled the small/marginal farmers to obtain remunerative prices by making use of the operations launched by NAFED.

Outlook

The target for the production of foodgrains in 1990-91 was fixed at 176.5 million tonnes (99.1 million tonnes in the kharif season and 77.4 million tonnes in the rabi season). The good south-west monsoon, distributed well over time and space, would have exercised a favourable impact on foodgrains production in the kharif season. The confluence of favourable weather factors and the increased utilisation of inputs, such as seeds and fertilizers, suggests that the kharif foodgrains production may be around 100.8 million tonnes. While kharif rice output may be around 67 million tonnes, output of kharif coarsegrains and kharif pulses may be about 28 million tonnes and 5.4 million tonnes respectively. There may be some set-back in kharif oilseeds production in view of the absence of initial rains in the Saurashtra region of Gujarat resulting in a sizeable reduction in area under groundnut cultivation. However, other kharif oilseeds, and even groundnut, have fared better in other regions producing oilseeds. In total, the loss of kharif oilseeds production may, therefore, be marginal. The shortfall in the kharif season, if any, is expected to be offset by an increase in rabi oilseeds output. Cotton, sugarcane and jute have also benefitted from the conducive seasonal factors. The targets set for sugarcane and jute may be achieved. The output of cotton may register a small decline, because of a loss in production in the major cotton producing districts of Punjab and Haryana. Given the behaviour of post-monsoon rains, together with an increased level of water storage in reservoirs and the availability of inputs, foodgrains output in the rabi season would attain the expected level. The total production of foodgrains in 1990-91 is, therefore, expected to be slightly higher than the targetted level and may be about 177 million tonnes.

12

THE FOOD PROBLEM

"What we have to cure is not a crisis but a chronic malady and the remideis for it are of a routine rather than of a spectacular nature."

—Professor M.L. Dantwala

Planning and economic development sold hollow indeed when the common man is not able to procure even the minimum amount of food at reasonable prices. As the Ford Foundation Agricultural Production Team stated, "Without food enough, India's hopes for improving human welfare, achieving social justice and securing democracy will become almost impossible of attainment." Probably the best description of food problem which India has been facing, not for one or two years, but for decades, is to regard it as a "chronic malady". As Professor Dantwala states: "It is my considered view that what we have to cure is not a crisis but a chronic malady..."

THE FOOD PROBLEM AND FOOD POLICY

India's food problem, properly speaking, dates back to the separation of Burma from India in April, 1937. With the separation of Burma India had to import 15 to 20 lakh tonnes of rice from Burma to augment internal production and to regulate prices. When Burma fell to Japan, Burmese rice supply was cut off and India had no alternative source of supply of rice. India faces its first serious food crisis in the form of the severe Bengal famine of 1943 when millions perished due to starvation. The Government of India appointed the First Foodgrains Policy Committee (in 1943) to advice it on the adoption of a proper food policy. The bring about equitable distribution of scarce food supplies, the Government introduced rationing on the recommendation of the Committee. By the time India became independent, nearly 54 million people were bought under statutory rationing and another 90 million people came under other forms of public food distribution.

The Partition of the country in 1947 further weakened India on the food front, as India received 82 per cent of the population, but only 75 per cent of acreage under cereals and 69 per cent of irrigated area. West Punjab and Sindh were surplus areas and used to supply about 10 to 12 lakh tonnes of foodgrains every year to other States. While separation of Burma forced India to import rice, the partition of the country made India dependent on foreign countries for wheat also. The important implications of the food policy inherited by the Government of India at the time of independent were as follows:

(a) extension of the rationing system made the Government committed to feeding all the consumers;

(b) procurement of internal supplies failed to keep with the expanding Government commitments making it necessary to import foodgrains;

(c) imports of foodgrains steadily increased from 1.6 million tonnes in 1944 to 2.7 million tonnes in 1947: and

(d) imports had to be paid for in foreign exchange; further, as imported foodgrains were more expensive than the locally produced foodgrains, the Government had to subsidies the distribution of foreign foodgrains.

All these four aspects of the food policy were not fully appreciated then. But they are highly significant even now.

Before the Era of Planning

In September, 1947, the Government appointed the Second Foodgrains Policy Committee to examine the food situation afresh. The 1947 Policy Committee clearly expressed the view that under conditions of overall shortage of foodgrains, it was impossible for any government to undertake commitments to consumers on any scale by means of procurement of internal supplies alone and that growing dependence on imports was built into the system of controls and rationing as practiced by India. Hence, the 1947 Committee recommended a new approach to the food problem, to be undertaken immediately and on a war footing : (a) dependence on imports of foodgrains to be liquidated by orderly and planned stages, and (b) the commitments undertaken by the Government to the consumers to be similarly liquidated in equally planned stages. According to the 1947 Committee, only a substantial increase of internal food production within the earliest possible time could solve the Indian food problem and as a first step the Committee recommended decontrol. The Government accepted the recommendations of the Committee and introduced decontrol. Prices rose in the wake decontrol and the Government got frightened and reintroduced control. The recommendation of the Foodgrains Policy Committee (1947) were not given proper trial.

The Government of India appointed the Foodgrains Procurement Committee (1950) to examine the food question again. According to the 1950 Procurement

Committee, the food policy faced many difficulties because public procurement did not match with the commitments of the public distribution. The 1950 Committee argued that the failure to procure from the producers their entire marketable surplus was directly due to the existence of a free market with a free market price ruling above the procurement price; this, in turn, led to increased difficulties in procurement. The Committee recommended a logical, consistent and complete system of controls which would leave no room for a free market whatsoever—monopoly procurement at the village level, abolition of free market and of free movement and complete statutory or near-statutory rationing everywhere. The commendations of the 1950 Procurement Committee were rejected as the Government believed them to be administratively and politically impracticable.

The 1947 Policy Committee and the 1950 Procurement Committee debated the food question of Government commitments through the public distribution system and the difficulties of public procurement and recommended exactly opposite policy measures; viz., total control with complete or near complete abolition of free market for foodgrains and total decontrol with complete free market in foodgrains. The Government found both the alternatives "politically" not feasible and found it expedient to have partial procurement, some rationing, some informal rationing, and some degree of free market. Such a food policy gave the impression that the Government was able to satisfy the conflicting interests of the consumers and producers and was somehow managing to solve the problem on an and hoc basis. But it was clear that the Government's food policy was undeniable and ineffective because (a) the policy was incapable of controlling the prices of foodgrains when heavy pressures were built up to push up prices; and (b) the policy forced the country to bet for food aid from abroad and make India dependent on large food imports.

Food Policy Since 1950-51

The Government of India was eager to reduce our dependence on foreign food and food aid, and self-sufficiency in food was thus made one of the goals of the First Five Year Plan.

The end of the Korean war as well as the end of the worldwide commodity boom and successively good harvests in the country with decline in foodgrains prices afforded an opportunity for the Government to relax the controls and later remove them altogether. The policy was successful at least for the time being and there was a general sigh of relief that the food problem was almost solved. The favourable position on the food front was clearly reflected in the movement of food prices during the period—food articles registered a decline of 23 per cent during the First Plan period, as compared to a decline of 3.5 per cent by the manufacturers. The Government was extremely satisfied with the food position at the end of the First

Plan and the Planning Commission stated : "The increase in agricultural production had helped to end inflation, stabilise the economy and prepare the way for a higher rate of development during the Second Five Year Plan."

The food policy of decontrol was successful at least for the time being and there was a general sigh of relief that the food problem was almost solved. But this feeling was short-lived, as the basic problem of food shortage and rise in prices again emerged in 1955. In that year, the millet crop was poor and even from October, 1955, the prices of foodgrains took an upward turn. In the beginning, the rise in prices was slow but soon it was rapid. There were reports of local scarcity in certain parts of Rajasthan, Bombay, Bihar, Orissa and Tamil Nadu, mainly caused by drought, floods and cyclone. Throughout 1956 the rise in prices gathered momentum. By 1958-59, the food problem became acute and assumed the characteristics of a crisis But the strange aspect of the problem was that throughout this period there was increase in foodgrains production. In previous periods the food problem consisted of shortages of internal production to meet internal demand and the necessity to import foodgrains to fillip the gap. But foe the first time, during 1958-59, a new situation arose Internal production and imports together were considerably higher at 77 million tonnes in 1959 as compared to 67 million tonnes in 1955. There was no food problem in 1955, but veritable "food crisis" developed in 1959. The basic difference, however, was in the matter of foodgrains prices which rose by 41 per cent in the first three years of the Second Plan. So long as food was available at reasonable prices people believed that there was no food problem, but when the prices of foodgrains rose rapidly, they felt panicky even though there were adequate stocks.

The basic causes for the rise in foodgrains prices during this period were the increased purchasing power in the hands of the people short-fall in the production of major crops, extensive hoarding and speculation by the producers and the wholesalers, etc. The Government sought to control the food situation through various steps. Selective credit controls were used by the Reserve Bank of India for the first time to check hoarding of foodgrains. But the main thrust was through larger imports and the use of the public distribution systems.

New Food Policy Since 1956

The Government of India entered into an agreement in 1956 with the U. S. A—known as the PL-480 agreement—for the import of 3.1 million tonnes of wheat and 0.19 million tonne of rice for the next three years. The Government found the PL-480 food imports a good tool to stabilise food prices in the country, since it could release unlimited quantities of imported wheat at the lowest possible price to create a psychological feeling of abundance and to bring down the price of foodgrains in the country.

The Government of India appointed the Foodgrains Enquiry Committee in 1957, which completely endorsed the Government's food policy. Recognising that the "food problem was likely to remain with us for a long time to come", the 1957 Enquiry Committee asserted categorically that "assurance of continued imports of certain quantities of foodgrains will constitute the very basis of a successful food policy for some years to come." The 1957 Enquiry Committee discussed the two possible alternatives of food policy—full control and complete decontrol—and rejected both and concluded, "The solution to the food problem, in our view, lies between complete free trade and full control." Thus, a stable and long-term food policy based on heavy imports emerged and it promised to be a successful one on account of the following factors:

(a) It was politically acceptable.

(b) It did not cost the Government anything; in fact the Government could acquire rupees for development from food imports.

(c) Administratively, it was easy to operate, importing foodgrains and releasing the same through the public distribution system of fair price shops at extremely low prices to the consumers.

(d) In practice, there would be complete free trade within the country.

(e) The vulnerable sections were fully protected from the high prices of domestic foodgrains.

The Government's food policy was regarded so "successful" that the import of 3 million tonnes for 3 years under the 19567 PL-480 agreement was used up in one year. There were two more agreements in 1958 and 1959 and India acquired the capacity of absorbing nearly 4 million tonnes of imported foodgrains every year. In 1960, India signed a spectacular agreement with the U. S. A. to import 16-million tonnes of wheat and one million tonnes of rice for the next 4 years. Price of foodgrains would be keep constant; and even if prices rose, due to shortfall in internal production, the vulnerable sections of the population would be fully protected from such a rise-they could draw their requirements from the fair price shops at low and nearly fixed prices.

For a whole decade, between 1956-66, the food policy of the Government of India was based on imports from U. S. A. None of the policy makers could foresee the dangerous implications of our humiliating dependence on foreign food. The irony of the whole situation lay in the pious hope of the policy makers in New Delhi that the food surplus of the U. S. A. would continue for years to come and that India could plan its industrial growth on American food surplus.

Failure of the New Food Policy during the Third Plan

The food policy formulated in 1956 and strongly endorsed by the Foodgrains Enquiry Committee, 1957 crumbled during the Third Plan, particularly after 1962-63. In that year and in the next, domestic production of foodgrains fell. Through there was a record crop during 1964-65, production of foodgrains fell to the worst possible level during the last year of the Third Plan. The production figures are given in Table 12.1.

TABLE 12.1

Production of Foodgrains

	1960-61	1964-65	*(million tonnes)* 1965-66
Cereals	69.3	76.6	62.2
Pulses	12.7	12.4	10.0
All foodgrains	82.0	89.0	72.2

Source: *Indian Agriculture in Brief*, 1971.

TABLE 12.2

Imports, Procurement and Public Distribution of Foodgrains Since 1956

Year	*Net Production of foodgrains (million tonnes)*	*Imports (million)*	*Procurement (million tonnes)*	*Public Distribution (million tonnes)*	*Imports as percentage of public distribution*
1950	60.7	1.4	0.1	2.1	66
1961	72.0	3.5	0.5	4.0	88
1962	72.1	3.6	0.5	4.4	82
1963	70.3	4.6	0.8	5.2	88
1964	70.6	63	1.4	8.7	72
1965	78.2	7.5	4.0	10.1	74
1966	63.3	10.4	4.0	14.1	74

Source: Economic Survey, 1973-74.

As the figures show clearly, the output in the last year of the Third Plan (1965-66) was a little more than 72 million tonnes, less by 17 million tonnes as compared to 1964-65 and less by 10 million tonnes as compared to 1960-61. While food production fell, factors affecting demand, specially rise in the volume of purchasing power in the hands of the people, black money with unsocial elements, hoarding and blackmarketing by traders and middlemen, etc., were pressing very hard on the food front. The result was severe shortages and inflationary pressures within the country. While general prices rose by 38 per cent during the Third Plan, food prices rose by 50 per cent, and many particular items of food rose by still higher percentages. Bihar, U. P. and many other parts of the country came under the grip of one of the worst famines this country had ever witnessed. The Government tackled the food problem, almost on a war footing,' imported huge quantities of foodgrains from abroad and saved precious human lives. As was to be expected, more and more people left the free market and consequently, the demand on the public distribution system increased rapidly. The relevant figures are given in Table 12.2.

The public distribution of foodgrains amounted to 5.2 million tonnes in 1963, 8.7 million tonnes in 1964, 10.1 million tonnes in 1965 and 14.1 million tonnes in 1966. To meet the growing demand of the public distribution system, imports too were increased-4.6 million tonnes in 1963, 6.3 million tonnes in 1964, 7.5 million tonnes in 1965, and 10.4 million tonnes in 1966. Procurement of foodgrains within the country was also increased as is clear from Table 12.2. However, between 75 and 80 per cent of foodgrains distributed through the public distribution system came via imports. For the first time, however, the Government realised that even huge and assured imports failed to halt the rise in the prices of foodgrains or to protect the vulnerable sections of the society. The Government further realised the dangerous implications of our humiliating dependence on imported food and the necessity to formulate a food policy of distribution what is produced within.

Integrated Food Policy Since 1966

The Government of India set up yet one more Foodgrains Policy Committee in 1966 to review the food problem afresh. The Committee found India's dependence on food imports as anything but reassuring, since imports from the U. S. A. were not likely to be either large or easy in further but would have to be paid for in dollars and not rupees. The Committee took a serious note of the fact that the food aid was used openly to influence the internal economic policies and foreign affairs policies of the Government. The 1966 Policy Committee did not like India's continued dependence on foreign imports and recommended the preparation and implementation of a National Food Budget—involving a national plan of supply and distribution of foodgrains. The four aspects of the National Food Budget which constituted the integrated food policy were :

(a) procurement to ensure necessary supplies; (b) control on inter-State movement to facilitate procurement and keep prices at a reasonable level; (c) a system of public distribution to ensure equitable sharing; and (d) the building up of a buffer stock to provide against difficult years.

It is interesting to observe that the so-called intergrated food policy was more or less the same policy which was in operation between 1943-47 and with some experiment with decontrol, up to 1952. This policy, as shown earlier, came into difficulties because public procurement did not match the commitments of the public distribution. While the 1950 Procurement Committee ascribed this failure to the existence of free market and free market prices and, therefore, recommended the abolition of the free market altogether, the 1966 Policy Committee did not accept this diagnosis and, accordingly, recommended only partial procurement and partial distribution but permitted clearly private trade with a free market price. The integrated food policy (1966) faced the same problems as those of the previous policy, viz.

(i) Would public procurement by adequate to meet fully the requirements of the public distribution system? and

(ii) Would not the existence of the free market create the same difficulties as it had earlier secreted?

If imports had to be ended, public procurement would have to be sufficient to meet the requirements of the public distribution system, or the public distribution system should make only such commitments which could be satisfied through public procurement.

Besides, the essence of the integrated policy is the preparation and implementation of the National Food Budget. For this purpose, the requirements of the deficit states and the amounts that suplus states could procure should be estimated in advance. The success of the National Food Budget would obviously depend upon whether:

(a) the deficits of the deficit states would be matched by the surplus of the suplus states; and

(b) the surplus of the surplus states; would be mobilised into central pool for distribution to deficit states.

If the National Food Budget did not balance in any year, the Government of India would be naturally forced to import foodgrains or beg for food, as it had done in earlier years.

Annual Plans and the Fourth Plan

The Government of India accepted the recommendation of 1966 Policy Committee. Fortunately for the Government, internal production of foodgrains picked up very well from 1967-68 onward. Imports were gradually reduced; internal procurement was increased; and large buffer stocks were built up. The production of foodgrains during the Annual Plans and the Fourth Plan is given in Table 12.3.

TABLE 12.3

Production of Foodgrains

Year	*Cereals*	*Pulses*	*(million tonnes)* *Total Foodgrains*
1967-68	83	12	95
1968-69	84	10	94
1969-70	88	12	100
1970-71	96	12	108
1971-72	94	11	105
1972-73	87	10	97
1973-74	95	10	105
1974-75	90	10	100
1975-76	108	13	121
1976-77	100	11	111
1977-78	114	12	126
1978-79	119	12	132
1979-80	101	8	109

Source: Economic Survey 1974-75 and 1980-81.

As compared to the output of 76 million tonnes of 1966-67, the production of foodgrains was a record of 95 million tonnes in 1967-68. The output of foodgrains during the first two years of the Fourth Plan (i.e., 1969-70 and 1970-71) was quite encouraging viz., 100 million and 108 million tonnes respectively. Visions of food self-sufficiency and even the possibility of exporting foodgrains were openly discussed. But the successive bad crops of 104 million tonnes and 95 million tonnes of 1971-72 and 1972-73 respectively created panic again on the food front. The decline in agricultural production in 1972-73 was particularly severe — the output in that year was 8 million tonnes less than that in the previous year and 11 million tonnes less than the year before. The revival in 1973-74 was only marginal. But the food situation has been fluctuating from year to year. Revival in 1973-74, decline in 1974-75 and big bumper crop in 1975-76 (121 million tonnes), a decline again in 1976-77, bumper crops in 1977-78 as well as in 1978-79 and a steep decline in 1979-80 and this has been the picture in our food situation in recent years.

Table 12.4 brings out clearly the gradual decline in imports till 1972, corresponding increase in internal procurement and the increasing role of the public distribution system. Since 1972, imports of foodgrains have been speeded up and public distribution system has been put on a permanent basis. As the prices of foodgrains have been rising considerably in the free market, even the higher income groups in urban areas leave the free market and enter the public distribution system. It was at this time that the Government of India took a major step in the form of take-over of wholesale trade in wheat.

TABLE 12.4

Net Availability, Procurement, Imports and Public Distributions

(in million tonnes)

Year	*Net production of foodgrains*	*Imports*	*Net availability of foodgrains*	*Procuretent*	*Public Distribution*	*Percentage of Procurement to Public Distribution*
1967	65.0	8.7	73.0	4.5	13.2	34
1968	83.2	5.7	86.8	6.8	10.2	67
1969	82.3	3.9	85.7	6.4	9.4	68
1970	87.1	3.6	89.5	6.7	8.8	76
1971	94.9	2.1	94.3	8.9	7.8	113
1972	92.0	0.5	96.2	7.7	10.5	73
1973	84.9	3.6	88.9	8.4	11.4	73
1974	91.0	4.9	96.8	5.7	10.8	53
1975	87.3	7.4	89.4	9.6	11.3	85
1976	105.9	6.4	102.1	12.9	9.2	140
1977	97.3	0.4	98.9	9.9	11.7	85
1978	110.6	—1.0	109.6	11.1	10.2	110
1979	115.4	—0.9	114.1	13.8	11.7	120
1980	95.2	—0.4	100.5	11.1	14.8	80

* Net availability = Net production+Net imports-Change in Govt stocks.

Source: Economic Survey,. 1976-77 and 1980-81.

Take-over of Foodgrains Trade, 1973

The take-over of wholesale trade in foodgrains was first recommended by the Foodgrains Enquiry Committee (1957) for the purpose of stabilisation of prices. But this recommendation was ignored by the Government. The decision to take over the wholesale trade in foodgrains was taken by the Congress Party in its Bombay session in December, 1969 and was reiterated at the Plenary Session of the Congress Party at Calcutta in December, 1972. The Central Government set up a Committee of Ministers of surplus and deficit States headed by the Union Agriculture Minister to suggest ways and means of reforming administration—financial, transport, storage and other facilities—in the wake of the take-over of the wholesale trade in wheat and rice. It was agreed that the wholesale trade in wheat should be taken over from the rabi season of 1973 and that of rice from the kharif season of 1973-74. Accordingly, wholesale trade in wheat was taken over from April 1, 1973.

The major objectives of take-over were (i) to assure regular supplies of essential commodities to consumers at reasonable prices; (ii) to stabilise the price level of foodgrains at a level which covered the cost of production and provided adequate incentives to producers to increase their production; (iii) to remove large fluctuations in prices in response to local or seasonal variations in supply and demand; (iv) to prevent the unsocial activities of the wholesalers and middlemen; and (v) to eliminate middlemen so as to help the Government in procurement (for the farmers would have to sell their surplus grains to Government agencies) and to maintain the flow of foodgrains through the public distribution system.

There was, however, considerable opposition to the Government take-over of foodgrains trade. The All India Foodgrains Dealers' Association requested its members to suspend their trade in other commodities also with effect from May, 1973, so that the whole system might fail miserably. Important criticism to the Government takeover of wholesale trade in foodgrains were as follows:

(i) The State take-over was not related to its fight against monopolies or the attainment of socialistic objectives,

(ii) Lakhs of wholesalers operated in hundreds of mantis all over the country and it was impossible to manipulate prices to the disadvantage of farmers and consumers,

(iii) It eliminated from trade 3,500 mandis and nearly 5 lakhs of people working for them, and the Government did not have the experience or suitable machinery for such a complicated, specialised and far-flung commercial operation,

(iv) The price rise was due to general inflation, severe drought, a psychosis of shortage and retention of stocks by the cultivators themselves. These factors would continue in the future, and therefore, the take-over would not solve the problem of providing adequate supply to the consumers at reasonable prices,

(v) The take-over of wholesale trade would further aggravate inflationary pressures in the country. For one thing, the Government would raise nearly Rs. 3,000 crores for buying suplus wheat and rice from the market. For another, the usual wholesale traders in foodgrains would be left with their funds to hoard and speculate coarse grains, spices, etc.

(vi) Nationalisation would prove expensive to the consumers as Government handling of foodgrains would be much more expensive,

(vii) The Government would pay only fixed prices for wheat or rice. Naturally, good quality grains would not be supplied to the Government but would somehow pass into the black market,

(viii) The Central and State procurement and distributive agencies were not equipped to handle anything between 15 and 20 million tonnes of foodgrains when private trade was removed from foodgrains trade in general,

(ix) The major attack was that the Government did not have the necessary machinery to handle trade of such huge dimensions, that the past record of government's performance as regards procurement and distribution of essential foodgrains through the public distribution system was extremely poor, and that this measure was found to affect the poor income groups adversely, for whom it was primarily meant.

The State take-over of wheat from April, 1973 resulted in utter chaos and confusion in the wheat-growing States. Despite a good rabi crop of the order of 30 million tonnes, business had come to a standstill and wheat prices had scaled new peaks. The public distribution system failed in most parts of the country. Strong opposition by wholesale traders and the rich farmers, with support given by the political parties and internal dissensions in the ruling party itself, forced the Government of India to scrap the take-over of wholesale trade in wheat and announce its new policy.

The New Food Policy, 1974

In April 1974, the Government of India announced a new food policy under which the Government fixed the wheat procurement price at Rs. 105 and the issue price of Rs. 125 per quintal in the fair price shops. Private traders were allowed to enter the market along with the Food Corporation of India (FCI) and the State Food and Civil Supplies departments. The Government fixed the maximum price of wheat at Rs. 150 per quintal and permitted private traders to sell within a state or outside. The new food policy provided a good price to the farmers so as to discourage their hoarding. Besides, the policy was expected to create favourable conditions for equitable distribution of foodgrains in the country as the availability of wheat particularly in the deficit areas would improve. Finally, it was felt the public distribution system would be able to meet the reasonable requirements of the people.

Even though the new food policy of the Government was criticised by different sections of the people, it was relatively successful. The prices of foodgrains started falling after September 1974 under the impact of Government's monetary and fiscal measures. At the same time, the anti-hoarding and anti-smuggling measure taken by the Government both before and after Emergency were also responsible for the reduction in the prices of foodgrains.

Present Food Policy

Soon after its assumption of power in March 1977 at the Centre, the Janata Party announced a change in food policy. The salient features of this policy were as follows:

(i) The precument price of wheat was raised to Rs. 110 per quintal; but the issue price of wheat to the consumers continued to be Rs. 125 per quintal.
(ii) Price support was also extended to barley and gram as Rs. 65 and Rs. 95 per quintal respectively.
(iii) No procurement target for wheat was fixed and the system of bonus as incentive to surpass the procurement targets was scrapped.
(iv) All inter-state restrictions on wheat movement were removed.

The new policy was a compromise policy. The price at which the consumer bought wheat in the ration shops would remain the same. In turn, other prices, the movement of which was determined largely by food prices, were expected to remain stable. With the abolition of interstate movement restrictions, the price of wheat would even out, with rise in surplus areas and fall in deficit areas. Prices would reflect the demand supply relations. There would be no scope for smuggling too.

FACTORS AFFECTING FOODGRAINS PRICES

The foodgrains prices in India are influenced by production of foodgrains, marketed suplus, import of foodgrains, growth of population and urbanisation, changes in consumer's preferences, level of income and income elasticity of demand for food, stock holding and hoarding and speculative activity. All these factors can be conveniently grouped under three headings: factors affecting demand for foodgrains, factors affecting supply of foodgrains and impact-of Government policy.

1. Factors on the Demand Side

Four general factors have been responsible for rise in foodgrains prices. They are : increased tempo of investment since 1955, deficit financing, expansion of bank credit and the rise of population.

(a) *Planned investment.* Though planned investment started much earlier, yet it was only after 1955 that it had a positive influence on the price level. For instance, from about Rs. 340 crores in 1953-54, annual investment rose to nearly Rs. 2,000 crores during the Third Plan period and to Rs. 3,000 crores during the Fourth Plan period, Rs. 7,000 crores during the Fifth Plan and is likely to be Rs. 20,000 crores during the Sixth Plan period. Increased investment led to increased income and purchasing power in the hand of the public, increase in demand for foodgrains, among other things, and consequent rise in prices of foodgrains.

(b) *Deficit financing.* While the volume of investment was increasing, there was an increased volume of deficit in planned expenditure. Deficit financing was of the order of Rs. 333 cores during the First Plan, about Rs. 950 cores during the Second Plan, about Rs. 1,135 crores during the Third Plan and Rs. 2,740 cores during the Fourth Plan. Thus, there has been considerable amount of expenditure which has been financed by new money. Deficit financing has had tremendous inflationary effect, through its influence on money supply.

(c) *Bank credit and money supply.* Since 1951 there has been a continuous expansion of bank credit in the country. Bank credit of all commercial banks rose from Rs. 580 crores in 1951-52 to Rs. 1,500 crores in 1960-61, Rs. 4,800 crores in 1970-71 and Rs. 16,000 crores in 1978-78. Deficit financing and expansion of bank credit led to increase in money supply with the public. In 1951-52, money supply was Rs. 1,850 crores. But by 1960-61 it rose to Rs. 2,870 crores and by 1977-78 to Rs. 18,100 crores. Expansion in money supply and bank credit push up prices.

(d) *Increase in population.* India's population has been rising rapidly. From 361 million in 1951, India's population rose to 439 million in 1961 and 547 million in 1971. The population in 1981 is 684 million. The rate of increase in population was 2.2 per cent, 2.5 per cent and 2.4 per cent per year during the last three decades respectively. An additional population of about 10 to 11 million people has to be fed every year, in addition to providing more and better foodgrains to the existing population. The continuous and persistent increase in population poses the most serious threat to the country.

Other factors on the demand side which have been pushing the foodgrains prices continuously upward since 1961 are as follows:

(i) High income elasticity of demand for foodgrains of the bulk of the Indian population who live on the marginal level,

(ii) Change in the volume and pattern of food consumption of many sections of people and a clear shift from coarse grains to superior grains,

(iii) Increase in population and growing urbanisation due to the tempo of industrialisation and increase in the demand for rice and wheat as against millets,

(iv) Hoarding of foodgrains at all levels, viz., by producers, traders and even consumers,

(v) Speculation in foodgrains by the traders because of the conditions of instability in the markets.

(vi) Extensive use of black money by wholesale traders to corner the stock of foodgrains.

2. *Factors on the Supply Side*

While factors operating specifically on the demand of foodgrains are responsible for the rising trend in foodgrains prices in the country, changes in the supply position are also important. In fact, sometimes the changes in the supply position of individual crops usually initiated changes in prices. The supply factors which contribute to the food problem are :

(i) *Production.* A very important factor for the continuance of food problem in India is the failure of food production to increase side by side with increase in demand. After record food crop of 80 million tonnes in 1964-65. there were two consecutive famines, considered to be the worst in the 20th century. The output of foodgrains was 72 million tonnes and 74 million tonnes in 1965-66 and 1966-67 respectively. While the demand for foodgrains had been rising rapidly, the production of foodgrains had declined in this period. Naturally, prices of foodgrains rose, creating a veritable food crisis. Production of foodgrains since 1967-68 had been consistently rising and touched a record output of 108 million tonnes in 1970-71. Since then, output of foodgrains fluctuated badly touching as low as 97 million tonnes in 1972-73 and as high as 132 million tonnes in 1978-79. Decline in food production and difficulty to augment internal supplies through imports are basic factors for our food problem.

(ii) *Marketed surplus.* More important than the output is the marketed surplus, i.e., the amount which the farmers put in the market. It is generally known that the farmers have now greater holding power and that they stock their output with the idea of selling at a higher price at a later date. The failure of wheat take-over in 1973 and the subsequent wheat policy is due to the refusal of the farmers to sell their surplus foodgrains in the market. Compulsory procurement or a system of compulsory levy will be useful if it can be enforced. Generally, however, increase in production of foodgrains has led to increase in market arrivals as well.

(iii) *Imports.* Imports of foodstuffs help to overcome the shortage of foodgrains. As a result of inadequate output within the country, the Government increased imports of foodgrains. For instance, as against net import of 3.5 million tonnes in 1961, the Government increased imports to 10.3 million tonnes in 1966, 8.7 million tonnes in 1967 and 5.7 million tonnes in 1968. Since then, imports came down considerably partly because of the Government's efforts at self-reliance and partly because of good crops within the country. Food imports touched the low level of 0.5 million tonnes in 1972. The food problem became really acute since the middle of 1972 because of the failure of internal production, difficulties of importing foodgrains from abroad and the failure of public distribution system. With increase in domestic food production and large accumulation of food surpluses, imports have been cut off since 1976-77.

3. *Government Food Policy*

As we have shown earlier, the Government food policy—or the lack of it or the failure of it was responsible for the acute food shortage and consequent rise in foodgrains price. At no time since Independence, the Government of India had a proper and well-thought-out long-term food policy, but adopted ad hoc measures to solve the food problem. The basic objective of India's food policy has been to supply foodgrains at reasonably low prices to the vulnerable sections of the community through fair price and ration shops. Imports and procurement were used to supply the public distribution system. Till 1966-67, the major plank of our food policy was import of food from abroad under PL 480 aid. An attempt was made at self-reliance since 1966. The food policy of the Government started crumbling almost from the middle of 1972 with the failure of food crops in that year. The Government announced the take-over of the wheat trade at a wrong time and the opposition of the traders and the farmers was responsible for wrecking the Government's decision. Much of the chaos, confusion and uncertainly on the food front was definitely due to the absence of clear-cut food policy of the Government.

MEASURES TO SOLVE THE FOOD PROBLEM

India's food problem is older than our independence but it is a pity that no permanent solution has been found all these years; in fact, the food problem is becoming more critical, year after year. The basic reason for our failure to solve the food problem is the adoption of ad hoc and compromise measures. From the beginning our food policy was based on two objectives : equitable distribution and reasonable prices. There is nothing wrong with our objectives. By 1950, two clear-cut but opposite policies were placed before the Government and the nation: (i) full free market conditions and complete decontrol of food trade, and (ii) complete elimination of market forces and full control of distribution of food. The Government rejected both the alternative on political and administrative grounds and, instead chose a compromise solution—some control, some procurement, some import and some public distribution system. Till today, this type of compromise solution has been continued. The steps, the Government of India has taken to solve food problem, are:

A. *Measure to Increase Production*

(i) *Technological change.* Among the measures meant to increase the production of foodgrains, the least controversial are the technological changes. To increase food production, it is essential to provide improved seeds, more fertilisers, more irrigation to the farmers, and so on. The intensive agricultural district programme consisting of package use of hybrid varieties of seeds, liberal use of irrigation and fertilises, is being vigorously extended in the country ushering in green revolution. Besides intensive cultivation through these methods, extensive cultivation has also been adopted through land improvement and land reclamation.

(ii) *Institutional changes.* The second method of increasing agricultural production is through necessary institutional changes, viz., through land reforms. It is asserted that the present agrarian structure is such that there are no incentives for increased production. With tiny holdings, which are found scattered all over the village, with a system of land holding in which the tenant has no security of tenure and which results in rackrenting and with extreme inequalities in land holdings, it is impossible to expect the tiller to put his heart into the business of cultivation.

The Government of India was putting much emphasis on land reforms which included consolidation of holdings, ceiling of holdings, control and regulation of tenures and the formation of co-operative farms. Doubts have been expressed in our own country about the success of land reforms to bring about increased agricultural productivity but tenancy reforms and land ceiling are a necessity in the agricultural climate of today.

(iii) *Organisational approach.* The third approach to agricultural development is the organisational, approach. It is pointed out that our efforts to increase agricultural production through improved seeds, fertilisers, irrigation, pesticides, etc., have not been successful mainly because of inadequate and defective organisation, which includes, not only the government administrative system, but the entire framework of official and semi-official institutions and agencies like the co-operative societies, the panchayats and the community development.

B. Distributional Changes

In the last few years, the Government has expanded the public distribution system considerably. From about 2.1 million tonnes in 1956 the public distribution system handled over 11 million tonnes in 1973. The Government has undertaken to satisfy the requirements of deficit states. It has, therefore, found it necessary to control wholesale trade through licensing fixing of margins, etc. Procurement is done in the surplus states by state governments which, in turn, hand over the procured surplus to the Union Government for sale in deficit states. The Food Corporation functions in every state and is allowed to undertake procurement operations on a large scale. In April, 1973, the Government nationalised the wholesale trade in wheat and announced its intention to take over rice trade from the coming kharif crop. When the take-over was a failure, the Government reversed its policy and brought back the wholesales.

C. Stabilisation of Foodgrains Prices

The main object of food policy in recent years is to hold the foodgrains prices in check. The Government has been adopting such short-term measure as the maintenance of foodgrains imports at a high level, extension of internal procurement and stepping up of government purchase of foodgrains for release through fair-price shops, measures to curb hoarding and profiteering, fixation of maximum control prices and the use of selective control measures by the Reserve Bank of India. These measures did have some influence in keeping prices in check but the past experience shows that the price stability has not been fully achieved. The Foodgrains Enquiry (Ashok Mehta) Committee as well as the National Council of Applied Economic Research (NCAER) also recommended the setting up of a National Price Stabilisation Board to suggest suitable agricultural, industrial and wage policies and also to suggest suitable administrative, monetary, fiscal and commercial measures needed for establishing foodgrains prices. Indeed, the Union Government has set up the Agricultural Prices Commission for recommending procurement prices for foodgrains.

The buffer stock operations by the Government is the key to the problem of stabilisation not only of food prices, but also of the general price level in the country. If managed with wisdom and flexibility, it would go a long way to protect both the farmer and the consumer against violent fluctuations in prices. The idea behind buffer stocks is that in good years when prices fall, the Government steps in and buys; and in bad years when prices tend to rise, in unloads a part of what it had purchased in the past. It is through this process of contra-cylical purchases and sales that price stability is sought to be maintained in the country. Besides, buffer stock operations will also involve government buying foodgrains in surplus areas and selling them in deficit areas. This would help in better distribution of the available foodgrains, besides bringing about regional equality of prices. The buffer stock operations should be so planned and executed that the Government is not involved in heavy financial losses.

The Government of India decided to build a buffer stock of 5 million tonnes of foodgrains by 1973-74, but substantial progress had been achieved and the actual stock with the Government at the end of June, 1978 was 20 million tonnes. This sizable increase in stocks was made possible by massive procurement of wheat whose production registered a spectacular rise, particularly in Punjab, Haryana and U. P. This achievement is highly commendable. Keeping in view the interest of the economy as a whole, the size of buffer stocks, if it is to be really effective to keep the prices under check has to be really effective to keep the prices under check, has to be substantially large. In the first place, there are considerable fluctuations in agricultural production in the country, for example, an increaser of 13 million tonnes in 1958-59 over the previous year, a decrease of 17 million tonnes in 1965-66 over 1964-65 and increase of 21 million tonnes in 1967-68 over 1966-67, a fall of 13 million tonnes in

1972-73 as compared to 1970-71 and an increase of 21 million tonnes in 1975-76 over the previous year. There is every possibility of such short-falls occurring in the future, necessitating large imports and extensive use of the pubic distribution system in the country. Likewise, if there is a big jump in production due to favourable monsoon and weather conditions there would be large increase in market arrivals and the Government will have to purchase substantial quantities to assure reasonable prices to the farmers and to prevent any setback to production.

D. Price Incentive to Agricultural Producers

The real difficulty centres round prices of foodgrains. On the basis of the experience of U. S. A. and other countries which have achieved "industrial revolution in agriculture" a team of American experts suggested that highly incentive prices were the key to any scheme for intensive agricultural development. But under the single State Zone System prices of foodgrains in the surplus states remain depressed. This means that the farmers of suplus states cannot take the risk of developing agriculture into an industry by investing heavily in the introduction, of modern techniques. The government has abandoned the policy of single state zones and has permitted free movement of foodgrains within the country.

Till now, the Government has been fixing foodgrains prices at "slightly higher than the average of immediate post-harvest prices during the previous three years." The principle of fixing farm prices is indeed wrong as the immediate post-harvest prices are nothing but distress prices' and if there are to be fixed for the whole year they would act as a disincentive to the farmer who has to invest heavily so as to produce more. The well-do-do farmers on whom the main responsibility for increasing production rests, have been in the habit of getting a much higher price by holding on to their suplus stocks till the lean season. It is imperative that the concept of remunerative prices has to be built nearer the lean season prices than the immediate post-harvest prices if food self-sufficiency is to be achieved.

Again, the pricing and procurement policy may offer the advantage of price stability to the producers in year of number crops, but ordinarily they have been much lower than those which would have been available in free markets. A certain amount of compulsion has been exercised on the farmers to part with a portion of their output at procurement prices. This indirect compulsion has been in the form of zonal restrictions on the movement of foodgrains. Consequently, the surplus areas have not been able to take advantage of the opportunities of a national market. Normal trading in foodgrains has become fragmented and restricted. As a result, although the extent of procurement is only 6 to 7 per cent of the total production of foodgrains, the impact of procurement on the farmers has been disproportionately heavy. The farming community feels strongly, and probably justifiably, that it is being prevented from taking legitimate advantages of the market situation. In this sense, the pricing and procurement policy of the Government is acting as a disincentive to the expansion of agricultural production.

Food Problem: Prospects

In the last 30 years, India's food problem has changed fundamentally. In the beginning India's food problem was one of scarcity, shortage of rice after the separation of Burma and shortage of wheat also after the partition of the country in 1947. Initially, the major concern of the Government was to increase the domestic supplies either through increased production or through imports or through both. In the second half of 1950's and during 1960's the major concern of the Government shifted to control of foodgrains prices, for the Government was assured of regular imports by U. S. A. to supplement internal production. In fact P. L. 480 imports were the basis of our agricultural and industrial development. In the 1970's and particularly after 1974 the there has been growing surplus stocks—from a target of 5 million tonnes the Government carried 20 million tonnes of buffer stock in foodgrains in 1978. The food problem is not any more one of shortage or of high prices but how to enable the lower income groups to purchase the available foodgrains. In other words, the present concern of the Government is how to make use of the huge food stocks to help accelerate the process of economic growth.

An important programme introduced for the first time in 1977-78 is the food-for-work programme. It is designed to offer gainful employment to the rural through the development of infrastructure, the creation of durable community assets by utilising suplus food stocks available with Government-payment to workers is made partly or wholly in the form of wheat or rice. The existence of 20 million tonnes of food stocks valued at nearly Rs. 2,600 crores may create a feeling of complacency that the food shortage is a thing of the past. True, agricultural output has been quite large in the last two years and there is every possibility of the output becoming larger with the expansion of irrigation, fertilisers, rural electrification, etc. But it should be clearly understood that the highly fluctuating monsoons and the consequent ups and downs in food output can always spell danger. Naturally, efforts should continue to keep the population in check. Any amount of increase in production of foodgrains under the green revolution will not and cannot solve our problem. The ultimate solution to the problem is control of population and there is no way of getting away from it—and the sooner the measures of population control are undertaken on a war footing, the better it would be.

13

AGRICULTURE LABOUR

Present Position of Agricultural Labour in India

Agricultural Labour is provided mostly by economically and socially backward sections; poor sections from the tribes also swell their ranks. It may be divided into four types:

(a) Landless labourers who are attached to the landlords;

(b) Landless labourers who are personally independent but who work exclusively for others;

(c) Petty farmers with tiny bits of land who devote most of their time working for others; and

(d) Farmers who have economic holdings but who have one or more of their sons and dependents working for other prosperous farmers.

The first group of agricultural workers have been more or less in the position of serfs or slaves; they are also known as bonded labour. They do not normally receive wages in cash but are generally paid in kind. They have to work for their masters and cannot shift from one to another. They have to provide *begar* or forced labour. In some cases, they have to offer cash and also supply fowls and goats to their masters. Among the other groups mentioned above the second and the third are quite important. The problem of landless labourers is the most serious problem in the rural sector.

Magnitude of Agricultural Labour

Accurate figures about the number, income, standard of living, etc., of rural labour are not available. But some information is available in the form of the reports of committees and commissions. According to the Second Agricultural Labour Enquiry published in 1960, agricultural labour families constituted nearly 25 per cent of all rural families. According to this, more than 85 per cent of the rural workers are casual, serving any farmers who is willing to engage them and only 15 per cent of agricultural labourers are attached to specific landlords. More than half of the workers do not possess any land, and even the rest of them own only very little of land. The number of those who have adequate economic holding but who work for others because of surplus manpower in the family is very small indeed. Agricultural labourers predominantly belong to the scheduled castes, scheduled tribes and other backward classes. According to one estimate, between 75 and 80 per cent of all agricultural labourers belong to the scheduled castes.

Recent studies about the mangitude of landless labourers have shown that agricultural labourers who numbered 315 lakhs in 1961 have increased to 475 lakhs in 1970-71 and have further swelled to 580 lakhs in 1977-78. On an average, the increase of agricultural labour during 1960-61 and 1977-78 was of the order of 15 lakhs per year. This has to be seen in the background of the increase of total rural labour force which increased from 1,740 lakhs in 1964-65 to 2,040 lakhs in 1974-75 and 2,230 lakhs in 1977-78. As a proportion of the total rural labour force, the percentage of landless labourers has increased from 18 percent in 1964-65 to 24 percent in 1977-78.

This implies that marginal farmers have been swelling the ranks of landless labourers. On account of the increasing economic compulsions in terms of past debts, they have been forced to sell their lands. This supports the popularly held view that the benefits of millions of rupees invested in irrigation, tubewells, fertilizers or roads have gone to the big farmers. It also implies that co-operatives and rural banks have also helped the rich farmers.

Another finding of the recent inquiries is that the days on which an agricultural labourer is employed have declined. He worked on an average for 277 days in 1964-65, but for only 250 days in 1974-75. It may be noted that work does not mean a full day work, but part-time work is also computed as gainful employment for a day.

During this period (1964-65 to 1974-75), the average daily earnings of agricultural labourers have gone down from Re. 1 to 88 praise, a fall of 12 per cent. Obviously, there is chronic underemployment among the agricultural labourers.

Official data about land distribution indicate that 61 per cent of the rural households either own no land or small fragments of land or marginal uneconomic holdings of less than 1 hectare. Taken together, 61 per cent of households own barely 8 per cent of cultivated area. Out of them, 22 per cent households own no land at all: another 25 per cent own less than half a hectare (1.2 acre). Thus, these marginal farmers are the recruits in the army of landless labourers, since they subsist at the border of the poverty line and have been gradually slipping below poverty line.

Agricultural wages in India

The average annual income of an agricultural labour household in India in 1950-

51 was Rs. 447, it fell to Rs. 437 in 1955-57; it increased to Rs. 600 in 1963-64 and increased significantly to Rs. 1,671 during 1974-75. Agricultural wages have been the lowest in every part of the country. This has been so far a long time. When the Minimum Wages Act was passed soon after Independence, the average wage fixed ranged between 62 paise and Rs. 1.50. The adoption of efficient farming methods and the introduction of mechanisation have led to "green revolution" and have been responsible for the prosperity of the middle and rich farmers. At the same time, they have been responsible for reduction in demand for labour and reduction in real wages. In a recent study of agricultural labour in Rohtak district in Haryana. Sunil K. Dhawan found the average wage between Rs. 2 and Rs. 3 in certain villages at the lowest level and between Rs. 4 and Rs. 5 at the highest level : the average money wage is around Rs. 3 per day per worker. The highest range of money wages prevails in Kerala where a male worker gets Rs. 10 and a female gets Rs. 7. According to a study by Professor Pranab Bardhan for the period 1960-71 to 1967-68, agricultural production rose by 6 per cent but it was accompanied by a decrease in real wages in all states except in Kerala where the agricultural workers are unionised. Professor Bardhan comes to the conclusion that in the Indian context, the bargaining strength of the landless labourers might be considered as an important determinant for high real wages than the spread of technological progress. The real problem is to organise uneducated and unskilled farm labourers in trade unions.

Causes of the Poor Economic Condition of Farm Labour

The following important reasons explain low wages and poor economic conditions of agricultural labour:

1. *Low Social status.* Most agricultural workers belong to the depressed classes which have been neglected for ages. The low caste and the depressed classes have been socially handicapped and they had never the courage to assert themselves. They have been like dumb-driven cattle.
2. *Unorgnised.* Agricultural workers are illiterate and ignorant. They live in scattered villages. Hence, they could not be organised in unions. He urban areas workers could generally organise themselves and it was convenient for political parties to take interest in trade union activity. This is almost difficult in case of farm labour. Accordingly, it is difficult for them to bargain with the land owners and secure good wages.
3. *Seasonal employment.* The agricultural workers do not have continuous work. On an average, a farm labourer finds employment for about 197 days in a year and for the rest of the year he is idle. Apart from under-employment there is also unemployment in rural areas. Unemployment and underemployment are two important factors responsible for low income and consequently low economic position of the agricultural workers

is India. But then the nature of work in agriculture is such that a farm labourer cannot get work continuously. In most cases, work on the farms is seasonal and intermittent. In many cases, there is single cropping which means work only for six to seven months in the year.

4. *Paucity of non-agricultural jobs.* Paucity of non-agricultural occupations in village areas is another important factor for the low wages and poor economic condition of the farm labourers. For one thing, the growing pressure of population is increasingly felt in rural areas and the number of landless labourers is steadily increasing. For another, the absence of another occupation in rural areas and lack of inter-regional mobility have been responsible for worsening the pressure of population on land.
5. *Rural Indebtendness.* Agricultural labour is heavily indebted. Normally, the farm labourers borrow from the landowners under whom they work. Since they have no security to offer, they pledge themselves to the money-lenders and rich landlords and become bonded labourers in many areas. Naturally, they are forced to accept lower wages.

Thus, partly because of factors beyond their control and partly because of their inherent bargaining weakness, the farm labourers have been getting very low wages and have, therefore, to live a miserable sub-human life.

Suggestions for the Improvement of the Condition of Agricultural Labour

The following suggestions have been made for the improvement of agricultural labour:

(i) Removal of serfdom. Agricultural serfdom which exists in many parts of the country should go. In fact, according to the Constitution of India, the practice of slavery in any form is prohibited. But agrarian serfdom which has been in vogue for centuries cannot disappear so easily. This is so because the labourers are helpless, ignorant and illiterate. Education of the rural masses and better opportunities are some of the remedies for the removal of the system of slavery. The abolition of bonded labour and control of the activities of the money-lenders, provision of house sites for landless labourers, etc., under the 20-point economic programme were desirable steps.

(ii) Better implementation of minimum wages in agriculture. Agricultural workers have been getting very low wages except in Punjab and Kerala. Measures should be taken to raise the wages of farm workers. Unless this is done it is impossible to raise the economic condition of the agricultural workers. Minimum wage legislation alone is not sufficient but should be taken to enforce it.

(iii) Rehabilitation of landless agricultural workers. In order to improve the conditions

of agricultural labour, the landless workers should be provided with land. This can be done in many ways. One way is that the new reclaimed land may be allotted only to them. The Bhoodan movement is another method through which those who have land contribute voluntarily for those who have not.

(iv) Improvements in agriculture. The agricultural labourers do not have full-time employment because of seasonal character of Indian agriculture. Both intensive cultivation and extension of irrigation are very much needed to increase agricultural work. By these methods, there will be double cropping and employment will be available for the whole year round. Besides, there will be increase in the productivity of labour which will also increase the wage of workers. There is very great scope for the setting up of village and rural industries which will use the raw materials readily available in the villages and which will provide occupation for the rural population. There is no necessity for setting up of big industrial units. There can be small size industries employing modern techniques which can be easily set up now-a-days because of the extension of electricity to rural areas. The setting up of such industries will help agricultural labour in many ways, as for instance : (a) seasonal unemployment will be reduced: (b) surplus labour power will be diverted to these industries; (c) the pressure of population on agriculture will be reduced and this will help to raise agricultural productivity and also raise wages of agricultural labour; and (d) those unable to find employment in agricultural may be readily employed and thus help to supplement the income of the family.

(v) Trade Unions for Agricultural Labour. As referred to earlier wherever trade unions have been organised for agricultural workers, their wages have been protected and their exploitation by powerful landlords and moneylenders can be eliminated. Hence, greatest emphasis should be placed on the formation of unions of farm labourers and the Government should help actively in the formation of such unions.

(vi) Public works programme. One very good method of providing employment to rural labour and of utilising it fully is the construction of public works by the government. The government can plan its projects in the rural areas carefully, so that the workers who may be unemployed during the off-seasons may be gainfully employed. Such projects include the construction of roads, the digging and deepening of tanks and canals, afforestation, etc. The setting up of small industries and public works programmes have great importance in the mobilisation of manpower resources in villages and raising the wages of the rural masses and, indirectly, the income of the country.

(viii) Co-operative farming. Ultimately it may be necessary to fashion a society in which all the rural folk will be more or less equal. The final solution probably lies in the creation of co-operative village system in which all, including the present labouring class, will have equal rights, duties and opportunities.

Government Measures Pertaining to Agricultural Labour

Since the Independence of the country in 1947, the Centre as well as the State Governments have taken some measures to improve the economic condition of agricultural labour. They include the passing of legislation to fix minimum wages for agricultural labour, the removal of disabilities, the ceilings on holdings and the redistribution of land among the landless labourers, etc.

(1) *Indian Constitution.* The Indian Constitution has declared the practice of serfdom an offence. It has abolished agrarian slavery including forced labour by law but it will take some time before it is removed in practice.

(2) *Minimum Wages Act.* The Minimum Wages Act was passed in 1948, according to which every State Government was asked to fix minimum wages for agricultural labour within three years. The minimum wages are fixed keeping in view the total costs and standard of living. Since conditions in various parts of the country are different and since even within the state the law allows different rates of wages to be fixed, the minimum wages very from Re. 0.62 to Rs. 1.50. In practice minimum wages are very difficult to enforce effectively. In many states the rates are fixed even below the current rates of wates. In practice, it has failed to increase the wages and earnings of agricultural labour.

(3) *Other legislative measures.* The zamindari system has been abolished by law in all the State and with that all the exploitation associated with the system has been removed. Besides, tenancy laws have been passed in most of the states protecting the interests of the tenants and labourers, and enabling them to acquire the lands they cultivate. Many states have passed legislation fixing ceiling on agricultural holdings by which the maximum amount of land which a person can hold has been fixed by law. According to these laws, the surplus land of rich land-owners is to be distributed to the landless labourers.

(4) *Organisation of labour co-operatives.* During the Second Five-Year Plan, efforts were made to encourage the formation of labour co-operatives. These co-operatives whose members are workers undertake the contract of government projects, such as, construction of roads, digging of canals and tanks, afforestation, etc. They provide employment to agricultural workers during off-seasons and also eliminate the possible exploitation of workers by the private contractors. The basic idea of the movement is commendable. The movement has yet to gain momentum in the rural areas.

(5) *Employment Guarantee Scheme.* The Government of Maharashtra introduced in 1977 the '*Employment Guarantee Scheme*' under which any able bodied person in rural areas can apply for a job to the Collector of his district or to his authorised subordinate officials and the latter will provide him employment within 5 kilometres

of his place of residence. For this purpose, the Government has to prepare and keep in readiness various public works, such as irrigation works, road construction, etc. The rate of wages will not be such as to attract agricultural workers from their normal employment in agricultural operations. This was a pioneering scheme which the Government has been trying to improve upon with the experience gained. Under the scheme, the State assumes the responsibility to provide work on demand. If work is not provided in a fortnight from the date of demand, a payment of allowance at the rate Re. 1 per day is to be made. The Maharashtra Employment Guarantee Scheme is being recommended for adoption by other states.

(6) *20-Point Programme.* In July, 1975, soon after the Emergency was declared, the Government introduced the 20-point economic programme which included a number of measures to improve the economic condition of the landless workers and other weaker sections of the community in our villages. These measures were :

(a) Speedy implementation of ceiling legislation and distribution of surplus land among landless labourers and small peasants;

(b) provision of house sites for landless labourers and conferment of ownership rights of the houses if they have been occupying them for a certain period;

(c) abolition of bonded labour;

(d) liquidation of rural indebtedness and moratorium on recovery of debts from landless labourers, artisans and small peasants; and

(e) review of the minimum wage legislation for agricultural labour and introduction of suitable enhancement of minimum wages wherever necessary.

The landless labourers who form nearly 24 per cent of the total population have been living a miserable existence, with highly inadequate food, clothing and housing. Their wages are low; they are forced to work for the same landlords for generations; and they do not own even the huts in which they live. The landless labourers are indeed the worst exploited section of the community. The 20-point programme was the first concrete step to ameliorate their economic condition and promote their welfare and a sincere and serious attempt was made to implement the programme vigorously. The Janata Party which came to power after the Emergency abandoned the 20-point programme. However, the Government convened a conference on rural unorganised labour in January 1978 to consider the problems facing agricultural workers. A Central Standing Committee has since been constituted to advise the Government on measure to ameliorate the socio-economic conditional of un-organised rural labour. Politics apart, the various points of the 20-point economic programme are worth implementing.

Abolition of Bonded Labour

One item of the 20-point economic programme was the abolition of bonded labour. It came as a great surprise to most urban dwellers that millions of landless labours were still being treated as bonded labour or contract labour and that they have been living like serfs or slaves. Bonded labour has existed in India for centuries. It is a peculiar phenomenon of our agricultural economy. The system grew out of extreme poverty and helplessness of scheduled castes. Adivasis and semitribal communities, who have always depended on wage-income which too they received only for a part of the year. When they needed foodgrains during the lean agricultural season or when they needed money for special occasions such as marriage or festival or for medical treatment, they had to borrow but they had no security to offer in the form of land, jewels or any other property except their own labour. Accordingly, they 'pledged' themselves. The moneylenders and the high caste landlords took advantage of their helplessness and ignorance, and entered into repressive contracts. The wages were so low and the rate of interest so high-often 25 per cent to 50 per cent— that the total loan got accumulated with the passage of time. In fact, the total burden of the loan became so huge that it could not be rapid even with the labour of many generations of the debtor's family. Once a landless labourer borrowed money, he and his descendants were doomed to perpetual slavery. Even in modern times, the scheduled caste agricultural labourers who could secure employment for hardly five or six months a year had to subsist for the rest of the time on money and foodgrains borrowed from their landlords. Bonded labour is known by various names in the country, as for example, 'Paniyas' in Madras, 'halia' and muliya' in Orissa, 'baramasiya' in North Bihar and 'Kemia' in South Bihar, 'harwaha' in Madhya Pradesh', 'sewak' and 'haris' in Uttar Pradesh, etc.

There are no reliable statistics of bonded labour even though the system has existed for centuries. No attempt was made to collect reliable information about bonded labour. What was worse, many states even denied the existence of bonded labour. But in many parts of the country, lands and houses belonging to the poorer farmers have been progressively passing into the hands of the usurious moneylenders and thus swelling the ranks of the bonded and contract labourers. During the period 1961 and 1971 alone, the number of landholders declined from 98 million to 78 million but the number of landless labourers shot up from 27 million to 47 million. This shows the clear trend of growing concentration of property with a few and the growing pauperisation of the rural community.

Bonded labour (Bandhua Mazdoor) generally borrowed foodgrains during the lean agricultural season from the rich landlords. The repayment of these loans is generally in kind and the extent was 1.25 to 1.5 times the original amount. As a result of this type of loans, the rich cultivators remained almost assured of regular supply of labour for their agricultural operations. The bonded labour incurred double burden in the form of paying high interest rate and selling their labour cheaply to their masters. Most bonded labourers worked as hereditary serfs in the houses, gardens and farms of the upper-class people. Their work ranged from tilining the land to menial domestic chores. In Maharashtra, Andhra, Bihar and other States, the Adivasis have suffered severe humiliation at the hands of the high caste cultivators and money-lenders.

Effects were made in the past to abolish bonded labour. The earlier effort was the law passed in 1933 by the British Parliament to abolish slavery throughout the British empire. Unfortunately it did not have any impact in India as the different states felt that there was no slavery in India. After Independence, the Constitution of India prohibited slavery, *begar* and other forms of forced labour. Soon after some states passed laws to abolish bonded labour. But the system persisted since the laws were not enforced and the urban people were either not aware of the evil or did not take notice of the evil. A series of agitations by the bonded Adivasi labourers against their exploiters were organised by the communists. They focussed attention of the Government and of the general public to their terrible exploitation by the high caste land-owner and rapacious money-lenders. Often these movements were condemned as communist uprisings and were cruelly suppressed. At the same time, the Government was influenced by the regular reports of the Commissioner for Scheduled Castes and Scheduled Tribes highlighting the exploitation and the humiliation of the backward communities. Accordingly the Government included the abolition of bonded labour as one of the basic points of the 20-point economic programme. In October, 1975, the President of India promulgated an Ordiance declaring bonded labour illegal and cancelling their indebtedness. For the first time, the Government took swift action against moneylenders, building contractors, quarry-owner and others who have been found practising some form of bonded and contract labour.

Future of Bonded Labour

Abolition of bonded labour is a positive step in the emancipation of agricultural labour. But the real abolition of the age-old system cannot be done through ordinances and legislation and use of police methods alone. As indicated earlier, several states had enacted laws to abolish the evil much earlier but the system persisted. The basic difficulty is the extremely weak economic condition of landless labour in our rural areas, particularly those belonging to the Scheduled Castes tribals and others. In fact, in the tribal and more backward areas, where the economy is non-monetised, there is in effect a certain amount of connivance between the bonded workers and their masters because the former have no alternative means of livelihood. According to Mr B. Siveraman of the Planning Commission, thousands of workers if freed from bondage over-night, would simply starve unless they could be immediately absorbed in other gainful jobs. Conditions in rural areas are such that unemployment would recur and make the landless labourers vulnerable to exploitation by the money-lenders and the rich farmers. It is, therefore, of utmost importance that gainful employment is created in rural areas through nationwide programme of accelerated rural development. This should also include the setting up of industries based on agriculture and forest resources. The Maharashtra Government's "Employment Guarantee Scheme" is a step in the right direction and it should be adopted by all other states. Under this scheme, every able-bodied worker is assured of employment near his place of residence. Such a scheme will help to harness the labour of millions of unemployed and under-employed workers in our villages for productive purpose, for capital formation and for raising national and per capita incomes. Regular employment will also obviate the necessity to approach the money-lenders. In this connection, the recently started regional rural banks as well as nationalised banks and cooperative banks in rural areas have an important role to play. But present indications are that they are not going all out to help the erstwhile bonded labourers.

Abolition of bonded and contract labour, redemption of past doubts, enforcement of minimum-wages,— all these are meant to help the exploited millions in our villages. All right thinking people would accept the correctness of the steps taken by Indira Gandhi and would extend maximum cooperation.

Agricultural Labour and Minium Wages

After the Minimum Wages Act, 1948 was passed, pressure was brought upon the Government to extend the Act to cover agricultural labour. The Act was subsequently extended but, there are still some. Scanes which have not brought large sections of agricultural labour within the ambit of the Act. Even where the Act is implemented, the machinery for the fixation and enforcement of minimum wages is not uniform. In some States, it is the Revenue Departments which implements the Act and in some others this is done by Labour Department. But generally speaking:

(i) The Act has been a dead letter in every State; (ii) minimum wages in agriculture have not been revised over long periods; (iii) almost everywhere, the actual wages rule higher than the minimum wages during the peak season and tend to fall in slack seasons; (iv) the machinery for enforcement is hopelessly inadequate to cope with the task of effective implementation, the institution of proceedings under the Act is almost negligible and successful prosecution even more disappointing; and (v) other difficulties in the implementation of the Act arise mainly from poverty and illiteracy of agricultural labour, absence of knowledge of the existence of legislation, scattered nature of agricultural farms, casual nature of employment, unorganised character of agricultural labour, etc.

There is strong opinion in favour of suspending the operation of the Minimum

Wages Act in rural areas. The basic argument is that at the peak of the agricultural season, labour can dictate its own terms and naturally accrue wages tend to be much higher than the minimum wages, under these conditions, the enforcement of minimum wages has no meaning. On the other hand, during the slack season, the supply of agricultural labour is much in excess of demand and, therefore, the enforcement of the minimum wages may be difficult. The National Commission of Labour gave due weight to this view, yet it strongly recommended the continuance of the Act on the ground that in the long run, with greater awareness of their rights on the part of agricultural labourers, the fixing of minimum wages would be useful. As regards enforcement, the Commission suggested that the Panchayats might be asked to enforce minimum wages, even though this has obvious defects. The Commission also suggested wide publicity to the fixing of minimum wages and the notification of wages at a public place which would make the employers cautions in denying to their workers the notified wages.

There are considerable disparties in wages between regions, between different crops as well as between wages paid to men, women and children. The National Commission on Labour had sufficient evidence before it to conclude that between 1956 and 1963 the incomes of agricultural labour increased faster than the wages of industrial workers. This was mainly due to the fact that the average wage in rural areas was increasing significantly. This was specially true in the case of Punjab and in some selected areas of Tamil Nadu.

Under the 20-point economic programme, the Government at the Centre had promised to review the minimum wages legislation for agricultural labour and to introduce suitable enhancement of minimum wages wherever was done in this direction. The abolition of rural poverty largely depends upon the guarantee of minimum wages.

14

IRRIGATION AND AGRICULTURAL INPUTS

Agricultural efficiency and production depend largely upon the inputs and investment in agriculture and the methods of production used. Progressive agriculture will demand, among other things (i.e., favourable institutional and organisational structure), improvements in inputs and methods. Irrigation, better seeds, better manures and fertilisers, land reclamation and soil conservation, plant protection, use of mechanisation, etc.—these are various aspects of agricultural inputs which have to be considered here.

IRRIGATION

Water is indispensable to agricultural production. In areas where railfall is plentiful and well-distributed over the year, there is no problem of water. But rainfall in certain areas is very scanty as well as uncertain. This is so in Deccan and Central India. Punjab and Rajasthan. In these areas, artificial irrigation is absolutely essential, for without it cultivation is almost impossible. In certain regions, railfall may be abundant but it may be concentrated in a short period of the year, the rest of the year being dry. As a result, cultivation may not be possible for the whole year. In these regions, provision of irrigation will facilitate growing of more than one crop in the year. Finally, there are certain food and cash crops such as rice and sugarcane which require abundant, regular and continuous supply of water. In short, continuous supply of water is essential in order to protect various regions from famines and to ensure double and, if possible, triple cropping and rapid increase in agricultural production (which is an essential condition for rapid industrialisation in India.)

Net cultivated area in 1976-77 was 1,402 lakh hectares and artificial irrigation was provided to 325 lakh hectares of land in 1975-76 i.e., 25 per cent of the net cropped area was irrigated. In 1950-51, about 18 per cent of the net cropped area was irrigated. In it clear that the progress of irrigation has been quite slow, despite the enormous importance given to it during the five year plans.

TABLE 14.1

Area Irrigated by Sources in India

Source of Irrigation	*1950-51* Area (lakh hectares)	*1950-51* Per-centage	*1976-77* Area (lakh hectares)	*1976-77* Per-centage
Canals	83	39.7	138	39.6
Wells	60	28.7	148	42.7
Tanks	36	17.2	39	11.2
Other sources	30	14.4	23	6.5
Total	209	100.0	348	100.0

Source: Indian Agriculture in Brief, 18th ed. (1980) p.26.

Source of Irrigation in India

Since 1950-51, considerable importance had been attached to the provision of canal irrigation and well irrigation. Even though nearly 10 per cent of irrigation is supplied by canals, now well irrigation has caught up rapidly; irrigation by tube wells has been expanded considerably. In the meantime, tanks and other sources of irrigation are declining in importance.

In India, irrigation works are classified into major and minor irrigation works. Since 1951, the "major" irrigation projects have been defined as those costing more than Rs. 5 crores, "medium" projects as those costing between Rs. 25 lakhs and Rs.5 crores, "minor" irrigation works costing less than Rs. 25 lakh each. Major irrigation works involve huge investment and take a long period for completion. But they have high capacity and can benefit large areas of land, sometimes running into millions of acres. They offer almost complete protection against famine. Moreover, in recent years, major works are multipurpose projects since they are designed not only to supply irrigation water but also flood protection, navigation, hydro-power etc. The main advantages of minor works are that they require small investment, that they can be completed within a short period and their effect on agricultural production will be felt immediately. If quick returns are desired, minor irrigation works are very suitable. Form 1978-79, the Planning Commission has introduced a new classification of irrigation schemes:

(a) *Major irrigation Schemes*—Those with culturable command areas (CCA) more than 10,000 hectares.

(b) *Medium Scheme*—Those with culturable command areas (CCA) between 2,000 and 10,000 hectares.

(c) *Minor Schemes*—Those with culturable command areas (CCA) up to 2,000 hectares.

When India started planned economic development in 1951, the irrigated area from major and medium irrigation was 83 lakh hectares and from minor irrigation works was 126 lakh hectares, totalling 209 lakh hectares. Between 1951-52 and 1973-74 (end of the Fourth Plan) over Rs. 3,330 crores were spent on major and medium irrigation works and on flood control and drainage works. As a result, additional irrigation facilities of 177 lakh hectares were created. Accordingly, at the end of the Fourth Plan, the total irrigation potential from major and medium works came to 130 lakh hectares and minor works conributed another 196 lakh hectares making a total of 326 lakh hectares. The proposed Fifth Plan outlay for major and medium irrigation projects as well as for flood control was Rs. 2,680 crores and the additional irrigation potential at the end of 1977-78 was 43 lakh hectares.

Irrigation and Flood Control in the Sixth Plan (1980-85)

Under the Sixth Plan, the main support to the agricultural programme is a large step up in the irrigation potential viz., 150 lakh hectares, consisting of 65 lakh hectares from major and medium irrigation works and 85 lakh hectares from minor works this is in sharp contrast with the irrigation potential of 79 lakh hectares created i n the first four years of the Fifth Plan (1974-78). In the Sixth Plan, an outlay of Rs. 12,000 crores has been projected for increasing irrigation potential which accounts for 12 per cent of the total plan outlay. To achieve this order of irrigation potential the Sixth Plan attaches the highest importance to:

(a) getting full utilisation of the potential already created;

(b) expeditions completion of on going projects;

(c) proper maintenance and operation of the existing irrigation systems;

(d) ensuring that the smaller farmers get their fair share of water; and

(e) settlement of inter-State river disputes so that projects utilising the water potential can be started expeditiously.

As regards minor irrigation, the Sixth Plan would make a careful assessment of ground water potential so as to minimise wasteful expenditure and prevent over-exploitation and co-ordinate all these programmes with rural electrification schemes. The Sixth Plan also attaches highest priority to expeditious completion of the on going schemes of flood control.

Multipurpose River Valley Projects Since Independence

When the country became independent, the national government gave highest importance to major irrigation works. But the new works were not for irrigation only. They aimed at controlling floods, generating electricity and providing navigation facilities. Hence, these projects came to be knows as multi-purpose river valley projects. During the Five Year Plan periods, many important projects have been planned and executed.

The execution of large works during the Five year Plan periods has brought out a number of problems and weaknesses related to these large projects. Firstly, irrigation facilities offered by large projects have been poorly utilised. During the First Plan only 48 per cent of irrigation facilities were utilised. During the Second Plan, the pace of utilisation was accelerated and about 76 per cent of the facilities were utilised by the end of the Second Plan. The reasons for the poor utilisation of irrigation facilities are; absence of synchronisation of various construction programmes, non-completion of field channels in time, lack of publicity and demonstration to the cultivators about the suitable cropping pattern, etc. Secondly, financial returns of large multipurpose works have been quite poor. They have to be improved through fuller utilisation, betterment levey, flood cess, water rates, etc. Thirdly, large works require continuous operation and maintenance. To large extent emphasis was placed on construction of new works rather than on the maintenance of old works. Irrigation facilities have to be expanded but maximisation of benefits can be derived only when the existing works are properly looked after.

Deffects of Irrigation

Irrigation, specially canal irrigation, is not an unmixed blessing. There are three special difficulties of irrigation which are highly significant and which currently attract the attention of the authorities. These are changes in the system of cropping, water-logging and salt-effervescence.

(i) Change in System of Cropping

The provision of irrigation has changed the system of cropping in various parts of the country. In some cases, even two crops in a year have been made possible where one crop was raised earlier. This, in fact, is good. But irrigation has led to excessive concentration on cash crops and has lowered the standard of crop rotation in many regions. In U.P. for instance, sugarcane, which remains on the soil for large part of the year, has altered the system of rotation which formerly included peas, arhar and fodder crops. In several parts of West Bengal and Bihar, jute has displaced the crops of peas, grams and oilseeds, all of which help to keep up the fertility of the soil. As a result of the changes in crop rotation due to extension of irrigation facilities, not only land is getting exhausted but in certain areas certain plant diseases have started appearing.

(ii) Water-logging

Whenever irrigation canals intersect drainage canals, rain and flood water may be help up and lands may be water-logged. Sometimes usual rains and floods in irrigated regions increase the supply of water so much that land remains under water for long periods Over-irrigation, which is becoming a common practice, is also one of the reasons for water-logging. This is becoming a menance in Punjab and U.P. where large areas of land are affected flood control measures and drainage of surplus water and proper planning of drainage and irrigation canals, etc., are being adopted in these States.

(iii) Salt-Effervescence

The soil contains certain salts which are bad for the plants. So long as these salts lie deep in the earth, there is no problem. But they are brought to the surface and the upper layers of soil, which are most helpful in the growth of crops, get soiled. The reason for the salt coming up to the surface—known as salt effervescence—is the seeping of the canal water into the lower layers thus forcing sub-soil water to come up. As the sub-soil water level rises, it is drawn up by capillary action and the salts of the earth come up to the surface. Lands suffering fro salt-effervescence become useless for cultivation, because these salts are destructive of all vegetation. The problem can be tackled best by lining canals and distributaries with concrete to prevent water from seeping in. Pumping out sub-soil water by tubewells will prevent the sub-soil water coming up.

Irrigation and New Challenges to Agriculture

An important constituent of the current strategy for raising agricultural production is the increasing reliance on irrigation. The strategy of Intensive Agricultural District Programme (IADP) and the High Yielding Varieties Programme (HYVI) have been introduced only in those areas which have assured rainfall and and irrigation facilities. As the scope for extending the cropped area is limited, greater reliance has to be placed on irrigation so as to have double or multiple cropping. The basic objective is to have a higher yield per hectare of the irrigated crop but also of producing a much higher yield of grain output per hectare by promoting the sowing of two or more crops on irrigated land.

There is still great scope for the extension of irrigation. The Planning Commission has always given much attention to this subject since the advent of planning. As against the estimated 107 million hectares of land which can ultimately be irrigated. India has developed so far irrigation potential for about 45 million hectares. During the Fifth Plan it was proposed to increase it by another 12 million hectares-taking the total to 57 million hectares. That leaves still 50 million hectares of potential irrigation for the successive five year plans. According to the Irrigation Commission, the cost of exploiting this potential would work out at Rs. 30,000 crores (at 1980-81 prices) and this might be accomplished before the end of this century. In order to meet the new challenges of agriculture we have to give maximum priority to the extension of irrigation so that all the available irrigation potential is exploited within the shortest possible time.

Under-Utilisation of Irrigation Capacity

Equally important is the better utilisation of the irrigation facilities already created. So far, we have failed hopelessly in getting the maximum benefits out of our investment in irrigation in the sense that irrigated land has not contributed its maximum to agricultural output. Theoretically, irrigation should make double cropping possible, if not multiple cropping, but the bulk of the irrigated area in India still continues to be single crop area. This is revealed by Table 14.2.

TABLE 14.2

Extent of Double and Multiple Cropping on Irrigated Land

Year	*Irrigated area (million hectares)*			*Percentage of double cropped area to net irrigated area*
	Gross	*Net*	*Area Sown more than once*	
1950-51	22.6	20.0	1.7	8.2
1960-61	28.0	24.7	3.3	13.5
1970-71	38.2	31.1	7.3	23.3
1976-77	43.1	34.8	8.3	23.9

Source: V.K.R.V. Rao: Dr. Panse Memorial Lecture, April 1974 and Indian Agriculture in Brief (1980).

In 1950-51, only 8.2 per cent of the net integrated area was sown more than once the percentage rose to 13.5 in 1960-61 and 23.9 per cent in 1978-77. In other words, of the 43.1 million hectares under irrigation, only 8.3 million hectares (or 23.9 per cent) were sown more than once. Either the bulk of irrigation gives security to only one crop or agronomic practices in the irrigated areas have not developed sufficiently to permit more than one crop.

If we assume that the entire irrigated area can be made to support two crcps, then on a single crop basis, over 76.3 per cent of irrigated land is under-utilised. This extent of under-utilisation is not met with in any other type of public enterprise. Scientists have explained the possibility of raising 10 to 12 tonnes of grain per hectare on irrigated land, subject to multiple of cropping and proper alternative rotations of crops. It is therefore, clear that full utilisation of existing irrigation resources alone can boost foodgrains production from the present 125 to 130 million tonnes to anything between 300 and 500 million tonnes. But what can be the causes for this colossal under-utilisation of irrigation resources? What practical remedies can we offer to remove this problem? The following are some of the important factors for under-utilisation and the necessary steps to remove them.

(i) At present Indian farmers, by and large, lack the necessary knowledge to optimise irrigation use. They do not know appropriate agronomic practices including suitable varieties of short-maturing crops, appropriate crop rotations, etc. In this connection, it is necessary to provide better extension services, suitably linked with research organisations of scientists, and adapted to multi-cropping farming practices.

(ii) Supporting facilities necessary for optimum use of irrigation,i.e., land levelling, land shaping, land consolidation, efficient land channels, etc. are absent in many parts of the country. To remedy this situation, it is necessary to have large-scale rural public works for land-levelling, land-shaping in irrigated areas, etc.

(iii) At present, even major and medium irrigation works have not been properly maintained. Minor irrigation works, particularly tanks and open well, have been largely neglected. To remove this important defect, it is necessary to undertake extensive rehabilitation, renovation and modernisation of existing irrigation system. Irrigation works, both major and minor, should be properly maintained to provide maximum benefit to the community. At the same time, canal irrigation should be reinforced and supplemented by well-irrigation at suitable points.

(iv) At present, faulty irrigation practices and absence of proper and adequate drainage facilities have not only been responsible for waste of water but for water-logging, salinity and alkalinity which have permanently damaged considerable portion of cultivable land. Education in water management and provision of drainage facilities will help to remove this defect.

In general, there has to be all-India co-ordinated programme to promote double and multiple cropping and optimal use of water. The objective, as we have emphasised earlier, is to raise the yield per hectare of the irrigated land through double and multiple cropping. This is the only way by which the new challenge in agriculture can be met.

FERTILISER AND MANURES

In any scheme for boosting agricultural output, the use of chemical fertilisers has an important role. India's soil though varied and rich is deficient in nitrogen and phosphorus—two plant nutrients which together with organic manure influence crop return. With population rising at a fast rate, the use of larger and larger doses of chemical fertilisers is the only way to augment our foodgrains production. The New Agricultural Strategy was based on increased use of fertilisers.

Production of Fertilisers. Generally speaking, fertiliser industry made rapid progress during the past decade. The production of nitrogenous fertilisers rose from 11,000 tonnes to more than 1 million tonnes between 1950-51 and 1973-74. During the same period, production of phosphatic fertilisers rose from 16,000 tonnes to 323,000 tonnes.

During the Fifth Plan, a total provision of Rs. 1,555 crores was made for fertiliser projects and pesticides. It was expected that by 1978-79, the installed capacity of nitrogenous fertilisers would increase by 4.7 million tonnes and its production by 2.9 million tonnes. Similarly, the installed capacity of phosphorous fertilisers would also increase by 1.3 million tonnes and its production by 0.7 million tonnes. Besides the new plants envisaged in the public sector; it wad expected that additional capacity would be promoted in the private sector as well. The Revised Sixth Plan (1978-83) has allotted public sector outlay of Rs. 1,680 crores for fertilisers and pesticides. The major stumbling block to the progress of fertiliser industry is of course the Government.

For one thing, there is extra-ordinary delay in the setting of fertiliser plants. Sometimes, it takes about 8 to 9 years to complete a project from the time of issue of the letter of intent, causing unnecessary increase in the cost of the project. Secondly, there is the problem of capital for the fertiliser programme. It has been estimated that between Rs. 925 and 1,400 crores would be required of which about Rs. 300 crores or more would be foreign exchange component. Recognising the need for increased foreign investment, the Government of India at the end of 1965 granted certain, concessions of attract foreign capital for the fertiliser industry such as majority equity participation, distribution rights, etc. But the foreign response to these offers has been very poor, mainly because (a) profitability of exports of fertilisers to India has been very high, and (b) there has been uncertainty with respect to the nature and ability of raw materials.

Imports of Fertilisers. As internal production has been short of continuously increasing demand, the government had to depend upon imports. The percentage of imports of fertilisers has been rather high. In 1951-53, the percentage of imports to the internal production of nitrogenous fertilisers was 45.5 per cent and it increased to 71 per cent in 1967-68. Such huge import of fertilisers through the use of scarce foreign exchange is the result of the absence of a clear-cut and vigorous policy of setting up fertiliser units within the country. During 1978-79, the total production of fertilisers was of the order of 39.4 lakh tonnes but total consumption aggregated to 51.1 lakh tonnes. Imports of fertilisers of the order of 19.9 lakh tonnes valued at Rs.

460 crores were made during 1978-79. The share of imports in total supplies of fertilisers was about 34 per cent in 1978-79. In the world market, fertiliser was becoming scarce and its price was shooting up since 1972-73 and particularly after the middle of 1973 in the wake of the oil crisis. Heavy pressure of demand the high cost of feed-stock of fertiliser factories, high rates of ocean freight—all these have reduced the international supplies of fertilisers to India and at the same time have raised their price.

Distribution of Fertilisers. On the whole, distribution facilities did not fare well during the past 15 years. The Sivaraman Committee Report on fertilisers has shown that the performance of co-operative agencies in distribution of fertilisers has not been up to expectations in many areas due to lack of marketing experience, inadequate distribution margins, etc. Fertiliser distributor has in many cases been a monetary drag on the co-operative system. The credit available to farmers from co-operative sector for purchase of fertilisers was also low—about 30 per cent of the total borrowing of rural households.

There has been heavy shortage of fertilisers in recent years and consequently a serious black market has arises in fertilisers. The Government has taken various steps to curb malpractices in fertiliser trade. Besides, the Government has asked the ports to give the highest priority to the handling of fertiliser ships in the interests of agricultural production. Likewise, fertilisers getting priority in rail movement.

TABLE 14.3

Consumption of Fertilisers (000 Tonnes)

Year	*Nitrogenous (N)*	*Phosphatic (F)*	*Potassid (K)*	*Total (N + P + K)*
1960-61	212	53	29	294
1965-66	525	133	77	785
1970-71	1470	541	236	2256
1975-76	2,149	467	278	2894
1978-79	3,420	1,108	591	5117

Source: Fertiliser Statistics.

Consumption of Fertilisers. Since the adoption of the New Agricultural Strategy in 1965-66, the consumption of chemical fertilisers has been growing rapidly. In recent years, special measures were taken to streamline distribution of available supplies through better transport regulated supplies of fertilisers to priority crops and areas specified by the State Governments ensuring adequate supplies of wagons through high level co-ordination with the railways, provision of short-term loans for fertilisers, promotion of balanced use of fertilisers (through a fertiliser promotion scheme), setting up of soil testing laboratories in various parts of the country for proper soil testing, and so on. In spite of rapid increase in the consumption of fertilisers in the country in recent years, India's position is much behind other progressive countries. Per capita consumption of fertilisers in India was estimated at 16 kg. per hectare in 1971-72 (it was only 8 kg in 1966-67) which compares very poorly with 690 kg. per hectare in Netherlands, 590 kg, in Belgium, 580 kg. in New Zealand and over 400 kg. in Japan.

Energy Crisis, Challenges to Agriculture and New Fertilisers Strategy

Till now, the new agricultural strategy has been based on heavy application of chemical fertilisers. India has been depending on imports of fertilisers or on import-based domestic fertilisers. But the energy crisis has given a new shift to the problem of chemical fertilisers. In the Panse Memorial Lecture (April 1974), Prof. V.K.R.V. Rao has outlined a few daring and un-orthodox methods to solve the problem.

The basic aim in the application of chemical fertilisers is to maximise output per hectare, even though it meant using fertiliser inputs at diminishing level of marginal productivity. So far large holdings with their large financial resources were able to go in for larger inputs of fertilisers that would maximise their individual monetary return. But India is now faced with the problem of limited amount of land as well as limited supply of fertilisers. Instead of maximising output per hectare, it would be more appropriate and economic to maximise the output per unit of fertiliser. According to Prof. Rao, a large number of holdings should be permitted to get limited supplies of fertilisers for use on their individual holdings. This may not maximise individual output, but this would definitely add to the national output. In support of his thesis, Prof. Rao refers to the research findings of the Department of Agricultural Research and Education (D.A.R.E). According to the DARE report, the "Management Mini-kit-programme" adopted in 1973 under the All-India Co-ordinated Rice Improvement Project resulted in a return of 50 kg. of grains for each kg. of nitrogen applied at moderate level with good management of non-cash inputs. According to the DARE report, "Good management of both cash and non-cash inputs resulted in a yield of 5,200 kg. per hectare which was 76 per cent of the yield obtained with four times more nitrogen applied with the sarac non-cash inputs." The simple and obvious conclusion which follows it this: better management with limited inputs of fertilisers can result in a large return per unit of fertilisers than where there is unlimited application of fertilisers.

While the current agricultural technology is based on the heavy use of fertilisers. Prof. Rao makes the unorthodox suggestion that India should try to minimise the use of fertilisers. It is true that HYV respond very well to the application of fertilisers but in general they have an inherent genetic capacity for giving higher yields than traditional varieties. The DARE report brings this out very clearly from actual field trials in different parts of the country for different crops.

TABLE 14.4

Average Response (Kg. Per Hectare)

Region State		*Variety*	*No. of yield trials*	*Average without fertiliser*	*Response N60*	*to N60 P30*
			Wheat			
All regions	HYV	2235	1988	695	1072	
	TV	1433	1650	496	755	
			Rice			
All regions	HYV	947	2913	748	1155	
	TV	861	2493	773	1157	

Source: Simple Fertiliser Trials on Cultivators' Fields 1967-68 to 1970-71, Report of DARE.

The DARE report come to the conclusion that average yields per hectare for HYV wheat and rice without fertilisers were higher than those of the traditional varieties without fertilisers. Prof. Rao's contention is that, in the context of the international energy crisis and high cost of fertilisers, India should go in for less use of fertilisers, but shift from traditional varieties of seeds to the new hybrid seeds. It would also be easier for the farmer with limited resources to adopt them a he does not have to go in for the costly cash input that is involved in the use of chemical fertilisers.

Finally, the fertiliser strategy in the Sixth Plan is oriented towards increasing use of organic manure, both farmyard manure nd urban and rural compost. On a rough estimate, over one-third of cowdung is not collected and one-third is used as fuel by the villagers; and the amount actually collected and used is about 349 million tonnes. At present, cattle urine which has valuable manurial properties is completely wasted. If cattle urine can be mixed with dung, the available manure will be about 400 million tonnes. The provision of alternative fuels to the rural population will help to increase the availability of farmyard manure. Besides the increasing use of gobar gas plants will help in making organic manures available to the cultivator. There is great scope for the manufacture of compost from urban waste, from forest litter and other waste materials and also for the use of green manures. All these will help in reducing our dependence on chemical fertilisers.

Ultimately the new fertiliser strategy, in the context of the energy crisis, will necessitate shift to coal-based fertiliser plants. At present, fuel oil is the feedstock for out fertiliser plants. But as India does not produce sufficient crude but has vast coal potentials-currently estimated at 200 billion tonnes—it is necessary to perfect the technology to increase the production of coal and decrease its cost of production and go all-out for a coal-based technology. The coal-based fertiliser factory fast coming up at Ramagundam in Andhra, will show the real possibilities of coal based fertiliser factories.

IMPROVED SEEDS

Indian farmers are aware of the importance of improved seeds in cultivation, for good seeds make possible an increase of 10 to 20 per cent in production. But they generally use seeds of very indifferent quality either because the special good quality seeds kept for sowing purposes are consumed away during the off-season or because good seeds deteriorate through bad storage. Agricultural Department, and the Indian Council of Agricultural Research have done much to evolve and pupularise improved and disease-resisting varieties of seeds suitable for different local conditions.

With the introduction of high-yielding varieties programme, the need for stepping up the production of better seeds has increased considerably. The Government set up the National Seeds Corporation in 1963 and the State Farms Corporation in 1969, to produce and supply quality seeds.

The National Seeds Corporation functions as an agency or production of foundation and certified seeds and promotes the use of improved varieties of seeds. The Corporation's sales increased from Rs. 3.6 crores in 1970-71 to Rs. 4.6 crores in 1971-72. With a view to undertaking the production and distribution of quality seeds of improved varieties, the State Governments have set up State Seeds Corporations. In this connection suitable schemes are being drawn up during the Sixth Plan.

The country enjoys self-sufficiency in seeds of major food crops and foreign markets are being exposed for exporting good quality seeds. A number of high yielding varieties of rice and wheat, hybrids of maize, jowar and bajra, have been introduced. The High Yielding Varieties Programme (HYVP) was started in 1966. Under the Fourth Plan, the coverage under high yielding varieties rose from the level of 9.2 million hectares in 1968-69 to 26 million hectares in 1973-74. By 1977-78 a total of 38 million hectares of land were covered by high yielding varieties. The Government is giving considerable attention to (a) research in evolving better seeds suitable to Indian conditions, (b) larger production and better distribution of quality seeds, (c) provision of adequate credit facilities to enable even the poorest farmers to purchase and use quality seeds and (d) expansion of irrigation, fertilisers and pesticides which are necessary for the efficient use of quality seeds.

SOIL CONSERVATION AND RECLAMATION

Although positive data are lacking, available tests show clearly that Indian soils have reached the lowest stage of deterioration. While heavy crops are grown year after year, very little is returned to the soil by way of manures. Apart from soil exhaustion and deterioration, there is also the problem of soil erosion. Soil erosion takes place when the surface soil is washed away through excessive rains and floods. Soil erosion occurs because of cutting of trees, removal of vegetation which exposes land to wind and rain, uncontrolled grazing and cultivation on hill slopes. In U.P. alone as many as 8 million acres have been lost through soil erosion. Other States to suffer heavily through soil erosion are West Bengal, Punjab, Maharashtra and Madhya Pradesh. Deserts of Rajasthan covering over 20 million hectares are posing a serious problem as they are creating conditions of soil erosion in nearby areas. It has been estimated that nearly one-fourth of the land area in the country—i.e. 80 million hectares—is suffering from soil erosion. Indian soils also suffer from the problems of salt effervescence and water-logging.

The remedies to soil erosion are preservation of forests and afforestation, contour building, regulation of land use, etc. Soil conservation measures were allotted Rs. 1.6 crores in the First Plan. Rs. 18 crores in the Second Plan, Rs. 72 crores in the Third Plan and Rs. 160 crores in the Fourth Plan. The Union Government set up the Central Soil Conservation Board in 1952 and similar boards were set up for each State. Over 24 million hectares of land have been brought under soil conservation measures. To check the advance of desert sands, the Government has been encouraging improved dry farming practices. During the Second and Third Plan periods, an area of about 9 million hectare was brought under improved dry farming. Besides, soil conservation programmes also include reclamation of water-logged saline and alkaline lands and of ravine lands.

LIVESTOCK SITUATION IN INDIA

Livestock can be in important source of income for a nation. In fact, pastoral production is far more resilient than crop production, and more remunerative too. the share of livestock production in the aggregate agricultural output in some of the West European countries is about 60 to 80 per cent in recent years. Livestock products like meat, milk and milk products have a prominent place in the dietary habits of people in the advanced countries. Besides, the general trend in these countries has been that the good articles of livestock origin are gradually relegating conventional food items like cereals and pulses to a position of secondary importance. As a result of this shift in consumer preference many European countries as well as the U.S.A. Canada, Australia and New Zealand are developing their agriculture on the basis of a prosperous, progressive and forward looking livestock enterprise.

Livestock in India and the World

It is a well-knows fact that India has the Largest number of livestock in the world.

Table 14.5 shows the livestock position in India according to the censuses of 1956 and 1966.

India, is at the top of the world having one-sixth of the livestock population—less than one-fifth of cattle, half of buffaloes, and over one-fifth of goats and sheep. China comes second with only 10 per cent. Countries which have well-developed livestock-based industries like Australia, Belgium, Denmark, Finland, Ireland, italy, Norway, the Netherlands, Sweden, Switzerland etc. have only small herds, that is, less than 1 per cent of the total.

TABLE 14.5

Livestock Resources in India

			(In crores)	
		1958	*1966*	*1972*
1.	Cattle	15.9	17.6	17.9
2.	Buffaloss	4.5	5.3	5.8
3.	Sheep	3.9	4.2	4.0
4.	Goats	5.5	6.5	6.8
5.	Others	0.8	0.8	8.8
	Total	30.6	34.4	35.5

Source: C.S.O. Basic Statistics Relating to the Indian Economy.

Livestock requires sufficient land area for their sustenance, particularly grazing land, consisting of permanent meadows and pastures. India has the lowest per unit amount of grazing land; even densely populated countries of Europe have much higher amounts of grazing land per unit of livestock. India has to many units of livestock depending upon too small a patch of land.

Livestock Wealth and National Income in India

A good measure of efficiency of livestock enterprise in any country is its contribution to the country's national income. The gross value of output from livestock was Rs. 1,476 crores in 1960-61. It rose to Rs. 1,608 crores in 1970-71 and further to Rs. 1,667 crores in 1972-73 (at 1960-61 prices). During the period 1960-61 to 1972-73, the index of gross value added rose to 112.9, signifying an annual growth rate of 1 per cent. This indicates the poor improvement of output from our livestock. On an average, income from livestock was about 16 to 17 per cent of the gross value of output from agriculture. This is extremely small when we consider the huge livestock resources of the country.

Out of the income contributed by the livestock in 1972-73 nearly 65 per cent was accounted for by milk and milk products. 9 per cent by meat and meat products, a little more than 5 per cent by eggs and poultry about 12 per cent by animal dung and nearly 3 per cent by hides and skins. The poor yield of Indian cows is proverbial. It is estimated that the average yield of cows in India is 220 kgs. per year, as compared to 3,000 to 5,000 litres per year in Western countries. As regards meat, about three-fourths of the meat produced in India is from goats and sheep. Beef accounted for only 6 per cent. Thus though cattle accounted for the largest proportion of livestock in India, their contribution to the national income of India was insignificant.

TABLE 14.6

Gross Value of Output from Livestock

(Rs. crores at 1960-61 prices)

		1960-61	*1970-71*	*1972-73*
1.	Milk and milk products	902 (61.1)	1037 (64.5)	1075 (64.5)
2.	Meat and meat products	128 (8.7)	141 (8.8)	150 (9.0)
3.	Hides and skins	44 (3.0)	45 (2.9)	46 (2.8)
4.	Eggs and poultry	84 (5.7)	88 (5.5)	90 (5.4)
5.	Dung	208 (14.0)	203 (12.6)	206 (12.3)
6.	Increment in livestock	84 (5.7)	59 (3.7)	63 (3.8)
7.	Other products	26 (1.8)	35 (2.1)	37 (2.2)
	Total	1447 (100.0)	1608 (100.0)	1667 (100.0)

Source: C.S.O., National Accounts Statistics 1960-61, 1972-73 Disaggregated table, March 1975.

Companies with other Countries

Table 14.7 shows clearly that the contribution of livestock to income from agriculture has been more than 50 per cent in all the advanced countries of the world. This is exceptionally high in some countries (75 to 85 per cent)—such as Denmark, Sweden, Norway, etc-for they enjoy exceptional environmental advantages for rearing quality livestock. India with the largest cattle population in the world contributes only 14 per cent of the aggregate agricultural income.

It has been observed that i n advanced countries like the United States of America, West Germany, England, etc. the share of livestock in the gross agricultural income has remained steady since the 1930s but at very high levels. On the other hand, in India, the share of livestock in the gross-agricultural income has also remained steady but at very low levels. The important conclusion from this fact is that with economic development, the share of livestock income in the gross agricultural income increases. A very interesting illustration may be given from Italy. In the early 1980s, about 36 per cent of the national income of Italy was derived from agriculture and the contributions of livestock was only 18 per cent. By the beginning of the 1960s Italy had already gone through the crucial stages of industrial development and the share of national income from agriculture had declined to 20 per cent but the contribution of livestock to agriculture had gone up to 37 per cent. A similar situation is seen in the case of Yugoslavia too.

TABLE 14.7

Contribution of Livestock i.e. National Income in some Leading Countries in 1958-59.

Country	*Share of Agriculture in National Income (Percentage)*	*Livestock Income as per cent of gross agricultural income (percentage)*
India	50	14
France	11	64
Denmark	16	83
Netherlands	11	69
U.S.A	5	60
U.K.	4	78
Swedan	6	79
West Germany	4	71

Source: Agricultural Situation in India, September 1966.

The reasons for the poor contribution of livestock enterprise to national income in India are not far to seek. The hopeless condition in Indian livestock position illustrates in a striking fashion what long years of utter indifference and positive cruelty in the name of religion can do to livestock. The reasons for the poor condition of livestock in the country are not economic but religious and social. In India, the large cattle population contains a sizeable proportion of aged animals, both made and female, which are capable neither of work nor of yielding milk. These animals are clearly a drain on available resources and a drag on the farm economy.

It has been estimated that a cross-breed cow can easily produce 2,500 liters of milk a year. About 22 million such cows can produce the total requirements of milk of the country where 55 million cows and 26 million buffaloes (according to the 1966 census) are at present unable to meet even half the requirements.

At the same time, expansion of agricultural machinery such as tractors, power tillers, pumpsets operated by power and oil, etc. is reducing the dependence of farmers on animal power. The only obstacle is that the small and fragmented agricultural holdings in the country necessitate the maintenance of too many heads of cattle.

It may be pointed out here that our fodder resources are sufficient only for about 60 per cent of our livestock strength. There is, therefore considerable malnutrition to which our animals are subjected. At the same time, death rate among cattle population has been declining due to better veterinary aid. The result is that cattle population has been increasing almost as fast as human population and today there is surplus livestock in India. About 100 million animals are surplus and are costing the country at least Rs. 1,800 crores a year to maintain (at the rate of 50 paise per animal per day). If the meat, hides, horns, etc. of these animals are used, they would bring in about Rs. 2,000 crores (at the rate or Rs. 200 per animal) much of it in the form of foreign exchange. These operations would also create additional jobs such as packers, tanners, etc. But unfortunately religion and superstition are standing in the way of the country's adopting a rational and humane policy of reducing surplus and useless cattle.

Five Year Plans and Animal Husbandry

The development programme of animal husbandry under the five year plans have a three-fold objective—to increase the supply of milk and milk products, to provide draught animals for farm, operations and to raise the output of such products as wool and hides which have commercial importance. The Fifth Plan outlay on animal husbandry and dairying was Rs. 438 crores; the Sixth Plan outlay is much larger—Rs. 670 crores. The Revised Sixth Plan was proposed to increase production of livestock products by improved methods of animal husbandry supported by a co-operative marketing structure. Planned cross-breeding for improving the productivity of cows and draught power of bullocks, improving the buffalo stock and promotion of poultry in rural areas, selective breeding and cross-breeding of sheep, goats and pigs, the production of high yielding varieties of louder—these are some of the policies to be vigorously implemented during the Sixth Plan.

Future Prospects

What then is the future of cattle wealth in the country? The future is indeed bleak, because of religious prejudices of the vast majority of people in the country. But the past experience of developed countries can serve as a guide for India. In other developed countries, industrialisation has meant increase i n national and per capita income. With increase in income, people change their consumption habits. They shift from cereal-based food to diversified and protein reinforced diet. This will necessitate a re-adjustment of agricultural and pastoral production in such a way that the latter is given greater importance. The developments in animal husbandry will help the livestock enterprise to catch up with the increasing demand and consumption and consumer preferences. These developments can be expected in India in the future but the only snag, as mentioned before is the religious prejudice against the killing of cattle, even the most useless.

MECHANISATION OF AGRICULTURE

The tools and implements used by the Indian farmers are primitive, crude and antiquated, as compared to the most up-to-date farm machinery used by the farmers of the West. As a result of progressive mechanisation of agriculture, these countries have been able to experience an agricultural revolution comparable to industrial revolution of 18th and 19th centuries. Mechanisation of agriculture has resulted in increased agricultural production and reduction of costs. Besides, agricultural machinery has been useful in reclaiming barren lands. Thus, the prosperity and richness of peasantry in the Western countries has been due, largely, to the extensive use of farm machinery. Naturally there is now a common belief that progressive agriculture is impossible without mechanisation of agriculture.

By mechanisation of agriculture we mean the replacement of animal and human power by machinery wherever it is possible. Ploughing is to be done by tractor; sowing and putting of fertiliser by the drill; and reaping and threshing by the combined harvest-thresher; and so on. The tractor will also be used i n transporting crops to markets. The old-fashioned wooden ploughs, bullocks, sickles, etc, are eliminated and the work is done by machinery. Thus, mechanisation of agriculture stands for the use of machinery in all farming operations, ranging from ploughing to the marketing of the produce. While some regard mechanisation as desirable and necessary and want Indian farmers to give up out-moded implements and primitive methods, there are others who believe that farm machinery is utterly useless for India in the context of India's economic and social conditions.

Case for Mechanisation of Agriculture

The case for mechanisation of agriculture is based primarily on the advantages of large-scale production made possible by machinery. Machines work faster and work accurately. Man by himself can produce only very little but with the help of machinery he can produce much more. What a farmer with a pair of bullocks can plough in 10 days, a tractor can do in one day, and far deeper too. Secondly, farm machinery has relieved man of much of the heavy work. For instance, land reclamation, digging and carrying of earth, ploughing etc. are all heavy jobs. Farm machinery helps to perform these tasks easily. Thirdly, from machinery has led to large-scale production. Huge plots of land can be ploughed; large crops can be harvested; huge amounts of produce can be taken to the market; all these can be done by machinery without any loss of time. Output per acre as well a per man can be increased. The cost of production is reduced. Finally, mechanisation will create employment opportunities in industry, transportation, etc. and, therefore , those thrown out of employment in agriculture will be employed. There is, therefore no fear of unemployment being created by farm machinery.

In Western countries, particularly the U.S.A. and the U.S.S.R, where extensive mechanisation of agriculture has taken place, agricultural productivity has been raised manifold. In the U.S.A., for instance, only 12 per cent of the population depends on agriculture. But agricultural production has been increased so much that the American farmers are able to send their surplus farm produce to other countries of the world. As India's present problem is to raise production, there is a good case for mechanisation of agriculture.

From the point of view of immediate applicability too, there is considerable scope for mechanisation. For instance, tractors are very useful for extensive jungle clearance, reclamation of waste lands, in drainage and soil conservation operations, etc. Besides tractors, there is good scope for pumping sets and electrically operated tubewells. Electronic motors and diesel engines can be very usefully adopted for other agricultural operations such as cane crushing, oilseed crushing, etc.

Case Against Mechanisation

While the case for mechanisation of agriculture has some validity in the context of the need to raise agricultural production as fast as possible within the shortest possible time, the case against mechanisation is very much stronger. In the first instance, mechanisation has no scope in India because of the extremely small size of holding between 3 and 12 acres. Even these small holdings are scattered over the village in tiny bits. A tractor cannot be used to plough a half-an-acre plot. The existence of a large farm is an essential condition for mechanisation. For proper and best utilisation of agricultural machines, holdings will have to be large and should be found together and not scattered in tiny plots as is the case in India. In U.S.A the average size of a holding is about 60 hectares, in Canada it is 90 hectares and in U.S.S.R. it is 600 hectares. Mechanisation is possible in these countries but not in India. But this argument ignores the fact that machines suitable for small farms have been developed and successfully used in Japan.

Secondly, mechanisation of agriculture will result in too many agricultural workers becoming surplus. According to one estimate if complete mechanisation is introduce, it is possible to cultivate the existing area of land in India with only 3 to

4 million persons. Millions of farmers will be thrown out of agriculture and will have to be provided alternative source or employment. It is impossible to absorb such a large number in non-agricultural occupations. This is the most serious objection to mechanised farming. In U.S.A. and Canada, the real problem is shortage of labour. To overcome this difficulty, machinery is used. But this is not the case of India where there is abundant labour. There is no denying the fact that agricultural mechanisation will displace labour. But the displacement can be kept to the minimum possible. Besides, expansion of rural industries will diversify employment opportunities in rural areas.

Thirdly, the adoption of mechanised farming will render the existing cattle population surplus and unnecessary. To cut down the existing cattle population will also be a difficult problem. However, reduction of our cattle population is necessary since feeding them is itself a huge burden. In many parts of the country, there is a serious competition among men and cattle for subsistence from land.

Fourthly farm machinery will require petrol, diesel and kerosene oil. There is already a great shortage of these mineral oils in the country as internal productions is highly inadequate to meet internal demand and the international price of oil has gone up due to the oil crisis. India cannot afford to use extensively oil-based farms machinery.

Finally, mechanisation may increase productivity of labour, since primarily it comes to help as well as displace labour. But it may not increase productivity of land. In India the crucial problem is to increase the productivity of land—land being a scarce resource of the country. This is more important than the problem of raising productivity of labour. In a country like Japan where mechanisation of farming is not adopted, productivity per acre has been maximised because of intensive cultivation.

Scope for Mechanisation

It is true that full-scale mechanisation is not possible in India, as is being adopted in other countries. All the same, it is wrong to argue that there is absolutely no scope for mechanisation in a labour-surplus developing country like ours. Machine suitable for small farms, and those suitable for large co-operative and State farms can be introduced. Besides, agricultural machines may be necessarily—or even inevitable—in reclamation of land, conversion of forest land for cultivation, ploughing of barren lands, construction of wells in rocky areas, etc.

Selective Mechanisation—The Approach of the Fifth and Sixth Plan

In view of the fact that the size of holdings is small but the size of the farming population in India is very large, it would be a grave mistake to push forth a policy of indiscriminate mechanisation of agriculture. In India, land is scarce, but labour is an abundant factor, and consequently, policies aiming at improving land productivity must be co-ordinated with those leading to utilisation of rural manpower. It is therefore necessary to pursue a policy of limited mechanisation so that the labour displacement effects are minimised.

Recognising the need for such a pragmatic approach, the Draft Fifth stated. "In the Fifth Plan, a policy of selective mechanisation is proposed to be adopted. The general objective would be to increase cropping intensity and farm productivity. The present availability in terms of power is estimated at 0.4 H.P. per hectare (of which machine power is only one-fourth) in the country. This is extremely inadequate and will need stepping up. There is also the allied consideration that the new agricultural technology requires faster farm operations as well as accuracy and timeliness in undertaking them. Finally, there is the emerging problem of increasing cost of purchasing bullocks and their recurring maintenance. While all these factors underlie the need for farm mechanisation, the rate and form of mechanisation will have to be determined keeping in view other relevant factors such as size of holdings as well as the impact of mechanisation on agricultural labour employment situation. The last consideration would be particularly relevant with regard to combine harvesters. Their introduction would be confined to areas where there is a perceptible scarcity of labour during harvesting season."[1] The Sixth Plan states: "Unrestricted mechanisation of agricultural operations will not however, be in the interest of out country as it severely worsens the rural unemployment problem. As such a policy selective mechanisation will be adopted."[2]

TABLE 14.8

Census of Agricultural Machinery (in thousands)

		1956	*1972*
1.	Iron Ploughs	1,376	5,359
2.	Sugarcane crusher-Worked by power	23	87
3.	Oil pumps for irrigation purposes	123	1,628
4.	Electric pumps for irrigation purposes	47	1,628
5.	Agricultural Tractors	21	148

Source: India, 1974, p. 175 and Indian Agriculture in Brief 42-43.

Progress of Mechanisation in India

Some idea of progress of agricultural mechanisation can be had from the periodical censuses carried out in our country about agricultural machinery.

TABLE 14.9

Rate of Growth of Agricultural Mechanisation

Per one lakh hectares of cropped land

	Tractors	*Oil engines*	*Electric pumps*	*Commission of electricity (lakh) per one thousand hectares of cropped area*
1951	7	63	20	1,530
1956	14	83	32	2,145
1961	20	151	105	5,453
1966	35	304	268	12,214
1971	126	—	1,003	22,900*

* for 1970.

Source: Commerce, Annual Number, 1972, p.4.

The figures of Table 14.8 and 9 show that there is growing mechanisation in India, but the progress was indeed quite slow. However, the rate of growth of mechanisation in agriculture has been sufficiently fast since 1961. These figures are self-explanatory. The use of tractors, oil engines for irrigation purpose and electric pumps has increased extensively. Likewise consumption of electricity for agricultural purposes has also increased considerably- from about 1,540 kwh per 1,000 hectares in 1951 to about 22,900 kwh per 1,000 hectares in 1970.

The production of tractors for agricultural purposes is another good indication of the growing use of agricultural machinery in the country. In 1956, there were 21,000 tractors in use in India, and the number increased to 218,000 in 1974.

To facilitate the purchase and distribution of tractors, power, tillers, pump sets and other items of agricultural machinery to farmers in their respective states, agro-industrial corporations have been set up in various States.

SELECT REFERENCES

Ministry of Agriculture: Report 1977-78.

V.K.R.V. Rao: Dr. Panse Memorial Lecture on "New Challenges before Indian Agriculture."

Planning Commission: *Draft Five Year Plan* 1978-79 Vol.I and Vol.III.

NOTES

1. Planning Commission. *Draft Fifth Five-Year Plan*, 1974-79,-Pt. II.p.14.
2. *Draft Five Year Plan*, 1978-85, Vol III, ch 1. data given in Table 14.10.

15

LAND REFORMS

> "There is only one fundamental method of improving village life namely, the introduction of a system of peasant proprietorship under which the tiller of the soil is himself the owner of it and pays revenue direct to the government without the intervention of any zamindar or taluqdar."
>
> —*Indian National Congress Resolution, 1935*

THE NEED FOR REFORM IN A DEVELOPING ECONOMY

Productivity in agriculture is mainly dependent on two sets of factors-technological and institutional. Among the technological factors are the use of agricultural inputs and methods such as improved seeds, fertilisers, improved ploughs, tractors, harvesters, irrigation etc. which help to raise productivity, even if no land reforms are introduced. The institutional reforms include the redistribution of land ownership in favour of the cultivating classes so as to provide them a sense of participation in rural life, improving the size of farms, providing security of tenure, regulation of rents, etc. In other words, the institutional factors, such as the existence of feudal relations, small size of farms, sub-division and fragmentation, insecurity of tenancy rights, high rents, etc. act as disincentives to the peasantry to raise production. They weaken the capacity of the farmers to save and invest in agriculture also to enjoy the fruits of their labour. Consequently, two schools of thought emerged: The Socialists believe that the existence of feudal or semi-feudal relations was the real cause of backwardness and poverty in rural communities. The emancipation of the peasantry from the bondages of institutional depressors will unleash forces which shall automatically raise levels of production in agriculture. The other school of thought believes that agricultural productivity is purely a technological phenomenon and that it can be raised by the application of superior agricultural methods. Thus, whereas the key to higher productivity lies in technological change according to one school, it lies in institutional reform according to the other, Quite recently both the schools of thought are converging and opinion has come to centre round the idea that land reforms and technological change are not mutually exclusive factors but are complementary in the process of agricultural development. It is held that technological change can work more effectively in a congenial agrarian structure and i n this way the process of development can be accelerated.

The purpose of land reforms is, therefore two-fold. On the one hand, it aims to make more rational use of the scarce land-resource by affecting condition of holdings imposing ceilings and floors on holdings so that cultivation can be done in the most economical manner i.e. without any waste of labour and capital; on the other, it is a means of redistributing agricultural land in favour of the less privileged classes and of improving the terms and conditions on which land is held for cultivation by the actual tillers, with a view to ending exploitation.

THE CONCEPT AND SCOPE OF LAND REFORMS

Land reforms aim at redistributing ownership holdings from the viewpoint of social justice, and reorganising, operational holdings from the viewpoint of optimum utilisation of land. Besides this, there is the problem of conditions of tenancy, i.e. the rights and conditions of holding land. Land reforms aim at providing security of tenure, fixation of rents, conferment of ownership, etc. The entire concept of land reforms aims at the abolition of intermediaries and bringing the actual cultivator in direct contact with the state. Along with this, the provisions of security of tenancy and rent regulation provide a congenial atmosphere in which the agriculturist feels sure of reaping the fruits of his labour. Need it be mentioned that in the rural economy, the agriculturist is the lynch-pin of the whole social chariot, and that in order to revitalise the rural economy, it is he who should be given a sense of participation. Thus, the scope of land reforms entails:

(a) abolition of intermediaries; (b) tenancy reforms i.e. regulation of rent, security of tenure for tenants and conferment of ownership on them; (c) ceilings and floors an land holdings; (d) agrarian reorganisation including consolidation of holdings and prevention of sub-division and fragmentation; and (e) organisation of co-operative farms.

THE ABOLITION OF INTERMEDIARIES

When India achieved independence, the land system of the country was a pandora's box of various tenures and it was believed that the land tenure system then obtaining was unsuitable both from the viewpoint of more production and from the viewpoint of social and economic justice. Consequently, the need for re-organising the tenurial system was considered urgent and imperative. But before we discuss the proposals for reorganisation it would be advisable to give a brief account of land tenure systems on the eve of independence.

Three Types of Tenures on the Eve of Independence

It is customary to classify the various categories of land tenures into three broad head: Zamindari, Mahalwari and Ryotwari.

(i) Zamindari Tenure

Under the Zamindari system, which was introduced by Lord Cornwallis in 1793 in Bengal, land was held by one person or at the most by a few joint owners who were responsible for the payment of land revenue. The system was introduce by the East India Company to create vested interests in land and thereby cultivate a privileged and loyal class. The various forms of tenure such a Zamindari, Jagirdari, Inam, the princely States, etc., were artificially created. The revenue collectors were raised to the status of landowners. Earlier they were responsible for collecting land revenue for which they received a commission. The Zamindari settlements made them owners of land, thereby creating a permanent interest in land.

The Zamindari settlements were of two types-permanent settlements and temporary settlements. The permanent settlement fixed land revenue in perpetuity. The system was introduced in Bengal, North Madras and Banaras. Under temporary settlement land revenue was assessed for a period ranging between 20 and 40 years in various states. Land revenue, therefore, was subject to revision. Temporary settlement was effected with remaining zamindars of Bengal, taluqdars of Oudh, etc. Since the period of assessment was fairly long, temporary settlement was temporary only in name. Land revenue was thus fixed an in doing so, the primary object of the East Indias Company was to fix responsibility for the punctual payment of land revenue.

The British Government pleaded that the zamindars represented the most enlightened section of the rural population and the conferment of tenurial rights would result in improvements on land and better agriculture. But these expectations were not fulfilled. With growing population and decaying village industries under the British rule, the demand for land grew and it was possible for the landlords to charge very heavy rents from tenants. Zamindari system supposedly introduced to foster progressive agriculture degenerated into absentee landlordism. Thus, between the State and the actual tiller, there grew an intermediary who was interested in land only to the extent of extraction of exorbitant rent. Historically, the landlords as a class are known for their extravagance on women, wine and vices. The landlords of India were no exception. Thus, the money extracted from the cultivators by these parasites did not result in capital formation but increased conspicuous consumption. The zamindari villages were thus divided into two agricultural classes-the absentee owners and non-owner cultivators. The absentee owners exploited the actual tillers. Absence of state interference of any type gave a free hand, to the exploiting classes to indulge in rack renting, evictions, begar[1] and many other social evils. The landlords symbolised oppression and tyranny. Indian agriculture was reduced to a form of subsistence farming. It was disincentive-ridden, but there was no escape from it, since it represented the principal source of livelihood for the masses.

(ii) Mahalwari Tenure

Under the Mahalwari tenure the village land were held jointly by the village communities the members of which were jointly or severally responsible for the payment of land revenue. The system was first introduced in Agra and Oudh and later on in Punjab. Under the system, the village common or Shamlat is the property of the village community as a whole. Similarly, the waste lands also belonging to the village community and it is free to rent it out and divide the rents among the members of the community or partition it to bring it under cultivation without any leave of the Government. The system is the product of Muslim tradition and development particularly in Punjab.

A certain sum is assessed as land revenue for the whole village for which the whole body of co-shares are jointly and severally responsible. The village lumberdar collected revenue for which he received panchotra, i.e. 5 per cent as commission.

(iii) Ryotwari Tenure

Under the Ryotwari tenure, land may be held in single independent holdings. The individual holders were directly and severally responsible to the state for the payment of land revenue. The first Ryotwari settlement was made in Madras in 1972. It was the product of Hindu tradition. This form of tenure was prevalent in Bombay, Berar and central India. The ryot is a liberty to sub-let his land and enjoys a permanent right of tenancy so long as he pays the assessment of land revenue. Some elements of zamindari tenure did appear in this system too, because the peasants in ryotwari areas could sublet their land.

The popular nomenclature of ryotwari, mahalwari and zamindari concealed the vast transformation that had taken place during 150 year practice. Mr. Venkatasubbiah emphasising the point mentions. "If Lord Cornwallis and Sir Thomas Munro, the respective protagonises of the zamindari and the ryotwari, were to look at the system in 1940 they would barely reorganise them as such."[2] The co-existence of zamindari ryotwari and inahalwari led to interesting of characteristics. But the three systems gravitated towards the tendencies of the zamindari system. Sub-letting. rack-renting became a common characteristic even in the ryotwari areas. The mahalwari system acquired the characteristics of the zamindari system in states like Madhya Pradesh and U.P. (Agra) where the Government laid emphasis on joint responsibility of the village for land-revenue assessment; at the same time, it acquired the characteristics of absentee landlordism of the the ryotwari in Punjab where emphasis was on several responsibility for the payment of land revenue. Similarly, in inams and Jagirdari areas, the zamindars demanded between a half and two-thirds as settlement. As there were no records, they could charge quit rents from the cultivators. Thus, on the eve of independence, on the one extreme, there were landless labourers and tenants-at-will an don the other, were big landlords owning huge estates. But a very disquieting feature of the situation was the absence of the proper revenue records which made the task of abolition of intermediaries more difficult. Consequently, the need for a complete census of holdings was felt. the intermixing of the various systems made it difficult to know the rentier class a defined by the earlier acts.

Abolition of Intermediaries-the Policy and Measures

After independence a spate of enactments were passed by state legislatures to abolish intermediaries. While the aim was to abolish intermediaries between the 'tiller and the State,' in actual practice the legislative enactment equated intermediaries with zamindars and, consequently, the legislation left a class of rent-receivers and absentee landlords under ryotwari untouched. Mr. Venkatasubbiah emphasising the point writers: "The declared initial agrarian policy of the Congress did not amount to the elimination of all the million and a half rentiers but probably a large majority of them. The Party and the Government at the Centre and in the States began to give thought to curtailing the power of the non-zamindari rentier only at a subsequent stage of their agrarian policy."

Although steps were taken earlier the actual abolition of intermediaries started in 1948 with the enactment of legislation in Madras. Legislation was passed in all states, but for a few minor tenures and inams as in Assam, Gujarat, Madras and Maharashtra. Incidentally, it may be mentioned that West Bengal-the state worst affected by the ravages of absentee landlordism-was among the late comers to adopt legislation in 1954-55. Madras, Bombay and Hyderabad effected legislation in 1949 and 1950; followed by Bihar, Madhya Pradesh, Uttar Pradesh, Madhya Bharat, and Assam in 1951; Orissa, Punjab, Saurashtra and Rajasthan in 1952; PEPSU, Vindhya Pradesh and Bhopal in 1953, and West Bengal, Himachal Pradesh, Mysore and Delhi in 1954-55. As a result of the conferment of rights, about 30.4 lakh tenants and share-croppers acquired ownership, rights over a total cultivated area of 62.2 lakh acres throughout the country.

Problems Relating to the Intermediaries

State after State was effecting legislation abolishing intermediary rights. Naturally, the landlord class which held sway over the destinies of millions of peasants in the pre-independence period was not expected to swallow the bitter pill of tenurial reforms quietly. Instead of waging a fight on the desirability or otherwise of legislation pertaining to intermediaries, the scene of struggle shifted to extracting more compensation for the expropriated rights; and to resume cultivation and eyict tenants by force, or pressure (social, legal or illegal). In several states, the zamindars could effect partitioning of land before the new legislations were brought into force.

Forced evictions took place on a massive scale on the plea of resumption of cultivation and tenants were reduced to the position of share-croppers. Co-operative farming provided another convenient method of by passing land reform legislation. It would be worthwhile to examine these issues in a little more detail.

Compensation to Intermediaries

Unlike Communist countries, abolition of intermediaries was, not done without compensation. In Russia, China, Yugoslavia, etc., Landlords were expropriated from land without any compensation. They were reduced to the position of wage earners at the collective farms. But the Congress Party which assumed power after independence was committed to a system of peasant proprietorship and this transformation was to be done after the payment of compensation to the landlords. The idea that no none should be permitted to hold more than a certain specified amount of land had entered the minds of the masses and the official policy was just a recognition of this idea. On the compensation question the controversy did not centre round whether to pay or not to pay compensation, but round the problem how much to pay as compensation.

Although the makers of the Constitution provided for compensation they did not clearly mention 'just and equitable compensation.' Consequently, the Zamindari Abolition Acts were challenged in the High Courts and later taken to the Supreme Court for adjudication. The Supreme Court while upholding the right of the legislatures to acquire land for a public purpose, ruled that compensation is a justiciable issue. As a result, Article 31 of the Constitution was amended. Thus, the Constitution makers who wanted to take the compensation clauses of fundamental rights out of the pale of justiciability, were not allowed to do so. The rates of compensation, the ceiling limit of compensation and even the principles determining compensation and even the principles determining compensation were revised and the landlords were quite successful in getting equitable and in some cases more than equitable compensation.

The basis and rate of compensation varied from state to state. Compensation was fixed as a multiple of net income of the proprietor at the time of expropriation. This multiple was high in the case of lower income brackets and declined in upper income brackets. In some States, a uniform multiple of net income was introduced as compensation, but proprietors with small incomes were, in addition, to be paid rehabilitation grant. In some States, compensation was a multiple of the revenue assessment. Yet in some other States compensation was correlated with the market value of land (e.g. in Kerala). It has been provided at specified amounts for classes of lands (the amount varying between Rs. 50 and Rs. 900 per acre).

The compensation was, however, to be paid in cash or in bonds. These bonds were to be redeemed in equal instalments spread over a long period ranging between 10 to 30 years in various states. The big proprietors were to be given bonds but the comparatively small proprietors were to be paid in cash. The ex-intermediaries have been given compensation amounting to Rs. 670 crores in cash and in bonds. The three big states, viz, Bihar, Uttar Pradesh and West Bengal, had to pay Rs. 239 crores, Rs. 198 crores and Rs. 70 crores respectively as compensation. Taking the three states together a total sum of Rs. 507 crores was paid as compensation.

TENANCY REFORMS

The Problem of Tenancy Cultivation

Under the Zamindari and Ryotwari systems, tenancy cultivation has been quite common in India. Tenancy cultivation may be done by small proprietors who find that they have a insufficient quantity of land or it may be carried on by landless labourers. Sometimes, the tenants holding land from an intermediary may sub-let it for cultivation. So in this way, tenancy cultivation has been going on. Broadly speaking, tenants are divided into three categories: (i) Occupancy or permanent tenants, (ii) Tenants-at-will or temporary tenants and (iii) Sub-tenants. The rights of tenancy of the occupancy tenants are permanent and heritable. They can also receive compensation from the landlord in case they make some improvements on land. They enjoy a fixity and security of tenure which makes them the virtual, owners of land. It can be said, the only difference between the occupancy tenant and the peasant proprietor is that the former is required to pay rent to the landlord and the latter to pay land revenue to the State. So for all practical purpose, occupancy tenants are treated as land owners.

The position of tenants-at-will and that of sub-tenants is extremely weak. They are subject to ruthless exploitation. Frequent enhancement of rent, evictions on minor pretexts of several kinds, extractions and begar[4] are some of the popular ways of exploitation. It is very desirable that this category of tillers be brought in direct relation with the state and their rights of tenancy be protected. In a country where the demand for land is much more than its supply on account of a growing population, exploitation of weak and unprotected tenants at the hands of reckless landlords is a widespread evil. Fifty per cent of the produce was the normal rent under the Batai or share-cropping. On several occasions, the peasants had to forego even two-thirds of the produce as rent. A situation like this coupled with insecurity necessitated tenancy reforms.

The Extent of Tenancy Cultivation

The National Sample Survey in 1953-54 (8th Round) made a estimate of the land held under tenancy and sub-tenancy in different parts of India. The percentage of area leased out varied from 11 to 26 per cent, though the all-India average was 20 per cent. It showed that about one-fifth of the total area was held under tenancy and thus it was not possible to ignore a problem affecting such a wide area. It would be desirable to note that lands held by occupancy tenants are not included because their right of tenancy is permanent and they are, therefore , treated as owners of land. Since then, evictions on a considerable area had taken place. To the extent tenants have acquired rights of ownership, these data do not reflect the correct position existing at present, but may be considered a rough indicator of the situation. According to 1961 census, 17 per cent of the total cultivating households were in the nature of ownership holdings, 8 per cent on pure tenancy and 15 per cent in mixed tenancy. Compared with this 78 per cent of the total cultivated area was under ownership holdings, 4 per cent under pure tenancy and 18 per cent under mixed tenancy. Inequality of land ownership can be judged from the fact that the top 5 per cent of rural households own more than what the bottom 80 per cent own.

Besides this open tenancy, there is a considerable amount of land leased out on the basis of oral or hidden tenancy which accounts for anything between 35-40 per cent of total cultivated area.

Measures of Tenancy Reform

The legislation for abolition of intermediaries is aimed at providing land to the tiller but it does not mean that it shall put an end to the problem of tenancy. Some leasing is bound to remain. The Panel on Land Reform recognised this fact and recommended the desirability of allowing persons who suffer from disability, namely, a widow or an unmarried woman, a minor or a person suffering from mental infirmity and members of the armed forces to lease out their lands. Moreover, even with the limit of ceiling, it may not be possible for a family to cultivate the entire land and so some sub-letting is unavoidable. Besides, in order to induce agricultural population to take over the non-agricultural occupations, some sub-letting to tenants may be allowed. A total ban on letting or sub-letting land would neither be socially desirable nor administratively practicable. It is, therefore more rational to take measures to minimise the evils of tenancy cultivate.

Measures of tenancy reform pertain to (i) Regulation of rent; (ii) Security of tenure; and (iii) Conferment of ownership on tenants.

Regulation of Rents

During the pre-Independence period, rents were fixed either by custom or were the result of the market forces of demand and supply. Supply of land being fixed, the demand for land growing with an increasing population, there has been a continuous tendency for rents to rise. The decay of handicrafts increased the dependence on land still further and thus pushed up rents. Rack-renting was a common feature in Indian agrarian structure.

It was, therefore, imperative that rents should be reduced by enacting legislation. The rates of rent prevalent were one-half of the produce or more. Considering the return on ivestment in other sectors of the economy, these rents were excessive by any standard of social justice. Consequently, the First and the Second Plan recommended that rents should not exceed 1/4th or 1/5th of the gross produce. Various States have passed necessary legislation regulating rents, but there are large variations in the rates of rents fixed in different states. In Gujarat, Maharashtra and Rajasthan, one-sixth of gross produce is fixed as maximum rent. In Assam, Karnataka, Manipur and Tripura, maximum rents vary between one-fourth to one-fifth of the gross produce. In Orissa and Bihar, one-fourth of gross produce has been fixed as rent. In Kerala maximum rent varies between one-fourth to one-half. In Punjab, one-third of produce has been considered as fair rent, while in Tamil Nadu, it is between 33.3 and 40 per cent of gross product. In Jammu and Kashmir, one-third of the gross produce, in Andhra Pradesh one-fourth of the gross produce for irrigated lands and one-fifth in other cases has been fixes as rent. In October-November, 1970, the Government of Uttar Pradesh decided to abolish rent on all the 1.4 lakh holdings below a size of 6.25 acres.

Owing to the weak position of the tenants and the prevalence of the widespread land hunger, the law regulating rents is observed more in its breach than in its compliance. The Third Plan rightly asserted: "When there is pressure on land and the social and economic positon of tenants in the village is weak, it becomes difficult for them to seek the protection of law. Moreover, resort to legal processes is costly and generally beyond the means of tenants. Thus in many ways, despite the legislation, the scales are weighed in favour of the continuance of existing terms and conditions."[5]

Another suggestion in this regard is to fix rents in cash rather than in kind. Historically, rents have been paid in kind in this country, but in view of the fact that the peasants have to make a good many payments in money while purchasing seeds, fertilisers, implements and other necessaries of life, it would be desirable to switch over to cash payments of rents. This is in fitness with the requirements of a rural economy changing rapidly, from barter to money exchange. The Fourth Five-Year Plan, in this connection, states: "The rents as fixed by law are still high in Andhra area, Jammu and Kashmir, Tamil Nadu, Punjab and West Bengal and should be brought down to the level recommended in the Plans-to one-fourth or one-fifth of the gross produce. Beside, produce rents which are difficult to enforce should be abolished and replaced by fixed cash rents so that the uncertainties arising out of annual fluctuations in rents may be eliminated and the tiller assured of the full benefits of his investment.[6]

Security of Tenure

Sir Arthur Young rightly observed: "Give a man the secure possession of a bleak rock and he will turn it into a garden; give him a nine years' lease of a garden and he converts it into a desert." This remark very pithily sums up the need for providing security of tenure. The personal interest of a cultivator in land with rights of temporary tenancy is very thin. Tenants, therefore, take much less care in preparing land, sinking capital in the form of wells or tube-wells on land, or putting up a permanent fence etc. The fear of loss of tenancy right saps all initiative to make improvements on land, reclaim waste-land or make long-term schemes of preserving soil, fertility. Consequently, the ends of social justice and maximum production both necessitate the adoption of legislation granting security of tenure. The purpose of such legislation should be to confer the rights of permanent occupancy.

While framing legislation pertaining to security of tenure, three essential aims have to be keep in mind-firstly, the large-scale ejectments of tenants do not take place; secondly, that resumption of land may be taken up by the owner for personal cultivation only; and thirdly that in the event resumption, a prescribed minimum area is left with the tenant.

The problem of resumption for personal cultivation and eviction, Experience of the implementation of zamindari abolition suggests that on the plea of resumption of land for personal cultivation, evictions of tenants have taken place on a considerable scale. It was mainly because in the initial years of land reform, several loopholes were left in the definition of personal cultivation. During the First Plan period, the definition of personal cultivation was based on three elements, namely, risks of cultivation, labour and personal supervision. But while defining 'labour', it was implied that work done by the 'owner' or a member of his family be treated as personal labour. Moreover, this condition was not obligatory and personal supervision was sufficient for justifying personal cultivation. Thus, under the garb of personal cultivation, evictions took place on a massive scale. During the Second Plan, it was suggested that 'personal cultivation' should be so defined as to provide for personal labour as a necessary condition but so far this suggestion has not found favour with the states. The obvious result of half-heartedness in providing a more rigourous definition of personal cultivation to that "in many cases lands, my have been resumed ostensibly on grounds of personal cultivation, but are cultivated through arrangements lacking some of the essential elements of personal cultivation."

The question that obviously arises is: Should resumption of cultivation be allowed? There is no doubt that in certain cases and categories of holders, resumption should be allowed, but the plea of resumption should not lead to large-scale ejectment of tenants. For this purpose safeguards are needed. During the Second plan states framed provisions for resumption broadly on the following three different patterns:

(i) All tenants have been given full security of tenure, without giving the owners the right for personal cultivation.

(ii) Owners have been given the right to resume a limited area (not more than a family holding in any case) subject, however, to the condition that a minimum area is left with the tenant.

(iii) A limit has been placed on the extent of land which a land-owners may resume, but the tenant is not entitled to retain minimum area for cultivation in all cases.

Uttar Pradesh, West Bengal and Delhi belong to the first category. Gujarat, Kerala, Madhya Pradesh, Maharashtra, Orissa, Rajasthan, Himachal Pradesh, Assam Punjab belong to the second category. Jammu and Kashmir, Manipur, Tripura and West Bengal (in case of sharecroppers) belong to the third category. Obviously, the second type of legislation has by and large been accepted by the states.

Voluntary Surrenders and Restorations

Dr. Khusro's study 'Economic and Social Effects of Abolition of Jagirdari and Land Reforms in Hyderabad' (1958) revealed that evictions of tenants took place on a massive scale. The results of this study are very disturbing, 42 per cent of the tenants suffered at the hands of landed aristocracy and every method, legal or illegal, was applied to compel them to surrender their tenancy rights, Similarly, a study conducted by V.M. Dandekar and G.J. Khundanpur in 1957 on the working of the Bombay Tenancy Act. (1948) revealed that in the five years from 1947-48 to 1952-53, the ratio of protected tenants to the total number of tenants declined from more than 60 per cent to little more than 40 per cent. Moreover, it was disclosed that while on paper 85 per cent of the area resumed by the owners had been voluntarily surrendered, the tenants concerned had in fact been under severe pressure from the owners in about two-thirds of the cases. The net result is that the tenant is reduced to an inferior position of a share-cropper and is subjected to the same risks of exploitation which he faced earlier.

To check the evil of voluntary surrenders, two recommendations were made in the Third Plan. First, the voluntary surrenders of land by tenants should not be considered valid unless they were duly registered by revenue authorities, and second, in case of voluntary surrenders, the landowner should be entitled to undertake cultivation on land only to the extent of his right of resumption by law. There is much leeway to be covered in implementation in this regard so as to save the poor tenants-the most vulnerable, though the most important section in rural India.

Informal or Oral Tenancy

Informal tenancy has been a common feature of traditional agricultural societies. Although attempts have been made to provide security of tenure, redistribution of land and fixation of fair rents, yet informal or oral tenancy has continued to exist even to this day. the term 'informal tenancy', loosely referred to as oral tenancy, refers to tenancy without legal sanction and permission. In other words, tenancy arrangements which are made without any written agreement in violation of the prevailing tenancy Acts are covered in this category.

TABLE 15.1

Study of Tenancy Rights in Hyderabad

		Out of every 100 protected tenants
1.	Those who still remained as protected tenants	45.5
2.	Those who purchased land and became owners	12.4
3.	Those who were legally evicted	2.6
4.	Those who were illegally evicted	22.1
5.	Those who voluntarily surrendered	17.5

In two case studies, one in a district of Eastern U.P. and the other in Western U.P. conducted by Mr. D.S. Chauhan, it was revealed that in Eastern U.P., the extent of subletting was of the order of 28.8 per cent of the net cultivated area and 16.6 per cent of landholders were involved in it. In Western U.P., 13.2 per cent of the net cultivated area was sublet informally and 27.6 per cent of landholders were involved in it. A very disturbing feature of the prevailing situation is that formal sub-letting was less than 1 per cent of the total area in the study pertaining to Eastern U.P. and 4.1 per cent in Western U.P. In other words, bulk of the land is leased out on informal tenancy basis. The overall magnitude of the problem demand serious attention.

The principal purpose of shifting to informal tenancy is to extract higher land rents from tenants. This is more to in view of the high yielding varieties program which has brought a realisation among landlords that land is a very valuable asset and promises high rate of return. In a country marked by land buyer, it is possible to take advantage of the situation by charging high rents. Secondly, informal tenancy arrangements are a convenient device with the landlords for nullifying tenancy reforms. Thus, unrecorded or clandestine tenancy masquerading under informal or oral tenancy perpetuates a semi-feudal land system which was sought to be abolished by measures of land reform.

Rights for Ownership or Tenants

A very important feature of land reform is the provision of the right of ownership for tenants. The Second Plan considered it very desirable to bring tenants in non-resumable areas in direct contact with the State. Earlier the right to purchase was optional to the tenants but this did not prove to be effective. Thus, the Third Plan suggested that the optional clause be removed and peasants be required to purchase land. Legislation for this purpose was to be enacted in various states. For instance, in West Bengal the tenants and sub-tenants have been brought into direct relationship with the state by the conferment of full ownership rights. In Punjab, the right to purchases is optional. Legislation has been enacted in Gujarat, Kerala, Madhya Pradesh, Maharashtra, Karnataka, Orissa, Rajasthan, Uttar Pradesh, West Bengal and the Union territories. It is quite disappointing that in Assam, Bihar, Jammu & Kashmir and Tamil Nadu, no provision exists even for an optional right of purchase. While the state can facilitate the transfer of ownership rights of land from the landlords to the tenants, no financial burden is imposed on the state.

CEILING ON LAND HOLDINGS

Land reforms in India had envisaged that beyond a certain specified limit, all lands belonging to the landlords will be taken over by the state and allotted to small proprietors to make their holdings economic or to landless labourers to meet their demand for land. Professor D. R. Gadgil justifying an absolute limit to the amount of land to be held by an individual observed: "Among all resources, the supply of land is the most limited and the claimants for its possession are extremely numerous. It is, therefore, obviously unjust to allow the exploitation of any large surface of land by a single individual unless other overwhelming reasons make this highly desirable. Moreover, in the context of the current socio-political climate, redistribution of land would rather appear to be imperative"[8] Thus, the case for pursuing a policy of imposition of ceiling rests on the following grounds:-

(i) In the rural sector, land is the principal source of income. If land-the fountainhead of income-benefits only a minor fraction of the rural population, i.e, the zamindars, the whole structure of land ownership fails to meet the ends of social justice. The best course of bringing a reduction in inequalities of income will be to bring about a reduction in inequalities of land-ownership.

(ii) A policy of application of capital-intensive methods in Indian agriculture will lead to unemployment on a massive scale. Consequently the Indian government desires to create a large number of small peasant proprietors. Fears have been expressed by critics that the policy of breaking big estates will transfer land from the resourceful zamindars to the resourceless peasants or tenants. It is alleged that such a policy may although enlarge employment, but will adversely affect production.

The whole argument of large-scale economies and maximising production is theoretical. Farm management studies reveal that gross output per acre is greater on small farms than on large farms. According to Professor Lewis, the size of farm is not very material in securing high yields. The experience, of Japan-a country of small farms-justifies that labour-intensive methods can result in higher productivity per acre. In Japan, the average size of holding is 1.2 hectare and the yield per hectare for rice is 52.5 quintals. In U.S.A, whereas the average size of holding is 124 hectares, the yield per hectare is 52 quintals, almost equal to that of Japan. Similarly, the large-sized collective farms of U.S.S.R. have not been able to produce yield rates comparable to those of Japan. The yield per hectare of rice in the U.S.S.R was 40 quintals. Consequently, historical evidence supports the imposition of ceiling because such a policy can enlarge employment and while meeting the ends of social justice does not in any way act as a hindrance to achieving maximum production.

Main Problems Arising out of Ceiling on Land Holdings

The imposition of ceiling on existing holdings is a more knotty problem. In this case a reorganisation of the present land system has to be effected. For this, a thorough verification of ownership rights has to be made. With it are connected a good many problems viz., mala fide transfers, exemptions and disposal of surplus land. Legislation for ceiling on existing holdings and unit of application has been enacted in two phases. During the first phase which lasted up to 1972, ceiling legislation largely treated landholder as the unit of application. After 1972, it was decided to have family as the basis of holding. It was also decided that the ceiling limit should be reduced so as to bring about a more equitable distribution of this scarce asset. (Refer Table 15.2).

Exemptions and Mala fide Transfer

The legislation pertaining to ceiling on holdings led to a large number of mala fide transfers. These transfers are principally of three types : (a) transfers among the members of the family, (b) benami transfers and other transfers which have not been made for valuable consideration and through a registered document, and (c) transfers made for valuable consideration through a registered document. Since it has been decided to apply ceiling legislation to the aggregate area owned by a family and not by individuals, transfers falling under (a) and (b) may be disregarded. However, transfers falling under (c) may need to be dealt with separately. It is necessary to protect the rights of such transfers, at any rate, up to a prescribed limit, say a family holding. Plugging of mala-fide transfers is quite essential because they go against the spirit of land reforms.

Compensation and Allotment of Surplus Lands

Ceiling legislation aims at obtaining surplus lands above a specified limit and then passing it on to small holders, evicted tenants or landless persons against the payment of a purchase price. Thus, this problem as two aspects-(i) Compensation that may be paid to be landowners for the acquisition of surplus land; and (ii) the price that may be recovered from the allottees of surplus lands.

TABLE 15.2

Level of Ceiling of Existing Holdings

States	*Level of Ceiling for a family up to 5 persons (in acres) after 1973*		*Level of Ceiling before 1973*	
	Irrigated land with two crops a year	*Irrigated land with one crop- a year*	*Ceiling limit (acres)*	*Unit of application*
Andhra Pradesh	10.18	15 to 27	27 to 234	Landholder
Assam	16.66[1]		25	Do.
Bihar	15 (Public Source) 18 (Private Source)		10 to 30	Do.
Gujarat	10 to 18	15 to 17	19 to 132	All members of family[2]
Haryana	17.9 acres	26.9	27 to 100	Landholder
Himachal Pradesh	10 acres	15 acres		Landholder
Jammu & Kashmir	9.1 to 16.8	14.4 to 22.2	22.75	Do.
Karnataka	10 to 13	15 to 20	27 to 216	All members of family.
Kerala	12 to 15[1]		12 to 15	Do.
Madhya Pradesh	18 to 54[1]		27 to 75	Landholder
Maharashtra	Not available		18 to 126	Do.
Orissa	10	15	20 to 80	Landholder
Punjab	17.3	27.2	27 to 100	Do.
Rajasthan	18	27	22 to 336	All members of family.
Tamil Nadu	15[1]		12 to 60	Do.
Tripura	9.9 to 29.6		25 to 75	Do.
U.P.	17.4	27.0	40 to 80	Landholder
W. Bengal	12.4[1]		12.4 to 17.3	All members of family.
Delhi	Not available		24 to 60	Do.

1. No distinction has been drawn in these states between irrigated land yielding two crops a year and land yielding one crop a year.
2. Family defined as husband, wife and minor children.

Source: Complied from *Economic Times*, March 31, 1975.

With regard to the price to be recovered from the allottees, it has been suggested that the purchase should be so fixed that the annual burden falling on the allottee on account of instalments of compensation and interest payable thereon, if any, and the land revenue should not exceed the fair rent recommended in the plan, i.e., 1/4th or 1/5th of the gross produce. The total amount payable as compensation should be recovered from the allottees so that there is no net liability on the state.

In the allotment of surplus land those tenants who have been displaced as a result of resumption for personal cultivation should be given preference. Along with the case of farmers with uneconomic holdings, landless labourers particularly those belonging to scheduled castes and scheduled tribes, should be kept in the priority list.

Progress of Measures Undertaken under Ceiling Legislation

Under the old ceiling laws, till 1972 only about 23.15 lakh acres were declared surplus in India, out of which only 12.55 lakh acres were redistributed.

It is distressing to note from Table 15.3, that in Bihar, Karnataka, Orissa and Rajasthan, no land was declared surplus on the imposition of ceiling legislation. Obviously, the partitioning of land or benami transfers had taken place before the imposition of ceiling. It implies that the Government thought of closing the stable doors after the horses had been stolen.

The Government of Kerala passed Kerala Land Reforms Act which reduced the ceiling limit from 15 to 36 acres to 12 to 15 acres.

The West Bengal Land Reforms (Second Amendment) Act, passed in February, 1971), marked an advance over even what the United Front Government had done. Under the Act, a family of five consisting of husband, wife, minor sons and unmarried daughters was allowed to hold a maximum that ranges from 17 to 24 acres. Exclusion of major sons, it was feared, was a loophole that provided for evasion. The lowering of ceiling would release as estimated 2.5 lakh acres of land. The Act was to be retroactive from August, 1969 in order to defeat mala fide transfers that might have been made since the state initiated steps to bring forth legislation. Recently, the ceiling has been further lowered from 17.3 to 12.4 acres.

The Latest Phase of the Ceiling Controversy (1972 Onwards)

The Central Reforms Committee appointed by Congress (R) suggested a rendition of ceiling on agricultural holdings from 30 standard acres to 10-18 standard acres in perennially irrigated double-cropped areas.

At the level of the Ministry of Agriculture, the explanation of "irrigated land" was inserted as "land under assured irrigation from government sources." This

TABLE 15.3

Distribution of Wasteland and Surplus Land on the Imposition of Calling up to 1972

State	*Surplus area declared or taken possession on imposition of Ceiling (000 acres)*	*Surplus Area Distributed (000 acres)*	*Wasteland distributed (000 acres)*
Andhra Pradesh	74	Nil	880
Assam	68	1	121
Bihar	Nil	Nil	202
Gujarat	50	25	250
Haryana*	170	65	23
Jammu & Kashmir	450	450	16
Kerala	10	1.8	20
Madhya Pradesh	84	13	1200
Maharashtra	271	123	409
Karnataka	Nil	Nil	290
Orissa	Nil	Nil	67
Punjab*	178	64	68
Rajasthan	Nil	Nil	1800
Tamil Nadu	25	17	125
Uttar Pradesh	241	121	49
West Bengal	694	375	378
Delhi	—	—	1
Goa	—	—	1
Total	2,315	1,255.8	5,900

* [a Haryana and Punjab in the non-PEPSU areas, there was no ceiling on ownership of land at the time in question.

Source: Planning Commission, *Report of the Task Force on Agrarian Relations*, 1973.

phrase, at one stroke, removed the application of the proposed ceiling legislation to lands cultivated by private sources, viz., tube-well and other private sources. This raised a storm of controversy and fears were expressed about the sincerity of the Congress Government to implement land reform measures due to the pressure of the `Kulak lobby'. To resolve the issue, a conference of the Chief Ministers held on July 23, 1972 approved the following guidelines for implementation.

(i) The best category of land in a State with assured irrigation and capable of yielding at least two crops a year should have ceiling within the range of 10 to 18 acres, taking into account the fertility of the soil and other conditions.

Allowance may be made for land irrigated from private sources and capable of growing at lest two crops in a year equating 1.25 acres of such land with 1 acre of land irrigated from public sources and capable of growing at least two crops in a year. The ceiling for such land irrigated from private sources shall not, however, exceed 18 acres.

The term 'irrigation from private source' shall mean irrigation from tube-well or lift irrigation from a perennial water source operated by diesel and/or electric power.

(ii) In case of land having assured irrigation for only one crop in a year, the ceiling shall not exceed 27 acres.

(iii) In case of owners with holdings consisting of different types of land, holdings after converting the better categories of land into the lowest categories shall not exceed 54 acres.

(iv) Regarding the Unit of Application, the following recommendations were made:

(a) The unit of application shall be a family of five members, the term family being defined so as to include husband, wife and their minor children. Where the number of members in the family exceeds five, additional land may be allowed for each member in excess of five in such a manner that the total area admissible to the family does not exceed twice the ceiling limit for family for five members. The ceiling will apply to the aggregate area held by all the members of the family.

(b) Where both husband and wife hold lands in their own names the two will have rights in the properties within the ceiling in proportion to the value of the land held by each before the application of ceiling.

(c) Each major child will be treated as a separate unit for the purpose of application of ceiling.

Exemptions have been permitted in the following cases:

(i) The exemptions in favour of plantations of tea, coffee, rubber, cardamom and cocoa should continue;

(ii) Lands held by industrial or commercial undertakings for non-agricultural purposes should be exempted from the ceiling law; and

(iii) Sugarcane factories may be permitted to retain an area not exceeding 100 acres.

Compensation

(i) Compensation payable for the surplus land on imposition of ceiling laws should be fixed well below the market value of the property so that it is within the capacity of the new allottees mainly comprising the landless agricultural workers who belong to Scheduled Castes and Scheduled Tribes.

(ii) The compensation may be fixed in graded slabs and preferably in multiples of land revenue payable for the land.

Distribution of Surplus Land

The amended ceiling laws were to be given retrospective effect from a date not later than January 24, 1971. The revised recommendations of the Chief Ministers' Conference were clearly a compromise between the pressures of the agricultural lobby and the socialist forces.

Following the guidelines laid down by the Chief Ministers' Conference, 17 state governments had revised ceiling legislation lowering the ceiling limits, (Refer Table 15.2), but the progress in acquiring surplus lands received a serious set back due to judicial intervention.

The Draft Sixth Plan (1978-83) giving the progress of land reforms stated: " As on 31st July 1977, the estimated surplus area was only 5.32 million acres, the area "declared surplus" was 4.04 million acres, the area "taken over" by governments was 2.10 million acres and the area actually distributed was only 1.29 million acres. Thus the distributed area remains less than one-fourth of the estimated surplus."[9] This was the position one year later.

TABLE 15.4

Progress Under the New Ceiling Laws up to 31st July, 1977

		In million acres	%
(i)	Area estimated to be surplus	5.32	100.0
(ii)	Area declared surplus	4.04	75.9
(iii)	Area taken over	2.10	39.5
(iv)	Area distributed	1.29	24.2

Although land reforms occupy an important place in the programme for removing poverty and unemployment, but measuring from the keenness and the speed of implementation, hopes of fulfilling the larger of completing the distribution of ceiling surplus land in the next few years to be very bleak.

However, the Draft Sixth Plan itself admits that the officially estimated surplus is a fraction of the area held in large ownership holdings. The Draft Sixth Plan mentions: "According to the National Sample Survey (26th Round, 1971-72) the area owned in holdings of 30 acres or more was 57.81 million acres. Allowing for self-cultivation by surplus owners the potential surplus would be 21.51 million acres. Similarly the area owned in holdings of 50 acres or more was 25.87 million acres and after the self-cultivation deduction the potential surplus would be 8.37 million acres."[10] The perfunctory manner in which land reform has been dealt with since 1977 leaves no doubt on the minds of observers that the approach to acquiring surplus lands, plugging loopholes in tenancy cultivation, reducing rents, preparation of land records and redistribution of land among the landless labourers and marginal farmers will continue to be casual and half-hearted.

The tardy progress of land ceilings may partly be explained by court judgments and litigation going on but as the Union Agriculture Ministry's review (1976) points out, while litigation has been an inhibiting factor, there is no explanation as to why thousands of acres, not affected by writ petitions, have not been even scrutinised.

The history of land reform implementation has been one of providing loopholes in legislation, winking at the evictions affected by big landowners of poor tenants, of yielding under pressures from the Kulak lobby. The implementation has fallen short of the national guidelines. Dr. W. Ladenjinsky's assessment is quite appropriate: "While officially the states accepted the ceilings programmes, they rejected them in practice." Unless the implementation of land reform is effectively transferred to grassroot level committees with a majority representation of marginal farmers and landless labourers, the much desired result may not be forthcoming.

BHOODAN (LAND GIFT) MOVEMENT

The land gift movement or Bhoodan Yajna was started by Acharya Vinoba Bhave in 1951. The movement was not a part of the land reform policy as conceived in the First Plan, but later on come states effected legislation to help the transfers of land offered in Bhoodan and Gramdan. Acharya Bhave's movement is a moral appeal to the land-owning classes in the country to donate a part of their holdings for the less fortunate landless labourers. The basic precepts of the movement are:

1. Land belongs to all and its use ought to be as universal as that of air and water. Individual ownership ought to be reduced to the minimum.
2. Possessors of land owning greater quantity than they could personally till ought to share it with the landless, desiring to cultivate it.

3. In order to create the necessary atmosphere for the elimination of economic inequality, small as well as great landholders ought to be appealed to share their lands in common with their less fortunate landless brothers.

A target of 50 million acres of land collection under Bhoodan was fixed. From Bhoodan, i.e. donations of small plots of land, the movement widened its sphere to Gramdan, i.e. donations of entire village. The purpose of gramdan was to create ideal villages with an ownership vested in the village community as a whole.

Up to March 1967, 42.7 lakh acres of land were donated under Bhoodan. Nearly half of it was donated by the people in Bihar alone. Out of this 12 lakh acres have so far been distributed. Besides 39,672 villages were donated under gramdan till August 1967.

It has been frequently observed that people sometimes donate lands whose ownership is disputed; or sometimes wasteland unfit for cultivation are donated. Total land collected under Bhoodan which is unfit for distribution is about 12.7 lakh acres, Bihar, the biggest donor state, provided 10.5 lakh acres, nearly 50 per cent of the total land donated was unsuitable for distribution.

Bhoodan Movement draws inspiration from the same ideals which fired the imagination of Utopian Societies like Robert Owen, Sismondi, Louis Blanc and Fourier. No one can doubt the sincerity of purpose and the tremendous amount of missionary zeal put in the movement. After the first spare of donations, the enthusiasm for competitive donations lost its fervour. Moreover three-fourths of the land collected so far is the personal achievement of Acharya Bhave. Unless the ideals of the movement are universalised, the chances of achieving th target of 50 million acres will recede further. But credit must be given to the tremendous missionary zeal of the reformer in mobilising a group of followers and devotees to the movement. the problem of land hunger has very big dimensions and besides Bhoodan, other more effective measures of social policy alone can help to solve it. Mr. D.D. Karve has rightly termed the movement a "Quixotic approach."

SELECT REFERENCES

H. Ventakasubbiah: Indian Economy Since Independence, Chapter III.

Planning Commission: Five Year plans.

Planning Commission: Progress of Land Reforms (1963).

Planning Commission: Reports of the Committees of the Panel on Land Reforms.

Minhas, B.S.: Rural Poverty, Land Redistribution and Development Strategy: Facts and Policy. Indian Economic Review, Vol. V (New Series) No. 1 April, 1970.

Sidhu, B.S.: Land Reforms, Welfare and Economic Growth.

Planning Commission: Draft Five Year Plan (1978-83).

Ruddar Datt: Giri's Plan for Co-operative Land Colonisation, Economic Times, September, 1970.

Ruddar Datt: Poor Implementation of Land Reform Programme, Economic Times. 27th November, 1973.

Ruddar Dutt: "Ceiling on Holdings", Economic Times, 28th November, 1973.

Khusro, A. M. : Economics of Land Reform and Farm Size in India (1973).

Planning Commission: Report of the Task Force on Agrarian Relations (1973).

NOTES

1. A service for which no payment is made.
2. Venkatasubbiah. H. Indian Economy since Independence, p. 51.
3. *Ibid.*, p. 57.
4. Service performed by a tenant for the landlord for which no payment is made.
5. Third Five-Year Plan, p. 223.
6. Fourth Five-Year Plan-a draft outline, p. 130.
7. Planning Commission, Progress of Land Reform, p.10.
8. Report of the Committee of the Panel on Land Reforms, p. 99.
9. Planning Commission, Draft Five Year Plan (1978-83) p. 12.
10. *Ibid.*, p. 12.

16

TREND OF AGRICULTURAL PRODUCTION IN INDIA—SOME FACTS AND GLIMPSES

Extent of Arable Land

Total geographical area of India is 328 mha. ' The extent of arable land (viz., net sown area plus current and other fallow lands) is only 160.6 mha. In the first few years of planning after independence the net sown area in the country increased mainly on account of largescale land reclamation and then it stabilised. Starting with 119 mha it reached 129 mha during the First Five Year Plan and 139 mha by 1971-72. By 1981-82, it had reached 142 mha.

Progress of Net Irrigated Area

The gross irrigated area in the country went on increasing—from 19.4 mha during the quinquennium ending 1949-50, to reach 63 mha by 1983, as under :

by 1950-51	20.9 mha	increases due to canals and well irrigation viz., major, medium and minor irrigation projects.
1960-61	24.7 mha	
1970-71	31.6 mha	
1979-80	56.5 mha	
1983	63.3 mha	

The spread of irrigation in various parts of the country, however, was uneven. Among the major states, Andhra Pradesh, UP, Haryana, Punjab and Tamil Nadu have been doing well. See some figures in Table 16.1.

Table 16.2 gives details of the (Statewise) position of irrigated (net) areas under principal crops and its percentage to total area thereof.

Depending on the availability of plan funds, the above mentioned states have gone ahead with irrigation, for with irrigation is directly connected the success of the

TABLE 16.1

Net area Cultivated and Irrigated—Statewise In 1967-68 and 1978-79

(In' 000 hectares and %)

	State	*1967-68* Net Area Cultivated	Net Area Irrigated	4 as % of 3	% Irrigated to Total Area (1978-79)	Productivity of Food-grains KG/HA (1981-82)
1	2	3	4	5	6	7
1.	Andhra	11367	3089	27.2	35.8	1238
2.	Bihar	8284	2011	24.3	32.6	872
3.	Gujarat	9802	1108	11.3	18.6	1162
4.	Haryana	3514	1132	32.2	53.9	1391
5.	Kerala	2129	411	19.3	12.3	1508
6.	M.P.	17797	1143	6.4	11.1	719
7.	Maharashtra	18267	1476	8.1	11.6	744
8.	Karnataka	9987	1082	10.8	15.4	984
9.	Orissa	5939	977	—	19.2	911
10.	Punjab	3992	2333	58.4	83.0	2667
11.	Rajasthan	15097	1865	12.4	19.7	550
12.	T.N.	6083	2629	43.2	49.7	1520
13.	U.P.	17465	5657	32.4	43.5	1191
14.	W. Bengal	5569	1478	26.5	19.6	1071
15.	India	139702	27523	19.7	27.5	1033

Source : (i) Report of Irrigation Commission 1972-Vol 1, Page 82.
(ii) Indian Agriculture in Brief 19th Edition, 1982, pp. 230-31, 256-257.

new technology in agriculture, including higher productivity per hectare. The miracle of the high yielding variety seeds—as a part of the green revolution—does not also work without irrigation. Even Bihar and U.P. have done slightly better in increasing their area under irrigation, and hence yields per hectare, compared with that of M.P. and Maharashtra.

Spread of Irrigation Cropwise

Total cropwise, 80 per cent of the gross irrigated area was under foodgrains, among which the area under wheat increased three fold between 1950-51 and 1971-72, whereas under rice it rose hardly by 50% in the same period. By 1978-79, it was

TABLE 16.2

Showing Statewise (i) Area Irrigated by Principal Crops incl. Total Area Under Foodgrains Crops (ii) Percentage of Irrigated Area to Total Area Under Foodgrain (1978-79)

State	*Area Irrigated under*				*% of Irrigated to Total Area*				*Total Area under crop*			
	Rice	*Wheat*	*Total foodgrains*	*Total area*	*Rice*	*Wheat*	*Total foodgrains*	*Total area*	*Rice*	*Wheat*	*Total foodgrains*	*Total cropped*
1	2	3	4	5	6	7	8	9	10	11	12	13
Andhra Pradesh	3751	15	4027	4698	94.3	65.2	41.7	358	3979	23	9648	13121
Assam	532	—	538	572	33.8	—	32.6	17.3	2241	68	2445	3311
Bihar	1941	1298	3472	3707	34.8	71.9	34.0	326	5579	1805	10209	11381
Gujarat	180	378	693	1936	39.2	61.0	15.1	18.6	459	620	4590	10389
Haryana	426	1320	2188	2979	93.0	89.1	51.2	53.9	458	1481	4273	5522
Himachal Pradesh	53	55	137	156	55.8	16.8	16.1	16.7	95	334	853	935
J & K	241	54	331	—	89.9	26.1	38.9	40.9	268	207	850	990
Karnataka	688	83	1172	1718	62.7	21.3	15.6	15.4	1093	389	7494	11133
Karala	255	—	255	354	31.9	—	30.4	12.3	799	—	840	2886
M.P.	821	980	2005	2413	17.0	25.9	11.2	11.1	4821	3788	17836	21858
Maharashtra	396	484	1357	2306	26.4	40.8	9.6	11.6	1498	1187	14090	19860
Orissa	1161	38	1293	1586	28.6	61.3	19.4	19.7	4372	62	6690	8275
Punjab	1014	2512	3986	5506	96.4	91.7	83.8	83.0	1052	2738	4738	6630
Rajasthan	80	1544	2369	3451	37.0	77.5	19.0	19.7	211	1991	12433	17496
T.N.	2586	—	2893	3819	92.7	—	56.1	49.7	2756	2	5156	7684
U.P.	1118	5928	8381	10570	21.7	80.2	41.8	43.5	5147	7390	20072	24300
West Bengal	1353	32	1433	1541	28.7	40.5	29.5	19.6	4765	521	5957	7878
All India	16847	14770	36865	48090	41.6	65.2	28.5	27.5	40487	22657	129353	175177

Source: Indian Agriculture in Brief 1982, pp. 226-238.

net 65.2 per cent for wheat and 41.6 per cent for rice, though only 28.5 per cent for all foodgrains. These percentages of cropwise irrigated area in relation to total irrigated area from 1960-61 to 1978-79 is given below :

TABLE 16.3

Area Irrigated—Cropwise, and Percentage to total Irrigated Area

(In mha)

Crop	*1978-79*		*1977-78*		*1970-71*		*1960-61*	
	Area	%	*Area*	%	*Area*	%	*Area*	%
Rice	16.8	35.0	16.2	35.2	14.3	37.5	12.5	44.7
Wheat	14.8	30.7	13.7	29.8	9.9	26.0	4.2	15.1
Sugarcane	2.5	5.2	2.6	5.7	1.9	4.9	1.7	6.0
Oilseed	1.7	3.4	1.6	3.4	1..1	2.9	0.4	1.5
Cotton	2.0	4.2	2.1	4.6	1.4	3.6	1.0	3.4
Others	10.3	21.5	9.8	21.3	9.5	25.1	5.2	29.3
Total Gross Irrigated area	48.1	100.0	46.0	100.0	38.2	100.0	28.0	100.0

Source: Statistical Outline of India 1984, Tata Service Ltd., Deptt. of Economics and Statistics.

Efforts to Increase Production under the Plans

Starting with a production base of a little over 50 m.t. of foodgrains in 1951, India reached a level of about 150 m.t. in 1983-84 and more in 1984-85. It was a remarkable achievement, thanks to the valiant efforts of its farmers, helped with the planning and financial assistance from the Government of India in conditions of freedom and the urge to become self reliant. Increasingly larger allocations were made for agriculture under the five year plans: private investment was also increasing with enhanced profitability from cultivation, depending on prices and per hectare productivity under improved methods of farming, including the green revolution ethos. For assessing the progress and performance of agriculture, the period can be divided in three spans viz., 1950-51 to 1964-65, 1965-66 to 1972-73 and 1974-75 to 1982-83. In the first span of 14 years, production increased more or less on traditional lines, with emphasis on irrigation and traditional inputs like improved seeds and fertilisers. During the next two spans, a new strategy of agricultural production, based on scientific advances in farm technology under the label of Green Revolution was adopted—to achieve almost a breakthrough in production and productivity in some areas. The highest rate of growth in agriculture was 4.1 per cent p.a. in the first plan, 3 per cent p.a. each in the second and third plans and 2.2 per cent p.a. in the fourth, and still higher in the subsequent plan periods.

TABLE 16.4

Area, Production and Yield of Crop—All India (1950-51, 1960-61, 1970-71, 1980-81 to 1983 84)

A : Area in lakh hectares *P : Production in lakh tonnes/bules* *Y : Yield in kgs. per hectare*

Crop		*1950-51*	*1960-61*	*1970-71*	*1980-81*	*1981-82*	*1982-83*	*1983-84*
1		2	3	4	5	6	7	8
Rice	A	308.10	341.28	375.92	401.52	407.08	382.62	409.90
	P	205.76	345.74	422.25	536.31	532.48	471.16	597.68
	Y	668	1013	1123	1336	1308	1231	1458
Wheat	A	97.46	129.27	18241	222.79	221.44	23567	243.95
	P	64.62	109.97	238.32	363.13	374.52	42794	451.48
	Y	663	851	1307	1630	1691	1816	1851
Jowar	A	155.71	184.12	173.74	158.09	165.99	163.76	162.63
	P	54.95	98.14	81.05	10431	120.62	107.53	119.34
	Y	353	533	466	660	727	657	734
Bajra	A	90.23	114.69	129.13	116.57	117.84	109.42	118.10
	P	25.95	32.83	80.29	53.43	55.37	51.31	76.24
	Y	288	286	622	458	470	469	646
Maize	A	31.59	44.07	58.52	60.05	59.35	57.20	58.88
	P	17.29	40.80	74.86	69.57	68.97	65.49	79.24
	Y	547	926	1279	1159	1162	1145	1346
Total Cereals	A	78230	920.18	1017.82	1042.10	1052.95	1022 62	1069.36
	P	424.14	693.14	966.04	1189.62	1217.88	1176 62	1388.88
	Y	542	753	949	1142	1157	1151	1299
Gram	A	75.70	92.76	78.39	65.84	78.63	7399	7308
	P	36.51	6150	51.99	43.28	46.42	52.90	47.55
	Y	482	674	663	657	590	715	651
Total Pulses	A	190.91	235.63	225.34	224.57	238.43	228.33	234.12
	P	84.11	127.04	118.18	106.27	115.07	118.57	126.55
	Y	441	539	524	473	483	519	541
Total Foodgrains	A	973.21	1155.81	1243.16	1266.67	1291.58	1250.95	1303.48
	P	508.25	820.18	1084.22	1295.89	1332.95	1295.19	1515.43
	Y	522	710	872	1023	1032	1035	1163

Source : Annual Report of the Ministry of Agriculture, Government of India, 1984-85, p. 175.

From 1950-51 to 1973-74 the compound rate of growth averaged 2.46 per cent p.a. During 1949-50 to 1964-65 the annual growth rate in production was 2.98 per cent in foodgrains and 3.5 per cent in nonfoodgrains.

Table 16.5

Total Area under Foodgrains, Net Irrigated area, Total Foodgrains Production, Area and Production of Rice and Wheat—Punjab, Haryana, UP and Bihar compared

(Years 1967-68 to 1981-82 ; Production '000 Tonnes ; Area '000 ha)

State	1978-79	Net Irrigation area under 1978-79			1978-79		Total foodgrain production			Rice			Wheat			
										Area		Production	Area		Production	
	Total area under food-grains	*Rice*	*Wheat*	*All food-grains*	*% Irriga-ed area to total area under grains*	*1967-68*	*1970-71*	*1981-81*	*1962-65 (Av)*	*1981-82*	*1962-65 (Av)*	*1981-82*	*1962-65 (Av)*	*1981-82*	*1962-65 (Av)*	*1981-82*
1	2	3	4	5	6	7	8	9	10	11	12	13	14	15	16	17
Punjab	4,754	1014	2512	3986	83.8	5407	7306	13326	263	1270	297	3755	1540	2919	2040	8553
Haryana	4,273	426	1320	2188	51.2	3992	4751	6040	167	506	212	1250	699	1562	881	3682
UP	20,072	1118	5928	8381	41.8	16779	19585	24220	4359	5307	3245	5726	3974	7849	3348	12883
Bihar	10,209	1941	1298	3742	31.0	8627	7881	8591	5276	5459	4568	4261	678	1744	455	2569
All India	1,29,353	16947	14770	36885	28.50	95052	108442	129586	—	40705	—	53590	—	22306	—	37883

Source: (i) Report of Irrigation Commission 1972. Vol. I, p. 43.
(ii) Report of National Commission on Agriculture, Vol. I, 1274, pp. 360-66.
(iii) Indian Agriculture in Brief 19th Ed., 1982, pp. 225-6 and 231.

Table 16.4 on page 290 shows area, production and yield of crops All India, comparing 1950-51, 1960-61, 1970-71 with the last four years upto 1983-84. Total foodgrains production had increased from 51 m.t. in 1950-51 to 133 m.t. in 1981-82 and to 151 5 m.t. in 1983-34, average all India yield having doubled from 522 kg per ha in 1950-51 to 1032 kg in 1981-82 and 1163 19 kg in 83-84. More remarkably, the production of wheat had increased from 6.5 m.t. in 1950-51 to 37.5 m.t. 1981-82 and to 45.1 m.t. in 1983-84 viz., more than seven times increase, with per ha yield having gone up from 663 kg in 1950-51 to 1630 kg in 1980-81 and 1851 kg in 1982-83. In rice, the production had increased from 20.6 m.t. in 1950-51 to 54 m.t in 1980-81 and to 59.8 m.t. in 1983-84, with per ha yield having gone up from 668 KG in 1950-51 to 1336 kg in 1980-81 and 1458 kg in 1983-84. Productivity in terms of 'indices, if the triennium coding 1960-61 = 100 be the base index, dropped down to 94 in 1966-67 (the year of the very severe drought) but it increased to 124.8 in 1970-71. A consistant increase in wheat crop was recorded however from 130.7 in 1968-69 to 164.7 in 1971-72. In the case of rice there was only a modest increase.

Table 16.5 on page 290 gives a Statewise picture of land utilisation, net irrigation, including percentage of irrigated to total cultivated area under foodgrains, area as well as production of wheat and rice, comparing the progress of the years 1962-65 (average) with 1971-74 (average) and 1981-82.

The above figures reveal the following for the the period 1962-65 Av. and 1981-82 :

(i) In regard to *rice,* Punjab, Haryana, UP and Bihar have shown per cent production increases of 1164, 490, 76.5 and—6.73 respectively, and in respect of wheat, the per cent increases are 379, 318, 285 and 464 respectively.

In the same period percentages of area increases in case of rice have been 383, 203, 21.8 any 3 and for wheat 89, 123,98, and 157 respectively. Let us juxtapose the information in Table 16.6 on the next page.

In rice production, Punjab and Haryana have done remarkably well; whereas UP has been so so, Bihar has deteriorated. In wheat production, Punjab and UP have done relatively better though even Bihar's progress is praise worthy. This is only relative progress ; see (ii) below and the last column in the table on page 293.

(ii) Comparing the performance of these states between 1967-68 and 1981-82 in respect of total foodgrains, the best scorers are Punjab and Haryana—followed by UP. Least impressive is Bihar having lost on even what it was actually producing in 1967-68. See columns 7, 8 and 9 in Table 16.5 above.

(iii) Whereas it would be interesting to go into the reasons for the remarkable success of Punjab and Haryana in the matter of foodgrains production, as well as the reasons inhibiting similar attainments in UP and Bihar, it would be useful to project in what ways could the Punjab and Haryana lessons or experience be applied to the other two northern states.

TABLE 16.6

Percentage Increase in Area and Production of Rice and Wheat between 1962-65 (Av) and 1982-83

State	*Rice*		*Wheat*		*Productivity Rise*		*% increase in total foodgrains production 1967-68/ 1981-82*
	Area	*Production*	*Area*	*Production*	*Rice*	*Wheat*	
Punjab	383	1164	89	319	162	121	+ 144.7
Haryana	203	490	123	318	185	87	+ 51
UP	21.8	76.5	98	285	45	95	+ 44
Bihar	3	- 6.73	157	464	— 38	120	- 1

(iv) The procurement of foodgrains in Punjab and Haryana was most outstanding, followed by that of UP. It would be a matter of interesting analysis to ascertain what contributed to such a phenomenal jump in the productivity of these two states and the resulting surpluses of foodgrains therein.

India's Future Food Needs

Today (1985) with a population of over 70 crores, and comparatively low standard of living of majority of our people, we may be satisfied with a food production level of 150 m.t. per annum—though even that level is of a rather fluctuating nature depending on the weather, flood and drought conditions and these four or five-yearly fluctuations usually, disturb our national food security balance. We must, however, assess the real dimensions of the present challenge faced by Indian agriculture (80 per cent of which consists of production of foodgrains) in the context of the next few decades. As it is, agriculture absorbs 75 per cent of our labour force. This may go down to 60 or 65 per cent with industrialisation and urbanisation in the coming one score years or more. But India's population itself may also touch one billion by 2000 AD and be around 13 or 1400 millions in the two decades thereafter. Let us not forget that the time lag between reduction in our death rate and reduction in birth rate is still noticeably present and the demographic transition is still to take place. The resultant rising population would be pressing against the available food supplies produced, by and targe, by traditional even obsolescent cultivation methods in a major part of the country.

Even at the present levels of consumption[1] Indian agriculture will be required to accomplish the task of doubling its foodgrains production within the next 25-30 years, merely to survive on the physical plane. In other words, the production level will have to rise to 300 m.t. per annum compared with only 150 m.t. at present. If higher levels of consumption are to be provided for and some exportable surplus catered, we would need to reach a production level of 400 to 450 m.t. in the next 40 years. Quite an uphill task, if not all that impossible looking. Hope lies in the central and eastern plains of the country in the shape of higher productivity.

Apart from reaching the outer limit of per hectare productivity growth in states at present doing well, the real growth in the next decades will have to be concentrated in areas where agricultural productivity has so far stagnated, as for example, in the eastern and central states of the country. The growth has so far been sustained by the north western and some southern states only. The next quantum jump would be possible only when the eastern and central states of eastern UP, Bihar, West Bengal, Orissa and Madhya Pradesh raise their productivity through a determined, well planned effort. They have very large unexploited soil and water resources and they can make big strides in grain production with the by now available and well tested technology. As an example, West Bengal's experiments with wheat production in the mid 70s were very successful and a high level of productivity had actually been attained. "Eastern UP is catching up fast; the productivity in eastern districts has picked up and met the deficit of the state, even declared a surplus of 2 m.t. wheat in 1983 for the central pool"![12] Can the results be repeated in Bihar, MP, Orissa and Assam with suitable adjustments in policy and methodologies?

The task of bringing one state to the level of the other in agricultural productivity even when the resources are more or less equal is not all that easy. The comparatively well-known problems of input supplies, irrigation, fertilisers and, crop protection can, perhaps, be dealt with smoothly but the problems that need to be tackled besides are much more complex and difficult viz., problems of smaller localised tracts and individual villages or individual holdings in villages. There are also problems like consolidation of holdings that cannot be solved unless the whole village, tract or the State is mobilised collectively. With such a mosaic of ostensibly unsolvable problems challenging us the question is : are we prepared to accept the challenge?

Let us put another vital aspect of this problem in the words of P.V. Narasimha Rao, then our Defence Minister, who had stated as under in his address to the Hissar Agricultural University on 11 May 1984:

> "In particular low agricultural productivity in a state is not necessarily a result of low land fertility, want of water or any such inherent causes. It lies often times in a complex web of extraneous conditions—social, economic and attitudinal. It is here that agriculture comes to have ramifications far beyond the known bounds and any attempt to tackle this problem would have to be of

a composite interdisciplinary character, well coordinated between the central and state governments as well as several other institutions at different levels."[3]

Gilbert Etienne, who has done a good study of Indian agriculture, holds that any all India production figure, any general judgement in agriculture is misleading because of the diversity of growth conditions, crop yields and frequently the farmer's behaviour. He cites the example of the eastern plains where the soils are fertile, the rains relatively heavy and the potential of irrigation enormous. Despite this we have poor nurseries, unlevelled fields, lack of weeding and excess water preventing the use of fertilisers and hyv rice— "and yet India's ability to feed its population lies in future in the eastern plains where between 20—27 mha can be irrigated, compared with, the total irrigated area in the country of about 60 mha. With water management, including drainage and flood control, enormous progress can be achieved. A Bihar farmer harvesting 700-1000 kg per hectare wheat and some pulses can, in the first stage, reach 2000 kg of rice, followed by a similar amount of wheat. In West Bengal and Orissa, the farmer can harvest 4000 kg of rice in two crops and a third crop of quick maturing pulses could be introduced between rabi and kharif—while nature is ready to help, the main constraints are manmade. First, several land-owning castes are not agriculture-oriented by tradition. They are 'dhile hain'—they are relaxed and not pushing. Then unlike the already advanced districts of UP, these areas are starting late when population densities are already high. Also water management problems are much more formidable in the eastern plains than say in Punjab. In the eastern plains it will not be enough to have more private tubewells. Also needed would be major irrigation works, massive drainage projects and flood control measures—all costly and complex."[4]

This is a heartening picture in a way. Mr. B. Venkateswarlu of the IRRI Philippines has the following view in the matter of comparative advantages of states:

"Punjab stands for non-problem and high-yield environment in agriculture. It is blessed with all the desirables viz.,

(i) favourable climatic composition, that is, moderately high radiation, less cloudiness, low and low to medium relative humidity. It is anti-pest and anti-disease, facilitating high input management;
(ii) application of 126 kg NPK per hectare is very high and helpful; compared with the 35 kg national average;
(iii) Irrigation facilities are highly improved, including exploitation of the ground-water for cultivation of rice and wheat;
(iv) high amount of credit is made available;
(v) its land reforms are production-oriented. Rice cash crop system grew impressively in a revolutionary manner due to its modern and non-traditional nature of farming;

(vi) It has no big problem of water submersion and drought.

Thus the Punjab syndrome of farming system is a low risk and high production one.

The situation in north eastern states is quite opposite which could be said as the syndrome of low-production and high-risk environment—: It has (enumeration added):

(i) problems of excess and inadequate water condition for rice;
(ii) low input management;
(iii) low fertiliser consumption, (10-20 kg/ha);
(iv) limited non problem irrigation;
(v) monsoon season is characterised by high winds, low light, high humidity and waterlogged conditions favourable for high pest complex and low productivity;
(vi) modernisation faces resistance — traditionalism being still prevalent;
(vii) non-availability of finance/credit presenting a major hurdle to high level management."

It was particularly noted by Venkateswarlu that nontraditional wheat was spreading fast in West Bengal mainly because of availability of water conditions and better input management : "since it is a new system, the problems of traditionalism are limited."

"The southern India has prepared adequate and is ina `take off' stage but all the time the fight was against unfavourable climatic composition such as untimely rains, water submersion, high pest load and low fertiliser application (30 to 60 kg/ha). Nondevelopment of adequate irrigation, partial exploitation of ground-water potential and lack of adequate water facilities are major factors for a prolonged lull but the symptoms of advancement are steadily emerging. MP and UP have large problem areas (drought and water stagnation), their input applications are low, they have poor water delivery and drainage systems, poor credit facilities and very poor transport systems."[5]

This shows that transformation from traditional to modern agriculture under the Green Revolution concept is associated mainly with inputs, irrigation and credit facilities plus favorable climatic conditions. Where such facilities are lacking, productivity would be low. Low risk and high yield environments remain to be unexploited due to inadequacy of inputs rather than modern technology. This is no doubt an oversimplification of the problem; the question is how do we manage that all these facilities are made available to all concerned with farming, especially the

TABLE 16.7

Area & Production of Foodgrains in Different States During Last 2 Decades 1961-80 (Area in Million ha; Production in Million Tonnes, Yield in kg/ha)

State	*Area Total*		*Difference during the period*	*Average annual Area*		*Production Total*		*Difference during the period*	*Average annual production*		*1980-81*			*1981-82*		
	1961-70	*1971-80*		*1961-70*	*1971-80*	*1961-70*	*1971-80*		*1961-70*	*1971-80*	*Area*	*Production*	*Yield*	*Area*	*Production*	*Yield*
1	2	3	4	5	6	7	8	9	10	11	12	13	14	15	16	17
1. AP	92.40	91.00	—1.4	9.2	9.1	70.80	86.80	16.0	7.0	8.6	8.76	9.99	1143	9.14	11.38	1238
2. Assam	21.10	23.52	2.4	2.1	2.3	19.50	22.70	3.2	1.9	2.2	2.52	2.70	1072	2.50	2.43	969
3. Bihar	96.80	99.17	2.3	9.6	9.9	74.30	90.30	16.4	7.4	9.0	10.02	9.91	989	9.85	8.59	872
4. Gujarat	86.0	42.05	—44.0	8.6	4.2	27.90	37.70	9.8	2.7	3.7	4.47	4.47	1001	4.74	5.09	1162
5. HP	6.19	7.50	1.3	0.6	0.7	6.50	9.60	3.1	0.6	0.9	0.87	1.18	1359	0.85	1.06	1240
6. Haryana	17.60*	39.90*	—	1.7*	3.9*	16.10*	47.60*	—	1.6*	4.7	3.98	6.04	1519	4.34	6.04	1394
7. J & K	7.70	15.61	7.9	0.7	1.5	7.60	10.70	3.1	0.7	1.0	0.83	1.3	1563	0.83	1.27	1529
8. Karnataka	54.0	68.01	14.0	5.4	6.8	46.00	63.70	17.7	4.6	6.3	6.62	5.88	888	6.7	6.59	984
9. Kerala	11.90	8.90	-3.0	1.1	0.8	11.60	13.80	2.2	1.1	1.3	0.84	1.27	1541	0.83	1.28	1538
10. MP	158.20	130.42	—28.0	15.8	13.0	90.80	86.20	—4.6	9.0	8.6	17.80	12.41	697	17.69	12.72	719
11. Maharashtra	129.30	134.50	5.0	12.9	134	63.70	82.40	18.7	6.3	8.2	14.1	9.76	695	14.2	10.58	744
12. Orissa	53.90	62.80	8.9	5.3	6.2	54.30	47.80	—6.5	5.4	4.7	6.9	5.98	865	6.9	6.25	911
13. Punjab	49.80	43.50	—6.3	4.9	4.3	58.20	86.01	27.8	5.8	8.6	4.84	11.90	2458	5.0	13.26	2667
14. Rajasthan	90.70	132.27	41.5	9.0	13.2	52.20	64.50	12.3	5.2	6.4	12.34	6.5	526	12.29	7.15	550
15. TN	50.70	50.10	—0.6	5.0	5.0	58.30	80.70	224	5.8	8.0	4.11	5.49	1335	4.76	7.23	1520
16. UP	187.60	201.14	14.0	18.7	20.1	148.90	172.20	23.1	14.8	17.2	20.47	24.95	1219	20.33	24.22	1191
17. WB	62.30	63.10	0.8	6.8	6.3	60.5	77.3	16.8	6.0	7.7	6.1	8.28	1356	6.0	6.47	1071

Source: B—Venkateswarlu—"Dynamics of Green Revolution of India". Agricole Publishing Academy 1985, pp. 26-27.

poor ones, adequately and at the right time ? Is it then only an administrative matter? We will explore this question in later chapters.

In Table 16.7, area, production and productivity of foodgrains as a whole have been compared Statewise for the period of two decades 1961-1980. Disparities among states particulary in regard to yields (in kg/ha) are so glaring — even where similar common factors of production were many— that it is not difficult to conclude that the administration/extension services in the laggard states did not come upto the mark and thus failed in their duty to avail of the golden opportunity presented by the new 'miracle seeds', without the democratic processes or the federating centre having been able to take them to task or apply the correctives in time.

There is ample data to show (not added) that between 1960 and 1980:

(i) Land production ratio was less than 1.000 in MP, UP, Orissa and Bihar;

(ii) This ratio was more than one in Maharashtra, Gujarat, TN, AP, West Bengal, Karnataka and Punjab ; (Ratio=land in mha/production in m.t.);

(iii) Total rice area in India increased from 28 mha in 1950 to 40 mha in 1977; total production also increased from 21 to 55 m.t. in 1981;

(iv) The area under wheat increased from 10 to 22.5 mha and production from 7.0 to 36 m.t. in 1950 to 1980. In productivity Punjab recorded 2.75 to 3 tonne/ha, comparable to that in Italy, U.S.A. and the USSR.

NOTES

1. The per capita local availability of foodgrains In India was 131, 160, and 156 kg respectively in the years 1951, 1965, and 1976. The highest ever reached was 171 kg in 1972. If developed countries are any guide in the matter of per capita consumption of grain, we are lagging far behind.
2. *Times of India*, New Delhi, 16 Oct. 1984.
3. *Main Stream* New Delhi, 19 May 1984.
4. Gilbert Etienne "Greener Pastures", *India Today*—New Delhi, 30 April 1984.
5. Venkateswarlu "Dynamics of Green Revolution" Agricole Publishing Academy New Delhi 1985. pp 278-79.

17

INTRODUCTION OF GREEN REVOLUTION IN INDIA

I

BIOLOGICAL BREAK-THROUGH

General

The gap in the food needs of developing countries was made up most in the years 1965-75. Most important factor responsible for this advance was the development by agricultural scientists, of the short stiff-strawed, fertiliser responsive varieties of wheat and rice which essentially double the yield potential of mankind's two most important food crops.

The Paddock Brothers had predicted doom—a famine for hungry nations on the food front by 1975, in their book *"Famine 1975—America's Decision—who will survive?"* in 1967 and yet a "turn around" took place in the late sixties in the shape of dwarf varieties of seed—a turn around which has few parallels in human history.

Dr Norman Borlaug was incharge of the Wheat Development Programme in Mexico in the fifties. A down-to-earth visionary he, infact, was the genetical architect of the dwarf wheat. Earlier he and Dr Harrar had conducted in the late forties a most relentless breeding programme choosing the best of the wheat genes in the world. They worked round the clock, growing two generations of breeding material each year, saving time by half. By 1951, the efforts at breeding a suitable dwarf wheat variety were crowned with success in Mexico. Insisting on a package deal for the field, Borlaug saw to it that the farm acceptance was rapid also. Infact, a seed-fertiliser revolution had taken place. Mexico was self-sufficient in food by 1956. But Borlang's dream wheat had not yet been evolved. He continued his gene-hunting relentlessly and experiments continued. Mere treatment of rust was not enough. To meet the lodging: problem, the wheat plant "had to be stiffened and shortened so that it could stand erect with long heads containing plenty of kernels and holding them tighter until they ripen for harvesting."

The Japanese wheat variety NORIN-10 was crossed'with the Mexican improved varieties and the first break-through came in 1961 when the resulting Pitic 62 and and Penjamo 62 were released in Mexico and in result Mexican farmers were able to obtain yields as high as 7000 KG per hectare-2½ times of the previous varieties. These were non-lodging-varieties, heavy on fertilisers and irrigation but possessing heavier tillering and grain-filling qualities. In four years these varieties occupied 95 per cent of the wheat area in Mexico.

These experiments in Mexico led to the greatest practical contribution to world agriculture through the field of genetics and plant breeding, plant protection, soil science and general farm management etc. One was reminded of the following para in Swift's Gullivers Travels:

> "And, he gave it for his opinion: that whoever could make two ears of corn or two blades of grass to grow upon a spot of ground where only one grew before; would deserve better of mankind and do more essential service to his country than the whole race of politicians put together."
>
> *[A Voyage to Brobdingnag]*

Dr. Borlaug was honoured with the award of Nobel Prize in 1974.

For centuries the typical rice plant of the tropics was tall with long drooping leaves. It survived under conditions of inadequate insect, weed and water control and non-use of fertilisers but its yield was low, though dependable. When fertiliser was applied to these traditional varieties to augment the yield, the plant became taller, leafier and lodged before harvest. The situation which the tropical farmer faced was: if he did not apply fertiliser, yield would be low for lack of nutrition and if he used it he received no increase in grain because of loding and wastage. Earlier the response of traditional wheat plants to fertilisers was essentially the same; the same improvements to rice varieties were needed if yields were to improve beyond 1965 levels.

The Green Revolution in rice was triggered off by the intense upsurge in rice research resulting from the establishment of the IRR1 at Manila. Some work on rice had been done in Taiwan. Infact it was from Asia, the house of rice plant, that the germ plasma came which opened up new horizons of the rice crop. The IRRI was started by the Rocke Feller and Ford Foundations originally. Now it is receiving funds from a Consortium of over 30 foreign aid agencies called CGIAR—The Consultative Group on International Agricultural Research—which supports other research countries also around the world.

The story of the 'miracle seed'—dwarf rice—did not start till 1962; the path had been charted by wheat in which Borlaug and Co. had to spend many years evolving the ideal wheat plant. The procedure for rice also would be the same. Rice yield in temperate climate was 5000 kg per hectare, for example in Japan, whereas in the tropics it was only 1500. Why could not the latter catch up with the former?

In the beginning the mere application of fertiliser had led to decrease in yields by more lodging and less photo synthesis because of shade and less grain formation, including damage by pest attacks. The new plant, therefore, had to be short, sturdy, erect-leafed, nonlodging, disease-resistant and fertiliser-responsive, with a high grain building capacity. Many hundreds of rice strains were screened and experimented upon by the IRRI to locate the suitable genes for use in the breeding programme.

Success came in 1965 with IR-8-288-3 giving an outstanding performance in the field. This was the miracle rice, with a yield potential of 9000 kg per hectare— six times that of Indica rice varieties. IRS is a,

> "Short upright, heavy tillering, a great consumer of fertilisers, giving undreamt of yields—a great contribution from IRRI to tropical agriculture."

It was obvious that a combination of hyv of seeds and multiple cropping would raise the average hectare yield many times in tropical areas but in actual fact seven International Research Stations throughout the world and many national centres are experimenting with producing hyv seeds to fit the numerous physical environments. And even then the ground has to be prepared indigenously to receive the new seeds and capture their potential by socio-economic engineering. According to Sudhir Sen,

> "In short, biological engineering must be accompanied by socio-economic engineering. It is the "wide cross" between these two different types of engineering that can produce a new hybrid type of economy which will be truly high yielding."

Socio-economic engineering in this context would include all what needs to be done at the scientist, government department and the farmers' level to derive the optimum results from the new seed varieties.

It was in the drought years of 1964-65 that the seeds of Green Revolution were first field tested in India. Dwarf wheat seed from Mexico was quickly acclimatised by Indian scientists and produced in time to be planted in the Indian soil at record speed. Dr Borlaug was responsible for introducing these seeds to the Indian scientists in the year (1963) for a direct transplant. He had predicted still earlier in 1961 that India could double her wheat production in one decade. India received seeds of four dwarf and semi—dwarf varieties—100 kg of each and some other selections. These seeds were planted in different soils in Delhi, Ludhiana, Pusa and Kanpur. Lerma

Rojo and Sonora 64 yielded over 4000 kg per hectare—a four times increase over local varieties. Then a chain reaction followed, of experimentation, multiplication and evaluation etc., before the two varieties were released for general cultivation but after some national demonstrations were carried out by the scientists in about 100 different farmers' fields. To speed up cultivation, more seeds of Lerma Rojo—250 Tonnes—and Sonora 64 were imported from Mexico for distribution and multiplication at state seed farms and by farmers. In 1966 imports of seed went upto 16000 tonnes, enough for cultivating 4 lakh hectares of land. It was in 1967 that with intensive breeding local variants of dwarf wheat were released.

New Strategy of Agricultural Production

Around the time Dr. Borlaug's dwarf wheat seed varieties were on trial in India in the early sixties, C. Subramaniam became the Cabinet Minister for Food and Agriculture (1964-67). He took to these seeds in right earnest for he saw in them a great potential for solving India's food problem. He evolved a new strategy of agricultural production in consultation with the country's scientists and administrators working with him, convinced the Government of India of its efficacy and then went all out to put it on the ground with maximum speed. During the course of his tenure he availed of every opportunity to put his ideas in a most lucid and simple phraseology across to the audiences consisting of scientists, agronomists, administrators, field workers, agricultural universities, farmers' cooperatives and the peasantry. Simultaneously, he took all requisite steps at the ministry level to create conditions and facilities for implementing the new strategy. His message dwelt on all the components of the new approach, including a new deal for the scientists, priorities in agriculture, spread of hyv programme, agricultural prices, marketing, administration and research, package programme, fertiliser production and supplies, food distribution cum price policy, better agricultural statistics, agro-industries, role of voluntary agencies in agricultural development—and what not. His relevant speeches are collected by the ICAR in a volume called "A New Strategy in Agriculture" published in 1972.

In the year, 1978, C. Subramaniam gave a series of lectures on the invitation of the Australian National University as a Visiting Fellow in the Department of Economics, Research School of Pacific Studies and the Development Studies Centre Sydney. Sir John Crawford was the Chancellor of that university. These series of lectures relating to his experiences about the introduction of Green Revolution in India were given to the students of the Master's Programme in Agricultural Development Economics (MADE). These covered the New Strategy of agricultural production introduced from 1964 onwards. His contribution to the successful launching of the Green Revolution in this country and much of its development in the subsequent years were due to his continuing interest in the subject even when he was not directly responsible for it.

He was so sure of the new strategy that he made it an issue to give an economic and incentive price to the farmer for his grain. L.K. Jha Committee on prices appointed in 1964 recommended an ad hoc increase of 5 per cent in the procurement prices. Agricultural Prices Commission (APC) was set up in Jan. 1965 to emphasise the following in its terms of reference, interalia:

(i) the need to provide incentive to the producer for adopting improved technology and for maximising production;

(ii) the need to ensure rational utilisation of land and other productive resources;

(iii) the likely effect of price policy on the rest of the economy, particularly on the cost of living, level of wages, industrial cost structure etc.

The APC was soon followed by the Food Corporation of India (FCI) to make support price purchases when necessary. The APC and the FCI were to be the twin arms of the strategy of production which made a beginning at the marketing end. Psychologically this approach was understandable and highly imaginative. Simultaneously, the Minister had announced that production could be increased only with the introduction of science and technology into farming. To this end be encouraged the IARI scientists to introduce scientific methods of crop management, as against the approach of the traditional agriculturist. ICAR itself was reorganised and all the national level institutions brought under its direct control. An Agricultural Research Service was set up in the states. It was recognised that if India was to launch a new technology without scientific and technological competence, it might end in a disaster. Agricultural science had to be put on a firm foundation so that research work could be undertaken to meet the new challenges and problems.

Dr. Ralph Cummings of the Rockefeller Foundation in India introduced the new dwarf wheat and rice seeds to Mr. C. Subramaniam who intially set up three panels—a scientist's panel, a panel of agricultural economists and a panel of agricultural administrators. Though Senior scientists were conservative in outlook,

> "the younger generation took a more dynamic stand saying we should progress now that this technology was bound to yield results and, therefore, we should take a bold decision and proceed with the introduction of the new varieties on a large scale."[1]

The economists' panel wanted to study the proposal in detail. Political opinion was adverse to the proposal. High dose of fertilisers would mean bigger imports thereof and some of them believed "we would only mortgage our future to the foreign interests, especially American." Cabinet too advised caution. It was, however, decided to introduce new strategy and new varieties of seeds in the year 1965-66 sowing season; one year was thus lost but as a part of the tactical approach to the main goal in conditions of Indian democracy. This period was really gainfully employed in convincing the various interests about the utility of the new strategy. That its adoption would lead to greater social tensions was also mooted but it could not be an argument against the adoption and implementation of the new strategy. C. Subramaniam felt it was absurd to argue in that fashion; no one had even insisted on equality of benefits as a result of industrialisation. Why so in the matter of an agricultural development strategy? He said,

> "I asked: would you like to have such high production and attain selfsufficiency within the country and face these tensions by distributing these foodgrains on an appropriate welfare basis or would you prefer to continue dependence on foreign imports indefinitely?"[2]

Ultimately many came to the conclusion that perhaps it would be better to raise output in India even though it might create other problems.

After selecting new seed varieties, seed multiplication on a large scale was needed. From nucleus seed to foundation seed, then to controlled seed producers, before the actual farmer got his share of the new seed—all this had to be organised and coordinated. A National Seeds Corporation and State Seeds Corporations were set up. Big seed farms had to be organised also for multiplication, minimum of 200 hectare size or more, managed by a State Farms Corporation at the centre with smaller ones at the state levels, similarly. Seed, in short, had to be developed tested, multiplied and processed and stored to maintain its quality, purity and performance capability—an industry by itself.

In regard to the use of fertilisers—and the new wheat varieties required very intensive doses of fertilisers, even 60 to 70 kgs per hectare—the controversy arose whether it was better to use the available, indigenous and mostly imported, resources in an intensive manner or spread its use at the rate of 5 to10 kg/ha extensively, to secure bigger output in the end. The protagonists of the latter course were not aware of the potential of the dwarf wheat seeds which gave a much higher yield as the fertiliser dose went beyond 10 kg/ha upto 60 kg. They were correct in so far as the cultivation with traditional varieties was concerned. In the end it was agreed by all that if India went in for large scale fertiliser production, high dosage levels on an intensive area basis for the new varieties were bound to yield better results.[3]

Not only that there was the controversy over the relative merits of chemical and organic fertilisers—the latter improves the soil whereas the former burns it. It had to be sorted out. C. Subramaniam was a man of conviction, with a lot of patience and tact. He won in the end and got the necessary foreign exchange for the import of chemical fertilisers.

To apply these fertilisers in high doses, a more intimate knowledge of the soil was required, to avoid other problems later. Soil analysis, in the normal course, for the traditional cultivation was much simpler and less intensive. A new soil survey programme had to be formulated and organised for intensive field to field survey, a "horoscope" for each, as he called it.

The watering pattern for the crops from new varieties of seeds was also different; it was difficult to fit it in with the traditional irrigation systems meant for traditional varieties. Our normal system provided for water going from field to field, by flow irrigation within a water shed. Irrigation of any particular field could not be controlled unless the entire inlet flow was blocked. Here if high doses of fertilisers were applied in one field, it was bound to flow with the water into the adjoining one. Then for the paddy crop, water must stand in the field at least 150 mm, to curb the growth of weeds. No weedicide had come into use then. For new varieties of seeds controlled water was a must, anyway.

Expanded Extension Services and Demonstration Centres

Adequate extension services were required to convey the message of the new technology to the farmers. The Panchayati Raj Institutions were in a sense available with their VLWs, Block Samities, BDPOs, subject matter extension officers. Zila Parishads and so on, covering the entire country. The concept of a block as a compact area for agricultural development had been accepted much earlier. The extension agencies were located at the block headquarters. The VLW was the kingpin of the agency, playing a multi-purpose role. It was he who was coming in touch with the farmers. But he being multipurpose, and with a big area to cover, was thinly spread and in practice was a "jack of all trades and expert in none." He had naturally to be retrained for the new technology to act as an extension officer for only agriculture—new agriculture—and his jurisdiction was also to be curtailed, to become more effective. That was the initial change necessary in the concept and coverage of this extension agency of a VLW.

C. Subramaniam thought of big but attainable aims and objectives[4]—whether for the spread of wheat dwarf seeds or for rice 'miracle' seeds. The problem of entire area coverage being colossal, it was realised that small efforts would be self-defeating also. It was necessary to think and act reasonably big. Consistent with this approach, a lot of homework bad to be done by the departments, agricultural universities and others to produce the requisite inputs, create an adequate research and training base, commensurate with the needs of a dynamic programme of agriculture. Then for better soil management, weed control, irrigation techniques and implements had to be provided for.

Initially in 1965 only 200 tonnes of hyv wheat seed was available. When the success of demonstration farms came to be known there was great demand for it. A special team had to proceed to Mexico and bring 18000 tonnes more of it. It was this seed which set in motion the chain-reaction in 1966-67, not only in India but elsewhere also. A programme of national demonstration bad in the meantime been initiated in 1965. Mass communication media, radio and cinema (T.V. came later) were fully exploited for passing information about the new agronomy for higher yields. Farmers came ta see the demonstration farms as pilgrims.

The Government of India thus took the initiative in introducing the new strategy through the hyv programme, focussed mainly on research, demonstration and extension plus the rigorous supply of inputs and credit. The programme, as observed, had to be implemented by the farmers. The big farmers were naturally in the forefront but the small ones also followed as soon as their difficulties of credit etc., were to some extent provided for by the organisation of cooperative societies. Fertiliser supplies were short in the beginning and had to be rationed on the basis of area targets. Sometimes these had to be diverted from other areas. Soon the small and marginal farmers' programme had to be started to provide to them special credit facilities and various other essential inputs in a preferential manner.

Wheat production in the country in 1961 was only about 11 m.t. In 1971 it had crossed 23 m.t. Between 1970-74 there was some stagnation mainly due to deterioration in the quality of the seed and shortage of fertilisers (because of the oil crisis and all that). Once in five years at least the farmer must change the seed and the same must be made available to him in the shape of new wheat varieties, taking into account the many defects like rust, diseases and drought proneness of the previous ones. Fresh research and development was needed constantly.

On the ground, the Punjab and Haryana farmers acted as the pioneers of this new technology. They were used to innovations, new ideas and had the knack to survive even in adverse circumstances. Infact, they now set the pace for a new agricultural revolution in India and showed the way to increased agricultural production. The production of wheat multiplied by leaps and bounds literally. The farmer there took to rice growing as a cash crop under new hy varieties. Soon thereafter they even went in for a third "catch" crop, having gone in for mechanisation and modernisation. It was found in practice that after introduction of tractors and harvesters etc., more labour was needed for multiple cropping, particularly during post harvesting operations. Employment was provided indirectly in the shape of new servicing operations also.

The problems of coordination between agriculture and irrigation departments had to be tackled at the implementation level. Traditional agriculture was used to irrigation department's timetable for water release/closure and opening of reservoirs etc. New type of agriculture, however, needed as said before controlled irrigation and systematically according to needs. The installation of private tubewells served this new purpose of irrigation better. Where ground water was available, tubewells could be installed as supplemental to canal water. This, incidentally, helped the canal water-logged areas also. It was established that many adjustments and innovations were required, in concept and practice, to bring the new technology on the ground successfully. One thing led to another: tubewells required electricity for which a major programme of rural electrification was undertaken, in Punjab and Haryana particularly. Intensive programme of marketing of produce led to a big programme of link road construction and other means of transport—as a necessary concomitant of the new technology of agriculture.

The picture on the rice production front, especially in the traditionally rice-producing areas, was net as bright as on the wheat side. The new rice varieties provided by the IRRI and various rice research institutions had to face a wide range of conditions including moosoonish cloudy weather and moisture favourable for pests and diseases. An intensive programme of adoptive research was needed before a particular region could use hyv seeds developed elsewhere. Not only that: extension work for rice cultivation had to be more intensive and more intelligent, requiring a specialised training among the VLWs and a particular orientation among the farmers.

In 1965 India had approached IRRI for the supply of one tonne of TN-1 seed and airlifted it. More seed was obtained from Taiwan direct—which was promptly multiplied and distributed by the National Seeds Corporation. In 1965-66 TN-1 was sown in 60,000 hectares. About the same time came the news of IRRI developing the IR-8 seed which was obtained. The results were uniformly reassuring. The area under IR-8 rose from a token level of 12 hectare in 1966 to 1,00,000 hectares in Kharif 1967 and 1.5 million hectare a year later. The hyv programme in rice bad multiplied rapidly. This also helped the country's food supply position in the short run.

Problems arose; even the best rice varieties have to be adapted to local conditions. Rust was a problem with TN-1; then the monsoon damp also produced disease and leaf blight. Water supply was to be regulated both for TN-1 and IR-8— erratic flushing would not do. Of course, these varieties were insensitive to photo-period or day length and could grow in short period but in high humidity there was danger of the grain germinating. The contrast with Indian varieties was acute. "The tall indica rejects fertilisers and gives a pathetically low yield of fine quality rice—the dwarf avidly absorbs fertilisers and gives a fantastically high yield of coarse rice—not relished by the consumers."[5] Could the two seeds be married together?

The answer came towards the end of 1968 when hybrid Jaya and Padma varieties were evolved and released by Indian scientists. These could stand up well, in yield and other essential attributes against both TN-1 and IR-8.

Says Mr. Carrol P. Streeter in his "A Partnership to improve Food Production in India"—A Report from the Rockefeller Foundation, New York 1969:

> "India's wheat and rice programmes have been like two powerful jet planes at the Airport. Wheat rolled down the runway first and is now well aloft. Rice has only barely lifted off the ground but it is beginning its climb."

The use of hyv of wheat and rice led to food production and higher agricultural incomes. Not only that; before the end of the pipeline there were other benefits in the shape of changes introduced in the whole production process and the resulting numerous job and income opportunities like (i) a seed industry; (ii) a rapid development of tubewells and other irrigation facilities; (iii) a spurt in fertiliser consumption; (iv) increasing use of pesticides for plant protection: (v) a surge in the demand for farm machinery—viz., tractors, threshers, seed-fertiliser drills, driers etc., (vi) growing emphasis on multiple cropping; (vii) rural electrification; (viii) expansion of fertiliser and pesticide industries; (ix) increasing road construction for linking villages with main roads for marketing; *(x)* agro-business of various types, including expansion of agricultural credit—all these come as direct results of the adoption of dwarf wheat and rice—not necessarily in that sequence though.

All these results become clear as one sees the growth of foodgrains production in Punjab and Haryana in the post 1966-67 period. It is a remarkably heartening story which was the outcome of a well coordinated effort of all concerned—the scientists who evolved the dwarf seeds, the political bosses who gave the inspiration and the go ahead signal, the technocrats and the bureaucracy who spread the message and then the farmers who accepted the challenge and carried out the instructions for their own benefit and the welfare of the country. As simple as that, it looked.

C. Subramaniam observed as under

> "The Punjab and Haryana provide a story of agricultural development on a scale and speed which had not been witnessed before in any part of the world. If only we could repeat this miracle in other areas, the food problem and problems of rural unemployment and of agricultural stagnation in other areas would be solved. Unfortunately natural conditions do not permit this nor is such a human enterprise available in other areas. Our task, therefore, has to be to train the human material and to provide the incentive for participation in these development programmes."[6]

Let us develop this argument later, by studying the case of the four states of Punjab, Haryana, U.P. and Bihar, in their reaction to the Green Revolution technology.

REFERENCES

1. C. Subramaniam—"The New Strategy in Indian Agriculture" Vikas, New Delhi 1979, p. 23.
2. *Ibid.*, p. 27.
3. Towards the end of the Third Plan it began to be realised that spreading scarce resources thinly over large areas did not lead to their optimum utilisation and to the maximisation of production of foodgrains. In the Fourth Plan, therefore, the policy was to concentrate effort in areas with adequate irrigation facilities or assured rainfall which promised greater response to the use of improved seeds and fertilisers. As a matter of fact the new strategy had itself meant cultivation of hyv crops under a complete package of improved agronomic practices to exploit their maximum potentialities."

   See also K.S. Mann's, "Analysis of the Expected Shifts in Cropping Pattern in the Punjab resulting from hyv crops" P.A.U. Ludhiana, 1967.
4. When C. Subramaniam talked of covering 1/10 of the 130 mha of cropped area in the country with hyv seeds in 1965-66, and the four years following, through assured irrigation and selected progressive farmers, the experts from the Ford and Rockefeller foundation advised caution and the adoption of less ambitious targets. The experts from socialist countries advised collectivisation of small farms. The minuter was in a dilemma. He again conferred with his own panels of scientists, economists and administrators and thought of setting up demonstration farms, 1000 in number belonging to farmers, to prove that an additional yield of 200 to 300 per cent could be achieved by the adoption of new technology. The resistance of the traditional farmer could be broken only by actual field demonstration; how else would he 'fall in' quicker? This was an excellent idea which carried the day. In the midst of traditional agriculture these 2 hectare fields were cultivated with the new technology in 1965-66 kharif and rabi seasons. See "The New Strategy in Indian Agriculture", *Ibid.*, pp. 47-48.
5. Sudhir Sen 'A Richer Harvest', *Ibid.*, p. 53.
6. "The New Strategy in Indian Agriculture" *Ibid.*, pp. 59-60.

18

GREEN REVOLUTION: HOW, WHY AND WHY NOT?—AN ANALYSIS

We have so far dealt with the essentials of the Green Revolution and seen how the states of Punjab, Haryana, UP and Bihar were affected by it and reacted to it— one way or the other—in the last three decades. In that process we have seen how these states handled their problems of agricultural development in the period preceding and following the revolutionary introduction of dwarf wheat seed varieties on the Indian scene. In brief, Punjab made the best of the Green Revolution technology from 1965-66 onwards, mainly because the ground had been prepared for it earlier—not by design as such but in the ordinary course of successfully augmenting agricultural production in the traditional manner—and Punjab had built up a sound infrastructure in the process to an extent that when the miracle seeds arrived, these were received with open arms by its farmers who happily fulfilled Borlaug's dream that the production of wheat could be doubled in ten years—(1962—72). Haryana, initially suffering from a grievance that in the composite state of Punjab the interests of the Hindi region had been neglected, started its separate existence in November 1966 just when the Green Revolution was ushered into northen India and did not take long—just about six years—to catch up with the basics of science and technology push in agriculture and showed results in foodgrains production with a speed rarely experienced before anywhere in the country.

Punjab and Haryana were two small states—both not more than 2 crores of people together before they were separated towards the end of 1966. UP, with more than four times their population and with as many as 56 districts, tried its best to follow the path of the Green Revolution but succeeded only in its western region, firstly because of tome natural advantages but mainly because of its contiguity with Haryana State and the fact that with the partition of the country in 1947, some Clements of non-muslim population from Pakistan side of the Punjab had come to settle down in the western districts of UP and had introduced their outlook, habits of hard work and efficiency in the new areas of their domicile. As we go east from the western region of UP, down the Ganges, we still find only traditional modes of agriculture, by and large, as if the new technology of Green Revolution was something of not much concern to the residents of eastern UP and Bihar. Our examination in the earlier chapters has revealed that in the matter of food production not much solid work had been done in keeping with their natural potential, or better still, nothing useful had happened so far in the real forward direction in these areas. Taking into account the mounting population of these two states the per capita incomes, productivity and the general standards of living of their people have, if anything, deteriorated, even beyond where these stood in 1947. Disappointing, indeed.

Tables 18.1, 18.2 and 18.3 are added below showing (i) yield of principal cereal crops (Table 18.1), (ii) per capita income estimates of 15 important states of India, at current prices from 1960-61 to 1981-82 (Table 18.2) and (iii) some indices of interstate disparities in some spheres, in regard to the same states (Table 18.3)—in all cases comparing them with all—India averages. It will be noticed that Bihar state is at the bottom rung in all respects and the UP is only a shade better—but both are below the all India averages generally in all respects. Their backwardness in the agricultural sphere is borne out by their per hectare yield of cereals (Table 18.1).

Punjab stands at the top in this regard, followed by Haryana as No. 2—both far above the all India averages. UP just touches the all India average yield in wheat but in rice it is far behind. Both in wheat and rice, Bihar is at the bottom of the list, except for Madhya Pradesh.

One can go on multiplying these inter-state comparisons of growth in various spheres but the net result is not going to improve. Bihar and UP both come out lagging behind, especially in the matter of foodgrains production and productivity with which we are here concerned. It will be worth while to go into the matter at length, to find out what has been basically ailing these two important states or what are the factors responsible for disabling them from acting effectively in this regard. We can study the subject under the following broad categories of constraints in the coming chapters though not strictly in that sequence:

(i) Reasons of size, space, area and local resources at the macro level, reckoning any special advantages available to the states of Punjab and Haryana,

(ii) The extent of administrative involvement in intensive agriculture at the field level which showed beneficial results in some states, not in others,

(iii) A scrutiny into the common phenomenon in India of a yawning gap between promise and performance, between planning and implementation. Was Gunnar Myrdal right in calling India a soft state administratively?

(iv) Some human factors relatable to leadership, discipline and work ethos etc.

TABLE 18.1

Yield of Principal Cereal Crops, 1980-81 and 1981-32 (Some States Compared)

(In kg/ha)

State	*Rice*		*Wheat*		*Cereals*		*Foodgrains*	
	1981-82	*1980-81*	*1981-82*	*1980-81*	*1981-82*	*1980-81*	*1981-82*	*1980-81*
AP	2068	1947	—	—	1403	1310	1238	1141
Bihar	781	1015	1473	1314	920	1049	872	989
Haryana	2470	2602	2357	2359	1761	1743	1391	1519
MP	793	834	994	934	793	787	719	697
Punjab	2957	2736	2932	2730	2816	2598	2667	2458
TN	2261	1865	—	—	1744	1493	1520	1335
UP	1078	1053	1641	1650	1275	1273	1191	1219
All India	1317	1336	1696	1630	1160	1142	1033	1023

Source: Indian Agriculture in Brief, 19th Edition 1982. pp. 256-257.

TABLE 18.2

Per Capita Income Estimates of Various States and All India

(at Current Prices)

	State	*1960-61*	*1965-66*	*1968-69*	*1970-71*	*1973-74*	*1977-78*	*1981-82*
1.	AP	275	387	495	584 **	829 **	1002	1536
2.	Assam	315	399	514	538	644	932 *	N.A.
3.	Bihar	215	332	376	402	575	735	995
4.	Gujarat	362	498	605	829 *	1121 *	1452 *	2211
5.	Haryana	327	450	635	845 *	1174	1600	2574
6.	Kerala	259	380	496	584	800	987	1312
7.	MP	460	305	430	489 *	725 *	950 *	1217
8.	Maharashtra	409	534	700	809 *	1125 *	1637 *	2519
9.	Karnatak	296	448	588	685	973	1120 **	1458
10.	Orissa	216	329	464	482	697	799 *	1147
11.	Punjab	366	562	903	1030	1438	1991 *	3122
.12.	Rajasthan	284	373	414	623 *	824 *	964 *	1417
13	TN	334	403	499	595	830	1036	1373
14	UP	252	373	453	486	673	916	1309
15	WB	390	532	660	735 *	946 *	1268	1595
16	India	306	426	552	633	871	1189	1750
	Rank of UP	14	12	13	14	13	13	12
	Rank of Bihar	15	13	15	15	15	15	12

Source : (i) Statewise, Estimates of State Domestic Products. Natural Account Statistics, Govt. of India 1976, 1980.

(ii) Draft Annual Plan of UP 1984-85 vol. I p. 15.

(iii) P 56-See Annexure II UP Draft Sixth Five Year Plan 1980-81 (Review) vol. I Nov. 1980.

* Quick estimates.

** Provisional Estimates.

TABLE 18.3

Some Indices of Interstate Disparities

State	*% State of population*	*Per capita Income 1973-76 (Av) in Rs.*	*Per capita Plan outlay 1951-79 in Rs.*	*Per capita Central Assistance 1951-79 in Rs.*	*Per capita market borrowing 1970-79 in Rs.*	*Central Undertakings Investment upto 31.3.77 Interse %*	*Per capita Consumption of Electricity Kwh*	*Length of surfaced roads per 100 Sq/Km of area in Kms as on 31.3.75*
1	2	3	4	5	6	7	8	9
Punjab	2.63	1,576	1659	497	73.11	2.2	242	30
Haryana	1.95	1399	1671	529	107.33	1.6	174	34
Maharashtra	9.80	1,349	997	211	62.68	7.1	199	16
Gujarat	5.19	1,134	1032	261	81.66	6.0	192	19
Karnatak	5.70	1045	768	293	67.43	3.0	148	30
W.B.	8.61	1033	586	246	77.97	8.6	125	33
Kerala	4.15	948	709	347	79.86	3.1	93	50
T.N.	8.01	942	660	250	66.70	5.2	146	46
A.P.	8.46	928	669	283	42.67	4.4	77	19
Rajasthan	5.01	853	695	373	74.45	2.5	83	7
Orissa	4.27	793	696	385	50.03	7.3	112	9
M.P.	8.10	776	667	286	39.78	16.7	90	9
U.P.	17.17	715	636	251	35.17	4.2	86	11
Bihar	10.95	545	479	250	34.50	8.1	82	12
All India	100	930	734	279	55.35	100.0	121	17

Source: P. 37, Annexure III, Draft 6th five year Plan 1980-85 U.P: Review Vol. 1.

A recent study[1] of the problem of stagnant agricultural production in eastern India (Eastern UP, Bihar, Orissa and West Bengal) concluded that even though the bulk of this region was blessed with good soil, water and labour resources and historically its productivity in this field was the highest in the country, its pre-eminence got eroded in recent times for institutional, technological, economic and other reasons. When the Green Revolution technology came on the scene in the country in the mid sixties, it worked better in northern and southern regions because the conditions there were more favourable. Government did not create the necessary infrastructure in the eastern region nor did it bring the needed inputs within easy reach of the cultivators partly because of deficiencies of administration and management in the region. This neglect resulted in accentuation of the problem of unemployment and poverty in these states.

This study, however, expressed the view that some recent development measures taken by the Government had helped in reducing a few of the many constraints plaguing the region, and, in result, its agriculture was near, although not quite, on the threshold of accelerated progress requiring a big push in terms of investment, management and policy to get over more serious constraints blocking the way."

The study Report spelt out a number of programmes to be undertaken in a planned and coordinated manner. The spearheads of these programmes had to be in terms of massive provision of tubewells, pumps, drainage, micro water-shed development, power, inputs, custom services, investment and credit etc. Once these made an initial breakthrough, other suggested measures would become more effective and the resulting technological surge would help to weaken some of the institutional and structural barriers which had proved intractable so far. The Report expressed the hope, that conditions, thus, would be created for converting the eastern region into a granary that it was, but the effort required would be massive, with the government playing a key role in it. This study proposed an efficient and well coordinated management and simplified procedures as a precondition for a rapid and extensive adoption of modern technology by farmers of the region. Finally it expressed the view that the big investments suggested for the programmes in the Seventh and Eighth Five Year plans would be well justified in the shape of net returns from the programmes if these were to be properly implemented.

This was all well put and smoothly said. But there are many a "slip between the cup and the lip" in our present administrative system and the peculiarly difficult "environmental" situations in which the states of UP and Bihar are currently placed. The major question remains : how to put all these good ideas and technologies, so well researched and time tested in practice, on the ground in their soils? Merely providing for them materially and financially will not do. Merely putting them across in the shape of five yearly annual plans will not do. *Something else has to click all along and down the line in these areas before action starts in unison, coordination full steam, aggressively and fruitfully on the ground.* What follows in the rest of the chapters is an account of the possible ways showing how that situation can be improved—only improved—learning from the experience of Punjab and Haryana states over the last two to three decades. The minimum steps required for improving the present constraint of the yawning gap between promise and performance of the administrative machine on the one hand and by now the inevitable distance between the plan and its implementation, will become clear as the narrative develops. Virtually the overall objective aimed at would be to put the administrative machine in a proper working order which, if it is properly inspired, organised, disciplined, controlled and monitored, should be able to deliver the goods for which huge investments would need to be made. In a developing economy things cannot be left to private bodies and institutions in the hope that growth will take place automatically. Governments have to step into the fray directly with plans, investments, input supplies, education, research and extension as well as irrigation supplies, fertiliser distribution, hyv seed production, road building, power generation and distribution— in fact the whole gamut of supplies and services which go into the augmentation of agricultural production and productivity in a region the size of which is much bigger than many countries of the world. One must be prepared for many lapses, inefficiencies and wastefulnesses in the process but the effort would be to get the maximum result from the investments made. Proper planning and a well-coordinated implementation of that plan would be successful on the ground only if there is a well led team of workers of concerned disciplines in the field a team which is not only properly motivated, even inspired by the political or bureaucratic bosses but also is put in charge of a small compact area, with duties well defined and the scope of work properly thought out. Such a plan, the implementation of which could be properly coordinated, controlled and monitored must, for practical considerations, cover only a manageable area of the state or its regions. The bigger the area, the thinner is the spread of developmental administrative activity on its people. Problems arise with men and the land makes its contribution to those problems to the extent it affects men directly or indirectly. It is to help solve these problems that men have to be guided, prompted, cajoled and even pressured into taking some positive steps in their own social and economic interest, for which all facilities would be provided in the shape of supplies and services by an enlightened, well meaning government. The bigger the area to be covered by one team of workers, the more diffused will be the spread and effect of its efforts. What should be the size of such a unit of planning and its management, becomes a crucial question. Punjab and Haryana experience, interalia, will have to help in prescribing the guidelines. States with 12 or even 7 crores of population are, at any rate, too big to be developed economically in a short-time.

Let us now deal with the problem more specifically.

There is no doubt that Punjab right from 1947 had the initial advantage of being a small, manageable state in which people's problems connected with the resettlement of refugees, augmentation of agricultural production, industrial advancement, growth of education and people's welfare could be properly attended to by the administration, at political as well as bureaucratic levels with ease, promptness and in right earnest and workable solutions found to those problems. The political head of the state, the Chief Minister, then generally blessed with the political stability of his regime, knew his team of workers well, both at the political and the administrative levels, and could tackle the problems satisfactorily as they arose and with the requisite deliberation and circumspection. This top boss generally had the real feel of the situation; he could supervise and monitor, even personally, some of the developmental operations he had set in motion, with a depth of attention and apply correctives where and when necessary. In this respect the state was exceptionally lucky in having the late Partap Singh Kairon on the political scene for a little less than 18 years, first as Development Minister and then as Chief Minister (1956-65), fully in control of the situation, with a rare intelligence,' imaginative foresight and capacity for hard work, endowed with an iron will, a sense of humour and the knack to make others work effectively. He had the love of the state dear to his heart; he wanted it to grow politically, industrially and economically and to that end he was determined to do his best in every way. With example, persuation, threats and cajolery he was able to create an environment of genuine activity and progress all around, in which tangible results followed the earnest efforts on the ground as well as in offices. In short, all concerned were putting their best foot forward, some in the fear that their progress or lack of it was being watched and others in 'the hope that their earnest efforts would be amply rewarded. The 'whip and the carrot' were rarely used so well together, to produce top results in a remarkably short span of time.

As the chapter on Punjab shows there was a remarkable progress in the state in the field of agricultural production, not only in total production of cereals but also in productivity per hectare; all this success had placed it on top of the country and made this small state an envy of others, really an example to emulate. This performance came to international notice as well, particularly the success in the yelds per hectare of wheat reached by Ludhiana district— later in rice also. The secret of this success, next to inspiration from the top, was that the state government took quick decisions and timely action when any gap or laches came to its notices making sure that the pace of implementation of the programmes in the field did not suffer at the the same time. Since its overall objectives were clearly defined, one essential step after another came to be taken in proper sequence; when the broad strategy was abundantly clear, the inputs were arranged for supply to the farmers in the right place and at the right time. Decisions on problems were distinctly taken by the cabinet after due deliberations and within practical limits of resources but with a flair and imagination which would tend to bring the overall objective nearer attainment. Decisions relating to consolidation of holdings, augmentation of irrigation, first surface then ground water, rural electrification, flat rates for power consumption by tubewells, village link roads to help transport and marketing of produce and so on, fall in this category of decisions which were not taken on the spur of the moment, nor lackadaisically in a huff, but after proper consideration to fit in with the overall scheme of economic advancement, the progress of which itself was an open book. In fact, if one had the time and the patience to spare, it would be a useful study to go into the cabinet notes/minutes and the departmental files, to mark precisely the dates and the nature of decisions taken by the state government—decisions which tended to put the train of progress on its proper rails at the correct time, to obviate any possibility of drift, waste or procrastination.

Even though India's is a mixed economy, in the background of a socialist democratic society in the making, government actions still have an overwhelming primacy and influence on economic activity both in urban and rural areas. In the broad framework of economic development starting from a lowly base, the position remains that people move forward generally within the parametres prescribed by the government and the economy does, over time, take a turn in the direction desired by it. In fact, the government provides the main spur to economic activity as it allocates financial resources and material inputs for various departmental programmes, directly or indirectly, besides providing the technological and administrative expertise in the earlier stages. That being the pattern generally, it is not difficult to visualise how remarkable progress can be made in the desired direction if the political leadership at correct levels, confident of its own stability and supported by an experienced and confident bureaucracy, wishes it that way and is vigilant enough to take vital decisions boldly in suitable measure and at the correct time. Being in such an effective position potentially it becomes the responsibility of the chief minister not only to carry the political collegues with him but also to inspire the administrative machine adequately, to enable it to deliver the goods in the field. Even at the cost of repetition, a few words are quoted below about the morale of the bureaucracy, as recorded by the author elsewhere a few years back:

> "Of equal, if not greater, importance is the question of morale of the official administrative machine itself—the inevitable bureaucracy—which has lately been on the decline for reasons one need not go into here. This is an aspect of a serious crisis in our current socio-political life, as well as a decline in our individual and national character. The gross and persistent misunderstandings between the political ministers and top echelons of the services during the last about a decade or two have tended to create an atmosphere in which the forms of action are considered more important than their reality and where more time is spent in appearing correct than in improvising solutions and producing results on the grounds. As a matter of fact and of real importance, the experience shows that very little progress is possible in any sphere of governmental activity involving the people unless there is a complete understanding, based on mutual respect and confidence, between the political and administrative wings of the government machinery. The mutual goodwill for the common crucial task of agricultural cum economic development at all levels—particuarly between the ministers and the services—is a must if the overall objectives are to be achieved in a speedy manner and the results are to be commensurate with the funds expended. It has to be remembered that

between the political will of the government and the needs of the farmers in the field, the vital link is provided by the administrative cadres, the political intermediaries being only peripheral and advisory in the context of the present day realities".[2]

As said already, Punjab was fortunate in having the late Partap Singh Kairon as the chief minister long enough to leave an indelible mark of his capabilities on the state—a chief minister who had clearly visualised the type of a viable border state he wanted, economically flourishing, contented, forward looking, upcoming agriculturally and industrially. He decided not to leave any stone unturned for the attainment of this objective, come what may, by hook and if necessary by crook. And in a way he did achieve his aim but in the process, human nature being what it is, left a little bad taste in the mouth in the shape of the Das Commission enquiry in the mid 60's which did not establish much against him but indirectly showed that the chief minister was somewhat tolerant of unscrupulous friends and relatives and did not suffer from any unnerving confusion of ends and means in the pursuit of hit goals etc. Though it was not the function of the Commission to establish what good turn the policies and deeds of this chief minister had done to the state, and what salutary direction he had given to the state's economy, it was abundantly clear that his foresight and endless pursuit of the goals had led the Punjab state to reach the pinnacle of agricultural production in the post 1947 period and made the state capable of deriving optimum advantage from the Green Revolution, thus becoming the real bread basket of India.

All the same, it will be pertinent to state here that this type of personalised but effective administration would be possible to have only if the state happens to be of a manageable size, as Punjab and Haryana happened to be.

Haryana followed in the footsteps of Punjab as soon as it came to have an independent existence as a full-fledged state of the Indian Union. Enough has been said already about her efforts at shortening the time span of Punjab's experience and attain similar goals to an extent only next to the parent state and become number two in the field of agricultural productivity in the country. Here again the credit must go to the foresight, efforts and the unstinted devotion of its chief minister Bansi Lal (1968-75), to attain the goal of a quick economic uplift of the new state. He succeeded in abundant measure to show that even against ostensibly insurmountable difficulties and odds of finance, resources and backwardness, the gaps in foodgrains production and productivity could be made up in a short-time if the political will was available at the top to inspire, cajole and persuade—if necessary pressurize—the concerned wings of the bureaucracy to go to the farmers and convince them, with example and reason, of the advantages of adopting the new technology in agriculture. A contemporary write-up on the progress of Haryana state has the following to say:

> "In five years even though a small state, it made an all-round advance—a virtual social revolution—giving a peep into the making of supercharged minds dedicated to a particular cause——Rather dry in rain, its people hell bent on deriving the maximum benefit from their meagre resources——they are stoics, disillusioned men and women content to make the best use of what little they possess. Already in the southern districts they equate Bansilal, the miracle working CM who brought water to their villages and fields, with spirit-raising holymen given to alleviate misery—the state government is bending over backwards to introduce development projects likely to pull the long suffering villager out of his traditional backwardness. The results achieved in seven years are astounding in their magnitude. The state today enjoys the second highest per capita income in India."[3]

A small state—only 44,000 sq. kms, with a small population, now about one crore—one of the smallest; this fact helped rather, than retarded progress. The comparative smallness of resources was more than offset by the intensive personal supervision "the men on the bridge" were able to bring to bear on the execution of development projects.

> The same author continues about the Haryana leadership and experiment:—
> "Was Bansi Lal a saviour or a heartless tyrant?—Bansilal calculated that political stability in the state comes about only as a result of rapid economic growth. Without letting any grass grow under his feet he cut the bureaucratic red tape to smithereens and embarked upon ambitious projects of giving people what they needed most—water, electricity, roads transport, agricultural inputs— Ever since he came to power in 1968, the wheels of Green Revolution have turned faster and faster. Not even the vagaries of nature and the selfishness of man could slow down the speed of their onward thrust—His bulldog tenacity in the pursuit of his objectives is a characteristic that brings him in line with such pragmatic administrators as Sardar Patel and Rafi Ahmed Kidwai".[4]

The ends would appear to justify the means if the ends are successful. Both Partap Singh Kairon and Bansilal had the courage, fearlessness and the gift of steadfast pursuit of purposes they held so dear and bring them to a successful end.

Was Bansilal a saviour or a tyrant? Was Partap Singh Kairon a benefactor of Punjab or a self-seeking 'trickster'? There can be many explanations and many philosophical dissertations possible in answer to these pertinent queries and yet the answer may not be all that truthful or even correct. In our current work-a-day world, full of socio-political tensions, full of emphasis on quick economic advancement, a leader with a forceful personality gets a chance of his life time to make a mark on it to the best of his lights, ability and awareness; if he gets involved in the controversy of ends justifying the means or vice versa all the time, he may end up his political career as a big zero. But if he is a 'go-getter', sure of his destiny, with a result-oriented bias by nature and design, clear about the paths leading to a successful end of his efforts, he does somehow try to leave behind his footprints on the sands of time and space. These two remarkable chief ministers tried to do precisely that and brought their respective states on top of the Green Revolution ladder in as best a manner as the circumstances permitted them. In the process, they came many a cropper but they got over the hurdles, not losing sight of their long range objectives and eventually succeeded. In the overall bargain they earned a bad name of sorts, unsavoury sobriquets like 'a tyrant, 'a self seeker", a 'crook', even an 'unscrupulous politician'. To how many people is given the privilege of achieving positive results for the good of society, with the use of good means only and without earning a bad name in the bargain. The peers of Mahatma Gandhi are rare to come by.

If the chief minister of a state is strongminded and clearheaded about the objectives he wants to achieve and has the capability of pulling along with his colleagues as a leader, at the same time inspiring the bureaucratic machine, he can get away with the use of pressures at the right place and at the right time, with a view to bringing the socio-economic goals nearer. Such pressures—even the use of discrete force against recalcitrant elements—need not be looked down upon but accepted as an indispensable instrument of state craft in some difficult situations. In the implementation of land reforms, for example, some 'pressures' may have to be applied to get the vested interests either cooperating or out of the way, in pursuit of the "land for the tiller" policy objective. Douglous Ensminger, once consultant to the Ford Foundation on Rural Development and its representative in India for many years In the sixties, had observed as under:

> "Following World War II only 3 countries went all the way in creating conditions for small farmers to change their values from producing to survive to producing to improve family living, viz. Japan, S. Korea and Taiwan. In all these cases, land reform programme provided incentives for those who tilled the soil to increase production for family gain. They made land and water more equitably available to all who tilled the soil. They oriented their agriculture policies to small farmers. This facilitated the transition from traditional to modernised agriculture. It be noted that revolutionary land reform programmes were implemented in Japan and S. Korea under military rule and the Taiwan under semi-dictatorial powers.
>
> "Other newly independent nations however remained dominated by elite power structure, lacking the political muscle and the will to pasa or implement land reform legislation favourable to the small farmer. US PL 480 facilities greatly reduced pressure on the political leaders for reforms or policies to

> modernise small farm agriculture. The Indian case was not that hopeless. It passed legislation in favour of land reforms but the implementation was colourless. Some gains were made; for example, Zamindari was abolished, land tenure was secured to a considerable extent, small holdings were consolidated and land ceilings were prescribed, though not so rigidly enforced and so on."[5]

To conclude, the success of Punjab and Haryana in the sphere of Green Revolution etc. in the general sense has been due not only to the small size and the relatively favourable resources of these states but also to the fact that both were fortunate in having bold, energetic and imaginative chief ministers to guide their destinies at the crucial times. They performed remarkably well and made sure that the goods were delivered in time and in ample measure. Should one hazard the statement that the UP and Bihar states were not so lucky in having for long the top leadership of a calibre which could properly exploit their local resources with which the two states were endowed in no small measure? Experience of Punjab and Haryana shows that such a statement is not wide off the mark. These states of UP and Bihar have suffered for lack of adequate top level guidance and pursuit which, if available, could change the tenor of their economic life (Late Govind Ballabh Pant was on top of the political scene in UP but only for a few years after 1947). If the bureaucracy/administrative machine did not come up to the needs of the task, the responsibility for that also must ultimately fall on the poor political leadership. What, after all, is administration in a Democracy, except an instrument in the hands of the political leaders elected by the people? An individual official may have a conscience and a duty to fulfil, but the administrative machine as a whole is naturally to be at the beck and call of the chief minister. Other things being equal or similar an inadequacy of performance in UP and Bihar states can mainly be ascribed to the inadequacy of political leadership which failed to inspire that machine in rising to the occasion nor could it awaken the farmer towards the pursuit of enlightened "self-interest" that was his due in a socialistic democratic set up. In the instant case one cannot even say that the political leadership got inextricably involved in a conflict between ends and means in which means were beyond reproach and the ends were lost in the long run. An adequate effort was simply not there; political and administrative energies seem to have been squandered in some ineffective pursuits.

The present situation in the socio-political sphere, perhaps, requires in a chief minister—or in all political leadership at crucial levels—a certain brashness, coupled with dynamism which is willing to sacrifice principles, once accepted as sacred in its desire to "go ahead" with the job of economic advancement, sometimes going crashing as a steam roller to gain his objective.

REFERENCES

1. Report of the "Committee on Agricultural Productivity in Eastern India" 1984 (published by RBI in 1985 called S.R. Sen Report).
2. Author's *Evolution of Food Policy in India*...Macmillan (India) New Delhi 1981-Pages *265-66*.
3. Muni Lal, *Haryana, High Road to Prosperity*, Vikas, 1974, p. 7.
4. Muni Lal, *Haryana, High Road to Prosperity*, Vikas, 1974—pp. 29-30 and 45.
5. Douglous Ensminger.

19

THE GREEN REVOLUTION

"What the new strategy can be expected to achieve is not so much a sharp accleration in the over-all rates of growth of agricultural (or, even of foodgrains) outputs as in giving considerable stimulus to private investment in the bigger-sized farms in regions which have the necessary climate and soil conditions, as well as sources of water supply, for adopting the new technology."

—*K.N. Raj*

New Strategy in Indian Agriculture

The increase in agricultural production that took place during 1967-68 and 1970-71 revived new hopes and it was felt that a breakthrough in Indian agricultural has been made. Some have gone so far as to proclaim that a green revolution is round the corner. Our purpose is to examine the true position as it stands in the light of the latest information gathered.

The new strategy in Indian agriculture was tried as pilot project in 7IADP districts in 1960-61. After the reports of initial success, the programme was extended to other areas. By 1967-68, 6 million hectares of land were brought under High-Yielding Varieties (HYV) and on the basis of official estimates, it led to an increase of about 7.7 million tonnes in the production of foodgrains. By 1968-69 the coverage of HYV was extended to about 7 million hectares. During the Fourth Plan, it was decided to bring 25 million hectares of irrigated area under HYV Programme. As a consequence, it was expected that production of foodgrains would increase to about 129 million tonnes.

To realise this target the new agricultural strategy proposed to make a technological breakthrough in Indian agriculture which comprised the introduction of new and high-yielding varieties of improved seeds, increased application of the recommended dose of fertilizers and extension of the use of pesticides so that the crop produced could be saved from destruction by insects. For this purpose, research had to be directed to finding new high-yielding varieties in each crop, stepping up the production of fertilizers and making the peasant familiar with the use of pesticides. In the initial stages, the programme was restricted to areas which enjoyed better irrigation facilities and which were immune to natural hazards. But as the coverage of the programme has to be extended further, irrigation facilities will have to be expanded. Need it be emphasised that water is a basic input in agriculture and better seeds and fertilizers can produce desired results only if adequate quantity of water is made available. Not only that, to enable the poor peasants to make use of better agricultural implements, install rubewells, apply heavy doses of fertilizers and use improved seeds, agricultural credit has to be provided. Steps were taken in all these directions. It would be appropriate to examine the achievements of the strategy.

Achievements of the New Strategy

The major achievement of the new strategy is the boost in the production of major cereals, viz. wheat and rice. Table 19.1 gives the areas, production and yield of the principal crops during the last few years. The year 1979-80 was an abnormal year

TABLE 19.1

Area, Productions and Average Yield of Principal Crops

		1960-61	*1964-65*	*1970-71*	*1973-74*	*1978-79*	*1979-80*
Rice	A	34	39	38	38	40	39
	P	35	39	42	44	54	42
	Y	1,013	1,073	1,123	1,151	1,339	1,082
Wheat	A	13	14	18	19	22	22
	P	11	13	24	22	35	32
	Y	851	913	1,307	1,171	1,570	1,437
Jowar	A	18	18	17	17	16	16
	P	10	10	8	9	12	11
	Y	533	503	466	544	718	688
Bajra	A	11	12	13	14	11	11
	P	3	4	8	8	6	4
	Y	286	380	622	540	486	380
Maize	A	4	5	6	6	6	6
	P	4	5	7	6	6	6
	Y	926	1,009	1,279	965	1,076	970

Source: *Indian Agriculture in Brief*, 1975 and *RBI Report on Currently and Finance*, 1979-80 and *Economic Survey*, 1980-81.

A: Area million hectares, P: Productivity million tonnes

Y: Yield per hectare in Kgs.

with extensive crop failures in many parts of the country. For purposes of comparison, let us, therefore, take 1978-79. A close look at the table reveals that the production of wheat which stood at 11 million tonnes in 1960-61 rose to about 35 million tonnes in 1978-79, i.e., an increase of 218 per cent in total wheat production. Part of this increase can be attributed to an extension of the area, but the yield per acre rose from 851 kgs. to 1,570 kgs. per hectare, signifying an increase of about 84 per cent. Increase in production of maize during 1960-61 and 1978-79 is estimated at 52 per cent. The increase in rice production from 35 million tonnes in 1960-61 to 42 million tonnes in 1970-71 and further to 54 million tonnes in 1970-79 signifies a break-through in this major crop of India. The yield per hectare has also recorded an improvement from 1,013 kgs. in 1960-61 to 1,339 kgs. in 1978-79. In the initial phase, the new strategy was able to make a big impact on wheat output, but since 1973-74, production of rice has also responded to a significant extent under the new strategy.

From Table 19.2, it can be inferred that the impact of the new strategy could best be observed during the period 1967-68 and 1970-71, the years which were, by and large, free from any drought or floods or any other natural hazards. Wheat, sugarcane and bajra responded exceptionally well to the green revolution. Some breakthrough has been noticed in maize. Rice was below the average of all foodgrains. But it is very disappointing to note that the growth rates were negative in gram, jute and cotton. During 1970-71 and 1977-78, groundnuts and bajra have shown negative growth rates. There has been a significant improvement in cotton. However, the over-all growth rates of foodgrains which was 4.8 per cent during 1967-68 and 1970-71, declined to a bare 2.1 per cent during 1970-71 to 1977-78. All this leads us to the conclusion that the influence of green revolution which was felt during 1967-68 to 1970-71 started slowing down, despite the fact that more area is now covered under HYV.

As a result of the new strategy, area under improved seeds has gone up from about 2 million hectares during 1966-67 to 26 million hectares during 1973-74 and further to 41 million hectares in 1978-79 and is likely to touch 48 million hectares in 1980-81. Besides this, new varieties take a short-term duration and consequently, instead of growing one crop in several irrigated areas, two crops are being grown. In the case of wheat, unprecedented enthusiasm prevailed among farmers Punjab, Haryana, Delhi, Rajasthan and Western U.P. for the new Mexican varieties like Lerma Rojo, Sonara-64, Kalyan and P.V. 18 and a situation developed in which the demand for seeds by the farmers exceeded the supply. But in the case of rice, the new varieties like T.N. IR-8, Tainen-3 and ADT-17 which were tried and found successful on a laboratory scale could not be successfully applied on the field. The peasants in rice areas which account for a big chunk of the total farming families have not accepted the new varieties with favour and enthusiasm. Sime breakthrough, however, has been achieved in rice areas by developing better and more acceptable varieties like IR-8.

TABLE 19.2

Growth Rate of Production of Selected Crops

Crop	*Growth rate of Production during* 1949-50 to 1964-65	1964-65 to 1966-67	1967-68 to 1970-71	1970-71 to 1977-78
Wheat	4.1	113	12.1	+4.0
Rice	3.0	1.0	3.9	+32
Bajra	2.3	10.2	15.5	—7.2
Maiza	3.5	7.6	4.8	—3.3
Gram	2.7	—1.4	-4.0	+0.7
Sugarcane	4.8	0.8	40.4	+5.4
Cotton	5.4	—3.0	—5.2	+5.9
Jute	4.2	—3.0	—7.0	+1.1
Groundnut	4.1	0.2	2.0	—0.1
All foodgrains	2.8	3.3	4.8	+2.1
All non-food crops	3.9	0.5	32	

Source: Commerce Annual Number, 1974.

TABLE 19.3

Production of Cash Crops in India

Commodity	*Unit*	*1960-61*	*1970-71*	*1973-74*	*1978-79*	*1979-80*
Sugarcane (gur)	million tonnes	11	13	14	16	13
Cotton*	million bales	6	5	6	8	8
Jute**	million bales	4	5	6	6	6
Oilseeds	million tonnes	7	9	9	9	8

* Bale of cotton = 170 kgs.

** Bale of jute = 180 kgs.

Source. Economic Survey, 1980-81.

After the initial success achieved by the green revolution, it was expected that the trend of a rising output of foodgrains will continue. This culminated in a record achievement of 108 million tonnes of foodgrains in 1970-71. It was hailed that green revolution has resulted in stoppage in imports, it has led to substantial accumulation of buffer stocks, but the euphoria was cut short in 1972-73 which happened to be a drought year. Production of foodgrains slumped to 95 million tonnes. The total output of foodgrains in 1973-74 improved to just 105 million tonnes. Dr. V.K.R.V. Rao, therefore, concludes "The increase in output between the first and final years of the Fourth Five Year Plan is thus no more than about 8 to 9 million tonnes or an average of less than 1.8 per cent a year, while the difference in output between the second year of the Fourth Plan, also a good year, is practically nil; and this in spite of the green revolution which has continued throughout the period with its expansion of the area under HYV and increase in the volume of the fertilizer input." Sharp fluctuations in foodgrains output were observed during the Fifth Plan period as well. From a low level of about 100 million tonnes in 1974-75, foodgrains output rose to 121 million tonnes in 1975-76 and then again fell to 11 million tonnes in 1976-77 and further rose to 131 million tonnes in 1978-79. Obviously the difference between the output of a good monsoon year, i.e., 1975-76 and 1978-79 indicates an increase of 4 per cent which is not significantly large in view of the fact that additional 4 million hectares were brought under HYV programme during this period. There was a steep decline in production in 1979-80 due to adverse weather conditions-foodgrains output in that year was 109 million tonnes only, almost the same as 1970-71 output. Two conclusions can be drawn from the sharp fluctuations of output of cereals in India since the introduction of new agricultural strategy: (a) Output of cereals is subject to weather conditions as in the past; and (b) the maximum and minimum outputs, however, are now much higher than in the past.

Since the green revolution was mainly directed at the improvement of foodgrains, it is not reasonable to expect a sharp increase in the production of commercial crops. "An analysis of table 3 reveals that the production of sugarcane, cotton, jute and oilseeds did not record any significant improvement during 1960-70 to 1973-74. Dr. Dharam Narain described this situation as "near paralysis in the output of cash crops." Significant improvement in the output of sugarcane took place after 1973-74 and its output rose to 19 million tonnes in 1977-78. Likewise, there was some improvement in the production of other cash crops but the improvement was not such as to be called a revolution.

Taking an overall view, the green revolution did not cover pulses which accounted for 11.8 million tonnes of the total output of foodgrains in 1970-71. The output of pulses has further declined to 10 million tonnes in 1973-74; it rose to 13 million tonnes in 1975-76 but declined to 12 million tonnes in 1978-79 and an all time low of 8 million tonnes in 1979-80. Neither did green revolution cover barley, ragi, and minor-millets which accounted for 7.1 million tonnes in 1978-79. Thus, the green revolution was confined only to High Yielding, Varieties (HYV) cereals, mainly wheat, maize and bajra, while rice output increased at relatively slower rate. The singular crop which showed a continuously rising trend was wheat. Maize and bajra showed substantial fluctuations. The very fact that the cash crops and pulses have to far been not brought within the ambit of new technology, forces the conclusion that quite a substantial part of the agricultural output has not been even touched by the green revolution.

In 1968-69, total area under HYV programme was of the order of 9 million hectares. In 1977-78, it shot up to 38 million hectares. In terms of fertilizer input (NPK), the total application of fertilizers was of the order of 23 lakh tonnes in 1970-71 and it touched 43 lakh tonnes in 1977-78.

TABLE 19.4

Percentage Distribution of Cereals and Pulses in Foodgrains Output

Year	*Cereals*	*Pulses*	*Total*
1949-50	85	15	100
1960-61	84	16	100
1970-71	89	11	100
1973-74	90	10	100
1978-79	91	9	100
1979-80	92	8	100

Source: Economic Survey (various Issues)

TABLE 19.5

Percentage Distribution of Cereals Output

Year	*Rice*	*Wheat*	*Coarse grains*	*Total*
1949-59	50	14	36	100
1960-61	50	16	34	100
1970-71	44	25	31	100
1973-74	47	23	30	100
1978-79	45	29	26	100
1979-80	39	29	32	100

Source: Economic Survey (various Issues).

As a result of the green revolution, the crop pattern in Indian agriculture has undergone two significant changes. This can be discerned from Table 19.4. Firstly, as a consequence of the new strategy the output of cereals has risen at the rate of 3 to 4 per cent per annum but the output of pulses has remained stagnant or even declined. This has resulted in a decline in the importance of pulses in foodgrains output from 15 per cent in 1949-50 to 9 per cent in 1978-79 and 8 per cent in 1979-80. As against it, cereals have risen in importance from 85 per cent in 1949-50 to 92 per cent in 1979-80. Secondly, among cereals, there is a decline in the proportion of rice from 50 per cent to 45 per cent during 1949-50 to 1978-79 (and 39 per cent in 1979-80). But during the same period, the importance of wheat has more than doubled, i.e., from 14 per cent in 1949-50 to 29 per cent in 1978-79. Another noteworthy change is the decline in the share of coarse grains from 36 per cent in 1949-50 to 26 per cent in

1978-79. The growing production of wheat indicates a substitution of coarse grains with wheat, both on the side of production and consumption. This trend has begun even before the green revolution was ushered, but it has now been strengthened.

3. Assessment of HYV Programme Under the New Strategy

Case for High Yielding Varieties Programme

The advocates of the new strategy consider the intensive approach as the only means of making a breakthrough in Indian agriculture in the shortest possible time.

Secondly, it is admitted on all hands that agricultural inputs are scarce and it is not possible to meet the needs of the entire country in this respect. The choice is to have a thin layer of the inputs spread over the entire country or to apply concentrated doses in selected and promising areas. The latter choice, according to the apologists, is more rational as it ensures maximum production in the short period.

Thirdly, it was held that the application of high doses will be accompanied by increasing returns and, consequently, the application of high doses can be justified on economic grounds.

Fourthly, large-scale demonstration of better cultivation in actual field conditions will induce farmers in other areas to adopt improved techniques. Thus, the intensive development programme sets in motion a benefit circle which will move on and thus gather momentum. In this way the spread effect of the programme will raise, overall level of productivity in Indian agricultural production in a certain region will produce secondary and tertiary effects. For instance, the availability of more food in the country will decrease our dependence on food imports and thereby release scarce foreign exchange resources for other sectors of the economy. Similarly, increased production of commercial crops will enable the expansion of agro-based industries.

Criticism of HYV Programme

Main points of criticism are:

1. Growth of Capitalistic Farming in Indian Agriculture

The high yielding varieties necessitate heavy investment in fertilisers and water. These heavy investments are beyond the capacity of small and medium farmer. In Indian, there are about 81 million farm households but only 6 per cent of the big farmers account for 40 per cent of the land; they alone are making heavy investment in the installation of tubewells, pumping sets, fertilizers and agricultural machinery required for the purpose. Consequently, the new agricultural strategy has helped the growth of capitalist farming in India. The agricultural revolution is lacking in spread effects and, thus, is fostering an 'enclave-type development' in Indian agriculture. The poor peasantry is not being benefited thereby, but has led to concentration of wealth in the hands of the top 10 per cent of the rural population.

TABLE 19.6

Capital expenditure on Some Items on the Big Farms in Punjab (1967-68)

(Rs. perform)

Size group in acress	*Tractors*	*Tube-well and pumping set*	*Other equipment*	*Buildings* land improvement and repairs*
20-25	1,900	2,000	500	1,100
25-30	1,600	1,600	300	1,250
30-40	4,400	2200	900	1,850
40-50	4,900	2,000	1,000	1,600
50-75	7,600	4,700	1,900	3,500
75-100	7,500	7,400	1,400	5,200
100-150	21,000	..	7,000	6,750
150 & above	34,000	21,100	11,500	13,850
All groups	4300	2,800	1,000	2,100

*Figures are average of 1966-67 and 1967-68.

Source.: Ashok Rudra, "Big Farmers of Punjab," *Economic and* Political *Weekly,* December, 27, 1969 (Compiled).

Ashok Rudra, Majid and Talib conducted a study of the farmers in Punjab so as to analyse capitalist farming. They defined capitalist farmers as those who have at least 20 acres of land. The study revealed that the total number of farms above 20 acres was 67,000 covering over 26 lakh acres of cultivated area. Land owned by the big farmers increased by about 9.5 per cent between 1955-56 and 1967-68. A further breakup of big farms reveals that farms of size group 20-25 acres expand only by 4 per cent whereas those of the size group 100-150 acres increased by 40 per cent and that most of the additions took place through purchases. Thus, there is evidence of the growth of gentlemen farmers, who comprise ex-servicemen, retired civil servants, urban-based businessmen deriving their income from industry and commerce and who have recently taken up agriculture as an industry. This group finds investment in agriculture quite profitable. In Punjab they constitute about 3 per cent of total number of farmers. Capitalist farmers who include progressive farmers, besides the gentlemen farmers command 8.5 per cent of the total number of farms and cover 27 per cent of the total cultivated area. It is this group of farmers who are able to make investment in the form of tractors, tubewells and pumping sets and other equipment Similarly, these farmers are in a position to incur capital expenditure on buildings, improvement of land and other repairs on their holdings. This is evidenced by the data in Table 19.6. It is obvious that larger farms show a higher capital investment per acre which is an index of the extent of mechanisation.

Francine Franknel, USAID expert, undertook a study of the impact of new, strategy on the socio-economic relations of the peasantry. The main conclusions of this study are:

(a) Overwhelming majority of the cultivators having uneconomic holdings of 2—3 acres have managed to increased per acre yield from the applications of small doses of fertilizers, but aggregate gains in output have been insufficient to create capital surpluses for investment in land development.

(b) In cases where small farmers also take part of their holdings on lease, or are pure tenants, rising rentals in recent years (in response to the sharp spurt in land values), and/or the tendency of landowners to resume land for personal cultivation (with the introduction of more profitable techniques), has actually led to an absolute deterioration in the economic condition of the small owner-cum-tenant cultivator class.

(c) Only the small minority of cultivators with holdings of ten acres or more have been inter position to mobilise surplus capital for investment in land development, especially minor irrigation, as an essential pre-condition for the efficient utilisation of modern inputs. Moreover, this class has pyramided its gains by using increased profits to buy more land, improve land already under cultivation, and purchase modern equipment.

(d) Farmers with twenty acres or more have made the greatest absolute and relative gains, partly by mechanising farm operations to take up double or multiple cropping, but also by diversifying their cropping pattern to include more profitable commercial crops.

(e) The majority of farmers-probably as many as 75 per cent to 89 per cent in the rice belt—have experienced a relative decline in their economic position; and same proportion, representing unprotected tenants cultivating under oral lease, have suffered an absolute deterioration in their living standards.

2. *Sidetracking the Need for Institutional Reforms in Indian Agriculture*

The new strategy does not recognise the need for institutional reforms in agriculture. Among the institutional changes, the most important is the land-relations so that the technological change sought to be achieved can be effected with maximum social benefit. It is a truism that in India agriculture is carried on by millions of cultivators and it is the absorption capacity of fertilisers by these millions of cultivators that will determine the total agricultural production. Not to speak of providing ownership rights to the bulk of the peasant population, we have failed to provide even fixity of tenure and large-scale evictions have already taken place. As a result the tenants are being forced to accept the position of share croppers. Minhas and Srinivasan have studied the effect of crop sharing arrangements in fertiliser use. Their basic assumption is that the cost of fertilizers is met by the cultivator by borrowing and interest charges amount to 10 per cent of the cost. Basing their judgement on the capitalist principle of profit maximisation, the owner-farmers reap a profit of 180 per cent on irrigated lands in the case of wheat and 183 per cent in the case of rice. Contrasting it with the tenant cultivation (50 per cent basis) this return is reduced to 65 per cent in wheat and 67 per cent in the case of rice. The return is further reduced to a level of 42 per cent in the case of share-cropping on a 40 per cent basis. Thus, the conclusion is inescapable that tenancy cultivation poses itself as a big obstruction in the way of fertilizer use. Profit maximisation criterion so very familiar to capitalist economics clearly indicates that larger dosages of fertilisers will be absorbed by owner-farmers than by tenants.

3. *Green Resolution and Farm Income Distribution*

The study of the effect of green revolution on farm income distribution by Srivastava, Crown and Heady revels: "The income disparity even among adopters, will grow over time. The ratio of large-medium farms net income has grown from 2.39 to 2.48 or by about 4 per cent. An even more dramatic growth in disparity is seen between the large and small farms. The ratio of net income grew from 10.03 to 11.50, or by about 10 per cent. Further the small farmer is worse off relative to his own former position (net income falls by about 2 per cent), so that one can conclude that the small farmer will be even less (than he was formerly) to employ the new technology in an attempt to catch up. The income gap, therefore, will continue to widen in successive production cycles."

TABLE 19.7

Ratio of Large and Medium Farm Net Revenue with Respect to Farm Revenue

Net Revenue Ratio	*Original*	*First approxi-mation*	*Second approxi-mation*
Large farm to small farm	10.03	10.29	11.04
Large farm to medium farm	2.39	2.40	2.48
Medium farm to small farm	4.20	4.29	4.45

Note: Small farms have average acreage of 2.5 acres, medium farms have 7.5 acres, and large farms have 12.5 acres.

Source: *Economic and Political Weekly*, December 25, 1971, p. A. 170.

4. *Problem of Labour Displacement*

Very few studies are available to assess the impact of the mechanisation introduced under the garb of green revolution in terms of displacement of labour. Uma K. Srivastava, Robert W. Crown and Earl O. Heady have examined the effect of two types of technological innovations introduced under the Green Revolution—(i) biological and (ii) mechanical. The term 'biological innovations' refers to the changes in inputs that increase productivity of land. The introduction of better seeds, popularly called as high yielding varieties, and use of fertilisers, fall in this category. In this sense, green revolution is described as transformation or seed-fertiliser technology. The mechanical innovations refer to the introduction of new appliances which displace human or bullock labour. Thus, whereas biological innovations are labour-absorbing, mechanical innovations are labour-saving. It is therefore all appropriate to describe the green revolution as a biological-mechanical revolution. It is the net effect of the labour-absorbing and labour-saving innovations which will determine the extent to which mechanisation need to be introduced to check further displacement of labour. The study concludes: "Since mechanisation may dampen the increase in labour demand, resulting from the expanding factor of seed-fertilisers, the policies that encourage premature mechanisation in surplus labour economies, such as India's do not seem conducive to solving the problem of growing unemployment." Consequently, policies aimed at indiscrete subdivision of huge labour-displacing machinery through cheap credit availability need to be re-examined.

Martin H. Billings and Arjun Singh have made an attempt to quantify the labour displacement effects in the wheat growing areas of Punjab with different degrees of mechanisation. Given the total strength of agricultural workers on Punjab farms according to 1961 census, the share of energy supply used for farm operation has been calculated on the assumption that the average worker will be able to devote 20 days in a month to crop production. The rest of the time he will have to attend to animal husbandry, private and social activities, or he may not be able to work due to bad weather.

Farm management studies were used for estimating per acre labour requirements. The study assumes that by 1983-84 (a) nearly 100 per cent of the wheat will be mechanically threshed; (b) 100 per cent of the corn mechanically shelled; (c) 20 per cent of the gross cropped area tilled by tractors; (d) 50 per cent of the wheat will be mechanically reaped; (e) pumpsets will feed 60 per cent of irrigated area and irrigable areas will be largely irrigated.

Given these assumptions, and also assuming that no radical changes occur in public policies, and the relationship between farm costs and prices remains sufficiently favourable, as against 5.5 per cent displacement of labour in 1968-69 it is expected by 1983-84, that labour displacement will be of the order of 17.4 per cent. About 55 per cent of the total labour displaced is expected to be caused by tractors and pumpsets and 37 per cent by threshers and reapers. The analysis of Martin H. Billings and

Arjun Singh is very instructive from the point of view of future economic policy. It emphasises the fact that the technical changes that will successively gather momentum are bound to cause labour displacement. In other words, 'green revolution' unaccompanied by a massive programme of "rural industrialisation" will cause nothing but frustration amongst the large mass of Indian peasantry. Mr. M.S. Gurupadaswamy, a former Minister of State for Food and Agriculture, gave a timely warning, "if new openings for their productive involvement in alternative occupations are not created simultaneously (with the new agricultural strategy), unemployment or underemployment of agricultural labour might land us in a situation which might become explosive both politically and economically."

Lessons of Green Revolution

On a balance of considerations, it would be too optimistic to assert that the agricultural revolution is round the corner. Much leeway has to be covered before an assertion of this type can be made. It would be, therefore, proper to analysis the lessons of the experience of the Green Revolution embodied in the seed-fertilizer-irrigation package.

Firstly, the agricultural revolution initiated by the new strategy is limited to wheat, maize and bajra only. The major crop of India, i.e., rice, has only recently responded to the impact of the high-yielding varieties. Besides this, progress in major commercial crops, viz., oilseeds, cotton and jute is very slow. Agricultural research has not been diverted to the development of new seeds in them. In addition to all this, pulses which account for about 10-12 per cent of the total food production have not registered any significant spurt in production. It would therefore, be very premature to speak in terms of an agricultural revolution unless the projected upward trend in production in a few crops becomes pervasive in most of the major crops.

Secondly, the rise in food production has taken place in Punjab, Haryana, Western U.P. and in some selected districts of Andhra Pradesh, Maharashtra and Tamil Nadu. But these areas cannot claim to cover the bulk of India. All that can be said is that they have shown the way to a big take-off in agricultural production. In other words, the already better off areas have made their economic position still better. This has initiated a process of unbalanced growth in India. The regions which have lagged behind have to catch up with those that have marched ahead. Unless all the major States enter the take-off stage, it would not be fair to speak of an agricultural revolution.

Thirdly, the adoption of new technology depends upon the level of literacy of the farmers. Dr. D.P. Chaudhri in his study Education and Agricultural Productivity in India has shown that large farmers, invariably are always early adopters of yield-raising, cost-reducing innovations, while small farmers, who are generally illiterate, are among the laggards. The spread of the new technology, therefore, depends upon the extent of information and this in farm depends upon the level of literacy. A programme of removal of illiteracy can, therefore, become the chief vehicle of the spread of green revolution.

Fourthly, it has been observed that in the present rural set-up of co-operative societies and rural banks, it is the big farmer who is able to secure a loan at 6 to 10 per cent interest. The small farmer who wields very little influence in the village has to borrow from the moneylender or other such sources of unorganised money market at interest rates varying from 12 to 75 per cent. The net result of the prevailing situation is that whereas official agencies provide less than 30 per cent of the total credit, and that, too, mostly to the big farmers, it has been usually found that the tenant farmer and the small farmer who need credit at lower rates of interest, get the costliest credit while the credit-worthy large farmers are able to get it cheap. This introduces a difference in the real price of inputs to the large and the small farmers, obviously to the disadvantage of the latter.

Fifthly, the adoption of new technology depends upon control over water supply and ability of the farmer to regulate its timing. It requires lumpy investment in tubewells or diesel pumpsets. It has been observed that the ownership of tubewells was limited among farmers with ten acress or more. Obviously, there is a need to develop a rental market for water supply. This market is generally developed with a time lag. It is the large farmer who is able to adopt the new technology in the early phase, when such a rental market is being developed. In order to accelerate the pace of new technology, availability of water should be ensured.

Sixthly, the new strategy has created three kinds of conflicts, namely, between large and small farmers, between owners and tenant farmers and between employers and employees on agricultural farms. The holders of large farms are capable of making heavy investment in the form of fertilizers, pumping sets, tubewells and agricultural mechinery. They are also able to procure credit from co-operatives, as also obtain fertilizers and better seeds. For all practical purposes the large farmers enjoy the right of pre-emption of agricultural inputs and, thus, the small farmers are deprived of the much-needed inputs. This has, therefore, widened the inequalities of income and fostered the growth of capitalist agriculture in the country.

In India quite a significant group of peasants have small size holdings and consequently, they hire land on tenancy from large owners. Since the landlords pay for the new agricultural inputs, such farms are experiencing a dualism in agricultural production technique. The part of land hired by the tenants is provided with modern techniques and the small fragments of land owned by the tenants continue to be worked by traditional techniques. The conflict is the cause of social tension, more so, when the landlords demand exploitative rents on the land leased out by them.

The application of new technology in large farms has led to the substitution of human labour with mechanical processes. In other words, the back-wash effects of new strategy in terms of displacement of human labour have also to be taken note of. The greatest sufferers in the process of agricultural revolution are landless labourers. Unless alternative opportunities of employment are provided to this most vulnerable section of the rural community, agricultural revolution will be meaningless to the millions of landless peasants in this country.

It is, therefore, essential that landless labourers and tenants are organised and a trade union mechanism is developed, so that non-exploitative forms of tenancy can be developed in rural India and the wages of landless labourers can be improved. The gains of high-yielding varieties are being pocketed by the rich landowing classes and the new strategy has further tilted the scales of distribution in their favour. Organised peasant resistance alone can help the small peasant, the tenant, the share-cropper and the landless labourer, i.e., the agriculture proletariat, in maintaining its relative share in agricultural income intact.

Former U.N. Secretary-General U. Thant, rightly warned: "The green revolution may indeed prove to be the Pandora's box rather than a 'corncopia' unless developing countries urgently push through land reforms." He further added: "There are many observers who contend that if left to market forces, the green revolution is likely to benefit primarily those farmers who are already engaged in commercial production rather than subsistence farmers, and among commercial farmers, big ones more than small producers."

From an analysis of the socio-economic relationships that are developing with the introduction of new technology euphemistically termed as green revolution, one may conclude that if an effort is not made to provide credit to the small farmers, provide security of tenure to the cultivating tenants, reduce the rents charged from tenants/sharecroppers or to provide an institutional framework which takes care of the econonomic inequities, there is every danger of the green revolution turning into a "Red Revolution." It will be a strange paradox of rising production and rising frustration.

Optimisation of the Benefits of Green Revolution

A very relevant question that needs to be examined is: Does the present strategy of intensive agricultural development lead to optimisation of the benefits from the package of agricultural inputs? Professor V.K.R.V. Rao in his Panse Memorial Lecture has argued that the present strategy does not use the scarce inputs in an optimum manner and as such needs serious modification. The main arguments of Professor Rao are the following:

1. Enlargement of Double Cropping Area Vitally Necessary

Intensive agricultural development was directed in the first stage to those areas which had assured supply or irrigation or which were free from natural hazards. Such areas should have been available for double cropping. But ironically, in India in the year 1969-70, the total cultivated area sown more than once was 24.8 million hectares, out of which rain-fed area sown more than once was as much as 17.9 million hectares or 72.3 per cent of the total double cropped area. Irrigated areas are more suited to double or multiple-cropping. In India, however, out of 40.2 million hectares under irrigation in 1973-74, only 7.7 million hectares or 19.2 per cent of the net irrigated area was sown more than once. Thus, a serious modification of our policy should be not only to try for a higher yield per acre of the irrigated crop but also to produce a much higher yield per acre by promoting the sowing of two or more crops on irrigated land. Such a policy can increase the effective supply of good land and thus help in the maximisation of output. Professor V. K. R. V. Rao rightly mentions: "If we assume that it is legitimate to consider one unit of irrigated land as having the capacity to produce two units of land on a single crop basis, it means that we have today an unutilised capacity of 75 to 80 per cent in the use of irrigated land, an extent of under-utilisation that for surpasses the unutilised capacity in most public sector enterprises which is so much a matter of concern to Government and the Planning Commission."

In addition to this, there is a strong need to utilise the irrigation potential of the country. On present indications, it has been estimated that 107 million hectares of land can be ultimately irrigated both from surface and ground water sources, but only 44.7 million acres have been brought within its ambit. The Fifth Plan aimed to bring about 11.1 million hectares of additional land under irrigation. It is very necessary to extend minor irrigation and thus tap the private savings of the people to enlarge the area under irrigation. The effort of the State coupled with that of private sources can help to cover the entire irrigable area within the next 20 years. Since water is a basic input in the use of seeds and fertilizers, irrigation facilities must expand simultaneously. Moreover, this shall also reduce the disparities between regions. It is commonly believed that green revolution has benefited the better-off regions, i.e., those possessing irrigation facilities. The extension of irrigation shall certainly enable the backward regions also to make use of the latent potential. The example of Haryana, now a granary of India's is a case in point. The touch of irrigation, reinforced by higher dose of agricultural inputs, transformed the region to such an extent that it is considered to be one of the most prosperous belts of India.

2. The Problem of Better Allocation of Fertilizers

The planners have recommended very high doses of fertilizers per hectare on the land selected for intensive agriculture. Recent research has indicated that such an allocation is wasteful became it is inappropriate to assume a linear relationship between increase in the quantity of fertilizers applied and the increase in yield per acre. A realistic view is that response of productivity per unit of land to fertilizer use is subject to the law of diminishing returns in terms of physical quantities. The Department of Agricultural Research and Education (DARE) on the basis of the All-India Coordinated Rice Improvement Project (1973) demonstarted that an additional 50 kgs. of grains resulted from each kg. of nitrogen applied at moderate level and with good management of non-cash inputs. It is, therefore, imperative that government should modify the criteria of allocation of fertilizers so as to optimise their use with a view to obtaining larger return per unit of fertilizer applied.

3. Extension of Area Under HYV, Even When Fertilizers Cannot be Supplied

So far the Government has been treating IADP as a package program and as such High Yielding Varieties were to be provided along with doses of fertilizers. This has improved the productivity in areas selected for IADP but has again widened disparities between IADP and non-IADP areas. This policy is based on the assumption that without fertilizer a High Yielding Variety will not yield any additional output as compared to the traditional variety. But field trials have shown that this is not true. The High Yielding Varieties have an inherent genetic capacity to give higher yields than traditional varieties. On the basis of "Simple Fertilizer Trials on Cultivators' Fields" for the period 1967-68 to 1970-71, it has been established that average yields per hectare for HYV wheat and rice without fertilizer were higher than those of the traditional or tall varieties without fertilizer. The conclusion is, therefore, obvious that the State should at least extend the programme of provision of HYV seeds even to non-IADP areas, though it may not be possible to provide fertilizers at the present stage.

4. A New Fertilizer Strategy Should be Developed in View of the Present Energy Crisis Which is Not a Short Period Phenomenon in Character

There are two elements of the new strategy. Firstly, emphasis should be shifted from oil-based fertiliser plants to coal-based fertiliser plants. For this purpose, it would be necessary to augment the supplies of coal as also to perfect the technology in this area. This approach also has its own limitations. It would, therefore, he appropriate to supplement it by a second line of defence, viz., the programme of development of organic manure, both farmyard and urban and rural compost. Wet-cattle dung and cattle urine can be made use of in producing organic manure. In this connection, the scheme of gobar gas plant has a special significance because it serves a dual purpose—on the one hand, it supplies gas as a fuel and on the other it converts cattle-dung into organic manure to serve as a fertilizer nutrient. If out of 1,355 million tonnes of wet cattle-dung, nearly 1/3rd or 400 million tonnes are harnessed in this project, it would be possible to provide 11.7 lakh tonnes of fertilizer. Besides this, urban waste can also serve as a good source of fertilizer supply. A programme of suitable rotation by planting leguminious crops can also enrich the soil in fertilizer and thus reduce the need for fertilizer use for obtaining higher yield.

5. HYV Programme with Particular Reference to Commercial Crops Like Cotton, Jute and Oilseeds and Food Crops Like Rice, Millets and Pulses Should be Launched

In order to achieve a sustained rate of growth of 4 per cent in foodgrains and 6 per cent in commercial crops, it is vitally important that research be directed in the crops not covered so far so that a breakthrough in the output of these crops is also achieved.

Dr. C.H. Hamumantha Rao has studied the impact of technological change and distribution of gains in Indian agriculture. His study draws some very important conclusions for future policy considerations: (a) "Technological changes have contributed to widening the disparities in income between different regions, between small and large farms and between landowners on the one hand and landless labourers and tenants on the other. In absolute terms, however, the gains from technological change have been shared by all sections. This is indicated by the rise in real wages and employment and in incomes of small farmers in regions experiencing, technological change."

(b) ".... participation in the modernisation process is usually greater in regions which have shed the feudal and semi-feudal structures and attitudes and where the inequalities in wealth and status are not conspicuous." Rao cites the example of Punjab, Haryana and Uttar Pradesh where the technological change did not result in perceptible social tensions because real wages have risen in these regions. But such was not the case in Thanjavur. The landlords tried to counter the demand for higher wages by bringing in farm labour from outside.

(c) Technological change, in general, and mechanisation in particular, have been taking place among the large farms in high-income and high-growth pockets.

(d) "if the Green Revolution is regarded as a package consisting of HYV and fertilizers its contribution to employment has been substantial. Also, tubewells seem to have contributed significantly to the employment of labour.... The highest labour co-efficient relates to operated area followed by HYV and irrigation."

"The net employment effect of tractor—use may turn out to be negative when tractorisation of farm operations is complete. A harvest combine would displace farm labour on a large scale while its land augmenting effect would be negligible."

From the findings of Professor C.H. Hanumantha Rao, certain important policy implications emerge.

Firstly, for the success of the Green Revolution and its spread without social tensions, it would be advisable to make a frontal attack on feudal and semi-feudal land relations.

Secondly, growth would have been less costly from the social point of view and the prices would have been lower if public investment (e.g., in the major and medium irrigation projects and public tubewells, which account for lower unit costs for the services rendered because of economies of scale) were stepped up and technological changes were widespread in the labour abundant sectors such as backward regions and the small farms.

Lastly, in order that the effect of mechanisation does not become negative or negligible, it would be desirable to permit partial tractorisation or mechanisation.

To sum up, the strategy of green revolution needs to be modified. While doing so, another important consideration should be to help the small farmers. So far, the principal beneficiaries of the green revolution have been the big farmers who are able to grab the superior quality inputs and credit facilities to their own advantage. But the need for modifying policies focussing special attention on small farmers and backward regions is evident. Professor V.K.R.V. Rao, emphasizing the need for meeting this challenge of widening inequalities with the spread of green revolution, states in a very forthright manner, "It is now well-known that the so-called green revolution which helped the country to raise its output of foodgrains has also been accompanied by a widening of the range of inequality in rural incomes, the loss of their status as tenants by a number of small farmers and the emergence of social and economic tensions in the countryside ... the challenge which Indian agriculture faces is not only of production but also that of distribution; and in our anxiety to concentrate on production problems, we should not forget the human and social implications of agricultural development."

Dr. D.P. Chaudhri concluding his survey of green revolution states: "Land reforms without changing the working of the capital market and other rural institutions may create a conflict between the reduction of inequalities of income distribution and maximisation of growth. But land reforms with appropriate changes in the capital market and rural institutions would make possible maximisation of output and productivity growth completely consistent with reduction of inequalities of income distribution."

ENCYCLOPAEDIA OF
TEACHING OF AGRICULTURE

ENCYCLOPAEDIA
OF
TEACHING OF AGRICULTURE

[VOLUME – TWO]

Editor-in-Chief

Arun Kumar

INSTITUTE FOR SUSTAINABLE DEVELOPMENT
LUCKNOW

&

ANMOL PUBLICATIONS PVT. LTD.
NEW DELHI - 110 002 (INDIA)

ANMOL PUBLICATIONS PVT. LTD.
4374/4B, Ansari Road, Daryaganj
New Delhi - 110 002
Ph.: 3261597, 3278000
Visit us at: www.anmolbooks.com

Encyclopaedia of Teaching of Agriculture

First Edition, 2002

ISBN 81-261-1316-2 (Set)

PRINTED IN INDIA

Published by J.L. Kumar for Anmol Publications Pvt. Ltd., New Delhi and Printed at Mehra Offset Press, Delhi.

CONTENTS

PREFACE

If 'food' is a must for survival of man besides the air and water, it is AGRICULTURE which provides FOOD. Since the cave-dwelling stage to the present day journey of mankind Agriculture has seen many changes, ups and downs. Modern agricultural science include agriculture, horticulture, floriculture, pisciculture, animal husbandry, sericulture, veterinary science, soil science, irrigation and water technology, paultry, dairying, forestry, food processing and technology etc. etc. Thus from an occupation it has grown into an industry in modern sense of the term. Hence Agriculture education.

Agriculture, art, science, and industry of managing the growth of plants and animals for human use. In a broad sense agriculture includes cultivation of the soil, growing and harvesting crops, breeding and raising of livestock, dairying, and forestry. Modern agriculture depends heavily on engineering and technology and on the biological and physical sciences. Irrigation, drainage, soil conservation, and sanitation each of which is important in successful farming—are some of the fields requiring the specialized knowledge of agricultural engineers. Agricultural chemistry deals with other vital farm problems, such as uses of fertilizers, insecticide, and fungicide, soil makeup, analysis of agricultural products, and nutritional needs of farm animals. Plant breeding and genetics contribute immeasurably to farm productivity. Genetics has also placed livestock breeding on a scientific basis. Hydroponics, a method of soilless gardening in which plants are grown in chemical nutrient solutions, may solve additional agricultural problems.

Many universities, institutes and colleges are there imparting education of Agriculture in many of its specialized branches.

The present work (in two volume) is an effort to bring together encyclopaedic information on the subject. All these issues in the context of teaching Agriculture have been discussed in the present book in detail and therefore, it is hoped that it will be a very useful work for all concerned.

The authoritative information is organised in thirty seven chapters. To make the work more authentic, opinions and research of many scholars in the field are liberally cited and borrowed. We are deeply beholden to all those scholars of eminence and agricultural scientists. We are also thankful to all those who have assisted/guided us while preparing the draft of the work. Our colleagues and friends at the ISD deserve our appreciation. Finally we owe a deep sense of gratitude to Mr. J.L. Kumar, Managing Director, Anmol Publications Pvt. Ltd., New Delhi for his constant moral support and also for bringing out this work.

—Editor-in-Chief

20

Agriculture: Productivity Trends and Crop Pattern

"If you are late in doing one thing in agriculture, you are late in all things."

—*Cato, 2nd Century B.C.*

THE PLACE OF AGRICULTURE IN THE NATIONAL ECONOMY

Agriculture forms the backbone of the Indian economy and despite concerted industrialisation in the last over two decades, agriculture occupies a place of pride. Being the largest industry in the country, agriculture is the source of livelihood for over 70 per cent of population in the country. The significance of agriculture in the national economy can be best explained by considering the role of agriculture under different heads.

(i) *Share of Agriculture in the National Income.* Figures supplied by the National Income Committee and the Central Statistical Organisation show clearly that agriculture and allied occupations (viz., animal husbandry, forestry, etc.) contributed 52 per cent of the national income in 1960-61 but contributed about 44 per cent of the national income in 1976-77.

Two important facts must be emphasised here. Firstly, agriculture contributes a high share of the national income in India. Secondly, the share of agriculture in national income has been decreasing steadily. It is interesting to note here that national income estimates made in the past indicate not only the dominance of agriculture but also its gradual decline. For instance, the share of agriculture in national output up to the World War I (1914-18) was of the order of 65 per cent. Then the share of agriculture in national income declined to 57 per cent for the period 1925-29, and further 'to 53 per cent during the middle of the Great Depression (1931). Since then, the ratio has been steady round about 50 per cent. But after. 1960-61, the share of agriculture in national income has been declining.

Comparison can be made between the position of agriculture in India with that in the other countries as regards the share of agriculture in national income. In the United Kingdom, agriculture contributes only 3.1 per cent of the national income; in U.S.A. it is 3.2 per cent; in Canada it is 5 per cent; in Australia it is 7.6 per cent; and so on. The more developed a country, the smaller is the share of agriculture in national output and income. India, having not yet reached the stage of advanced economy, has an agricultural sector which is the dominant one in the country.

(ii) *Agriculture as a Source of Livelihood.* It is not surprising that in the Indian economy, with agriculture as the dominant activity, the main source of livelihood is agriculture itself. Seven out of every ten persons in India depend on agriculture. This proportion, indeed, is very high and what is remarkable is that between 1901 and 1971 this proportion (68 per cent of people depending on agriculture) has remained almost constant. In countries like U.K., U.S.A., West Germany, Canada, New Zealand, etc., people dependent on agriculture are less than 20 per cent of the total population; in some cases they are less than 5 per cent. What is still more significant is that in advanced countries like U.K., U.S.A. etc., the proportion of people dependent on agriculture is declining while in India it hai remained almost stationary.

(iii) *Indian Agriculture and Pattern of Employment in the Country.* Agriculture dominates the economy to such an extent that a very high proportion of working population in India is engaged in agriculture. According to 1971 census figures, 68.7 per cent of working population in India, is engaged in agriculture as compared to 69.5 per cent in 1961 census. But in the United Kingdom, only 3 per cent of the working population is engaged in agriculture; in U.S.A., the proportion is about 5 per cent; and in Australia, this is about 8 per cent. It is only in backward and underdeveloped countries that the working population engaged in agriculture is quite high. For instance, it is 55 per cent in Egypt, 64 per cent in Burma and 70 per cent in Indonesia.

There is another significant fact about the employment pattern in India. Census figures for the last 60 years show that the percentage of workers engaged in agriculture has been steadily increasing. For instance, in the beginning of the century about 62.5 per cent workers were engaged in agriculture; by 1921, the figure rose to 68.5 per cent; by 1961, this had increased to 69.5 per cent; in 1971, this had come down slightly to 68.7 per cent. This implies that alternative industries have not come up to provide employment for the rapidly growing population in the country and agriculture continues to be the main source of employment.

(iv) *Importance of Agriculture for Industrial Development.* The significance of Indian agriculture arises from the fact that it has been the source of supply of raw materials to our leading industries Cotton and jute textile industries, sugar, Vanaspati and plantation—all these depend on agriculture directly. There are many other industries which depend on agriculture in an indirect manner. Many of our small-scale and cottage industries like handloom weaving, oil crushing, rice husking, etc., depend upon agriculture for their raw materials.

But then, in recent years, the significance of agriculture to industries is going down as many new industries have come up which are not dependent on agriculture. Under the Five - Year Plans, iron and steel industry, chemicals, machine tools and other engineering industries, aircraft, etc, have been started. Despite this, agriculture provides raw materials for a number of industries, e.g. sugar, tea, cotton and jute textiles, hydrogenated oils, food products, soap and some other agro-based industries, which together account for 50 per cent of income generated in the manufacturing sector in India.

(v) *Role of Agriculture in the Field of International Trade.* Importance of Indian agriculture also arises from the role it plays in India's trade. Agricultural products—tea, sugar, oilseeds, tobacco, spices, etc.,—constitute the main items of exports of India. Roughly speaking, the proportion of agricultural goods which are exported may amount to 50 per cent of our exports, and manufactures with agricultural content (such goods as manufactured jute, cloth and sugar) contribute another 20 per cent or so; and the total comes to 70 per cent of India's exports. This has great significance for economic development. For, increased exports help the country to pay for the increased imports of machinery and raw materials.

It is significant to note here that it is the failure on the agricultural front which has upset the whole system of planning. For one thing, shortage of foodgrains within the country raised the prices, and led to rise in wages and other costs. Secondly, the Government has been forced to import foodgrains with foreign currency which can be used for importing more essential goods than foodgrains. Importance of agriculture in the national economy is also indicated by many other facts. For example, agriculture is the main support for India's transport system, since railways and roadways secure bulk of their business from the movement of agricultural goods. Internal trade is mostly in agricultural products, Further, good crops implying large purchasing power with the farmers lead to greater demand for manufactures and, therefore, better prices. In other words, prosperity of the fanners is also the prosperity of industries. Likewise, bad crops lead to a depression in business. Finally, finances of the government, especially of the State Governments, depend, to a large extent, upon the prosperity of agriculture.

It is clear, therefore, that agriculture is the backbone of the Indian economy and prosperity of agriculture can also largely stand for the prosperity of the Indian economy. At the same time, it is true that per capita productivity in agriculture is less than in industry. Naturally, most scholars of under-developed economies observe that this dominance of agriculture in India's economy is responsible for the low per capita income in the country. In their opinion, so long as the Indian economy is dominated by agricultural activity, per capita income will not rise to an extent which is considered desirable or necessary.

Agricultural Development Essential for Economic Growth

The significance of agriculture in India arises also from the fact that the development in agriculture is an essential condition for the development of the national economy. Ragnar Nurkse argues that the surplus population in agriculture should be removed and used in the newly started industries. Nurkse's thesis is that agricultural productivity will be increased on the one hand and on the other new industrial units would be set up with the use of surplus labour.

The Nurksian thesis has been questioned recently. Firstly, industrialisation does not consist of transference of workers from agriculture to industries. Industrialisation requires a particular set of motives and values which an agricultural economy cannot supply. A change in agriculture itself is: essential before such motivations and values are evolved. Secondly, the marketed surplus will have to be increased considerably to feed the growing urban population and to pre - vide raw materials to industries. Thirdly, the new industries, however fast they may develop, will not be able to provide adequate employment for the ever-growing millions in India. There is a limit to the capacity of employment in industries in the short period. Necessarily, therefore, increased employment will have to be found, not in the new industries, but in agriculture itself or in rural industries. This will, then, necessitate improvement in agriculture.[1]

It may, therefore, be concluded that general economic development will require agricultural development either to precede or to go hand in hand with it. Indian planners have learnt a bitter lesson during the Second and Third Five - Year Plan periods that failure of the agricultural sector to deliver the goods would spell disaster to the entire planning process.

It is interesting to observe here that at least at the initial stage, agricultural development may be—or should be—given greater emphasis in India. In the first instance, the capital-output ratio is not very high in agriculture. Hence, a small input of capital will bring in a large output of agricultural goods. To achieve a rapid increase in incomes at least in the initial period, a greater proportion of investment should, therefore, be made in agriculture. Secondly, the rate of saving and investment can be high for the country only if the rate of saving and in vestment is high in agriculture too. Thirdly, foreign exchange requirements are not so necessary for agricultural development as is the case with industrial development. In any case, agriculture contributes over 40 per cent to national income, unless there is development in agriculture, and increase in income in the agricultural sector, per capita income in the country as a whole cannot increase.

We reach two conclusions: (a) Agriculture is the most important sector in the Indian economy; and (b) development of agriculture is essential for general economic development of the country.

AGRICULTURAL PRODUCTIVITY AND PRODUCTIVITY TRENDS

Agricultural productivity can be analysed from two different angles—from the point of view of productivity per acre (i.e., land productivity) and from the point of view of per worker employed (i.e., labour productivity). At the outset, it may be stated that Indian agriculture is characterised by miserably low land and labour productivity Let us see, in greater detail, trends in agricultural productivity.

Land Productivity

Table 20.1 brings out two interesting points. First, agricultural productivity (i.e., yield per hectare) is miserably low in India, as compared to other countries.

Table 20.1

Land Productivity in Selected Countries

(Annual Average for the period Yield per hectare (Kgs.)

Commodity	*Country*	*1951 to 1955*	*1961 to 1965*	*1976 to 1978*
Rice	World	1145	1365	1692
	India	809 (70.6)	987 (72.3)	1244 (73.5)
	China	1744	1853	2349
	USA	1920	2917	3384
	Japan	2653	3341	3977
Wheat	World	1152	1209	1781
	India	714 (62.0)	835 (69.0)	1425 (80.0)
	China	935	882	1372
	France	2078	2930	4346
	Germany	2778	3275	4547
Cotton (Lint)	World	210	330	404
	India	90 (42.8)	120 (36.4)	154 (38.1)
	China	160	250	458
	USSR	N.A.	700	885
	Mexico	330	640	940

Note: Figures in parentheses represent land productivity in India as percentage of world annual average productivity during the period.

Source: *Production Yearbook*, published by FAO of the United Nations.

As regards rice, our major crop, productivity per hectare was only about 31 per cent of that of Japan during 1976-78. In case of wheat, average annual yield per hectare in India was 1,425 Kgs. during 1976-78, while it was 4,346 Kgs. per hectare in France and 4,547 Kgs. per hectare in Germany. In case of cotton, our major cash crop, India's per hectare productivity was just one-sixth that of Mexico. Similar is the position in other crops too. In fact, yield per hectare in India is below the world average in all the crops. As compared with the advanced countries of the world, land productivity was much lower.

Secondly, Table 20.1 shows that India is trying to catch up with other countries. In other words, land productivity has been steadily improving in India, more particularly after 1961-65 as a result of the impact of the green revolution. During 1961-65 and 1976-78, average output of rice per hectare increased from 987 Kgs. to 1,244 Kgs. — an increase of 26 per cent. As against it, the rise in productivity of wheat per hectare showed a much greater increase from 835 Kgs. during 1961-65 to 1,425 Kgs. during 1976-78, i.e., an increase of about 71 per cent.

As compared with the average productivity, the situation has improved. While land productivity in the case of wheat was only 69 per cent of the world average during 1961-65, it has improved to 80 per cent during 1976-78. In case of rice, t is 73.5 per cent, but in case of cotton (lint), it is just 38 per cent. Obviously, in case of crops in which the green revolution has made a greater impact, land productivity has risen much faster. For instance, yield per acre in wheat in India during 1976-78 was higher as compared to China, but in case of rice, the yield was just 53 per cent and in case of cotton just 34 per cent of that of China. However, there is a trend towards a steady increase in land productivity, though productivity continues to be low judged by international standards.

Regional Variations in Land Productivity

From Table 20.2 it can be observed that land productivity shows marked regional variations. It works out to be Rs. 1,037 per hectare for India as a whole. Land productivity is as low as Rs. 461 in Rajasthan and Rs. 2,716 for Kerala. Land productivity is dependent on the quality of the soil, extent of irrigation facilities available, favourable natural factors like rainfall, climate, etc. It b also dependent on the extent of Government effort in improving agricultural inputs and infrastructure. Differences in natural endowments can be narrowed down by human effort but it is not possible to completely eliminate them.

Labour Productivity in Agriculture

Output per worker in agriculture is also low. According to the National Income Committee, average productivity per engaged worker in agriculture for the year 1950-51 was Rs. 500 as against Rs. 1.700 in large establishments and Rs. 1,500 in commerce, transport and communications; in other words, labour productivity on land was the lowest in the country. Labour productivity is calculated by taking the average yield per hectare of land and the average number of agricultural workers employed on a hectare of land. Accordingly, the productivity of labour will reflect the variation in the pressure of population on land in different States.

Table 20.2

Agricultural Productivity of Land; Average for 1967-68 to 1968-70

Rank	*State*	*Net income per hectare (Rs.)*
1.	Kerala	2,716
2.	West Bengal	2,228
3.	Assam	2,102
4.	Himachal Pradesh	1,870
5.	Punjab	1,809
6.	Jammu & Kashmir	1,724
7.	Nagaland	1,702
8.	Uttar Pradesh	1,447
9.	Tamil Nadu	1,367
10.	Haryana	1,807
11.	Bihar	1,247
12.	Orissa	1,106
13.	Andhra Pradesh	1,068
	All India	1,037
15.	Karnataka	834
16.	Gujarat	774
16.	Maharashtra	683
17.	Madhya Pradesh	639
18.	Rajasthan	461

Source: *Commerce Annual* 1972, p, 14.

Table 20.3 brings out the differences in productivity of labour in agriculture in the different States but it does not show whether there has been any change in the labour productivity on hind. Some improvement in the productivity of land will be definitely taking place, because of the adoption of the new techniques of production. But it is generally accepted that the average productivity of labour working in Indian agriculture is very low.

TABLE 20.3

Agricultural Productivity Per Worker : Average for 1967-68 to 1969-70

Rank	*State*	*Average net output per worker at current prices (Rs.)*
1.	Punjab	3,195
2.	Haryana	2,922
3.	Kerala	2,072
4.	West Bengal	1,819
5.	Assam	1,707
6.	Gujarat	1,457
7.	Orissa	1,450
8.	Jammu & Kashmir	1,303
9.	Karnataka	1,321
10.	Himachal Pradesh	1,314
11.	Uttar Pradesh	1,236
	All India	1,213
12.	Rajasthan	1,129
13.	Andhra Pradesh	993
14.	Tamil Nadu	956
15.	Maharashtra	949
16.	Madhya Pradesh	859
17.	Bihar	755
18.	Nagaland	409

Source: *Commerce Annual* 1972, p. 14.

Labour productivity is put as Rs. 1,213 for India as a whole. It is as high as Rs. 3,195 for Punjab and as low as Rs. 755 for Bihar. Nagaland being a very small State should be excluded as an extreme case. Broadly speaking, differences in labour productivity are smaller (i.e., in the range of 1:4) at the inter-State level than differences in land productivity. Partly this is explained by the role of density of population. Whereas land productivity is highest in Kerala, productivity per worker is highest in Punjab. This is due to the fact that number of agricultural workers per hectare is much higher in Kerala than in Punjab. This supports the hypothesis that whenever land productivity is high density of population tends to be high.

CAUSES FOR LOW PRODUCTIVITY

We have explained that agricultural productivity in India—per hectare as well as per worker—is the lowest in the world. There has been some improvement in recent years, particularly during the plan period. But conditions in agriculture have not changed much. It will be useful to analyse the factors responsible for the backwardness of agriculture. An understanding of these factors will help to appreciate the lines of action adopted by the Government to tackle the agricultural problems. It is convenient to group the factors under these headings: (a) General Factors, (b) Institutional Factors, and (c) Technological Factors.

(a) General Factors

(i) *Overcrowding in Agriculture.* The real problem of Indian agriculture is that there are too many people who depend on agriculture. Since 1901 the proportion of people dependent on agriculture has almost remained constant—at about 70 per cent. Though the proportion is constant in terms of percentage, agricultural population in 1971 was about 383 million, as compared to 163 million at the turn of this century. The natural increase in population could not be absorbed in industries and even those who followed traditional handicrafts had to give them up and adopt agriculture. Overcrowding and the consequent pressure of population on land have led to sub-division and fragmentation of holdings, decline in the area of land per capita, disguised unemployment in agriculture, and marginal productivity of labour which may be zero or even negative. The pressure of population on land has been so heavy that between 1901 and 1971 the area of cultivated land per cultivator has declined from 0.43 hectare to 0.36 hectare, despite an expansion in acreage. It is clear, therefore, that not much can be achieved unless this continuing pressure of population on land is reduced

(ii) *Discouraging Rural Atmosphere.* The Indian farmers, generally speaking, are illiterate, ignorant, superstitious, conservative and bound by outmoded customs and institutions such as the caste system and the joint family. Superstition and belief in fate are the curses which keep the farmers fully satisfied with their primitive system of cultivation. The farmers are not motivated by considerations of economic progress. Unless this atmosphere which supports backwardness and stagnation is changed, there is no possibility of agricultural progress. But the important thing is that these conditions are changing fast.

(iii) *Inadequate Non-farm Services.* Indian agriculture has suffered because of the inadequacy of such non-farm services as provision of finance, marketing, etc. Either these facilities are non-existent or if they are available, they are quite expensive. For instance, till recently, farmers had to depend upon the village moneylender and had to pay rates of interest so high that once a farmer had borrowed, he was bound to lose his land and become a landless labourer. Other sources of finance did exist-such as cooperatives and the Government—but they were almost insignificant. Likewise, until recently the farmers could not secure storage facilities in mandi towns and were cheated by the wholesaler and commission agents there. Thus, a very important factor responsible for agricultural backwardness in India was the inadequate non-farm services.

(b) Institutional Factors

(i) *Size of Holdings.* The average size of holding in India is very low, less than 2 hectares or 5 acres. Not only agricultural holdings are small but they are fragmented too. In certain parts of the country plots of land have become so small that it is impossible to move even an ordinary plough. Since the average agricultural holdings are too small, no scientific cultivation with improved implements, seeds, etc., is possible. Small-sized holdings lead to great waste of time, labour and cattle power, difficulty in proper utilisation of irrigation facilities, quarrels and consequent litigation among farmers, wastage of crops in the absence of fencing, etc. The existence of small holdings is one of the causes for poor agricultural yield.

(ii) *Pattern of Land Tenure:* A very important factor for low agricultural productivity was the absence of proper incentives. Under the Zamindari system as well as under landlordism, the cultivator was not the owner but could be turned out of the land he was cultivating, by the owner. Even though the Zamindari system has now been abolished and tenancy legislations have been enacted in various States, the position of the tenants is far from satisfactory. The cultivator does not often own the land; he has to pay high rents for the land he cultivates; and he has no security of tenancy and may be turned out of his land at any time the landlord desires. Under these difficult conditions, it is impossible to expect the tiller to increase agricultural productivity.

(c) Technological Factors

(i) *Poor Techniques of Production.* The Indian farmers have been using old and inefficient methods and techniques of production. Since they are tradition - bound and also poor they have not adopted the modern methods which are so widely adopted in the countries of the West and in Japan. Only in recent years and that too to a limited extent, the farmers have started adopting improved implements like steel ploughs, sugarcane crushers, small pumping sets, water-lifts, barrows, hoes, seed drills, fodder cutters etc. Increase in production is possible only if proper and adequate manures are used. To revitalise fertility and to utilise fallow lands, the use of various kinds of manures is urgently called for. But in India, the use of both farm yard manure and chemical fertilisers is extremely inadequate. The importance of good quality seeds to increase agricultural productivity needs no emphasis. Between

10 and 20 per cent of increase in productivity can be brought about through the use of better seeds. But Indian farmers have been using seeds of indifferent quality, either because they do not have the means to purchase good quality seeds, or because the seeds they kept apart for sowing purpose have deteriorated due to bad storage conditions. Recently, the National Seeds Corporation has been set up to organise on a countrywide scale, the production and distribution of high-yielding disease resistant varieties of seeds.

(ii) *Inadequate Irrigation Facilities.* One of the basic causes for the weakness of Indian agriculture has been that most of the farmers throughout the country have to depend upon rainfall and very few of them can avail the facilities of artificial irrigation. For instance, before Independence and partition of the country, about 24 per cent of the land was irrigated. This proportion declined to 19 per cent after partition. Despite a vigorous programme of major and minor irrigation works during the Plan period, the ratio of irrigated land to total cultivated land is now about 22 per cent. This shows that there is great scope for artificial irrigation in the country.

The causes of low productivity which we have briefly outlined above suggest their own remedies too. At one extreme, attempts are being made to find alternative employments for rural population and to change the occupational structure in such a way that only 60 per cent of people will depend cm agriculture. On the institutional front, Government is tackling the agricultural problem through land reforms. On the technological front, a very good beginning has been made in converting farmers to the use of improved implements, seeds, chemical manures, etc. Irrigation facilities are being increasingly made available. Double cropping, better rotation of crops, fighting plant diseases and pests etc., are being looked into. In course of time, therefore, land and labour productivity in agriculture may be expected to go up. The sooner this is achieved, the better it will be for the national economy.

New Challenges to Indian Agriculture

Indian Agriculture has been in doldrums for a long time. Attempts to rejuvenate it were made so many times in so many different forms but without any spectacular success. The introduction of the new agricultural strategy which goes by the name of the green revolution was hailed to have solved India's food problem, but this hope was only short-lived. Despite five five-year plans and three annual plans since 1950-51, India has failed to solve its basic and number, one problem i.e., the feeding of its teeming millions,

In recent years, the agricultural problem has become much more severe and intense. The tremendous growth of population from 350 to 630 million since 1951 with further possible rise of the population to touch 1,000 million before the end of this century, increase in demand for foodgrains and agricultural raw materials due to economic growth, however slow and tardy it may have been, the persistent and often help less it - sort to deficit financing by the Government leading to rise in money demand for foodgrains and other basic necessities, the revolution of rising expectations in the minds of the common man consequent to the slogans of "garibi hatao" and the direct attack on the new technology itself by the oil crisis which has raised the price of imported fertilisers and import-based domestic production of fertilisers-all these have posed some serious challenges to Indian agriculture. These challenges, of course, were not new; only they have grown in intensity.

All the above causes will have to be tackled simultaneously. But it is of paramount importance that we raise the rate of growth in agriculture. According to Prof. V.K.R.V. Rao, India requires a minimum growth rate of 4 per cent in foodgrains and 6 per cent in non-food crops during the rest of this century to meet the basic requirements of the economy for food and agricultural raw materials. But, as pointed out earlier the rate of growth in agriculture has been only 2.8 per cent (compound) between 1949-50 and 1971-72. The Fourth Plan target for growth in foodgrains was 5.6 per cent per annum but the actual growth rate is now put at 3.4 per cent per annum. The Fifth Plan target of 4.2 per cent growth rate in foodgrains which was quite modest also could not be achieved. The new challenge to the Indian agriculture lies her: how to push up agricultural production in the shortest possible time so as to catch up with mounting demand for agricultural goods?

The growth rate in agricultural output is composed of growth rate in area and growth rate in yield. The growth rate in area has been declining in India, from 1.9 per cent during the first decade of planning to 0.61 per cent during the second decade of planning. There may be no more scope for increasing area under cultivation in the coming decades. Naturally, increase in agricultural output will have to come wholly or mainly from growth in yield. This will mean raising agricultural productivity through multiple cropping (so as to increase the effective supply of land) and relying on irrigation, high yielding varieties, etc. The new challenges will have to be boldly met and tackled; otherwise, the country will be doomed.

GROWTH RATES IN AGRICULTURE

Productivity Trends — Before Independence

Available agricultural statistics - even though defective and incomplete - indicate that over the first half of this century (roughly between 1901 to 1947) agricultural production declined (as compared to the increase in population). Table 20.4 proves this point.

From this table, it should be clear that before independence population rose by 38 per cent but cultivated land rose by only 18 per cent. Average annual output of foodgrains and pulses remained almost constant but that of non-food or commercial crops increased by 53 per cent. It was because of the increase in the latter that the index of all crops showed an increase of 13 per cent. Thus, increase in population had over—taken increase in agricultural production by a considerable extent.

TABLE 20.4

Productivity Trends in Indian Agriculture

Period	*Population Index*	*Index of Area Cultivated*	*Index of Average Annual Output*		
			Food	*Non-Food*	*All crops*
1900 - 01 to 1904-05	100	100	100	100	100
1940-41 to 1944-45	138	118	101	153	113

Source: J.P. Bhattaoharjcc (Edited), *Studies in Indian Agricultural Economics.*

Dr. Panse of the Indian Council of Agricultural Research (I.C.A.R.) made a special study of the yields of principal crops in India for the period between 1910-11 and 1945-10 and came to the conclusion that the yield per acre of cereals did not show any consistent decline or increase but there was a positive increase in the yield per acre of commercial crops and foodgrains. He did not agree with the general belief that there had been a deterioration in fertility or in the standards of agriculture.[2]

Productivity Trends—After Independence

The process of decline in productivity which had taken place between 1900 and 1945 seems to have continued in the post - war period also. The First Five-Year Plan estimated that as compared to the pre-1939 period, the average yield of cereals per acre during the period 1946-47 to 1949-50 had declined from 619 lbs. to 565 lbs.[3] According to Dr. Rangnekar the volume of output in India was about 0.86 metric ton per hectare in 1951, as compared with 0.90 metric ton before Second World War.[4] Similar conclusions have been drawn by I.CA.R. studies and the Grow More Food Enquiry Committee.

But broadly speaking, the decline in agricultural productivity in general and foodgrains productivity in particular which was so clear in pre-independence period and to a certain extent in the three years following it, was definitely reversed after 1950-51. Under the impact of Five-Year Plans and intensive method of cultivation, an upward phase was initiated. This is illustrated in Table 20.5.

Table 20.5 shows trends which are the exact opposite of the trends observed in the pre-Independence period. Between 1950-51 and 1960-61 foodgrains productivity had increased by 30 per cent, but the productivity of non-foodgrains had increased by about 10 per cent only. The difference between foodgrains and non-foodgrains productivity is due to greater attention being given to the former. The overall index of all agricultural productivity had, however, gone up by 24 per cent during the period. While agricultural productivity rose gradually since 1950-51, it received a severe jolt during 1965-66 and 1966-67 because of severe drought over vast areas of the country. The upward trend was renewed in 1967-68. Between 1968-69 and 1977-78 productivity of foodgrains had increased by 36 per cent while productivity of non-foodgrains had increased also by about 36 per cent; the productivity of all crops had also gone up by 36 per cent.

TABLE 20.5

Index Number of Agricultural Productivity

(Boat : Triennium ending 1969-70 = 100)

	All Crops	*food grains*	*Non -food-grains*
1960-61	87	86	89
1965-66	81	76	91
1970-71	112	113	100
1978-79	138	139	135
1979-80	117	114	122

*Provisional.

Source: *Reserve Bank of India Bulletin*, April-May, 1970 and *Economic Survey 1980-81*, p. 08.

The increase in agricultural production is partly brought about by increase in area under cultivation and partly by rise in productivity.

Table 20.6 reveals that during the first decade of planning, the new strategy of agriculture had not even been conceived but the second period coincides with the application of the new strategy to the major foodcrops. During the first decade of planning, production had increased at the annual rate of 3.3 per cent. This was composed partly of increase in area and partly of increase in yield; but the increase in area was much more—1.9 per cent as against 1.3 per cent increase on account of per acre yield.

But during the second decade of planning the rate of growth in output declined to 2.64 per cent per annum (from 3.3 per cent) but the interesting point is that the rate of growth in yield has increased by 44 per cent over the previous period. But this order of increase in the yield was not sufficient to compensate for the fall in the rate of growth of area by 2.5 per cent.

TABLE 20.6

Compound Rates of Growth in Food-grains since 1949-50 (in per cent)

Period	*Area*	*Yield*	*Production*
1949-50 to 1959-60	1.90 (57.6)	1.30 (39.2)	3.30 (100.0)
1960-61 to 1971-72	0.61 (23.1)	1.87 (70.8)	2.64 (100.0)
Combined (1949-50 to 1971-72)	1.08 (38.4)	1.61 (57.2)	2.81 (100.0)

Note: Figures in brackets represent percentage of total.

Source: Dr. V.K.R.V. Rao's "Dr. *Panse Memorial Lecture*", April. 1974. (Mimeographed).

TABLE 20.7

Compound Growth Rates of Production (P), Area (A) and Yield (Y) of Selected Crops in India

Crops		*1961-62 to 1965-66*	*1966-67 to 1968-69*	*1969-70 to 1973-74*	*1974-75 to 1977-78*	*1961-62 to 1977-78*
All Crops	P	—1.0	+6.9	+3.1	+4.7	+3.0
	A	+ 0.3	+0.9	+1.2	+0.5	+0.7
	Y	—1.9	+5.4	+1.7	+3.0	+1.6
Foodgrains	P	—2.0	+9.4	+2.8	+5.6	+3.2
	A	—0.1	+1.6	+1.0	+0.2	+0.6
	Y	—2.2	+6.9	+1.5	+4.2	+2.0
Non-foodgrain	P	+ 0.1	+2.4	+4.0	+3.3	+2.7
	A	+ 1.7	—1.5	+1.9	+1.4	+1.1
	Y	—1.3	+2.9	+2.1	+1.0	+1.0
Rice	P	—1.7	+9.4	+2.4	+6.2	+3.3
	A	+ 0.8	+1.4	+0.8	+1.1	+1.0
	Y	—2.5	+7.8	+1.6	+4.5	+2.2
Wheat	P	—0.1	+20.2	+5.2	+9.7	+7.4
	A	—0.5	+8.4	+3.2	+3.5	+3.1
	Y	+0.3	+10.3	+1.7	+6.1	+3.8
Pulses	P	—3.7	+4.4	—	+5.3	+0.9
	A	—0.6	—2.1	+2.1	+0.4	+0.1
	Y	—2.8	+5.9	—2.0	+4.1	+0.6
Sugarcane	P	+2.9	+2.8	+2.7	+7.1	+3.8
	A	+3.6	—2.1	+2.3	+4.2	+2.3
	Y	—0.7	+4.5	+0.3	+2.7	+1.3

Sources: *Reserve Bank of India Bulletin*, April-May, 1979.

The lower rate of growth in output of foodgrains during the second decade of planning (2.64 per cent) coincided with two drought years—1965-66 and 1966-67—which are claimed to be so exceptional that they are said to occur only once in 50 years or more. Even if an allowance is made for this factor, the rate of growth is not satisfactory.

Trend of Agricultural Production

Table 20.7 gives an analysis of growth trends in agricultural production during the 17-year period beginning with the Third Plan. Growth rate of all crops showed an average increase of 3.0 per cent per year during the entire 17-year period (1961-62 to 1977-78). Out of this, the increase introduction due to increase in area was of the order of 0.7 per cent per annum and that of increase due to a rise in yield per hectare was of the order of 1.6 per cent per annum.

Plan-wise analysis indicates that agricultural production followed an erratic trend during the Third Plan period. During the first three years of the Plan, stagnation in agricultural production was witnessed. During 1964-65, however agricultural production increased by 11.1 per cent over the previous year. Severe drought during 1965-66 resulted in a steep fall in agricultural production. This was followed by another year of drought during 1966-67. However, agricultural production picked up during 1967-68 and 1968-69. During the period of the three Annual Plans, agricultural production index rose at an average annual rate of 6.9 per cent. Part of this was due to increase in area but the major part was due to increase in yield. The index of yield improved at an average rate of 5.4 per cent annually during this period, whereas the index of area under crops increased at an annual rate of 0.9 per cent.

During the Fourth Plan (1969-70 to 1973-74), a steady rise in agricultural production, though with a lower annual average growth rate i.e., 3.1 per cent, took place as compared with the Annual Plans. Growth rate of yield fell to 1.7 per cent while the growth rate of area under all crops marginally improved to 1.2 per cent per annum.

Agricultural production recorded a much higher growth rate during the Fifth Plan. Average annual growth rate of agricultural production was 4.7 per cent during 1974-78; the lion's share in this increase being contributed by increase in yield at an annual average rate of 3.0 per cent.

Trend of Foodgrains Output

During the 17-year period, the growth rate of foodgrains output was of the order of 3.2 per cent per annum, the contribution of area was 0.5 per cent and that of yield was 2.0 per cent per annum. However, the trend of foodgrains output followed more or less the same pattern as agricultural production. During the Third Plan, severe drought conditions resulted in a negative growth rate of 2.0 per cent per annum. The situation improved during 1967-68 and 1968-69. During the Annual Plans, growth rate of foodgrains production touched a record level of 9.4 per cent, major part of this growth rate being contributed by increase in yield (6.9 per cent per annum), the contribution of area was barely 1.6 per cent per annum.

During the Fourth Plan, the production of foodgrains increased at an annual average rate of 2.8 per cent—much lower than the growth rate of the period of the Annual Plans. The situation improved considerably during the Fifth Plan. Annual average production of foodgrains rose by 5.6 per cent per annum—double the growth rate of the Fourth Plan. This was the result of a very high growth rate recorded in rice (6.2 per cent) and wheat (9.7 per cent) during the Fifth Plan.

Trend of Production of Non-Foodgrains

Non-foodgrains consist of oilseeds, cotton and jute fibres, plantation crops like tea and coffee and some miscellaneous crops like sugarcane etc. Over the 17-year period, non-foodgrains showed an average growth rate of 2.7 per cent per annum. Increase in area accounted for 11 per cent and increase in yield 1.0 per cent. In other words, the impact of high yielding varieties did not have as strong an effect on non-foodgrains as they had on foodgrains. Non-foodgrains fared better than foodgrains during the Fourth Plan inasmuch as the production index of the former increased as an average growth rate of 4.0 per cent per annum while it was 2.8 per cent in the case of the latter. But during the Fifth Plan period (1974-78), non-foodgrains showed a growth rate of 3.3 per cent as against 5.6 per cent of the foodgrains. The impact of increase in yield was to the extent of 1.1 per cent per annum and that of area to the extent of 1.4 per cent. This obviously shows that in case of non-foodgrains, a greater part of the increase in production was accounted for by the extension of area and a relatively small part by increase in productivity per hectare.

Productivity Trends of Some Important Crops

It would be of interest to study productivity trends of some important crops.

Rice. Growth rate of production of rice started showing an increase after 1967-68. Rice production touched a record level of 52.7 million tonnes in 1977-78 as against the earlier peak of 48.7 million tonnes. During the 17-year period, (1961-62 to 1977-78), production of rice increased at an average rate of 3.3 per cent per annum. It is only during the Fifth Plan that a significant breakthrough was achieved in rice production This is evidenced by the fact that whereas production index of rice showed a growth rate of 6.2 per cent per annum during 1974-78, a major proportion was accounted for by rise in yield per hectare (4.5 per cent).

Wheat—Growth rate of production of wheat showed the highest impact of high-yielding varieties programme. Production of wheat rose from 10.4 million tonnes in 1965-66 to 34.9 million tonnes in 1978-79. During the 17-year period, production index of wheat showed an annual average growth rate of 7.4 per cent per annum — 3.8 per cent was accounted for rise in yield and 3.1 per cent by extension of area. The high profitability generated by HYV programme induced fanners to divert a greater proportion of their area towards the production of wheat.

Pulses—During the 17-year period, pulses recorded a very poor growth rate of 0.9 per cent per annum. This highlights the fact that no significant breakthrough could be made to improve their production under the HYV programme.

Sugarcane—In case of sugarcane, during the 17-year period, production index showed an annual average increase of 3.8 per cent, but a major portion of this increase was accounted for by increase in area (2.3 per cent) and the yield effect accounted for only 1.3 per cent per annum. This trend was largely observed during the Fourth and Fifth Plan periods.

The analysis dearly highlights the fact that the possibilities of raising production by bringing in more area (i.e. extensive cultivation) are extremely limited. Even when in certain crops this is achieved, it is at the cost of other crops. Obviously, the country has to concentrate on improving yield per hectare (i.e. intensive cultivation) as a future course of policy.

CROP PATTERN IN INDIA SINCE 1951

Crop Pattern Before Independence

By cropping pattern is meant the proportion of area under different crops at a point of time. A change in cropping pattern implies a change in the proportion of area under different crops. We have agricultural statistics—though defective and not so reliable—to explain the cropping pattern before Independence. The additional area brought under cultivation in the first 45 years of this century had not been proportionate to the increase in population. While population had increased by about 38 per cent, the gross sown area had increased by a little more than 18 per cent.[5] The area under food crops had increased at a still lower rate, viz., by about 14 per cent. However, the land under non-food crops (i.e., oilseeds, fibres, etc.) had increased very much more - that is by 41 per cent.

Among the food crops, wheat had registered the highest increase in acreage - 34.6 per cent - while increase in jowar was the lowest (about 11 per cent). Except for wheat, all other crops had been increasing rather gradually; only wheat had recorded a steady rise throughout the period.

In the case of non-food crops like oilseeds, fibres, plantation crops and sugarcane, increase in area had been over 40 per cent during the five-year period ending 1939-40, after which there was a decline. In fact, the five-year period ending 1944-45 actually marked a reversal of the earlier trend of steady increase. Only sugarcane among cash crops registered an increase in acreage in this period. Thus, between 1935 and 1945, the decline in area under non-food crops had been accompanied by a large increase in area under food crops. There could be two possible reasons for this change in the crop pattern: (a) The Grow More Food Campaign started by the Government in 1943 with the object of increasing food production within the country; and (b) the partial loss of export market of cotton, oilseeds and jute led to the shifting of area to food crops.

It is also interesting to study the pattern of distribution of the area under different crops during the period 1900-1945. There was a gradual shift from food crops to non-food crops. At the beginning of the century, more than 83 per cent of land was put under food crops and about 17 per cent under non-food crops. By 1944-45, area under food crops had come down to 80 per cent and area under non-food crops had increased to 20 per cent. This shift in crops from foodgrains was mainly due to the higher price of non-foodgrains, commonly known as cash crops. It reflected a change from subsistence cropping to commercial cropping.

Crop Pattern After Independence

Table 20.8 shows the change in area under foodgrains and non-foodgrains since Independence.

TABLE 20.8

Changes in Area in Different Crops Between 1950-51 and 1975-76 in India

Year	*Area under foodgrains*	*Area under non-foodgrains*	*Area under All crops*
1950-51	100	100	100
1960-61	120	127	121
1967-68	126	139	128
1972-73	121	134	123
1975-76	132	144	134

Source: Indian Agriculture in Brief (1974), Ministry of Agriculture, pp. 28-29. Basic Statistics relating to the Indian Economy (1950-51 to 1975-76) Table 20.8.

Between 1950-51 and 1975-76 total acreage rose by 34 per cent and as in the pre-independence period, the increase in non-food crops has been much more than the increase in food crops. Non-food crops registered an increase of 44 per cent, while food crops registered an increase of only 32 per cent. Table 20.8 gives only the broad changes in major crop categories. Let us see the position of some selected crops during this period.

TABLE 20.9

Changes in Area Under Different Crops Since 1950-51

Crops	*1950-51*	*1960-61*	*1970-71*	*1979-80*
Rice	100	110	121	125
Wheat	100	132	184	219
Jowar	100	118	112	106
Pulses	100	128	117	106
Oilseeds	100	127	137	212
Cotton	100	130	131	143
Sugarcane	100	141	159	146

Source: Indian-Agriculture in Brief (1980) and Basic Statistics relating to Indian Economy (1950-51 to 1975-76) Table 20.8.

Among foodgrains, the largest increase in area has been recorded by wheat, with an increase of 106 per cent. While the increase in the case of rice has been quite modest, small millets and opther minor crops have actually declined, showing a general shift from minor to major crops. Increase in acreage under pulses has been modest. The largest increase in food crops has actually been in the case of potatoes an increase of 166 per cent. The traditional commercial crops, viz., oilseeds, cotton, jute, sugarcane, tobacco, etc., have made impressive increase in acreage, much more than food crops.

TABLES 20.10

Nature of Crop Distribution Since 1900

	Crops	*1900-01 to 1904-05*	*1940-41 to 1944-45*	*1950-51*	*1970-71*	*1978-79*
(a)	All Crops	100	100	100	100	100
(b)	Food crops	83	80	74	78	78
(c)	Non-Food crops	17	20	26	22	22

Source: Figures based on "Basic Statistics Relating to the Indian Economy" (Planning Commission) and RBI *Report on Currency and Finance*, 1978-79.

Table 20.10 shows the share of different crop categories in the total area sown. Table 20.10 seems to give a confusing picture of the cropping pattern in India. The trend since the turn of the century was a declining share of foodgrains in the total acreage and corresponding increase in acreage of non-food crops. From about 83 and 17 per cent under the two types of crops at the turn of the century, they changed to 80 and 20 per cent, just before Independence. The acreage figures for 1950-51 show a further shift in favour of commercial crops, about 74 and 26 per cent respectively. Almost the same position was maintained in 1960-61. But the acreage figures from 1967-68 show a reversal of the above trend, and a definite shift from commercial crops to foodgrains. Two important reasons may be given to explain the importance of foodgrains as compared to non-food crops. First, prices of foodgrains have been rising quite fast and the farmers have suited growing foodgrains for the market, in the same way they grow oilseeds, cotton and other commercial crops. In other words, the traditional classification between food crops and commercial crops is losing its significance. Second, the cultivation of foodgrains has become highly remunerative and productive under the impact of the new technology.

TABLE 20.11

Area of Important Crops in (000 Hectares)

	1960-61	*1970-71*	*1973-74*
Rice	227	390	520
Bajra	123	207	147
Maize	327	565	567
Wheat	1400	2299	2338
Barley	66	57	110
Gram	838	358	352
Mash	19.3	25.9	38.7
Moong	6.8	6.9	9.8
Massur (lentil)	29.8	13.3	18.7
Groundnut	67	174	156
Rape & mustard	107	103	179
Sesamum	7.8	14.0	20.5
Linseed	2.9	2.8	2.7
Sugarcane	133	128	110
Potato	9	17	
Cotton	447	397	523

An idea of shift in cropping pattern in India can be had from the experience of Punjab. The technological breakthrough in agriculture, higher production, and higher net return per unit area per time—and in general economic gains - are the basic causes for the shifts in cropping pattern by the farmers.

The area under rice, maize and wheat increased by 129 per cent, 73 per cent and 67 per cent respectively between 1960-61 and 1973-74. The sharp rise: in area under these three crops, was due to the introduction of high yielding varieties in these crops, advances in farm technology, and the economic gains by the farmers. The area under barley and gram declined rapidly by 1971-72 and 1972-73 but showed an upward trend during 1973-74 resulting in a decrease in the area under wheat. Tin's shift in the area was the result of high cost of inputs, particularly fertilisers and a lower procurement price for wheat than those for barley and grain. The area under oilseeds also showed an upward trend, particularly under rape, mustard and sesamum. The area under sugarcane showed a decline during this period; likewise barley, gram and cotton showed a decline till 1972-73. The shifts in cropping pattern by the farmers are mainly based on the economic gains.

FACTORS AFFECTING CROPPING PATTERN IN INDIA

At one time many believed that cropping pattern in India could not be changed. S. N. Sinha, for instance, gave expression to such an opinion when he wrote: "In a audition-ridden country with a very low level of knowledge, the peasants are unwilling to make experiments. They accept everything with a spirit of resignation and a sense of fatalism. For them, agriculture is a way of life rather than a commercial proposition.... In an agricultural community where the members are illiterate and tradition - ridden, there is hardly any possibility of crop shifts."[6] This opinion is not correct any more as is clear from the change in cropping pattern in Punjab. It is widely agreed that the crop pattern of a country like India can be changed and should be changed.

Physical and Technical Factors and Cropping Pattern

Cropping pattern of any region depends upon physical characteristics as the soil, climate, weather, rainfall, etc. For instance, in a dry area where the rainfall is scanty and where there is high uncertainty of monsoons, there will be a greater dependence on jowar and bajra, as these crops can be managed with a small quantity of rainfall. Water-logging in parts of Ludhiana and Sangrur districts in Punjab has led to an increase in area under rice; for rice can stand the extra water better than other crops. In the newly reclamed lands of Madhya Pradesh, millets are grown for a few years after which rice is cultivated.

Apart from soil and climatic conditions, the cropping pattern ofa region will depend upon the nature and availability of irrigation facilities. Wherever water is available, not only can a different crop be grown, but even double or triple cropping will be possible. When new irrigation facilities are provided the whole method of cultivation may change. A superior crop can be grown; a new rotation of crops, where there was none, or a better rotation over what prevailed may be possible. One of the important factors responsible for increase in the cultivation of sugarcane, tobacco, etc., is the extension of irrigation facilities. It ispossible that because of lack of capital, agricultural pre-requisities, better implements, improved seeds and finance for getting fertilisers, it might not have been the right crop that wasbeing grown; but given these facilities, the cropping pattern may change.

Economic Factors and Cropping Pattern

Economic motivations are the most important in determining the cropping pattern in a country. Whatever may have been position in India in the past there are very clear indications that Indian farmers are being clearly influenced by economic factors now. Among economic factors affecting crop pattern, the following are important:

(i) *Prices and income maximisation.* Many empirical studies have brought out clearly the relation between price movements and crop pattern. A study of inter-crop price parities undertaken by the Ministry of Food and Agriculture shows how price variations exert an important influence on acreagre shifts. "It seems that prices influence the acreage under the crops in two ways. One is that the variations in the intercrop price parities led to shifts in acreage as between the crops. Another is that the maintenance of a stable level of prices for a crop.... provides a better incentive to the producer to increase the output than what a very high level of price does, if there is no uncertainty of this level being maintained over a number of years."[7] Fixed procurement price of wheat and rice and other Government controls have induced farmers to shift to cash crops like sugarcane.

According to some authorities, income maximisation pull has greater influence in changing the crop pattern, that is, the farmer would choose that combination of crops which would give him maximum of income. Dr. Raj Krishna, however, argues that relative profitability per acre, is the main consideration which influences the crop pattern. In any case, the farmer is influenced in the choice of his crop by such consideration of price parities between different commodities, or maximisation of income or relative profitability per acre.

(ii) *Farm Size.* There is a relationship between farm size and the cropping pattern. The small farmers are first interested in producing foodgrains for their requirments. They would go in for cash crops only after they have met their requirements of foodgrains. Small holders, therefore, devote a small relative acreage to cash crops than large holders. This point has been brought out in many empirical studies. But a recent study of Deoria district of uttar pradesh brins out clearly the fact that almost all farmers, big and small, try to grow some cashcrops. In fact, in recent years it is the samll farmers who have been increasing their sugarcane area more than large farmers.

It is true that the need for subsistence has traditionally dominated the cropping pattern of the small farmer. But his marginal need for money income cannot be less than that of the large farmer. And, as the economy grows, we should expect the small farmer to make very significant adjustments in hit crop pattern in order to maximise his income.

(iii) *Insurance against risk.* The need to minimise the risk of crop failures not only explains diversification but also some specific tea-lures of crop patterns. For instance, the persistence of millets in many regions which puzzles many economists can be understood mainly at insurance against bud seasons in dry areas.

(iv) *Availability of Inputs.* As already Indicated, crop pattern is also dependent upon the availability of such inputs as seeds, fertilisers, water, storage and marketing, transport, etc. Of the additional facilities, the most rewarding would be irrigation. The NCAER estimated that, "if additional irrigation facilities were provided in Punjab, the cropping pattern on as much as 3.4 million acres could be changed, of which nearly 1.6 million acres now under gram could be put to more paying crops."[8] The availability of groundnut seed was one of the important factors which induced many farmers to increase the area under this crop in Madhya Pradesh. Another reason why farmers prefer groundnut to cotton is that the former is quick. yielding, while cotton is on the field for a long time, and does not easily satisfy the need for quick cash.

(v) *Tenure.* Under the crop sharing system, the landlord has a dominant voice in the choice of the cropping pattern; and this helps in the adoption of income maximising crop adjustments.

Government Action and Crop Pattern

Government can influence crop pattern through legislative and administrative measures. Steps may be taken by the Government to ease or subsidise the supplies of the farm inputs and knowledge. The government may attach the provision of some service or facility with a particular crop pattern. The provision of irrigation facilities or the supply of seed and fertilisers, etc., may be related to the adoption of a given crop pattern by the farmers. Food Crop Acts, Land Use Acts, intensive schemes for paddy, for cotton, for oilseeds, etc.,—all these bring sharply into focus the possibility that while each individual measure may push the crop pattern in the direction intended in that measure the overall effect of all the measures taken together on the entire crop pattern may not be in accordance with national requirements. In India, the *Grow-More-Food* and other campaigns prove this point.

Apart from the personal prejudices, and inadequate financial and other resources of the tanners, there may be factors like recurrent drought or pest infestation that prevent them from opting for a more remunerative set of crops. In those situations if more of irrigation, fertilisers or pesticides are made available, it would be possible for them to change the crop - structure and so earn larger returns from their land. To the extent it is not possible for a farmer to acquire all these by himself, the Government could come to his help and procure these for him. Yet other possibilities of helping the farmer to improve the cropping pattern are building of new roads which will improve the flow of commodities to the market where they will fetch better prices and help to establish industries or townships nearer to their land.

Policy to Induce Improvements in the Cropping Pattern

If plan targets indicate a pattern different from the one which the peasants may desire to have, then the Government must enforce some type of control of acreage, but such control in an economy like India where there are millions of farmers may be difficult, if not impossible. Naturally, the Government may have to rely, in the main, on the manipulation of product prices and input prices, discrimination in the supplies of controlled inputs like seeds, water, fertilisers, etc. In certain cases, the use of taxes and subsidies may also be called for.

For implementing the necessary programmes for a better cropping pattern, the NCAER has made the following suggestions :[9]

(i) The present way of haphazard allocation of area to different crops by individual farmers is not in the best interests of the farmers or of the country. As persuasion is a slow remedy and often not so effective, the NCAER would like legislative compulsion to bring about a desired cropping pattern in different areas. There are many countries like the U.S.A., U.A.R., etc., where legislation prescribes the extent to which different crops can be grown in particular regions.

(ii) District Planning officers may be appointed who could plan the crops for each season according to changes in the seasonal and other factors like prices, yields, etc. The farmers should work in close co-operation with departments of agriculture, irrigation, etc., so that the chalked out programmes may be implemented.

(iii) An Agriculture Mechanisation Corporation may be set up for States like Madhya Pradesh where the size of holding is too large and where the average farmer is unable to manage with hired labour, the Corporation should advance the farmer the requisite machinery, preferably on loans and the value repayable in easy instalment.

(iv) Government should give the greatest importance to the promotion of transport and marketing facilities and consolidation of holdings.

Conclusion

The most important consideration affecting cropping pattern is the economic consideration. Even in a country like India which is dominated by farmers teeped in poverty and conservatism and where farmers hold tiny bits of land, cropping pattern can be changed through appropriate change in economic motives. Experience in recent years has been that the farmer does accept the logic for a change wherever he is shown a better cropping pattern. The real difficulty in adopting a better cropping pattern is that the farmer may not have the requisite capital to invest now or possess the know-how that may be necessary for changing the crops. It is here that the Government may come to his help.

NOTES

1. Here is, however, a possible contradiction. Improved agriculture based on improved techniques, seeds, etc., will lead to increased productivity, but also to the displacement of labour.
2. *Agricultural Situation in India,* June, 1952.
3. *First Five-Year Plan,* p. 155.
4. Rangnekar: *Poverty and Capital Development in India,* p. 42 and 298-99.
5. Refer to J.P. Bhattacharjee (ed.): *"Studies in Indian Agricultural Economic",* pp. 10-15.
6. S. N. Sinha : *"Economics of Cropping Pattern" in AICC Economic Review,* Vol. XV, January, 1964.
7. *Indian Journal of Agricultural Economics,* Vol. XVIII, No. 1. Conference Number, *"Agricultural Price Policy in India"* (Mimeo), Directorate of Economics and Statistics, Ministry of Food and Agriculture.
8. NCAER, *Cropping Pattern in the Punjab.*
9. NCAER: *Cropping Pattern in Madhya* Pradesh, pp. 49-51.

REFERENCES

J. P. Bhattacharjee, *Studies in Indian Agricultural Economics,* Chapters I and II.

Indian Journal of Agricultural Economics, Volume XVII, No. 1. Conference Number.

V.K.R.V. Rao, *New Challenges before Indian Agriculture* (Dr. Panse Memorial Lecture) .

Kartar Singh Chela, *Crops Rotations,* (Punjab Agricultrual University Publication).

Commerce Annual Number 1972, Regional Profile of Indian Agriculture.

21

AGRICULTURAL REGIONALISATION OF INDIA

One of the approaches to study the general level of development of an economy might be to assess the annual rate of growth of national income in relation to the annual growth rates of other undeveloped, developing and developed economies. The annual rate of growth of an economy is mainly a determinate of the growth in the goods producing sector of an economy less the annual rate of growth of population. Here, if the increase in prices during the stipulated period of growth are allowed for, real rate of growth per annum could be worked out. Another approach for assessing the growth of an economy relates to the development of various input-output technical co-efficients with the help of various surveys and linear programming techniques. Yet, another approach for assessing the level of development of an economy might consist of studying the growth pattern of certain selected physical indicators, which might be considered sensitive enough to reflect the growth and development of an economy.

Among the various approaches used for assessing the level of development, the selected indicators approach has special relevance for a developing economy. In the case of developing economies, the impact of the various developmental measures undertaken for accelerating the level of social and economic development cannot be quantitatively measured so as to reduce them into some sort of a composite index which might be considered to reflect the level of development at a particular point of time. This approach again has a special relevance for rural areas where planning is more broad-based including as it does in its scope not only increasing employment opportunities which might reflect the levels of per capita income but also the creation of social overhead facilities in various sectors, like, Education. Health, Transport and Communication, Rural Water Supply, Rural Works programme, etc.

It is realised that the agricultural development should be assessed not only by the trends in the agricultural production and productivity which reflects the essence of various agricultural developmental measures but also on the basis of various physical inputs, like, labour (human and animal), irrigation, fertilizers, improved seeds, extent of cultivated area, etc. Assessment of the level of development of each of these physical inputs can reflect broadly the potential level of agricultural development which may or may not be reflected in the average productivity levels. The discrepancy in the agricultural sector between the potential production and actual production is often due to the weather conditions. Besides, the impact of the various agricultural programmes comes up against a formidable difficulty in the existence of million of farmers with diverse sizes of holdings. Any quantitative assessment with regard to the motivation and reactions of the various classes of workers to various programmes is still very hazardous. It is in this context that it is felt that perhaps the approach of selected indicators is preferable in the fields of agricultural development. It should, however, be emphasised that selected indicators should be sensitive enough so as to explain quite a sizeable proportion of the total variation in agricultural and/or general development.

The selected indicators used here for assessing the potential level of agricultural development are:

(i) Gross area irrigated as percentage of gross area sown; (1959-61)

(ii) Average annual rainfall; (1959-61)

(iii) Net area sown as percentage of total geographical area (Reporting area); (1959-61)

(iv) Intensity of cropping = $\dfrac{\text{(Gross are swon)}}{\text{[Net area sown (1959-61)]}} \times 100$

(v) Gross area sown per capita = $\dfrac{\text{[Gross area sown (acres)]}}{\text{[Total rural population (1961)]}}$

(vi) Soil index rating:

The indices of soil rating for various districts have been taken from a study undertaken by Shome and Ray Chaudhari. (1960) The authors, in their study, have tried to develop soil evaluation which may be defined as the capacity of soil to produce crops and is the determinant of crop productivity rating. They have attempted to evaluate soil index rating of Indian soils using three main factors—

Factor A—Character of the soil profile

Factor B—Topography, texture and structure

Factor C—

(i) Degree of climatic suitability

(ii) Salinity

(iii) Stoniness

(iv) Tendency to erode.

Each of these factors is evaluated on the basis of 100 per cent for the most favourable conditions. The soil index rating is obtained by the product of the factors, A, B and C and the final index rating is expressed in percentages.

By the very nature of the factors involved in the calculation of soil index rating, the rating cannot be final and infallible and may have to be modified in the light of new data on the subject. It may, however, be emphasised that the figures of the soil index rating are average approximations for a district/group of districts which have different kinds of soil.

The consumption of fertilizers, manure and improved seeds, though quite a sensitive index for assessing the level of agricultural productivity, have not been considered here due to the non-availability of necessary data at the district-level.

In order to develop a composite index which reflects the composite effects of the six indicators enumerated above, the following methodology has been used:

After working out the necessary calculations for each of the 300 and odd districts of India, the values for each indicator have been arranged in the ascending order so as to work out the 'deciles'.

The relevant decile values for each indicator have been given in table 21.1.

On the basis of 9 decile values, ratings ranging from 1 to 10 have been assigned to each district for each indicator separately. These ratings based on decile values broadly reflect the physical level of a particular indicator. The co-efficient of variation has been worked out for each indicator and for each decile separately so as to ascertain the extent of variations prevalent within a particular decile. The relevant data are contained in table 21.2 which shows as expected, relatively greater variation in the lowest and the highest deciles. The intermediate deciles do not conceal any sizaeble variation. On this basis, one might say that the decile ratings do not conceal much variations so as to give a distorted picture of the realities.

In order to arrive at composite decile rankings for a district, some weightage seems necessary for each indicator. It may be clarified that the problem of relative weightage as between indicators is an arbitrary exercise based on personal judgement and trial and error method.

TABLE 21.1

Decile Values of Six Selected Agricultural Characteristics

Decile rank	*Average annual rainfall (inches)*	*Gross area sown per capita (acres)*	*Gross area irrigated as % of gross area sown*	*Net area sown as % of total geographical area (Reporting)*	*Double cropped area as% of net area sown*	*Soil productivity*
Lowest	72.6	0.4	1.3	23.3	0.7	50.4
Second	27.7	0.5	2.9	30.1	2.4	54.4
Third	31.7	0.6	4.8	38.0	3.9	*
Fourth	34.7	0.7	8.3	42.7	6.6	57.6
Fifth	38.9	0.8	12.5	50.0	9.4	60.8
Sixth	43.9	1.0	19.3	58.2	15.4	61.2
Seventh	49.6	1.3	24.5	63.3	22.6	64.6
Eighth	56.5	1.5	31.6	69.3	29.5	68.6
Ninth	74.6	1.7	43.3	75.4	38.6	72.7
Highest	74.7+	1.8+	43.4+	75.5+	38.7+	72.8+

* As the decile values for second and third deciles were of the same order, only second decile has been considered while allotting the decile rank.

The weightage problem will differ from region to region. For the sake of simplicity, uniform weights for the different states and districts have been assumed. Uniform weights have also been used for the sake of comparing the composite indices over districts and states. The relevant weights and their basis are given below.

Indicators	*Weights*	*Remarks*
Area cultivated as % of total geographical area (Reporting area)	1.0	
Gross area sown per capita	1.0	No weightage has been given for the amount of area sown per capita. The quality of area has been taken into account under the relevant rainfall categories as also under the level of irrigation.
Average annual rainfall	Upto 30":1, Between 31" to 74":2, 75"+1.5	Areas having rainfall upto 30" have been considered to be dry areas. Areas having rainfall between 31"-74" have been considered to be assured rainfall areas good for double cropping and hence the

cont....

Indicators	*Weights*	*Remarks*
		weightage of 2.0 has been assumed. Rainfall beyond 74" has been considered relatively less beneficial and therefore 1.5 has been assumed as the weightage.
Gross area irrigated as % of grass area sown	2.0	Based on the order of difference in the per acre yield rates between irrigated and unirrigated areas.
Intensity of cropping	1.0	No separate weightage for double cropping has been assumed because the necessary account for difference in the quality of land has already been taken care of under irrigation and rainfall categories.
Soil productivity	1.5	In order to isolate the impact of the quality of soil reflecting the fertility and chemical contents, a weightage of 1.5 has been assumed.

TABLE 21.2

Coefficient of Variation in Deciles

Indicator	*Lowest*	*Second*	*Third*	*Fourth*	*Fifth*	*Sixth*	*Seventh*	*Eighth*	*Ninth*	*Highest*
1. Gross area irrigated as percentage of gross area sown	53.09	20.85	10.33	13.02	11.90	11.63	6.03	6.70	8.23	28.39
2. Net area sown as percentage of total reporting area	44.99	12.97	7.42	3.59	4.71	18.91	1.98	2.56	2.05	4.46
3. Double cropped area as percentage of net area sown	56.06	18.02	15.05	11.05	12.53	16.84	7.58	8.06	6.87	2.57
4. Gross area sown per capital	21.83	—	—	—	—	5.38	5.74	—	3.07	25.66
5. Average annual rainfall	26.98	6.70	4.25	3.13	3.29	2.97	4.17	3.86	26.62	16.92
6. Soil productivity	11.08	2.57	*	1.52	2.22	—	2.05	2.92	2.33	2.94

* See footnote in Table 21.1.

(-) Indicates no variation

On the basis of the weights indicated above, decile ratings for each of the 6 indicators considered here have been combined for each of the districts studied in order to arrive at composite decile ratings. The composite decile ratings based on the selected 6 indicators have been assumed to reflect the land resources in each of the various districts. In order to assess the relative level of land resources various districts of India, average all-India composite decile ratings were worked out and have been taken as the base (=100) for constructing composite land resources index for all the districts in the 15 major states of India. Thus, land resource indices have been worked out for each district of the major states on the basis of their relative composite ratings. On the basis of the land resource indices, the agricultural regions of India have been delineated and the results are summarised below.

While agricultural regions have been delineated on the basis of land resource indices explained above, account was also taken of the relative valuation of the per acre gross value productivity based on 20 crops for the triennium 1959-60 to 1961-62. The 20 crops considered here for the purpose of studying the relative productivity levels are rice, wheat, jowar, bajra, maize, barley, ragi, gram, tur, linseed, rapeseed and mustard castorseed, groundnut, seasamum, sugarcane, cotton, jute, sunnhemp, mesta and tobacco. For the purpose of valuation of production of 20 crops, weighted all-India harvest prices were used, the State-wise production being the weights for each of the various crops.

In the case of crops where harvest prices were not available, wholesale prices in the peak marketing season, were used. Uniform all-India weighted prices were used for valuation of a particular crop over districts. For getting the gross value of production for 20 crops for each of 300 districts, the average production during the triennium 1959-60 to 1961-62 was multiplied by the all-India weighted prices for each of the 20 crops in a district and this was divided by the average gross area sown (1959-60—1961-62) under these 20 crops for getting the per acre gross value productivity. Thus—

$$Y_d = \frac{\sum_{j=1}^{20} q_j \dfrac{\sum_{i=1}^{15} P_{ij} W_{ij}}{\sum_{i-1}^{15} W_{ij}}}{\sum_{j=1}^{20} a_j}$$

Where Yd—Per acre gross value composite productivity (20 crops) for a district.

qi—average gross physical production for a crop (j) during the triennium 1959-60—1961-62 for a district.

Pij—Annual harvest price for a crop (j) in a State (i) for 1960-61.

Wij—Average gross production for a crop (j) in a State (i)

ai—Average gross area sown (1959-62) for a crop (j) in a district.

The advantage of this formulation of per acre gross value productivity is that it gives due weightage to the importance of various crops as reflected by the relative per cent of area under a crop to total cropped area under 20 crops in a district. At the same time by adopting uniform price weightage for each of 20 crops in various districts, it has been possible to aggregate the per acre physical productivity for each of 20 crops into a composite per acre value productivity for 20 crops for various districts.

After working out the per acre gross value productivity, district by district, the indices of per acre productivity, like the land resource indices, were worked out taking the average all-India per acre gross value productivity as the base (=100).

For the purpose of evaluating the economic status of different regions, the indices of land resources and productivity have been classified under the following groups:

Land Resource Indices		*Crop productivity indices*
Very low:	Less than 50	Less than 50
Low:	50-75	51-100
Medium:	76-100	101-150
High:	101-125	151-200
Very High:	126-150	201+

Before we proceed further with the study of agricultural regions, it may not be out of place to give here a table indicating the maximum and minimum levels of land resource indices and composite crop productivity indices obtained in various districts of India.

Agricultural Regions Within States

It may be pertinent to mention that land resources may not always be reflected in composite crop productivity because of the heterogeneity in the utilisation of resources and their availability as also the risk and uncertainty involved in climatic conditions. The actual pressure of population on land, untimely rainfall, non-availability of credit facilities and late ploughing or similar disturbances in some agricultural operations may lead to the deviations in the results. Keeping in mind this broad limitation, we may proceed for a study of agricultural regions within administrative boundaries of States. The agricultural regions have been considered on the basis of 1961 Census zones. These zones are as follows:

1. Northern Zone — Jammu & Kashmir, Punjab (including Haryana), Rajasthan, Delhi, Himachal Pradesh.

TABLE 21.3

Maximum and Minimum Levels of Land Resource Indices and Crop Productivity Indices

	Land resources		*Crop productivity*	
	Maximum	*Minimum*	*Maximum*	*Minimum*
1	2	3	4	5
Andhra Pradesh	126 (West Godavari)	84 (Chittoor)	216 (East Godavari)	66 (Adilabad)
Assam	128 (Nowgong)	69 (Sibsagar)	203 (Nowgong)	118 (Kamrup)
Bihar	136 (Shahabad)	84 (Saharsa)	174 (Dhanbad)	81 (Palamau)
Gujarat	91 (Surat)	60 (Kutch)	194 (Surat)	4.37 (Banas kantha)
J & K	104 (Jammu)	63 (Doda)	160 (Anantnag)	67 (Punch)
Kerala	123 (Trivandrum)	70 (Kozhikoda)	238 (Kottayam)	162 (Kozhikode)
Madhya Pradesh	134 (Balaghat)	58 (Jhabua)	127 (Bastar)	57 (Rajgarh)
Tamil Nadu	137 (Thanjavur)	79 (Tirunelveli)	255 (Kanniyakumari)	127 (Ramanathapuram)
Maharashtra	118 (Bhandara)	69 (Buldana)	194 (Kolaba)	48 (Wardha)
Mysore	126 (Shimoga)	70 (Gulbarga)	299 (Coorg)	31 (Bijapur)
Orissa	126 (Balangir)	76 (Koraput)	151 (Ganjam)	105 (Baudh Khondhmals)
Punjab	148 (Firozepur)	36 (Lahul & Spiti)	158 (Ludhiana)	45 (Lahul & Spiti)
Rajasthan	117 (Bharatpur)	44 (Jaisalmer)	99 (Udaipur)	11 (Bidaner)
Uttar Pradesh	138 (Saharanpur)	81 (Almora)	214 (Muzzaffarnagar)	70 (Bahraich)
West Bengal	128 (Birbhum)	86 (24-Paraganas)	223 (Hooghly)	127 (Malda)
15 States	148 (Firozepur)	36 (Lahul & Spiti)	229 (Coorg)	11 (Bikaner)

2. Central Zone — Uttar Pradesh, Madhya Pradesh.
3. Eastern Zone — Bihar, Orissa, West Bengal, Assam, Manipur, Tripura, Arunachal Pradesh, Nagaland, Sikkim.
4. Western Zone — Gujarat, Maharashtra. Dadra & Nagar Haveli.
5. Southern Zone — Andhra Pradesh, Mysore, Kerala, Madras (Tamil Nadu), Pondicherry, Laccadive, Minicoy & Amindivi Islands.

Areas which have not been included in any Zone — Andaman & Nicobar Islands and Goa. Daman & Diu.

NORTHERN ZONE

Jammu & Kashmir

TABLE 21.4

Agricultural Regions of Jammu & Kashmir

State	*Region*	*Ranking*	*District*	*Land resources (All India = 100*		*Per acre gross value Productivity (20 crops) (All India=100)*	
				Indices	*Rank*	*Indices*	*Rank*
Jammu &	Western	AI(I)	Baramulla	91	4	119	3
Kashmir	Srinagar			82	6	148	2
			Anantnag	93	3	160	1
			Udhampur	89	5	75	6
			Punch	76	8	67	9
	Western	AII(2)	Jammu	104	1.5	77	5
			Kathua	104	1.5	71	8
	Eastern	BI(3)	Ladakh	81	7	80	4
	Eastern	BII(4)	Doda	63	9	74	7

On the basis of land resource indices, the State of Jammu & Kashmir can be divided into 4 broad agricultural regions. The region consisting of Jammu and Kathua, is showing high land resources. Barring Doda, the rest of the districts fall in the medium category of resources. As against this, the districts of Anantnag. Srinagar and Baramulla are having relatively high productivity indices compared to resource indices. Anantnag falls in the high productivity group and Baramulla and

Srinagar in the medium category. The rest of the districts are having productivity levels in the low group of productivity.

Punjab (including Haryana)

TABLE 21.5

Agricultural Regions of Punjab (Including Haryana)

State	*Region*	*Ranking*	*District*	*Land resources (All India = 100)*		*Per acre gross-value productivity (20 crops) (All India=100)*	
				Indices	*Rank*	*Indices*	*Rank*
1	2	3	4	5	6	7	8
Punjab (including Haryana)	Northern	A(I)	Gurdaspur	119	10	116	10
			Hoshiarpur	112	14.5	111	12
			Kangra	108	16	103	15
			Lahul & Spiti	36	19	45	18
			Simla	88	18	100	14
	Western	(B)2	Amritsar	137	5	133	5
			Kapurthala	132	7.5	131	7
			Jullundar	132	7.5	157	2
			Ludhiana	142	3	158	1
			Firozepur	148	1.5	147	3
			Bhatinda	148	1.5	144	11
			Sangrur	131	9	119	9
			Hissar	141	4	78	16
			Rohtak	134	6	109	13
	Eastern-Central	C(3)	Patiala	144	12.5	140	4
			Karnal	144	125	121	8
			Ambala	102	17	125	6
	Southern	D(4)	Mahendragarh	117	11	45	19
			Gurgaon	112	14.5	67	17

Four broad regions seem to emerge from the above table. It may be observed that the agricultural regions formed by the group of districts, like. Amritsar, Ludhiana, Firozepur running through Bhatinda down to Hissar is perhaps the best agricultural region of Punjab and perhaps of India and high value productivity crops, like, wheat and cotton are grown here. In fact, the highest productivity levels are observed in this set of districts in the whole of North Zone. The hilly districts of Lahul & Spiti and

Rajashthan

TABLE 21.6

Agricultural Regions of Rajasthan

State	*Region*	*Ranking*	*District*	*Land Resources (All India = 100 (20 crops)*		*Per acre gross value (All India = 100)*	
				Indices	*Rank*	*Indices*	*Rank*
1	2	3	4	5	6	7	8
Rajasthan	Western	AI(1)	Jaisalmer	44	26	22	24
			Bikaner	54	25	11	26
	Western	AII(2)	Barmer	60	24	21	4.5
			Jodhpur	73	21	18	25
	Western	AII(2)	Nagpur	72	22	28	21
			Churu	64	23	21	22.5
			Ganganagar	92	17	61	12
	South-Western	BI(3)	Jalore	94	14	33	20
			Sirohi	94	14	67	11
	Southern	BII(4)	Pali	106	5	43	17
			Udaipur	100	9	99	1.5
			Chittorgarh	103	6	79	5.5
	Southern	BIII(5)	Banswara	91	18	84	3
			Dungarpur	79	20	99	1.5
	Southern-Eastern	BIV(6)	Ajmer	112	3	53	16
			Bhilwara	114	2	80	4
	Eastern	C(7)	Jhunjhunu	98	12	37	19
			Sikar	100	19	38	18
			Jaipur	94	14	69	9.5
			Tonk	101	7	55	14.5
			Bundi	93	16	79	5.5
			Sawai Madhopur	100	9	74	7
			Kotah	80	19	58	14.5
			Jhalawar	99	11	58	13
		D(8)	Alwar	110	4	69	9.5
			Bharatpur	117	1	73	8

Simla come under low and very low resource group, though from the point of productivity. Simla falls in the medium group. Punjab is one of the fertile areas of India and has in its domain alluvial soil which is rich for the growth of wheat, gram and rice. In fact, in productivity. Punjab is richer than any other part of northern region and it so appears that the productivity levels observed in Punjab tend to decline as one proceeds towards South-east through the western and central U.P. plain. The districts comprised of Gurgaon, Mahendragarh, represent the backward region of Punjab (now in Haryana) and therefore, both resource and productivity indices are comparatively low in these districts. The extension of Bhakra Nangal irrigation to this region will perhaps change this part into a prosperous area.

It may be observed that Rajasthan has 8 agricultural regions on the basis of land resources. The most prosperous region is represented by the districts of Ajmer, Bhilwara, Pali, Udaipur and Chittorgarh as also Alwar and Bharatpur. From the point of view of productivity, practically all the districts are relatively backward excepting the districts of Dungarpur (99), Udaipur (99), Bundi (79), Banswara (84), Sawai Madhopur (74), Chittorgarh (79), and Bhilwara (84), Most of these districts fall in the productivity range 75—100 which represents low productivity group. In none of the districts, the productivity exceeds the all-India productivity. It touches very near all-India productivity in the districts of Dungarpur and Udaipur. In most of the districts of Rajasthan, the average size of holding is relatively high but the land is of poor quality. The availability of red and black; desert soil; brown: red and yellow soil, with relatively less rainfall and low per cent of irrigation, are good only for the growth of inferior grains and, therefore, both physical and the gross value productivity is relatively low. In some of the recently developed parts like Ganganagar, the yield is still below the all-India average yield. As it well still take some time to tap the irrigation potential created there.

CENTRAL ZONE

Uttar Pradesh

TABLE 21.7

Agricultural Regions of Uttar Pradesh

State	*Region*	*Ranking*	*District*	*Land resources (All India = 100)*		*Per acre gross value productivity (20 crops) (All India = 100)*	
				Indices	*Rank*	*Indices*	*Rank*
1	2	3	4	5	6	7	8
U.P.	Kumaon &	A(1)	Uttarkashi	86	52	N.A.	—
	Uttara Khand		Chamoli	95	46	N.A.	—

cont....

1	2	3	4	5	6	7	8
			Pithoragarh	86	52	N.A.	—
			Tehri Garhwal	86	52	N.A.	—
			Garhwal	95	46	N.A.	—
			Almora	81	54	N.A.	—
	Western U.P. Plains.	BI(2)	Naini Tal	120	17	149	6
			Bijnor	111	33	175	3
			Rampur	118	22	124	12
			Bareilly	119	20	125	11
			Budaun	120	17	86	42.5
			Moradabad	105	42	121	13
	Western U.P. Plains	BII(3)	Dehradun	120	17	127	10
			Sharanpur	128	1	164	4.5
			Muzaffarnagar	134	2.5	214	1
			Meerut	130	4.5	190	2
			Bulandshahar	122	135	139	7
			Aligarh	119	20	104	25
			Mathura	123	11	102	27.5
	Western U.P. Plains	BIII(4)	Agra	113	30.5	97	33.5
			Etah	123	11	98	32
			Mainpuri	125	8	164	4.5
			Farukhabad	113	30.5	105	20.4
			Etawah	124	9	105	22.5
			Kanpur	117	23	9	30
	Southern U.P. Hills	C(5)	Jalaun	106	40	79	45
			Jhansi	89	49	78	46
			Hamirpur	87	50	75	47
			Vanda	91	48	86	42.5
			Allahabad	98	44	93	37
			Mirzapur	95	46	92	39.5
	Central U.P. Plains	DI(6)	Pilibhit	101	43	128	9
			Shahjahanpur	107	38	107	22
			Kheri	109	36	116	15.5
			Sitapur	109	36	103	26

cont....

1	2	3	4	5	6	7	8
			Hardoi	106	40	92	39.5
			Unnao	106	40	92	39.5
			Lucknow	111	33	95	35.5
			Bahraich	115	28.5	70	48
	Central U.P. Plains	DII(7)	Faizabad	116	25.5	116	15.5
			Sultanpur	121	15	102	27.5
			Rae Bareli	116	25.5	99	30
			Fatehpur	111	33	95	355
			Pratapgarh	115	28.5	92	39.5
			Jaunpur	119	20	112	19.5
	Eastern U.P. Plains	E(8)	Gonda	127	6.5	84	44
			Basti	134	2.5	99	30
			Gorakhpur	130	4.5	109	21
			Deoria	123	11	133	8
			Azamgarh	127	6.5	112	19.5
			Barabanki	122	13.5	120	14
	Eastern U.P. Plains	E(9)	Varanasi	116	25.5	114	17.5
			Ghazipur	109	36	114	17.5
			Ballia	116	25.5	97	33.5

Madhya Pradesh

TABLE 21.8

Agricultural Regions of Madhya Pradesh

State	*Region*	*Ranking*	*District*	*Landresources (All India = 100)*		*Per acre gross Value productivity (20 crops) (All India = 100)*	
				Indices	*Rank*	*Indices*	*Rank*
1	2	3	4	5	6	7	8
Madhya Pradesh	North-Western	AI(1)	Gwalior	104	5.5	114	6
	(Madhya		Bhind	104	4	91	9
	Bharat low	AII(2)	Morena	69	35.5	75	20.5
	Lands)		Shivpuri	84	16	68	29
			Datia	66	40	68	20.5
	Northern	BI(3)	Tikamgarh	95	9	90	10
	Madhya		Chhatarpur	85	14	86	12

cont....

Pradesh -do-	BII(4)	Panna	73	31	75	20.5
		Satna	75	28	66	31
		Rewa	79	22	58	40
		Sidhi	73	31	78	15
		Shahdol	73	31	88	11
Eastern	BIII(5)	Surguja	86	12.5	98	8
Madhya		Bilaspur	103	5.5	116	4
Pradesh (Chhatisgarh Plains)		Raigarh	76	26.5	116	4
-do-	C(6)	Balaghat	134	1	125	2
		Durg	114	3	107	7
		Raipur	118	2	116	4
		Baster	96	8	127	1
Southern	D(7)	Betul	88	11	67	30
Madhya		Hoshangabad	86	12.5	58	40
Pradesh		Seoni	91	10	81	13
		Narsimhapur	99	7	75	20.5
		Chhindwara	82	18	73	34
Central	E(8)	Guna	67	38	61	37
Madhya		Sagar	67	38	71	27
Pradesh		Vidisha	77	25	77	18
		Raisen	73	31	73	24
		Sehore	73	31	71	27
		Raigarh	79	22	57	43
		Shajapur	80	20	65	32.5
		Damoh	79	22	76	16
		Jabalpur	84	16	76	17
		Mandla	81	19	80	14
Madhya	F(9)	Mandsaur	84	16	63	34
Bharat		Ratlam	72	34	58	40
Plateau		Ujjain	62	41	58	40.5
		Dewas	69	35.5	65	32
		Indore	67	38	58	40
		East Nimar	61	42	71	27
Madhya		West Nimar	75	26.5	73	24
Bharat		Dhar	78	24	62	36
Hills		Jhabua	58	43	63	34.5

On the basis of land resource indices, U.P. can be delineated into 9 agricultural regions as well be seen from the table given above. Relatively speaking, western U.P. represented by the districts of Saharanpur, Muzaffarnagar, Meerut Bulandshahar, Aligarh, Etah and Mainpuri represent the best agricultural region in whole of U.P. and this is reflected by very high productively levels and very high land resource in many of the districts in this tract of U.P. Another good agricultural region from the point of view of availability of land resources is found in eastern U.P. and is represented by the districts of Gonda, Basti, Gorakhpur, Deoria, Azamgarh and Barabanki. These districts are relatively better from the resource point of view, yet form the point of view of productivity there seems to be under utilisation of agricultural resources in this part. Broadly speaking it has been observed that both land resources and productivity decline from western U.P. to central U.P. and again in eastern U.P. though the resources in some of the districts, again go up. Productivity tends to decline in the Indo-gangetic belt. However, within the same region say eastern U.P. composite crop productivity tends to decrease from north to south. Thus, productivity tends to decrease from Muzaffarnagar and Shaharanpur towards south. One of the possible reasons might be the commercial bias represented mostly by sugarcanes cultivation. This region is also relatively prosperous because of agro-based industries which have developed around this region. Relatively speaking, eastern U.P. though has sugarcane cultivation, yet there are frequent floods and drought almost every year and because of relatively less commercialised cultivation, the region on the whole, happens to be backward and hence there is a great need for establishing agro-based industries in this part of Indo-gangetic belt.

It would be seen from above that Madhya Pradesh could be divided into nine agricultural regions, on the basis of availability of land resources. The region formed by the group of districts falling in South Eastern Madhya Pradesh—Balaghat. Durg and Rajpur is perhaps the best agricultural region. The crop productivity falls in the medium category in this region. The high productivity observed in these districts gets extended towards Bastar, Raigarh, Bilaspur, Surguja, etc. In these sets of districts the high productivity levels might partly be explained by the high valued crop of rice which is grown over red and yellow soil.

The most backward agricultural region of Madhya Pradesh seems to be located in the Central Madhya Pradesh where medium black soil is prevalent. In this part both land resources and crop productivity indices are relatively low. Low productivity is partly explained by low valued crops, like, Jowar. Wheat and cotton are other crops grown here but their area proportions are rather insignificant.

It will be seen from above that Gujarat could be divided into five agricultural regions. The Saurashtra region formed by the districts of Jamnagar, Rajkot, Surendranagar Amreli, Bhavnagar and Junagadh is relatively backward both in respect of land resources and crop productivity. Groundnut, jowar and Bajra, being low valued crops, perhaps deflate per acre gross value productivity index for these districts. The soil types in this region are deltaic alluvial, coastal alluvium and medium black.

WESTERN ZONE

Gujarat

TABLE 21.9

Agricultural Regions of Gujarat

State	*Region*	*Ranking*	*District*	*Land resources (All India = 109)*		*Per acre gross Value productivity (20 crops) (All India = 100)*	
				Indices	*Rank*	*Indices*	*Rank*
1	2	3	4	5	6	7	8
Gujarat	Kutch	AI(1)	Kutch	60	17	42	16
	Saurashtra	AII(2)	Jamnagar	64	12	61	14
			Rajkot	81	3	64	11
			Junagadh	76	5.5	89	5
			Amerli	69	-	61	-
			Bhavnagar	74	8	63	13
			Surendra nagar	62	13.5	51	15
	North-Gujarat	BI(3)	Banas Kantha	76	5.5	37	17
			Sabar Kantha	76	5.5	87	6
			Ahmedabad	62	13.5	75	9
			Mahesana	88	2	64	12
			Panch Mahals	61	15.5	82	7.5
	South Gujarat	BIII(4)	Kheda	76	5.5	139	1
			Vadodara	72	-	111	1
			Bharuch	61	15.5	82	7.5
			Surat	91	1	194	3
	Dangs	BIII(5)	The Dangs	67	11	92	4

Maharashtra

TABLE 21.10

Agricultural Regions of Maharashtra

State	*Region*	*Ranking*	*District*	*Land resources (All India = 100)*		*Per acre gross value productivity (20 crops) (All India = 100)*	
				Indices	*Rank*	*Indices*	*Rank*
1	2	3	4	5	6	7	8
Maharashtra	Konkan	AI(1)	Thana	72	25	173	3
			Greater Bombay	92	8	N.A.	—
			Kolaba	89	11	194	2
			Ratnagiri	93	7	127	4
	Deccan	AII(2)	Dhule (Dhulia)	91	9.5	72	10.5
			Jalgoan	88	13	88	6
			Nasik	103	2.5	62	15
			Aurangabad	86	16.5	56	20
			Ahmad-nagar	103	2.5	69	13
			Pune	101	4	64	14
	Deccan	AIII(3)	Satara	88	13	73	9
			Sholapur	95	5	51	24
			Sangli	76	23.5	81	7
			Kolhapur	88	13	216	1
	Marathwada	BI(4)	Buldana	69	26	71	12
			Nanded	78	22	58	18.5
			Parbhani	80	20	54	21.5
			Bir	79	21	61	16.5
			Osmana-bad	76	23.5	72	10.5
	Vidarbha	CI(5)	Akola	87	15	58	18.5
			Amravati	83	18	61	16.5
			Yavatamal	82	19	52	23
	-do-	CII(6)	Wardha	94	6	48	25
			Nagpur	91	9.5	54	21.5
			Bhandara	118	1	10.9	5
			Chandrapur	86	16.5	76	8

Regions three and four constitute the Gujarat area. The Northern districts of Banas Kantha, Sabar Kantha, Mahesana and Panch Mahals are relatively backward both from the point of view of land resources and crop productivity, as compared to Southern districts of Kheda, Vadodra, Bharuch and Surat. The medium black soil and coastal alluvium prevalent in Southern districts seem to be favourable for high value productivity crops, like, cotton, rice. Thus, per acre gross value productivity indices for Surat, Kheda and Vadodra are 194,139 and 111. Productivity in Bharuch perhaps gets deflated due to cotton-jowar combination.

The above table shows that Maharashtra could be divided into six agricultural regions on the basis of land resources. From the point of view of land resources, the region formed by the districts of Nasik, Ahmadnagar, Pune, (Dhulia), Jalgaon and Aurangabad, which fall in Deccan areas having medium and deep black soil is perhaps the best region. However, per acre gross value productivity index is not very high, because crop pattern is oriented towards jowar, bajra and ground-nut complex. From the point of view of productivity, Konkan area (Thana), Kolaba, Ratnagiri, Deccan (Kolhapur) and Vidarbha (Bhandara) seem to be the best.

Perhaps the cultivation of rice and cotton in these districts, as against jowar, bajra and ground-nut combinations in other districts explain their superiority in productivity;

SOUTHERN ZONE

Andhra Pradesh

TABLE 21.11

Agricultural Regions of Andhra Pradesh

State	*Region*	*Ranking*	*District*	*Land resources (All India-100)*		*Per acre gross value productivity (20 crops) (All India-100)*	
				Indices	*Rank*	*Indices*	*Rank*
1	2	3	4	5	6	7	8
Andhra Pradesh	Coastal	A(1)	Srikakulam	122	3.5	177	4.5
			Vishakhapatnam	116	6.5	159	7.5
			East Godavari	124	2	216	1
			Krishna	122	3.5	189	3
			West Godavari	126	1	208	2
	Central	BI(2)	Guntur	93	16.5	167	6
			Mahbubnagar	93	16.5	79	15

cont....

1	2	3	4	5	6	7	8
			Nalgonda	94	14	73	18.5
			Hyderabad	87	19	79	15
			Khammam	93	16.5	84	12
	North-Western	BII(3)	Medak	107	11	95	10
			Nizamabad	116	6.5	177	4.5
			Karimnagar	103	12.5	79	15
			Warrangal	118	5	83	13
			Adilabad	93	16.5	66	20
	Southern	C(4)	Kurnool	110	10	75	17
			Nellore	114	8	116	9
			Cuddapah	112	9	86	11
			Anantapur	103	12.5	73	18.5
			Chittoor	84	20	159	7.5

Four agricultural regions emerge on the basis of land resources. The best agricultural region is formed by the coastal districts which are having both high resources and high productivity. The relatively backward region seems to be located in the central part of Andhra Pradesh, formed by districts like Mahbubnagar, Nalgonda, Khammam, etc., Guntur though included in this region is an exception from the point of view of productivity. In the Southern Andhra Pradesh, though the resources are relatively high, yet these are not reflected in high productivity levels and, therefore, need investigation.

Mysore

TABLE 21.12

Agricultural Regions of Mysore

State	*Region*	*Ranking*	*District*	*Land resources (All India = 100)*		*Per acre gross value productivity (20 crops) (All India = 100)*	
				Indices	*Rank*	*Indices*	*Rank*
1	2	3	4	5	6	7	8
Mysore	South Eastern	A(l)	Bellary	93	7.5	79	13.5
			Chitradurga	94	10	73	15
			Tumkar	99	6	103	10
			Kolar	87	13	120	8
			Bangalore	82	15	90	11
	Central-Southern	B(2)	Shimoga	126	1	160	5
			Chikmagalur	121	2	136	7

cont,...

1	2	3	4	5	6	7	8
			Hassan	122	4	139	6
			Mandya	119	3	187	3
	Dry North	C(3)	Bidar	73	16.5	64	16
			Gulbarga	70	18.5	37	18
			Raichur	70	18.5	40	17
			Bijapur	73	16.5	31	19
			Dharwar	84	14	79	13.5
			Belgaum	97	9	88	12
	Malnad	D(4)	North Kanara	107	5	190	2
			South Kanara	93	11.5	163	4
			Coorg	93	11.5	299	1
			Mysore	98	7.5	113	9

The above table shows that Mysore could be divided into four agricultural regions. The best agricultural region from the point of view of land resources is formed by the districts of Shimoga, Chikmagalur, Hassan and Mandya having red and laterite soil (Central Mysore). These districts are also having relatively high value productivity due to rice-ragi crop complex. The best set of districts from the point of view of value productivity are located in South-Western portion, represented by the districts of South Kanara, North Kanara, Coorg and Mysore having rice as the major crop grown on laterite and coastal aluvium soils.

The districts located in Northern Mysore and South-Eastern portion are having relatively less land resources along with relatively low productivity. The crops grown in North Mysore are Jowar, Bajra, Ground-nut and Cotton on deep black, medium black, mixed red and black soils. In South-Western portion, Ragi, Rice, Ground-nut seem to be common crop combinations grown on red soil.

The above table shows that four agricultural regions could be formed on the basis of land resources. High land resources seem to be located in Trivandrum. Trivandrum, Kottayam and Palghat are having very high productivity levels as against relatively low productivity levels in Cannanore, Kozhikode and Trichur. High value productivity levels of rice might explain the high productivity in these areas, in the background of mixture of soil types like coastal alluvium lateritic, laterite and forest. Rich growth of coconut and tapioca is the special feature of these districts. It may, however, be pointed out that the land resources and land productivity cannot strictly be compared in Kerala, because of the inclusion of the various land resources on the side of resources, but on the productivity side the output on account of plantation crops has been excluded and since Kerala is an important plantation area, this exclusion might deflate the per acre productivity and, therefore, these indices are not as realistic in Kerala (as is also the case in Assam) as in the case of other States where plantation crops role is rather insignificant.

Kerala

TABLE 21.13

Agricultural Regions of Kerala

State	Region	Ranking	District	Land resources (All India = 100)		Per acre gross value productivity (20 crops) (All India = 100)	
				Indices	Rank	Indices	Rank
Kerala	Northern	A(l)	Cannanore	88	8	164	8
			Kozhikode	70	9	162	9
	Central	B(2)	Palghat	94	6.5	219	5
			Trichur	112	3	174	7
			Ernakulam	113	2	210	6
	Southern	CI(3)	Kottayam	94	6.5	238	1
			Alleppey	98	4	214	3
			Quilon	97	5	213	4
		CII(4)	Trivandrum	123	1	231	2

Madras (Tamil Nadu)

TABLE 21.14

Agricultural Regions of Madras (Tamil Nadu)

State	Region	Ranking	District	Land resources (All India = 100)		Per acre gross value productivity (20 crops) (All India = 100)	
				Indices	Rank	Indices	Rank
Madras	Coromandal Coast	A(l)	Chingleput	118	4.5	168	7
			North Arcot	115	6	201	3
			South Arcot	127	3	189	4
			Thanjavur	137	1	236	2
	Inland	B(2)	Salem	98	10	142	11
			Tiruchirappalli	107	8	159	9
			Madurai	106	9	179	5
			Ramanatha-puram	109	7	127	12
			Coimbatore	118	4.5	175	6
		C(3)	Nilgiris	78	11	153	10
			Tirunelveli	79	12	162	8
		D(4)	Kanniyakumaril	33	2	255	1

It would be observed that of the two agricultural regions high and resources are observed in Coromandal coastal districts like Chingleput. South Arcot, North Arcot and Thanjavur. Productivity also happens, to be high in this belt. In the inland Madras districts, land resources and productivity fall in the medium category barring Kanniyakumari which is having very high land resources and very high productivity.

On the three agricultural regions, the coastal region, though has relatively less land resources yet these are best productivity districts. Resources are high in Western Orissa, but the crop productivity in some of the districts (Bandh Khondamals, Kalahandi) is relatively low and, therefore, needs investigation.

EASTERN ZONE

Orissa

TABLE 21.15

Agricultural Regions of Orissa

State	*Region*	*Ranking*	*District*	*Land resources (All India = 100)*		*Per acre gross value productivity (20 crops) (All India = 100)*	
				Indices	*Rank*	*Indices*	*Rank*
Orissa	Coastal	A(l)	Ganjam	103	8	151	1
			Puri	107	5	146	2
			Cuttack	105	7	145	3
			Dhenkanal	106	6	140	5
	Northern	B(2)	Sundergarh	100	9	125	9.5
			Keonjhar	96	10.5	142	4
			Mayurbhanj	96	10.5	123	9.5
			Balasore	95	12	122	11
	Western	C(3)	Sambalpur	118	3	137	6
			Balangir	126	1	132	7.5
			Bandh Khondmals	111	4	105	13
			Kalahandi	121	2	112	12
			Koraput	76	13	132	7.5

Bihar

TABLE 21.16

Agricultural Regions of Bihar

State	Region	Ranking	District	Land resources (All India = 100)		Per acre gross value productivity (20 crops) (All India = 100)	
				Indices	Rank	Indices	Rank
Bihar	North-Western	AI(1)	Saran	102	8	97	14
			Champaran	113	5	127	5
	North-Eastern	AII(2)	Muzaffarpur	94	11	96	15
			Darbhanga	88	14	110	11
			Shaharsa	84	17	116	9
			Purnea	86	16	121	8
			Monghyr	91	13	92	16
	Eastern	B(3)	Bhagalpur	114	4	123	7
			Santal Parganas	111	6	142	3
	Central	C(4)	Shahabad	136	1	114	10
			Gaya	115	3	105	12
			Patna	118	2	129	4
	Chotanagpur	D(5)	Palamau	103	7	81	17
			Hazaribagh	92	12	126	6
			Singhbhum	98	10	145	2
			Dhanbad	87	15	174	1
			Ranchi	100	9	103	13

Five agricultural regions seem to emerge from the above table, relatively high resources have been observed in central Bihar (Shahabad, Gaya and Patna). But the productivity levels are relatively low in Shahabad and Gaya. These cases need careful investigation. On the whole, there seem to be a declining trend in productivity, with the increase in resources. How for this situation is explained because of annual recurrence of floods and drought in Bihar, needs detailed surveys and investigations.

West Bengal

TABLE 21.17

Agricultural Regions of West Bengal

State	*Region*	*Ranking*	*District*	*Land resources (All India = 100)*		*Per acre gross value productivity (20 crops) (All India = 100)*		
			Indi-ces	*Rank*	*Level*	*Indi ces*	*Rank*	*Level*
West Bengal	Northern A(l)	Cooch-Behar	101	12	High	181	7	High
		Jalpaiguri	101	12	High	202	3.5	Very High
		Darjeeling	108	10	High	207	2	Very High
		West Dinajpur	109	9	High	147	12	Medium
		Malda	101	12	High	127	15	Medium
	Eastern B(2)	Nadia	92	14	Medium	147	13	medium
		Twenty-four Parganas	86	15	Medium	146	9	High
	Western C(3)	Murshidabad	112	7	High	146	14	Medium
		Burdwan	112	7	High	202	3.5	Very High
		Hooghly	116	4	High	223	1	Very High
		Howrah	112	7	High	182	5.5	High
		Birbhum	128	1	Very High	182	5.5	High
		Purulia	113	5	High	159	11	High
		Bankura	117	3	High	179	8	High
		Midnapur	125	2	High	160	10	High

It may be observed that in practically all districts of West Bengal, the productivity indices exceed the land resource indices. High productivity indices are observed in Himalayan west Bengal as also in Gangetic west Bengal. The Western Bengal, through which flows the Ganges is, of course, having high land resources alongwith high productivity.

Assam

TABLE 21.18

Agricultural Regions of Assam

State	*Region*	*Ranking*	*District*	*Land resources (All India = 100)*		*Per acre gross value productivity (20 crops) (All India = 100)*	
				Indices	*Rank*	*Indices*	*Rank*
Assam	Western	AI(1)	Goalpara	109	3	158	6
			Kamrup	103	4	118	11
			Darrang	97	5	163	5
			Nowgong	118	1.5	203	1
	Western	AII(2)	Garo Hills	83	8.5	131	9
			Khasi & Jaintia Hills	111	1.5	138	7
	North Eastern	BI(3)	Lakhimpur	93	6	170	4
			Sibsagar	69	10.5	172	3
			UM & NC Hills	92	7	123	10
	South Eastern	BII(4)	Cachar	69	10.5	172	2
		Mizo	Hills	81	8.5	135	8

Four agricultural regions seem to emerge from the above table. High agricultural resources alongwith high productivity seem to be located in U.K. and J. Hills, Nowgong, Goalpara and Kamrup. On the whole, productivity indices seem to be exceeding the resource indices in practically all districts of Assam. This is despite the fact that we have not taken into account plantation crops in the productivity index.

REFERENCES

Shome, K.B. and Ray, (1960), "Ratings of Soils of India". In:

Chaudhary, S.P., *Proceedings of the National Institute of Sciences of India*, Vol. 26(2) Supplement.

22

COTTAGE AND AGRO INDUSTRIES

The meaning attached to the term "Cottage Industries" has undergone changes with the progress in technology and its definition no longer restricts it to industries using manual labour and simple tools only. The use of power and machines has been admitted in the definition of the term. The basic implications which differentiate the cottage industries from large scale and heavy industries are: (1) the small scale of production, and (2) usually the worker works on his own, although the employment of wage—paid labour is not ruled out. Attempts to define the term very precisely have not been quite successful. With technological progress the concept behind cottage industries has been undergoing modifications. However, the sense in which the term is employed is commonly well understood.

Cottage industries may be divided into four broad classes, *viz.*, handicrafts, home industries, small-scale industries and village industries. The dexterity of human hands and mainly their use in making the final product are the characteristics of handicrafts. Handicrafts, as the name indicates, always represent the best type of art for the art produced by hand can hardly be excelled.

In home industries, hand or power driven machinery may be used. But the chief characteristic of home industries is that raw materials are procurable locally and labour consists normally of family members. If there are any outsiders they are employed only as apprentices. The industry thus gets the most careful attention of the master.

Small industries are distinguishable by the place of work, the amount of capital invested and equipment used. They are carried on either in the house of the artisan or in a small workshop run by the entrepreneur. The capital invested may be theoretically large as is the case with ivory carving. But generally the capital is limited and this places restrictions on the employment of labour, the use of machinery, etc.

Hired labour may be employed both in the small-scale and village industries; but the village industries have necessarily to be located in the villages, while the small industries may even be located in tine urban or semi-urban areas. The number of employees in the small industries should also not be large and may not exceed nine persons.

PROGRESS AND DECAY OF SMALL AND COTTAGE INDUSTRIES IN INDIA

India has been well known for the products of her cottage industries. Thousands of years ago when the art of embalming was practised in ancient Egypt, the mumies were embalmed in Indian silk. Courts of Imperial Rome glittered with gold and silver brocades of Delhi. Muslins of Dacca (Dhaka) were famous for their superfine qualities. In the words of William Hunter "The industrial genius of her inhabitants even more than her natural wealth and her extensive sea-boards, distinguished India from the other Asiatic lands". When barbarians of Britain were painted as savages, Martin was amazed by the gossamer muslins of Dacca, the beautiful shawls of Kashmir and brocaded silks of Delhi adorning the proudest beauties at the court of the Ceasars.'

India continued to enjoy this proud position till the end of the 16th century. The decline in handicrafts began towards the close of that century but became more pronounced by the middle of the 19th century. The chief causes of their decay were the following factors.

Native Courts

The internal demand for fine arts and exquisite material was primarily from the nobles for the state occasions. Besides this, the patronage given to these industries by the Mughal emperors was with drawn when their own position became very weak. At the early stages even when the courts disappeared, the decline was not so rapid. All those places which were famous for certain products, continued to enjoy their reputation. But as a generation passed, the old splendours were quite forgotten. The result was a rapid decline and even disappearance of the whole industry in a number of cases.

British Rule

The advent of British rule affected this industry in a number of ways. The early policy of the British was not to encourage the Indian industry. Every possible hindrance was put in the way of workers. The rule indirectly weakened the power of guilds which regulated the trade in the cottage products. The new rich classes comprising mainly the landlords preferred to live in towns and retired from their estates. The educated classes developed new tastes following British fashions and traditions. This naturally had an adverse effect on the demand for the products of cottage and small scale industries.

The Government Policy

Dutt wrote long way back, "the invention of the powerloom in Europe completed the decline of Indian industries." The United Kingdom produced large quantities of textiles and was out for finding markets for them. A deliberate policy was thus adopted by which British machine-made fabrics could very conveniently compete with the Indian manufactures which were naturally costly as compared to the mass production of textile mills in England. Development of the railway system in India aimed at carrying Indian cotton to the UK and sending back to India manufactured cloth. The British East India Company was primarily a trading concern. Instead of rendering assistance to the struggling handicrafts of India the then Government went out of its way to give direct assistance to British manufactures so as to exploit the Indian market.

The whole policy of transport, tariffs and foreign trade was geared to help the British industry at the cost of India as well as other nations who were trying to compete with British trade.

India also had to fall in line with the other countries of the world in industrial progress. The development of manufacturing industries in India further helped in driving away the cottage made products from the market. In most cases of general utility articles, the mill produced goods are cheaper, more uniform and even better in quality as compared with hand-made goods. Gradually the cottage industry survived in those spheres in which its products did not complete with the mill products. In parts of Bihar, Bengal, Orissa and South India, people still prefer handloom textile because of tradition. Handlooms have, therefore, survived in these areas. Despite severe competition from internal manufacturing concerns and foreign-made goods, cottage industries still survive and make no small contribution to the production of goods in the country. The National Income Committee estimated in 1950-51 the value of output of small enterprises to be Rs 910 crores while the output of factory establishments was worth about Rs 550 crores only.

Importance of Rural Industries

The importance of village and cottage industries is emphasised by the fact that India, in spite of considerable progress towards urbanisation in the recent past, still lives in villages and her national economy is predominantly agricultural. Her agriculture, even when the required degree of intensity of cultivation is achieved through the development of irrigation, use of fertilizers and above all with the assistance of an all-prevading extension service, by its very seasonal nature and overcrowding in it, would not be able to free itself from the disguised unemployment and under-employment.

The industrial development, with its concentration in towns, is no solution to the problem of unemployment in villages. The small enterprises employed about 20 million workers as against about five million workers in the factory establishments in 1966. There has been a great rise in building and other construction activities and people have found new employment opportunities in them. But this is so only during the transitional stage of economic development and the tempo of construction programmes may slow down. Therefore a more stable source of employment has to be looked for. Rural industries provide great promise in this respect.

The greatest advantage of small and cottage industries is that employment is provided in the natural setting of workers' own place with numerous physical moral and other benefits that go with such employment. It helps them to develop inherent talents and aptitudes in occupations congenial to them. Another important reasons for employment to be found in the villages and not outside is that agriculture may not provide full employment, yet a large labour force has to remain in villages at certain periods to provide the needed labour for cultivation, weeding, harvesting and other operations. Mechanised agriculture, to any appreciable extent, does not seem to be a practical proposition for many years to come. The introduction and encouragement of well-chosen village industries in each region of India, on grounds of availability of raw material, local talent, the ability of the industry to fit in with the adjust within a limit with the peak labour demand in a agriculture, should add to rural welfare in many ways.

1. It will reduce considerably unemployment and underemployment in many regions.
2. It will provide useful outlet for many of the raw materials and waste material of agriculture, the transport of which, for reasons of their bulk and low value, is not economical over long distances. Such raw material will be converted into national wealth.
3. The farmer and his family members will get an additional source of income which is so needed to raise their standard of living and help them tide over difficulties caused by the failure of crops.
4. The country is facing an extreme shortage of foreign exchange. The restrictions on imports does not help very much to ease the problem. An all-out effort is required to encourage exports. The products of cottage industries have fairly reasonable prospects of commanding foreign markets where there is great demand for them. The opening of cottage industries emporia in some of the European countries has already brought to light the potentialities of these products as earners of foreign exchange.

   Phulkaries of Punjab, shawls, wood work and art pieces of Kashmir, are but a few items, which are much in demand in other countries and promise to fetch prices commensurate with the labour put in the production of these commodities.

5. The products of small industries which employ local raw materials and talent do not call for any heavy capital investment. This suits our austere economy.
6. There need not be any clash between the large and small industries. If properly organised, small industries will provide employment opportunities to enable India to achieve her cherished goal of full employment, but also act as feeders to large industries.
7. A developing economy depending upon deficit financing has to face a time lag between the commencement and completion of big projects particularly when the country lacks technical know-how and equipment. During this period, the available consumption goods cannot keep pace with the rise in the purchasing power. Products of cottage and small-scale industries would come in very handy under these circumstances. In addition to making available a variety of consumer goods, they will to a great extent be free from the distribution problem. Transportation over long distances would not be necessary and a major portion of the products would be consumed near the producing centres. These readily available goods will help to mop up the newly released purchasing power and fill the necessary gap, till the investment on capital goods starts producing consumer goods.
8. India is wedded to a democratic form of government and is committed to a socialistic pattern of society. Like peasant farming in agriculture, cottage and small industries in the industrial sector strengthen the foundation of democracy and socialist structure of society. For this reason also, cottage and small industries have to get a place in the national economy to usher in economic democracy and the desired social order.

Need for Rehabilitation

In view of the benefits conferred on the national economy and increase in rural welfare, cottage and small scale industries have a very strong case for their rehabilitation and development. The power projects completed and nearing completion under the Five Year Plans are adding to the supply of total electricity in the country at reasonable costs. The Government have already taken up the programmes of rural electrification. This should give further impetus to development of small industries in villages. This favourable development in the availability of cheap power in rural areas can greatly help to remove the stigma attached to cottage industries of being sweated ones with low returns. As technological progress in the form of the steam engine gave a staggering blow to the small scale production in the 19th century, the change from steam power to electric power, particularly in countries where electric power can be produced cheaply, can help in reversing the process from large scale to small scale production in the case of a number of products on economic grounds, considerations of rural welfare and decentralisation of economy in the interest of a democratic socialistic set up. It is just not a theoretical deduction but is in actual practice in countries like Japan and Switzerland. The various forms of help needed by the small scale industries are described below:

1. An organised extension service of technical guidance to the cottage workers in designing, use of tools and machines and production of quality goods or goods according to demand.
2. The extension service has to be supported with appropriate research institutions at central, regional and state levels.
3. Development of industrial cooperatives for supplies, services, finance, and marketing of products and in some instances even for production. If farming can be run on cooperative lines, there appears no reason why it cannot be done in the production of goods in certain cottage or village industries as well. For an understanding of the causes of failure of cooperative societies in the past, the findings of Sri Vishwanathan Tekumalla who surveyed some of the representative industrial cooperatives in Andhra would be of interest. These are:

(a) *lack of adequate capital.* Societies do not have enough capital either to give full time work to members or to enable them to manage until the products are sold. Consequently the members are obliged to produce for the local market or for the local *sahukar.*

(b) *Lack of good faith among the members.* The limited benefits conferred by the societies make the members disloyal. The artisans distrust one another.

(c) *Personal interest of office-bearers.* Some influential members as office bearers utilise the societies' funds to further their own personal ends.

(d) *Low purchasing capacity.* The consumers, majority of whom have low incomes, care more for cheap goods rather than for quality. The member artisans face keen competition from non-members who under-sell by producing interior staff for the market.

(e) *Lack of co-operation among departmental and non-official staff.* Instead of obtaining the cooperation of the supervising officials, the office bearers of some societies actually betray their own cause by hoodwinking them.

With cooperative education, proper training and supervision, these difficulties can be overcome but it must be clearly realised that efforts on a cooperative basis in all possible directions are necessary for the development of handicrafts and cottage industries. It is only through cooperation that these industries can reduce their costs of production and marketing.

4. There should be adequate and institutional credit made available to the cottage workers for meeting the financial needs of their business.
5. The State should help develop home and foreign markets for cottage goods.
6. A system of quality control must be imposed on the cottage products.
7. Government should give preference to cottage products in buying their requirements.

COTTAGE INDUSTRIES UNDER THE PLANS

The Industrial Policy Resolution of the Government of India, 1948, recognised cottage and small scale industries as an important sector of our economy. The Cottage Industries Board and the Cottage Industries Directorate were accordingly set up. Since each of these industries have their own complexities and problems, separate Boards were also created to look after individual industries such as coir, silk and handicrafts. The Small Scale Industries Organisation was also created to take care for those small enterprises which were outside the orbit of all such Boards. For purposes of definition, it was decided that a unit which has an investment of less than Rs 5 lakhs and employs less than 50 persons could be classed as small scale. The limit on employment has been removed since then.

In spite of the awareness of the importance of cottage and small scale industries and initiation of the organisational measures as mentioned above nothing much could be achieved during the First Plan period. Some protection was given to handloom and three or four other village industries. But a large number of the cottage industries faced strong competition from mill and factory-made goods.

The Karve Committee

In order to review the problems and programmes of such industries, a Village and Small Industries Committee was set up in 1955 under the chairmanship of Prof. D.G. Karve. The recommendations of the Committee deal with the following three objectives:

1. Further technological unemployment such as occurs in traditional village industries should be avoided as far as possible.
2. Employment opportunities for as large a number as possible should be created through different village and small industries.
3. The basis should be prepared for an essentially decentralised society and also for progressive economic development at a fairly rapid rate.

With these objectives in view, the Committee considered that the progressive expansion and modernisation of rural industry could be best brought about by the establishment of small industries units, along with the necessary services in large villages and small towns located all over the country. Industrial expansion on the periphery of large towns could scarcely be said to reduce the concentration of industry. What was needed, therefore, was a pattern of industrial activity in which a group of villages converging on their natural industrial and urban centre, form a unit, what the Committee called "a pyramid of industry broad-based on a progressive rural economy."

The Committee also visualised the introduction of technical improvements even in the traditional village industries. It was recommended that for the future there should be a regular programme of gradual transition to better techniques.

Industrial Policy Resolution of 1956

Necessitated by the acceptance of a socialist pattern of society as the national objective and to meet the requirements of the planned economic development, a fresh statement of industrial policy was made on April 30,1956. "The Resolution accepted the policy of supporting cottage as well as village and small scale industries by restricting the volume of production in the large scale sector by differential taxation or direct subsidies. The aim of state policy was declared as that of ensuring a decentralised sector acquiring sufficient vitality to be self-supporting and integrating its development with that of the large scale industry. The state was thus to concentrate on measures designed to improve the competitive strength of the small scale producer."

The Second Five Year Plan could then take bolder steps for the development of small scale and cottage industries, as a means of achieving dispersed industrial growth on the one hand and provide better employment opportunities on the other. The Development Commissioner's Organisation for Small Scale Industries was set up at the Centre as the apex administrative machinery which has since then rendered technical assistance through a network of 15 State Institutions, Branch Institutes and 36 Extension Centres already established for this purpose all over the country.

Intensive area and marketing schemes for khadi and village industries were introduced. The scheme aimed at the integrated economic development of selected contiguous areas with a population ranging between 20,000 to 30,000 with a view to developing village industries as an integral part of the rural economy. The number of such intensive areas was increased from 35 in 1955-56 to 200 at the end of the Second Plan.

A provision of Rs 10 crores was also made for setting up industrial estates. The object was to provide conditions favourable to working efficiency, maintenance of uniform standards in production and economic utilisation of materials and equipment. This was intended to enable the small scale units to have the advantage of common services and other facilities such as good site, electricity, water, gas, steam, compressed air, railway sidings, watch and ward, etc.

District Development Programme

It was felt in 1959 that the impact of the small industries programme was more confined to the cities and towns or to areas within close proximity of the headquarters of the Institutes and Extension Centres and had not permeated the countryside. Little advance had been made by this programme into rural areas. The objective of regional and dispersed development of industries as conceived by the Second Plan had not been achieved.

In order to carry the programme to the under-developed areas and make the people interested in the development of small industry, an intensive development campaign programme was chalked out.

The programme was primarily to be launched in such areas (districts), which though under-developed, offered maximum potentialities for the development of small industries. Closest cooperation was also intended to be provided among the Central Small Industries Organisation including the Small Industries Service Institute, the State Directorates of Industries, National Small Industries Corporation (P) Ltd and other official and non-official institutions and agencies. The underlying idea of the campaign was to carry the development programme to the very doors of the potential entrepreneurs and apprise them of the benefits that were likely to accrue to them from taking up small scale industries.

At the end of September 1960 the scheme was in operation in 65 intensive and 18 pre-intensive areas. The basic unit for working out the scheme was a selected area comprising 30-40 villages with a population of about 20,000. Warrangal District in Andhra Pradesh and Pali as well as Bhilwara Districts in Rajasthan were the first to receive the benefit of this programme.

Third Plan Programme

The Third Plan put forth a bold programme for the small industries. A total outlay of Rs. 264 crores as against an actual expenditure of Rs 180 crores in the Second Plan was made for these programmes.

The main objectives of the small industries programme during the Third Plan were to include the activity of the workers to enlarge the availability of institutional finance and to pay special attention to the growth of small industry in rural areas and small towns. The progress was encouraging during the first two years of the Plan; then slowed down for various reasons including the shortage of raw materials following the hostilities of 1962 and 1965; and has recently shown signs of revival. It was decided that if the small industries have to be developed in a massive way, the approach will have to be more fundamental and will have to be worked out along the lines of the reshaping of the structure of some parts of the industrial economy. The steps necessary to translate the policy of decentralisation into concrete programmes of action will have to be worked out and the following ten-point programme of assistance will have to be given special importance:

(i) *Reservation.* In certain selected industries, the entire, or the bulk of future development may be reserved for the small investor, artisan and worker, after a detailed study and investigation.

(ii) *Location.* A dispersed growth of small industrial units in towns will involve not only widespread provision of economic and social overheads but also greater attention to the planning of small towns so that the evils of congestion, growth of slum conditions and deterioration of sanitation are avoided.

(iii) *Industrial estates.* One way of meeting the problem of pressure on space and growth of industrial slums in urban areas is to construct industrial estates on specially selected sites and to provide workshed accommodation to the prospective small investor.

(iv) *Supply of raw materials.* Suitable arrangements will have to be made for the supply of adequate raw materials to small industries after an examination of the existing allotments and utilisation of scarce raw materials by large and small industries.

(v) *Training, management and marketing.* The industry will also have to be given support through training programmes in production methods, business management, salesmanship and market analysis.

(vi) *Preparatory and processing facilities.* It will be necessary to take steps to organise preparatory and processing facilities for the small industry, preferably on a cooperative basis.

(vii) *Finance.* Adequate finance will have to be provided for the rehabilitation and modernisation of equipment and for working capital so that the industry is placed on a strong and sound footing.

(viii) *Decentralised manufacture of equipment.* Manufacture of small machines or tools required by the small units will have to be undertaken, to the extent possible, on a decentralised basis.

(ix) *Research.* An important step would be the organisation of facilities for research in order to enable the small industry to adopt the appropriate level of technology and to keep pace with the advances of scientific and technical knowledge in a highly dynamic world.

(x) *Fiscal policy.* Taxation and fiscal measures will have to be so devised and adjusted as not to prejudice the programme of decentralised growth of industry.

Fourth Five Year Plan

The Fourth Five Year Plan has been formulated to improve progressively the production techniques of small industries so as to enable them to produce quality goods and to bring them to a viable level; to promote decentralisation and dispersal of industries; and to promote agro-based industries. In order to achieve these aim, it would be necessary to improve skills and provide a combination of incentives and disincentives for securing decentralisation and dispersal of small industries. Fiscal and other measures are required to enable these industries to stand competition with large industries. The operation of the industrial licensing system has not been effective in preventing competition from the large industries and in providing the required degree of initial protection. Nor has it has been possible to prevent concentration of industries in large cities and towns. Since a large number of industries is proposed to be delicensed during the Fourth Plan period, greater emphasis will have to be placed on a variety of positive measures of assistance including liberal credit facilities, adequate supply of scarce raw materials, provision of technical assistance and improved appliances, tax concessions and differential excise duties. It will be necessary to assist mechanised small scale industries to grow into larger and more viable units. Further, in order to protect small scale and traditional industries from undue competition, the existing reservation will be continued and modified in accordance with the requirements. This will have to be preceded by careful identification of industries, parts, components and processes in which the large size of operations or a high degree of mechanisation has no pronounced impact on economies. This identification would have to be followed by fiscal and credit policies and measures to accelerate their development in the small sector, evolution of appropriate technology for smaller units in different industries, introduction of quality control and formulation of well coordinated programmes of assistance. Outside the designated field for small industries, the small and large industrial sectors would be developed wherever possible, as complementary to each other so as to facilitate growth of ancillary industries.

The cooperative form of organisation will be encouraged wherever appropriate. In small scale industries and handicrafts with a tradition of an independent entrepreneurship and craftsmanship the "service" and "assembly" types of cooperative organisations and associations may be found more suitable than "producer" type which has taken roots in industries like the handloom, powerloom and coir. Recently, in pursuance of the recommendations of a working group appointed by the Reserve Bank to consider the question of ensuring adequate flow of funds for industrial financing through cooperative banks, for financing 22 broad groups of small industries including coir sericulture, handicrafts, tanning and flaying, leather goods, hand-pounding of paddy and cereals, oil crushing and general engineering. The Working Group has made other recommendations for stimulating the flow of funds to industrial cooperatives. These will be implemented and the position reviewed after two years.

The Fourth Plan envisages a total outlay of about Rs 295 crores in the public sector for the development of village and small industries.

TABLE 22.1

Outlay for Village and Small Industries in Public Sector

(Rs crores)

Industry	*Estimated expenditure*		*Outlay for 1969—74*			
	Third plan	*1966-69*	*Centre*	*Centrally sponsored*	*States and Union territories*	*Total*
Small scale industries	86.12	52.46	34.00	—	67.74	101.74
Industrial estates	22.15	735	—	—	18.15	18.15
Handloom industry	25.37	13.37	4.50	—	28.67	42.98
Powerlooms	1.52	0.46	—	—	9.81	—
Khadi and village industries	89.33	54.03	95.00	—	1.43	96.43
Sericulture	4.39	3.75	2.00	—	9.37	11.37
Coir industry	1.79	1.21	1.50	—	2.92	4.32
Handicrafts	5.30	4.80	8.00	—	6.52	14.52
Rural industries projects	4.79	6.70	—	4.50	—	0.60
Collection of statistics	—	—	—	0.60	—	0.60
Total	240.76	144.13	145.00	5.10	144.61	294.71

AGRO INDUSTRIES

Nobody quite knows who first propounded the thesis that India could stamp out poverty by setting up agro-industries in a big way. The term came into vogue no more than a decade ago. But since then it has become highly fashionable. Almost everything ranging from a giant nuclear power reactor to cellophane packing is now sought to be justified on the ground that it is necessary to promote agro industries. Most of the big business houses have already diversified into the field and others want to do so.

For the new entrepreneurs it is not merely a matter of keeping up with the established industrial Joneses. The opportunities are vast and highly attractive, considering the governmental incentives on offer. The authorities in their anxiety to industrialise the countryside have decided to extend the benefits to a large number of industries whose link with agriculture is remote if at all one exists. The 155-point list of "agro industries" issued by the Commissioner of Small-scale Industries, for instance, includes refrigeration plants for milk and factories for the canning of fish.

The Planning Commission goes even further. It says that any enterprise which processes farm produce supplies chemical inputs to agriculture (fertlizers, insecticides, etc.), manufactures improved agricultural equipment (hoes, bullock cart tyres, tractors, etc.), or helps raise farm output is an agro-industry. This means that the term not only covers the nation's biggest industry—cotton textiles—but also the whole range of plantations like tea, coffee and cashewnut and of course, jute.

Going only but statistics the result is somewhat ironical. They show that a much bigger segment of the Indian industry was agro-based at the time of independence than has been the case in recent years. In 1951 agro-industries accounted for 75 per cent of the total productive capital employed and 78 per cent of all value added by manufacturing in the organised sector. By 1963, however, these percentage had fallen respectively to 34 and 53. The steep decline in these figures does not mean that agro-industries are on their way out. But it does indicate that their growth has been slow in comparison with the phenomenal expansion of other industries like steel, oil refining of electronics.

The word "agro-industries" broadly means industries manufacturing "input" for agriculture, or processing agricultural "output." Agencies supporting agriculture by designing and manufacturing inputs are termed as "agro-industrials" and those supported by agricultural products are classed as "agro-based industries."

Generally most of the industries use agricultural output as basic raw material. But the concept of agro-based industries leave out of account "industrial product". It is confined to those industries that are engaged in processing the agricultural produce either for consumption or for the use of industry. They include wheat milling, rice milling, dal-making, oil crushing, cotton ginning, khandsari, fruit beverages, and so on. Thus, the agro-industries serve at both the ends of farming.

In the first instance agro-industrials provide the essential input and in the second phase agro-based industries function as an outlet for agricultural output. The purpose of helping these industries is to make farming a profitable and commercially viable proposition; since agro-industrials directly influence the cost structure of farming and agro-based industries, they have their impact on the receipts of the farmer. Agro-industries also serve as a means for providing better employment opportunities to the labour during off season.

Old System

The concept of agro-industrial is not new. Formerly the village smith, the carpenter, the cobbler used to prepare and repair the various traditional types of agricultural tools and implements. These industries were, and in some regions still are, the part and parcel of rural life. The agriculture implements are exchanged for a share in the harvest.

Likewise in its refined form modern agro-industrial manufacture, process and supply, agriculture inputs either on cash basis or realise the value in instalments. The distribution of fertilizers and seeds has been taken over by the cooperatives for quite a long time now. The various agro-industrial corporations in the states seek to supply agriculture machinery on cash, credit and hire-purchase system.

The idea of 'in-kind' assistance has been mooted from the simple fact that outright cash grant to the farmer for production is often misused and misdirected in consumption. Even those genuinely interested in buying implements and machinery, have faced numerous problems in purchasing them at reasonable prices. The farmer who is detached from urban life takes time in locating the supplier and picking up the operating technique. Because of "scarcity" of modern machinery, it is said that a "black market" has also grown in agricultural inputs.

The purpose of establishing the agro-industrial corporations, therefore, is not only to equip the farmer with modern implements but also to train him in the operation of machines—so essential at the early stages of mechanisation and rationalisation of agriculture. "In-kind" assistance given by the corporations, ensure proper use of funds, and protect the farmer from the difficulties, involved in purchasing right type of machinery and implements suitable for cultivation in his region. In India, the soil and climate conditions so sharply differ from one region to another that only regional corporations could take a better view and mould the tools according to regional requirements.

Specialised Agencies

Hitherto the emphasis was mainly on agro-based industries. The growth of these industries was no doubt important for providing a ready market to the agricultural outputs, to create employment opportunities in view of seasonal nature of agriculture and to stabilise the farmer's income. Several specialised agencies such as the State Finance Corporations, Small Scale Industries Finance Corporations Cooperative agencies, the Guarantee Organisation of the Reserve Bank, and the various technical guidance and assistance schemes of the Central Government. State Governments,

Reserve Bank and State Bank of India were directed mainly to sustain the growth of small scale and cottage industries.

It was, however, realised at a later stage that the institutional assistance at the 'output' end is only one part of the game. Indian agriculture does not face any serious problem from the demand side. It is with the improvement in the technique of production that the cost of farming can be reduced and the productivity of the land can be improved in order to meet ever growing demand of agricultural output from direct consumers as well as from industries. The institutional set up was, therefore, geared up to discover means for improving the technique of agricultural production and also to manufacture agriculture machinery.

At the highest level various agricultural research centres, headed by Indian Council of Agriculture Research, are performing important functions in the development and distribution of scientific knowledge of agriculture. There is also the Indian Society of Agriculture Engineers which provides wealth of information for the application of science and technology to farming. At the village level cooperative societies are propagating and, where possible, implementing, the programmes of the research centres.

Cooperative societies and various official organisations have performed important functions but because of their limitations, these agencies could not meet the requirements of mechanised agriculture which has recently gained importance for improving agricultural productivity.

Agro-Industrial Corporations

Thus with a view to meeting the rapidly growing demand of farmers for standard agricultural machinery equipment, *e.g.*, tractors, power tillers, power threshers, seed-cum-fertilizers, drills, power sprayers and dusters, diesel and electric pump-sets, encash or hire-purchase basis, the State Agro-Industrial Corporations have been established by the Government.

These corporations have been established in almost all the States. They are performing multifarious activities connected with not only the supply of agricultural machinery but also providing credit for the purpose. Some of them have ambitious programmes of designing and assembling tractors and manufacturing various tools and implements. But in fairness to the Corporations it must be recorded that none of them is more than three years old. What is more, some of them have the right ideas. In Gujarat, for instance, the AIC has decided to accord priority to the setting up of new cold storage plants in Banaskantha and Balsar, onion dehydration plants at Jamnagar and Bhavnagar and hydrogenated oil unit in the Saurashtra region. In West Bengal and AIC will supply diesel pumping set to poor farmers owning two to five acres of land on payment of Rs 70 cash down, the balance of the price—Rs

3,500—being recoverable in easy instalments. In Tamil Nadu the AIC is not only going into the processing and marketing of farm produce but will also set up factories for the servicing and repair of the 1,000 new tractors which are sold in the state every year.

It is evident that the growth of agro-industries cannot reverse the process of urbanisation. Most of the plants for processing farm products or "waste" materials released by the existing agro-based industries like textiles must be established near the consuming centres if they are to pave their way. All that can be done in the circumstances is to be establish new industrial units, both big and small, in district towns and away from the major cities. Any effort to confine the new agro-industries to the rural or semi-urban areas is bound to hinder their growth and will in the end be self-defeating.

Since there are quite a number of common problems which these institutions are likely to face, it might be worthwhile to establish a Central Agro-Industrial Corporation of India to coordinate the work of these state institutions. Besides coordinating the working of the state corporations, the central organisation can be entrusted with research and development of technology. Centralisation of research will eliminate waste and duplication. The State corporations will, however, continue to collect data from their regional offices to make them available to the central organisation. It will also be the task of the central agency to arrange the production of different components of machinery and implements in different areas according to the local facilities available on the 'comparative cost' basis. The proposed corporation can also serve as an effective agency in arranging the collection of technical knowledge from abroad and distribution of imported machinery.

Agro-service Centres

The massive expenditure that the Government has incurred on agricultural development has in one way or another gone to prop up a new rich class by passing millions of small farmers and landless labourers. There are growing indications of discontent among those who have found the cost of acquiring modern equipment and agricultural inputs soaring, on the one hand, and official aid somehow getting channelled away from them, on the other.

The Government's decision to establish 500 pilot agro-service centres in the country is a step in the right direction. The programme is being jointly sponsored by the Ministry of Agriculture and the Ministry of Industrial Development and is designed to provide a wide range of services from custom-hiring of tractors and other farm equipment and repair and maintenance of agricultural machinery to soil testing, crop spraying, water analysis and land and field services such as levelling and contour surveying. The scheme is intended to provide employment in the rural area as well as to accelerate the process of modernisation in the villages. It is particularly aimed at providing help to the small farmers and the weaker sections of the rural community in the relatively backward areas. The rapid increase in the use of tractors, pumps, mechanical threshers and other machines, not to speak of community radios, transistors and other personal gadgetry, has created a demand for easily available repair and maintenance facilities on a wide scale. Then agro-service centres can perform a valuable function provided they are suitably staffed and have the requisite managerial competence. The Government has negotiated World Bank assistance for tractor-cum-ground water development projects in Punjab, Haryana and Tamil Nadu, Agro-service centres could be established alongside these programmes for a start.

As far as mechanical equipment like tractors and pumps are concerned, it would be most desirable for the country to standardise on a few broad categories (not necessarily makes) so that problems of maintenance and the stocking of parts and components are not needlessly complicated by type difference. Little thought appears to.have been given to this matter so far. But it is obviously something that calls for an early policy consideration.

APPENDIX

Definitions of Cottage Industries

Cottage industries were first defined by the Industrial Commission as:

> "industries carried on in the home of workers, where the scale of operation is small, and there is but little organisation, so that they are, as a rule, capable of supplying only local needs."

The U.P. Industrial Organisation Committee, while discussing the scope of cottage industries did not agree with the above definition. They defined cottage industries as those "in which work is done, generally speaking in the homes of artisans, and occasionally in small factories run by small industrialists of the entrepreneur type, power-driven machinery being rarely used."

The question again came before the Committee constituted in Bombay in 1943 presided over by the well-known industrialist Sir Purshottamdas Thakurdas. The report defined cottage industries as follows:

> "Cottage industries are industries where no power is used and the manufacture is carried on, generally speaking, in the home of the artisan himself and occasionally in small *Karkhanas* where not more than 9 workers employed."

All these definitions have only raised controversies rather than solve them.

The definition by the Industrial Commission is rather restrictive. It assumes that the organisation in cottage industries is not important, and market, as a rule, is local. This is not true. The markets for the products of certain cottage industries extend far beyond the local boundaries. Another controversy is around the use of power. The definition of cottage industries has to take into consideration the stage of economic development and technological progress. Earlier, the concept of cottage industries was based on their being carried out in the place where the workers lived or in their cottage without any outside help. As a matter of fact, such a concept applied to all industries then. But with technological progress such a restriction on the meaning of cottage industries cannot be imposed any longer.

The use of power is another controversial point. Everybody agrees that the use of better tools is necessary to improve production and, if that is so, new tools or complicated machinery will have to be utilised by the cottage worker and in many cases no manual labour may be necessary. Electricity is a boon to humanity and when India reaches a stage in which electricity would be easily available at cheap rates in the villages, it would certainly be utilised by the cottage workers as well. In almost all Western countries power is used quite freely in all cottage industries.

The National Planning Committee agreed on a definition as follows:

> "A small-scale or cottage industry is an enterprise or series of operations carried on by a workman skilled in the craft on his own responsibility, the finished product of which he markets himself. He works in his own house with his own tools and material and provides his own labour or at most labour of such members of his family as are liable to assist. The workers work mostly by hand labour and personal skill, with little or no aid from modern power-driven machinery and in accordance with traditional technique, such supplementary energy as is provided by animal power may add to the economy and efficiency of the industry. He works finally for a market in the immediate neighbourhood that is to say in response to known demand with reference to quality as well as quantity."

As already mentioned, it is not necessary that cottage industry should possess all these attributes, as it is not necessary that labourers other than the members of the family should not help. But the necessity is that there should be no element of wage-slavery in the additional labour employed. Secondly, a large output is no barrier to a small-scale industry; so long as there is a potential demand for the produce the industry will continue to be a cottage industry. Thirdly, the machinery used should not render that industry a mass-producing one of standardised goods.

23

ANIMAL HUSBANDRY AND DAIRYING

In ancient India cattle tending was one of the items of vartha and it was entrusted to a section of the people who thoroughly understood the business. "When Prajapati created cattle, he made them over the Vaisya; and if a Vaisya is willing to keep them, it must not be kept by any other caste". (*Manu Samhita* IX, 326-327). Vaisyas were primarily agriculturists, formed a wealthy and respectable section of the community and produced fine breeds of cattle. There were experts and well-experienced cattle breeders in India who built fine and rich cattle herds—some of which existed in recent times in certain parts of India, such as Punjab, Bihar, Gujarat and Kathiawar.

Agriculture, cattle-rearing, trade and commerce constituted the four-fold vartha or pursuits suitable for making fortune. Cattle-rearing has been noted in the Epics as important and universal and occupation as farming (*Ramayana* II. 67.12; 100.45; *Mahabharata* II. 5.79; 13.2; XII.88.28), it has already been mentioned in the beginning how agriculture and cattle farming flourished simultaneously.

Contrary to the instruction of Manu that cattle rearing was the occupation of Vaisyas only, there are records in available literature that livestock and animal farming were not the business of a particular group or section of people. The kings themselves, the Ksatriyas, owned and reared the cattle and cattle-wealth was the mainstay of their household finances. The outstanding examples are the emperor of Kosala (Ramayana II. 100.50) or of the prince of Kasi (*Jataka,* 1.240).

Besides horses, elephants, cows, sheep, and goats they maintained buffaloes, camels, asses, mules swine and dogs for a variety of purpose (Arthasastra H. 29). In the Dhumakari Jataka (III. 410), the high-bred Brahmin is a goat keeper. "All the seventy families in Brahmana hamlet on the slopes of Grdhrakuta mountains near Rajagraha took to cattle breeding as the sole means of livelihood (Chinese Dharmapada by Beal)". The Setthis or merchants mentioned in Jataka (I. 388) were also keepers of cattle. The universality of cow-keeping and cattle trade is manifest in the common use of cows as a standard of value and medium of exchange in the transitional stage between barter and money transaction.

Animal husbandry was the occupation, either sole or part, of most of the people. Economic condition of the farmer classes of people was mainly dependent upon agriculture. Some of them were very rich and possessed extensive farms. This is best illustrated by the example in the Dhaniya, the son of a setthi in Vedaka (as mentioned in Suttanipata by Paramatthajotika). Dhaniya owned about 30,000 head of cattle, which 27,000 were milch cows, and luxuriant meadows for pasture. The Ganapati Mendaka enjoyed a bigger farm which had been managed by as many as 1,250 cow keepers (*Mahioagga* 34.19).

Every villager used to keep a few animals for draught purposes of to meet the supply to his own household. The village maintained on pay or on a share of produce, shepherds, who were entrusted with the work of taking the animals to the pasture ground in morning and bring them back in the evening (*Anguttaranikaya,* 1,205; *Rigveda* X 10 ef).

Cattle in India pose a difficult problem at a time when the economy is rapidly growing and agriculture is required to play a leading role. Mixed farming in India enabled products of land to be fully utilised for both human food and animal feed. This situation is now slowly changing. Finer rice and wheat for which demand is rapidly rising are poor yielders of fodder. On the other hand, milk consumption is below the nutritional level and with increasing income and population the total demand for milk tends to rise. The religious traditions discourage cattle slaughter. Weeding of cattle is therefore done by nature through malnutrition which is indiscriminate in its impact. This keeps down quality without adequately controlling the number. Without mechanisation and rural transportation, the farmer depends on bullocks and the use of cattle as a source of draught power and supply of organic manure still dominates. With expanding irrigation and double cropping, there is an increasing pressure of demand for cattle for which the present cattle population, despite its large number, is ill-equipped. A beginning has been made under the Five-Year Plans to improve systematically the quality of cattle with a long-term objective of increasing output of cattle products with proportionately less feeding costs.

Livestock play and important role in the national economy of the country which is predominantly agricultural in character. India Accounts for nearly 10 per cent of the world livestock population and 25 per cent of cattle and buffalo. Although the animal and human population are almost equal the contribution of India's livestock to the national income of the country is not very significant. "Animal husbandry" accounts for only 14 per cent of the national income from the agriculture sector and its contribution to total national income does not exceed 7 per cent, as shown in Table 23.1. The depressed position of the industry in the country is primarily due to low level of productivity in the sector.

TABLE 23.1

Gross & Net Value of Output in Agricultural Sector

(Rs. Million at 1960-61 prices)

	1960-61		1964-65		1970-71	
	Gross	Net	Gross	Net	Gross	Net
Agricultural	72,554		81,811		108,610	
Animal		65,740		71,982		90,598
Husbandry	12,090		13,225		17,173	
Total						
Agricultural Sector	86,828	67,827	97,260	74,105	129,185	95,820
Total Adjusted (Agricultural Sector)		68,846		75,124		93,839
National Income		141,400		167,100		241,300

Source: Draft Fourth Plan—Material & Financial Balances, September 1966, p.9.

Livestock products like hides and skins, fur skins, leather and leather products, dairy products and other crude animal materials are also an important source of foreign exchange in the country. The exports of these commodities during the past few years have steadily gone up from Rs. 40 crores during 1962-63 to Rs. 45.8 crores in 1965-66 and Rs. 88 crores in 1970-71. The major contributing item is "leather" which alone accounts for more than 50 per cent of the total foreign exchange earnings from livestock products. (Table 23.2.)

TABLE 23.2

Exports of Dairy and Livestock Products

(Rs. lakhs)

Items	*1962-63*	*1963-64*	*1964-65*	*1965-66*	*1970-71*
Live animals for eating	8.56	7.56	16.24	13.17	3.46
Meat and meat products	168.43	162.13	21.17	41.86	305.24
Butter	0.75	1.14	1.08	0.74	1.91
Eggs	5.18	0.01	neg.	neg.	0.46
Total DairyProducts, Eggs and poultry					
Hides and Skins-Undressed (Excluding fur)	1070.63	959.85	905.04	955.17	376.35

cont....

Items	*1962-63*	*1963-64*	*1964-65*	*1965-66*	*1970-71*
Fur Skin undressed	18.29	50.76	6.41	30.31	2.00
Crude Animals materials	524.74	436.85	638.97	692.76	913.42
Animal fats & oils	0.74	neg.	3.27	1.01	2.41
Leather	2245.23	2620.23	2716.23	2821.46	7158.77
Leather Manufacturer	17.37	17.60	18.92	23.05	58.15
Furs dressed	—	0.07	0.03	1.02	1.24
Total	4059.98	4256.37	4327.81	4581.34	8830.93

Source: *Monthly Statistics of Foreign Trade of India.*

Net foreign exchange earnings from the livestock sector are, however meagre as imports of livestock products (Table 23.3) have also gone up during the period under discussion from Rs. 12.8 crores to nearly Rs. 28 Crores.

TABLE 23.3

Imports of Dairy and Livestock Products

(Rs lakhs)

Items	*1962-63*	*1963-64*	*1964-65*	*1965-66*	*1970-71*
Live Animals for eating	1.53	0.61	5.60	10.11	11.11
Meat & meat products	0.84	0.94	0.69	0.77	2.78
Milk & cream-evaporated,	820.65	853.96	673.92	604.25	831.02
condensed or dried Butter	neg.	53.06	359.75	75.88	225.94
Total dairy products	880.35	938.85	1036.49	683.01	1061.39
Hides & skins undressed	276.25	335.71	312.34	148.20	154.44
Fur Skins — undressed	0.73	0.65	1.35	1.59	3.58
Crude animal materials	42.95	33.53	22.83	20.93	22.44
Animal fats& oils	69.32	53.49	68.23	573.24	1512.49
Leather	5.46	4.54	5.48	3.79	5.19
Leather Manufacturers	5.12	0.98	1.87	1.16	1.44
Furs dressed	neg.	-	-	0.68	
Total	1282.55	1369.30	1454.88	1443.46	2774.86

Source: *Monthly Statistics of Foreign Trade of India.*

Dairy products, animal fats and oils are significant in the import list. Such imports may be desirable in the context of improvement of Indian diet, at the same time they reflect the neglected development of the livestock industry. Along with the efforts to promote the export of livestock products, there would be need to reduce their imports not at the expense of consumption levels but through increased efficiency of the industry.

TABLE 23.4

Contribution of Cattle Products to Total Income from Animal Husbandry in 1958-59 at Current Prices

Items	*Quantity (thousand tonnes)*	*Value (in lakh rupees)*	*Percentage to total income from animal husbandry*
I. Milk and milk products			
(i) Cattle and buffalo milks	19,155	75,960	645
(ii) Others	561		
II. Cattle dung			
(i) Manure	226	11,361*	9.7
(ii) Fuel and other purposes	126	11,361*	9.7
III. Meat and its products			
(i) Cow and buffalo meat	167	1,040	0.9
(ii) Others			
(meat and meat products)	Lakh nos.		
IV. Cattle and buffalo hides	210	1,501	1.3
V. Goat and sheep skins	451	1381	1.2
VI. Wool and hair		1339	1.1
VII. Eggs and poultry		2,836	2.4
VIII. Bones		138	0.1
IX. Increment in stock		2,440	2.1
Total		117,468	100.00

* The value figures given here include in addition to the value of cattle dung, the value of manures etc., from intestines, oesophagus, animal blood, edible offals, heads and legs, animal fat, useless meat, tail stumps, glands etc. The total quantity of dung and all these items would be, 5,025,000 tonnes.

From among the various species of livestock,cattle enjoy a pride of place not only because they account for roughly two-thirds of the total livestock population but also because they constitute and indispensable adjunct to agriculture and contribute in various ways to the prosperity of the farmer in particular and the country in general: They are other main source of supply of draught power for nearly all the agricultural operations, viz., ploughing fields, carrying manures to the fields, lifting water for irrigation, threshing the produce, transporting the produce from the fields to the threshing floors and thereafter to the market as well as driving sugarcane crushers and oil ghanies. It is estimated that the working bullocks put in about 12,000 million working hours for various agricultural operations. According to one estimate, their contribution of draught power in monetary terms is worth about Rs. 400 crores. A more scientific evaluation of the draught power would raise this figure much higher. Even after mechanisation is introduced to the extent possible, cattle will continue to occupy an important position in so far as the requirements for farm operations are concerned.

Cattle are a predominant source of all livestock products except meat, skins, wool and hair and poultry products. Their contribution in the total value of livestock products is of the order of 88 per cent.

Cattle dung which is a very important source of soil nutrient, serves as a source of complementarity in the agricultural sector. Half of the cattle dung produced in the country is estimated to be burnt as fuel due to the scarcity of alternative source of fuel. With the increased use of chemical fertilizers the use of farm-yard manures will also have to be increased, to correct the imbalance caused in the soil due to the continued use of chemical fertilizers. Every effort has, therefore, got to be made to conserve and utilise dung for manurial purposes. According to rough estimates even under the present neglected state of affairs about two-thirds of the soil nutrients are being supplied by organic sources in which farm-yard manure figures prominently.

Indian diets are miserably unbalanced in so far as the supply of proteins and vital vitamins are concerned. The cattle help not only to raise crops but also provide protective foods in the form of milk and milk products which are the main source of fats of high biological value. Nutritional experts have repeatedly stressed the importance of animal proteins in the diet of the people. The role of livestock in improving the living standards of the masses is thus significant.

Livestock Population

The total livestock population in India consisting of cattle, buffaloes and other small animals like goats, sheep, etc. was 343.8 million according to the Livestock Census for the year 1966. The bulk of the livestock population consists of cattle and buffaloes, which together accounted for 67 per cent of the total livestock. Numerically the livestock population in India, judged on the basis of the number of cattle and buffaloes only, is one of the largest in the world. According to 1951 data, India had about one-fifth of the world's total cattle population and one-half of the total buffalo population. Though cattle in India were then three times more than buffaloes, India was probably the world's most important buffalo-rearing country. Indian cattle and buffalo population accounted for over 68.5 per cent of the total number in the Asian continent. This inevitably puts a heavy pressure on India's limited land resources. It may be mentioned here that countries like Denmark have a larger density of cattle and buffalo per square kilometre than India and countries like New Zealand, Argentina and Australia have a larger ratio of cattle population to human population.

Table 23.5 gives the increase in different categories of livestock since 1945.

Cattle, buffaloes, sheep and goats are the numerically important groups. The total livestock population increased by about 28.5 per cent during the 20 years. During the last five years the increase is more spectacular in the case of work animals, *e.g.*, mules (41.6%), cames (13.9%), goats (16.1%). Male buffaloes have increased by 6.6 per cent while the female buffaloes have increased by 4.5 per cent only. The cattle population has increased only marginally.

TABLE 23.5

Livestock Population in 1945-1966

(in thousands)

	1945	1951	1956	1961	1966
Cattle	135,960	155,099 (155,238)	158,651	175,557	176,057
Buffaloes	40,593	43,351 (43,401)	44,916	51,211	52,920
Sheep	37,729	38,829 (38,962)	39,246	40,223	42,014
Goats	46,286	47,077 (47,115)	55,405	60,864	64,566
Horses & Ponies	1,399	1,514 (1,514)	1,483	1327	1,149
Mules	45	60	40	53	75
Donkeys	1,132	1,239 (1,249)	1,055	1,096	1,054
Camels	654	629	776	903	1,028
Pigs	3,707	4,420 (4,424)	4,932	5,176	4,975
Others				22	30
Total livestock:	267,505	292,218 (293,595)	306,504	336,432	343,868

Note: Figures in brackets are comparable in coverage with those of subsequent Livestock Census.
Source: Tenth Livestock Census, 1966, January 1971, pp. 1-2.

TABLE 23.6
Age and Sex of Cattle and Buffaloes

(in thousands)

Cattle	*Males* (over three years)	*Females*	*Young stock*
1951	61,759	49,848	43,492
1961	72,528	54,204	48,825
1966	73,330	54,693	48,044
Buffaloes			
1951	6,778	21,842	14,731
1961	7,684	25,023	18,504
1966	8,192	26,144	18,584

Source: Ibid.

Male cattle are useful for work in the field or for transport while male buffaloes are used mainly for breeding, except in few states where they are used for work. Consequently we find that the number of male cattle is larger than that of cows, but in the case of buffaloes females outnumber males. Again, the ratio of young stock to cattle is relatively higher than that of young stock to total buffaloes.

Livestock in Rural and Urban Areas

Livestock population is proportionately much larger in rural areas, since the occupation of animal husbandry is largely combined with cultivation of land. However, owing to inadequate transport facilities for carrying fresh milk over long distances, a part of the livestock population is located in towns and cities. In the cities, horses are used for pulling carriages and bullocks for carts for local transport of goods. In 1956, of the total livestock of 306 millions in the country, 11.9 millions or about 4 per cent were in cities and towns, the rest were in rural areas. As will be seen from Table 23.7, the proportion was slightly less in the case of cattle compared to that for buffaloes. In both the cases, proportionately more female stock was found in urban areas. The proportion of sheep and goats in urban areas was as much as that of cattle. It was in the case of horses and ponies that a little larger proportion (8 per cent) was found in urban areas, as they carried the passenger traffic within cities.

The animals in urban areas are fed on fodder and other feed brought from rural areas. Since this involves cost, most of the dry animals are maintained in rural areas. On the other hand, proportionately a much larger number of female animals in milk is found in urban areas.

TABLE 23.7

Livestock Population in 1956

(in thousands) Rural Urban

	Rural	*Urban*
Cattles		
Male	63,021	1,846
Female	47,831	2,062
Young stock	42,122	1,681
Total cattles	153,060	5,590
Buffaloes		
Male	6,315	190
Female	21,069	1,266
Young stock	15,324	747
Total buffaloes	42,710	2,205
Sheep	38,123	1,123
Goats	53,047	2,358
Horses and ponies	1,360	123
Total livestocks	294,617	11,886
Total poultry	89,180	5,503

Distribution Among States

Uttar Pradesh with 14.5 per cent of cattle and 21.8 per cent of buffalo population is at the top of all the States so far as cattle population is concerned. With regard to cows and bullocks, Madhya Pradesh comes a close second with 14.3 per cent. Its position is fourth in respect of buffalo population. The second position with regard to buffalo goes to Andhra Pradesh. Maharashtra comes third with regard to the population of both cows and buffaloes. Rajasthan has 7.6 per cent of both cows and buffaloes. Punjab, while it possesses 8.5 per cent of buffaloes (males over three years being quite insignificant), has only 3.7 per cent of cows and bullocks. Between cows and bullocks the latter predominate.

Variations in the number of cattle from state to state are due to climate, grazing facilities, and the area under fodder crops which determine to a great extent the quality of cattle and the carrying capacity of land. Agricultural development, introduction of irrigation reclamation of land, etc., affect adversely livestock as the land available for grazing is reduced and the cultivated area under fodder and feed crops does not go up correspondingly, resulting in the reduction of total livestock feed. Both from the viewpoint of quality and number, the states for livestock importance are Rajasthan, Punjab, Uttar Pradesh, Madhya Pradesh and Andhra Pradesh.

As regards other animals, nearly 60 per cent of sheep are almost equally distributed in three States of Rajasthan, Andhra Pradesh and Tamil Nadu. A little more than 50 per cent of goats are found in Rajasthan, Maharashtra, Bihar and Uttar Pradesh, 65 per cent of horses and ponies are in Uttar Pradesh, Madhya Pradesh and Maharashtra. The mules are mainly in Uttar Pradesh and Punjab, camels in Rajasthan and Punjab and donkeys in Uttar Pradesh, Rajasthan and Maharashtra.

LIVESTOCK IMPROVEMENT

Livestock in India are in general of poor quality. Any improvement depends upon four factors—breeding, feeding, disease control and management.

The keynote of the genetic improvement policy for cattle is the extension of the programme of cross-breeding of indigenous animals with exotic germ plasm. To begin with, this programme was taken up in a few selected areas and exotic cattle of Jersey and Friesian breeds were imported from Australia and the USA for distribution among different states. Cross-breeding programmes were also taken up with foreign collaboration, *e.g.*, Indo-Danish Project at Karnatak for Red Dane cattle, Indo-Swiss Project in Kerala for Brown Swiss cattle and Indo-German Project in Himachal Pradesh. For the intensification of this programme, the Government of India is, presently, allocating free foreign exchange for the import of exotic cattle for distribution to State farms, agricultural universities and research institutes.

A big break through in milk production is expected to be achieved through a massive cross-breeding programme with exotic germ plasm simultaneously in different areas. A Frozen Semen Bank has been set up near Bangalore for the production and supply of frozen semen to different State Governmets. Establishment of four central projects for intensive cross-breeding of cattle with imported frozen semen is also being envisaged.

Feed and fodder development which is an important ingredient for cattle development, is, presently, being given close attention. A number of leguminous fodder crops like lucerne, berseem, some varieties of cow-pea and guar were being grown by our farmers even prior to the First Plan. Subsequently, giant hybrid Napier, Guinea grass, Rhodes grass and Para grass have also gained popularity in certain areas. Three regional forage stations have been set up to demonstrate improved practices of production, conservation and utilisation of forage. Two of them are receiving assistance under UNDP Grass Land Fodder Development Project in the form of equipment, training, fellowships and experts. Four more such stations are being set up. These regional stations would supply seed and planting material to livestock farms, State agencies and other centres for further multiplication and distribution. The National Seeds Corporation is also engaged in the production of improved fodder seeds in large quantities to ensure steady supply of quality seeds to farmers. The State Governments are running a few fodder seed production centres to meet a part of their seed requirements. In order to supplement feed and fodder resources in the country, attention is being given to the utilisation of agricultural and industrial by-products/wastes in cattle and poultry feeds.

Increasing attention has been given to provide adequate health cover for livestock development. Presently, a good network of veterinary institutions cover the entire country for the control of animal diseases. A massive programme for eradication of rinderpest disease was undertaken in 1954 and the incidence of the disease has come down from about 8,000 attacks per year.to 207 attacks during 1971-72. The disease is presently under complete control. Similarly, new Cattle disease which was a courage of poultry is no longer a serious problem in the country. Other animal disease like foot and mouth diseases are being taken care of.

There is a fairly strong machinery in the country for the diagnosis and investigation into animal diseases. All the vaccines required for protection of livestock are now being produced in India. The internationally reputed Indian Veterinary Research Institute at Izatnagar is undertaking research in several disciplines of animal sciences. It is providing training facilities both to students from within the country and abroad. This Institute is producing as many as 35 different biological products. Besides, the states have their disease investigation laboratories and biological product units.

Management

In livestock production, management occupies an important place as it embraces the fields of nutrition, feeding and the techniques acquired from the experience for handling livestock. This is reflected in proper housing and care of animals. The production can be considerably improved by improvement in the managemental practices. Recent researches have shown that buffaloes which do not breed during summer could be bred to obtain an even calving rate and flow of milk in case suitable modifications were made in the managemental practices. The calving interval of cattle which is normally 18-20 months in the village areas can be reduced by three months which will produce a net increase by about 15 per cent. In the case of sheep the breeding season can be telescoped to produce the lamb crop in suitable climates and environments.

Punjab has started an integrated programme of Agriculture, Animal Husbandry and Dairy Development. Schemes have been drawn up especially for training the farmers and the staff. The veterinary doctor is actively associated with all training camps for farmers arranged by the Department of Agriculture so that he could give practical advice to the trainees. Advice on cultivation of good fodder, importance of artificial insemination as well as after-care formed part of the syllabi of each training programme. Fodder development was also taken up on a campaign basis. The role of animal husbandry in encouraging rotational crops and importance of animal products in the context of rural economy was emphasised at meetings and conferences. Cinema shows were held on a campaign basis to explain the advantages of cross-breeding for improving the breed of the cattle and increasing milk yield.

Agencies for the development of small and marginal farmers and landless labour undertook to provide financial assistance (credit as well as subsidy) and technical know-how for those with marginal and small holdings to set up dairy, piggery and poultry units. The Dairy Development Corporation took up the onerous task of building up a marketing structure for the sale of milk and milk products through a network of milk plants and milk bars. Quality feed is provided by the Poultry Corporation.

The digestibility of different feed constituents in the rumen of the cow, the buffalo and the goat are being studied with the help of fistulated animals. The differences noted are being co-related with the types of micro-organisms present in the rumen of different species. This would enable greater use of agricultural and industrial waste materials, and non-protein substances like ureas to be used for routine feeding, increasing milk production and lowering the cost of milk.

A dairy animal takes 3-4 years to come into milk. Constant effort is being made to reduce the age at first calving, through better management practices, and to reduce the cost of rearing young stock by replacing milk used for feeding with equally nutritious "calf starters" prepared from cheaper materials like fishmeal, deoiled cakes etc.

Balanced feeding has been shown to be important not only for maximising production but also in maintaining the composition of milk to the same level throughout the year. Cow and buffalo milk through containing the same constituents, differ mainly in their properties. This leads to difficulties in utilising established manufacturing practices for preparing products like condensed milk, sterilised milk's cheese, etc.; with buffalo milk. Methods for studying in details reasons for observed differences were standardised and the components which give rise to these differences identified and isolated. This understanding has enabled the manufacture of as good products from buffalo milk as from cow milk.

A constant check is needed on quality. Rapid bacteriological and chemical tests have been devised to check hygienic quality as well as freedom from presence of extraneous substances, like water in milk, cheaper fats in ghee etc. With the help of the Hansa test, presence of as little as 3 per cent of buffalo milk can be detected in cow milk. With the same sensitivity, the freezing point of milk detects added water in milk and chromatograhic methods detect foreign fats in ghee.

In selecting the equipment for storage and processing, care has to be taken for their proper design to facilitate easy cleaning and eliminating areas where bacteria thrive. The NDRI has associated with the Indian Standards Institution in evolving of desirable practices.

Sheep Development

The wide range of environment in India extending from the Alpine climate of the Himalayas to the semi-arid areas of Deccan Plateau and desert areas of Rajasthan and Gujarat has given rise to a number of breeds and types of sheep suited to the agro-climate conditions of various areas. As mentioned earlier, the number of sheep and goats in India is fairly large—42 million in 1966. However, the productivity of most of the sheep stock is low in terms of wool and mutton. The bulk of wool produced is coarse and is used largely in the manufacture of carpets, druggets and coarse blankets.

Organised efforts to improve the quality and productivity of sheep have been carried out under the successive Plans. For improved breeding rams produced at the state sheep breeding farms are being distributed to the flock-owners through the sheep and wool extension centres established in sheep rearing areas of the country. These centres also provide guidance to sheep farmers in improved methods of sheep husbandry, protection against diseases, mass drenching of sheep and other facilities. Each such centre covers, on an average, 10,000 sheep. Sheep shearing and wool grading and marketing schemes are also being implemented.

Presently, seven large-scale sheep breeding farms with exotic breed of sheep are being set up in different parts of the country for supplying seed material for the cross breeding programme. An Indo-Australian sheep project has been established at Hissar with about 2,000 Corriedle sheep and the number of these sheep will go upto 7,000 in due course.

About five thousand sheep of superior exotic breeds like Ram-bouillet, Polwarth and Soviet Merino, have already been imported and another 7,000 are proposed to be imported during the Fourth Plan period.

Poultry Development

Poultry products are a valuable source of protein in the form of eggs and meat. Nutrition experts have laid a standard of one egg per day per head of population. The population of 500 million on this basis would require over 61 billion eggs, if it is assumed that 50 per cent of the population are potential consumers of this commodity. The total number of eggs available as against this is estimated at only 3630 millions. There is, thus a wide gap between production and requirements of eggs. This gap is likely to grow in a long run perspective ending 1975 and 1985 unless special production programmes are undertaken. As increase in egg production can be achieved both by extensive and intensive methods *i.e.,* increasing poultry population as well as yield per bird. The poultry population in the country increase from 73.5 million in 1951 to 114.3 million in 1961 *i.e.,* by about 50 per cent. According to the 1966 census poultry population increased marginally during the quinquinneum 1961-66, to 115.1 million.

Growth in Population

One reason for this change in the population growth pattern may be the severe draught conditions in 1965-66 to 1966-67 which led to an acute shortage of livestock feeds. A number of poultry units closed down during 1965 and 1966. The trend as indicated by 1966 census may, therefore, be considered as a passing phase.

In the past, table birds were largely from the called breeding stock and indigenous table birds upto the period of one year. Under the present practice where a larger proportion of the population is crossbreed, the male stock is sold for table before it attains the age of 6 months. It is thus likely that at the time of the enumeration, the male stock was not properly accounted for, indicating no visible increase in the poultry population. To this extent, it could be assumed that the 1966 poultry census was an under-estimate.

When improved breeds are introduced on a mass scale a large incubation capacity is necessary. In the First and the Second Plan period it was possible to create adequate incubation capacity as most of the poultry equipment had to be imported. During the Third Plan period, however, adequate capacity to manufacture larage-sized incubators has been created. It would now be possible to increase poultry population rapidly.

Yield Per Bird

Annual production of eggs per laying hen was estimated at 60 at the end of the second Five-Year Plan. This estimate was based on the performance of indigenous breeds under free range conditions. According to recent surveys conducted by IARS in a few states, egg production per bird is estimated at 89 in Andhra Pradesh, 96 in Kerala, 99 in West Bengal, 86 in Gujarat and 112 in Maharashtra. Table 23.8 gives the estimates annual production in these states.

On this basis we may assume an annual yield of 75 eggs per adult bird for the country as a whole. This reflects that the development of the poultry industry during the Third Plan period has resulted in major changes in the pattern of poultry production in the country. While in the past the entire production of eggs was based on free range, introduction of intensive rearing on deep litter system resulted in a change of pattern to the extent that a sizeable proportion of the urban supply of eggs was from commercial flocks raised under intensive system of poultry rearing. Under the Plan programme, a large number of birds of exotic breeds are being introduced to replace the indigenous stock. Intensive training and distribution programmes for rearing day old chicks have been taken up thereby making it possible to introduce a large number of exotic breeds in the rural areas through their distribution. The setting up of commercial units with foreign collaboration has further intensified the introduction of high-laying strains in the country. These programmes will need to be strengthened to effect a change over of the poultry industry from free range breeding of non-descript birds to the intensive rearing of improved stock.

TABLE 23.8

Egg Production Per Adult Female Four Per Year and Total Annual Production of Eggs

State	*Average egg production per bird per year*	*Total annual egg production in million*
Andhra Pradesh (1959-60)	84	188
Kerala (1959-60)	94	282
West Bengal (1962-63)	99	785
Gujarat (1963-64)	87	N.A.
Maharashtra (1964-65)	112	760

The development of poultry in India has received considerable attention during the last two decades. A modest beginning was made in the First Five Year Plan with the launching of a pilot project with 33 extension centres. During the Second Five Year Plan, a comprehensive All-India Poultry Development Project was launched under which five regional poultry farms were established in different agro-climatic regions. These farms aimed at production and propagation in the respective regions, of acclimatised birds and training of officers and farmers. Through the assistance of TCM, a wide range of equipment and a consignment of 30,000 chicks of White Leghron and Rhode Island Red breeds were imported to provide the necessary support to the project. Besides, 300 Poultry Extension-cum-Demonstration Centres were established in different parts of the country to introduce modern poultry farming in rural areas.

In the Third Year Plan, seventy-seven Intensive Projects closely linked with hatcheries, food mixing units, marketing centres and extension organisations were established and all the required inputs and services, including credit requirements were made available to the farmers. The Regional and State Poultry Farms were expanded. Two processing units, 68 feed mills and 18 marketing centres were also set up.

In the private sector also, four commercial hatcheries came up for the propagation of genetically superior stock of eggs as well as meat lines, developed in other countries. Many companies have come into existence to manufacture the entire range of equipment required for the poultry industry. Today, the country is not only self-sufficient in its requirement of equipment, but is also in a position to export the same to the developing countries.

In the Fourth Plan, emphasis has been laid on the production of better breeding stock, dissemination of latest scientific farming practices and strengthening of the existing marketing structure, besides taking up intensive production work in new areas. The Regional Poultry Farms are assisting 24 State Government Farms in implementing their breeding programmes. Attempts are also being made to exploit hybrid vigour through crossing of strains in various combinations. A Random Sample Laying Test Unit has been established to assess, from time to time, the genetic gains achieved through the breeding programme. An In-Service Training Institute in Poultry has been set up in Bangalore, where training courses in breeding, nutrition and management, including marketing will be conducted each year with the help of foreign experts under EPTA Programme of the FAO. Facilities for broiler production are being expanded. Two large processing plants at Poona and Chandigarh have started functioning. Some of the State Poultry Farms have also taken up a programme of raising one or two crops of broilers by utilising their existing facilities during the off-breeding season.

To ensure the farmers a reasonable price for their produce, it is proposed to encourage establishment of four egg collection and marketing centres at Calcutta, Delhi, Bombay and Madras in the cooperative sector with facilities for cold storage. An Egg Powder Plant has also been set up recently at Bombay.

Piggery Development

The pig is an important source for the supply of protein-rich meat. The indigenous pigs are un-economical for commercial exploitation because of their low growth rate, poor body weight, small litter size and late maturity. As against this, the exotic breeds like Yorkshire, Landraco and Hampshire are ideal for commercial exploitation. In order to develop pig production on systematic lines and organise it as a profitable industry, coordinated piggery development programme was initiated towards the end of the Second Plan. The salient features of this programme are: (a) setting up of pig breeding stations/farms for production of improved breeding stock for distribution especially to the weaker sections of the community who traditionally rear pigs; (b) organising pig production in compact rural areas by distribution of breeding stock at subsidised rates and making available other inputs; and (c) setting up of bacon factories/pork processing plants for providing ready and remunerative market to the producers.

Seven bacon factories have been established in different states to provide marketing facilities to the farmers for the quality exotic pigs produced by them. Around these bacon factories, piggery development has been stepped up through the supply of quality breeding stock and all health support and training of farmers in modern pig-raising.

INTENSIVE CATTLE DEVELOPMENT

The need for improving the quality of cattle and raising their productivity, particularly with regard to milk, has been fully recognised and this has been receiving increasing attention with the initiation of the First Plan. In percentage terms India has about 20 per cent and 54 per cent of the world's cattle and buffaloes populations respectively and together they form 23 per cent of the world's bovine population. In spite of such an impressive bovine population, India has been able to produce only 6.56 per cent of the world's milk production. This has been mainly due to low average yields of cows and buffaloes in the country. The average annual milk yield of cows and buffaloes in India has been estimated at only 382 lbs. and 1117 Ibs. respectively. With such low yield it is no wonder that we have a very unsatisfactory level of milk consumption in the country.

There are 25 well-defined breeds of cattle and eight breeds of buffaloes in the country, but vast majority of these cattle and buffaloes are non-descript and of poor productivity. It is estimated that 75 per cent of Indian bovine population falls under this category.

Key Village Scheme

Certain research projects aimed at cattle development were taken up from the thirties. Herd books were opened for eight improved breeds of cattle, namely, Haryana, Kankrej, Gir, Tharparkar, Red Sindhi, Sahiwal, Kangayam and Ongole and the Murrah breed of buffalo. Certain measures for encouraging selective breeding for improving the milk yielding capacity and relating or improving draught quality of cattle were laid down.

The first organised attempt to develop village cattle on an effective scale was initiated with the introduction of key village scheme during the First Plan. This scheme was introduced in the light distribution of a few premium bulls for upgrading village cattle and the production of a limited number of breeding stock in a few Government farms was not adequate for bringing about any substantial improvement in the quality of the stock. They key village scheme, therefore, embraced all aspects of cattle improvement, *viz.* controlled breeding, improved feeding, disease and sexual health control, better management and marketing and adoption of improved animal husbandry practices through proper extension methods. Since the First Plan, a number of key village blocks have been established in the different States and the total breedable female bovine population covered by these centres was spread over 529 blocks at the end of 1972. In concept of the key village scheme is indeed sound and this scheme has helped greatly in developing pockets of good quality cattle in different parts of the country. It has also given a sound approach for taking up programmes of cattle development in an integrated manner, but considering the vast magnitude of the problem this scheme has not been able to make the necessary impact for immediately making available the increased quantities of milk that are required to meet the demand in several places. It has been estimated that this scheme is at present handling only about three per cent of the breedable cow and she-buffalo population of the country. Moreover, these centres are generally dispersed all over the country, and in many places all the items of activities envisaged in the scheme have not been taken up. Each key village blocks is thus only a tiny area of well-organised activity surrounded all round by vast areas where indiscriminate breeding is continuing. As a result the effect of the good work done in the key villages gets obliterated by the uncontrolled breeding going on all round.

It is desirable that the key village programme should be continued and wherever necessary further extended and improved so that the general programme of cattle development is continued without break. It is at the same time very essential to close down those blocks which have been chronically lagging behind. In addition to the key village programme, it is essential to take up some large Intensive Cattle Development Projects on the lines of the package programme for agricultural production. These could be special projects for creating the necessary impact and confidence in cattle industry. These projects may be located both in the breeding tracts of the indigenous breeds of cattle and buffaloes which we would like to develop in the future and also in the milk shed areas of the large diary projects (even though they may be largely non-descript cattle) to enable the diary plants to procure milk to their installed capacities economically. In the former areas the policy of breeding should be selective. These areas should developed for the supply of quality breeding stock to other parts of the country to provide the necessary incentive to the producers. These areas will have to be linked up with the milk supply scheme but where this is not possible, dairy products in factories will have to be provided for the utilisation of milk. In the later areas, cross-breeding with suitable exotic breeds and/

or up grading with improved indigenous breeds may be practised. In determine the breeding policy for the various localities recommendations of the Working Group on Breeding Policy set up by the Ministry of Food and Agriculture and the Committee on Cross-breeding set up by the Central Council of Gosamvardhana will give helpful guidance.

Under each project envisaged to be taken up during the Fourth Plan, it is proposed to cover one lakh cows and she-buffaloes of breedable age so that it will have a significant impact on milk production in the area. The exact size of these projects will have to be determined after a proper survey of the areas where these centres are to be located. In this connection it should be borne in mind, as already experienced in the case of key village scheme, that good cattle cannot be raised everywhere and for development of cattle on right lines for improving their productivity expeditiously, certain natural conditions as well as incentives are necessary. It would, therefore, be desirable that these centres are located in places which will provide conditions conducive to good cattle raising. In the location of these centres this should be primary consideration. With a view to achieving concrete results within the shortest possible time, concentrated efforts will have to be made in all aspects of cattle development which have a breaking on increasing the productivity of the animals.

The programme popularly known as ICDP, was started in 1964-65 and there were 52 such projects in operation in 1971-72. The following programme of work is being taken up at each of the centres.

(i) Survey. A village to village survey of the area selected for the location of the project is undertaken and completed in a short time to ascertain the initial animal husbandry conditions, level of milk production, marketing and utilisation of milk, feeds and fodder resources etc. Initial data would be useful for periodical evaluation of the progress under the project.

The survey also introduces the staff to all the livestock owners in his area and gives an opportunity of acquainting them with the broad outline of the programme envisaged. When the survey has been completed a summary should be made which could be used all the year round for reference. The survey may be repeated preferably in the beginning of each year subsequently. Through this preliminary survey it will be ensured that the sites chosen for the location of projects offer suitable conditions for raising good cattle and are conducive for animal production.

(ii) *Breeding.* As controlled breeding alone would ensure progressive improvement in the genetic make-up of cattle for efficient production, adequate breeding facilities are provided for all the breedable bovine population both by artificial insemination and natural services. A cattle development officer is provided for the whole project who will be assisted by the officer incharge of the breeding operations in the four regions into which the project is divided for operational convenience. Each veterinary officer is incharge of the artificial insemination centre in his region and organises the breeding operation for the 25,000 breedable cows and buffaloes.

(iii) *Castration of scrub bulls.* For the successful operation of the project which envisages progressive breeding control it is very helpful if the Livestock Improvement. Act is enforced in the project area. Complete-removal of all scrub bulls from the area is necessary to ensure full breeding control. For this purpose a mass castration campaign is organised throughout the area initially by the project authorities as soon as the preliminary survey is completed. A mass castration team is then located in each Block to undertake castration of all scrub bulls in a planned and systematic manner.

(iv) Veterinary aid and disease control. During the preliminary survey information is gathered as to the contagious diseases which are prevalent in the area and necessary measures are immediately initiated for periodic prophylactic vaccination of livestock against these diseases. It would be advantageous if the Control of Contageous Disease Act is enforced in the project area.

Arrangements are made at each stockmen centre for the provision of veterinary aid. For this purpose the stockmen would be provided with required veterinary equipment and medicines to enable him to undertake this work.

(v) Registration and milk recording. With a view to ensuring proper breeding control and locating high yielding stock and utilising their progeny for breeding purposes for ensuring progressive improvement in the quality of stock registration and milk recording of milch bovines and their progeny is carried out. Every effort is made by the staff to get the subsidised calves registered in the Central Hard Books. The regular milk recording work in the respective stockmen units is carried out by the milk recorders.

(vi) Introducing high-yielding milch cattle. In order to encourage cattle owners to maintain high yielding milch animals they are advanced loans on usual terms for the purchase of good quality milch cattle of a sufficiently high productivity. The owners are required to maintain them properly and supply the milk to the local cooperative for the milk collection centres. The recovery of the loan is made in convenient instalments through deductions from the milk collection area of a large dairy project.

(vii) Subsidies and production incentives. In order to foster a spirit of healthy competition among the cattle owners and their cooperatives for breeding better-cattle and increasing milk production etc. production incentives and prizes for cattle owners are provided for maintaining highest milk yield in animals and also to cooperatives for organising and stepping up the collection of milk to make its

collection and handling economically. Subsidised rearing of selected progeny both young bull and calves on the lives of the calf subsidy under the key village scheme is also provided.

(viii) Feeds and fodder development. An adequate supply of feeds and fodder is essential to support production of increased quantities of milk in the intensive cattle-development areas. In fact, feeding constitutes the major portion of the expenditure on maintenance of cattle and, therefore, augmenting the available feeds and fodder resources locally to the extent possible is expected to be more economical than their procurement from other areas. Thus increased cultivation of high yielding and more nutritious feeds and fodder crops particularly crops like the hybrid maize and jowar, perennial crops like hybrid Napier grass para-grass and leguminous crops like berseem, lucerne, cow-pea, gram, guar, phillipeacra etc. could help in reducing the cost of maintaining cattle for milk production. This is achieved partly by increasing the area under fodder cultivation and partly by increasing per care yields through the cultivation of high yielding varieties of fodder crops and adoption of better cultural, manurial and irrigation practices. Emphasis is also to be laid on the cultivation of quick growing fodder crops as 'catch crops' thereby increasing fodder resources without effecting the existing crop patterns.

It would be equally necessary to popularise the conservation of green fodder during the flush season for the requirements of summer season when green fodder is generally in short supply and preventing the wastage of fodder by popularising the use of chaff-cutters in areas where these are not yet popular. In order to supplement the local resources, it is necessary to augment the supply of feeds and fodders by bulk purchase at economic rates from surplus areas and arrange for their supply to cattle owners through their cooperatives on no-profit no-loss basis.

(a) *Demonstrating cultivation of fodder crops.* In a programme of popularising cultivation of fodder crops and adoption of improved cultural, manurial and irrigation practices it would be necessary to demonstrate these on the cultivator's own field. Extensive demonstrations are, therefore, organised on small plots on the cultivators fields with regard to the cultivation of different fodder crops. In each intensive cattle development project area 400 demonstrations per year are taken up. In organising these demonstrations, seeds, fertilizers and technical assistance is provided free to the cultivator.

(b) *Distribution of seeds and planting material.* It is necessary to arrange for the timely supply of seeds and planting material of recommended varieties of fodder crops at subsidised rates for an initial period till the cultivation of these crops becomes a regular practice with the farmers. For this purpose the project authorities make necessary arrangements for the requirements of seeds and planting material from the fodder seeds farms and distribute the same at subsidised rates.

(c) *Irrigation facilities.* Adequate irrigation facilities are one of the essential requirements for extending cultivation of fodder crops, increasing the yields of green fodder and cultivating yearly summer fodder crops, necessary for maintaining milk production during the lean period. In areas where irrigation facilities are not readily available it would be necessary to assist the farmers liberally by way of loans and subsidies for wells and pumping-sets as provided for under the minor irrigation schemes.

(d) *Silage making.* Adequate supply of nutritious green fodder is essential for maintaining optimum milk production. To provide succulent fodder to cattle during summer months when green fodder is generally scanty, it is necessary to popularise conservation of green fodder during the flush seasons as silage. Silage making is, therefore, popularised by subsiding the construction of medium size silopits.

(e) *Popularising the use of chaff-cutter.* In a number of areas chaffing of fodder, particularly the thick stemmed fodder crops like maize and jowar, is not yet quite popular and this results in substantial wastage of fodder due to trampling the soiling. The use of chaff-cutters of improved types of choppers is being popularised in areas where this is commonly not used, by supplying chaff-cutters to the cattle owners at subsidised rates.

(f) *Dairy Extension.* With a view to making the farmer interested in better breeding, feeding and management of his cattle, it is necessary to provide him with a remunerative market for milk. For this purpose, these projects have been linked up with the milk supply schemes wherever possible and with the milk product factories in other cases. It is also necessary to support this programme by initiating dairy extension activities. Milk producers are organised into a co-operative society and they are given financial assistance for obtaining necessary dairy equipment. Under this programme loan are provided for the purchase of cattle feeds and fodder and milch stock. A suitable number of milk chilling stations is provided to facilitate the transport of milk from interior areas to the main dairy plant. At the project level, a dairy extension officer is incharge of this work.

DAIRYING AND MILK SUPPLY

Development of modern organised dairying and milk supply has been given attention since the inception of the First Five Year Plan in 1951-52. Dairy projects are being set up so as to provide an assured and remunerative market to the producers of milk in the rural areas and to make available depending quality to milk and milk products at reasonable prices to the consumers in the urban areas. Starting with a modest effort in the First Plan, the programmes have been substantially expanded in the subsequent Plans. Presently, there are 123 dairy plants in the country comprising 69 liquid milk plants, 10 milk product factories and 44 pilot milk scheme/rural dairy centres. Cooperatives are playing a significant role in dairy development in the country. Of the 123 plants, 34 are in the cooperative sector. In many States, cooperative constitute a major source of supply of milk to the dairy plants. The pioneering work done by the Kaira District Cooperative Milk Producers' Union, Anand, is being emulated widely. In addition, 26 dairy projects are under different stages of implementation. The average daily through put of milk from all the plants during 1971-72 was 2.4 million litres as compared to 0.7 million litres in 1960-61.

In order to make the country self-sufficient in respect of milk products as such as, better, cheese, milk powder, condensed milk, infant milk food and malted milk food, the private sector was also encouraged to set up factories for milk products in potential milk producing areas. The country is now almost self-sufficient in respect of all these milk products except milk powder which is imported to overcome the seasonal shortage in milk production and to maintain the level of milk distribution to the public in the major cities and towns.

Increase of milk production will be achieved by provision of such technical inputs as ready mixed concentrates and green fodder, artificial insemination, veterinary services and medicines. The milk cooperative movement is a new phenomenon in India which has in it the seeds of a social revolution. Not only will it inculcate the spirit of cooperative endeavour in the village community, but the cooperatives could become effective instruments for dissemination of information on modern farm techniques, dairy development, family planning, literacy drive and above all, healthy environment.

A national milk grid is proposed to be set up in India on the lines of water and power grid. This will be based in areas like Haryana where more milk is produced and will help sending milk to scarcity areas instead of sending milch cattle there.

Three methods of distribution of milk were now under consideration. They are the existing system of bottled milk, distribution by bulk vending units and distribution through tetrapack single service disposal packages. Distribution by bulk vending units was being organised particularly for the sake of the poor. Under this system milk will be available to the consumers seven paise cheaper per litre. An assured supply of milk through the development of the dairy industry and animal husbandry on a planned scientific basis is not only an urgent necessity to enable millions of the coming generation to develop a healthy body but is also imperative for stablising the income of rural folk. This is what Operation Flood stands for.

If a break-through in agriculture, ushering in the green revolution, was achieved by the use of high-yielding varieties of seeds evolved by Indian and international scientists, a white revolution of milk production, marketing and distribution is sought to be achieved smoothly by buttressing local milk supply with considerable quantities of imported milk powder and butteroil supplied by the World Food Programme, simultaneously with a determined and organised effort at improving the yields of the cattle and buffaloes.

In the field of dairy equipment manufacturing, the country has become almost self-sufficient. All equipment, except certain items of higher capacity and a few specialised and sophisticated items, required for the establishment of dairy plants, are now being manufactured in the country. This has enabled the country to save considerable foreign exchange.

Objectives and Outlays

As some of the existing dairy projects are operating at a loss, one of the principal tasks in the Fourth Plan is to review the work of the projects and to take corrective measures. This will include changes in the milk pricing policy introduction of modern management practices. The desirability of changing the public sector projects from departmental to corporate management will have to be pursued. It will also be necessary to establish a direct link between the small producers and the public sector milk plants through cooperative organisations. Dairy projects will need to be encouraged to take up extension work under their own auspices. At present, a number of dairy projects balance their operations by using imported milk powder. A phased programme is intended to be drawn up to increase production in the milk shed areas and gradually eliminate dependence on imported milk powder.

The organised sector of dairy industry will be extended to smaller towns with emphasis on milk production in the rural areas. Measures will be taken to ensure that dairy projects are economically viable and, as far as possible, organised in the cooperative sector.

The financing of dairy development will be based on three principal sources, namely Plan outlays, institutional finance and counterpart funds generated by the sale of commodity gifts under the World Food Programme. As far as institutional sources are concerned, the Agricultural Refinance Corporation has already entered the field and financed one dairy project. Further suitable schemes will have to be formulated for financing by ARC. As regard the commodity gifts from the World Food Programme, details are given later. Briefly, it is expected that funds of the order of nearly Rs. 95.40 crores will be generated and will form part of the Plan outlays. On this basis, the total outlay under the Plan will be of the order of Rs 138.97 crores. The break-up will be as given in Table 23.9.

TABLE 23.9

Outlay on Dairying

(Rs crores)

States	39.77
Union territories	1.95
Central schemes	97.25*
Total	138.97

*Includes an outlay of Rs 95.00 crores provided for Indian Dairy Corporation.

In the Fourth Plan, the first priority will be to complete the dairy schemes numbering 33 which split over from the earlier period. Organised dairy industry will be extended by taking up 24 new schemes in towns with a population of about 50,000. Four milk product factories are proposed to be established. In addition, 64 rural dairy centres will be organised in areas with a population of less than 50,000 with a view of providing chilling and marketing facilities is isolated pockets of milk production.

"Operation Flood" and White Revolution

With the cooperation of the World Food Programme (WFP) the Department of Agriculture formulated a project for stimulating milk marketing and dairy development in India. Under this project, known as 'Operation Flood' and launched in 1970-71 the WFP has agreed to supply, free of cost, during the five year period from 1970-71 to 1974-75, 126,000 tonnes of skimmed milk powder and 42,000 tonnes of butter oil, worth Rs 41.90 crores at international price. After recombination of the skimmed milk powder and butter oil into liquid milk at the public sector dairies at Bombay, Calcutta, Delhi and Madras, the milk is being sold and the sale proceeds from the quantity estimated at Rs 95.40 crores will be used for increasing milk processing facilities of the public sector dairies from 1.00 million litres at present to 2.75 million litres per day at the end of five-year project period. The generated funds will also be used for increasing milk production and procurement in the Union Territory of Delhi and the ten neighbouring states. This will be achieved by the provision of technical inputs which will include production of ready mixed concentrates and green fodder, artificial insemination, veterinary services and medicines, calf rearing assistance, development of improved milch animals and organisation of rural procurement of milk. The project will also provide for the resettlement of the city kept cattle and buffaloes in the adjacent rural areas.

For the implementation of this project, Government has set up an Indian Dairy Corporation with headquarters at Baroda. Up to the end of March 1972, the World Food Programme has supplied 16,500 tonnes of skimmed milk powder and 3,700 tonnes of butter oil. Funds to the extent of Rs 8.58 crores have been generated and advances aggregating to Rs 4.06 crores have been released to various States for expansion of milk plants, establishment of cattle feed mixing plants and for new dairies. In order to meet the requirements of the first phase of the expansion programme of the public sector milk plants, estimated to cost Rs. 1.91 crores, the Indian Dairy Corporation has placed indents amounting to Rs. 1.80 crores, on UNICEF for the import of dairy processing equipment and components. The balance of the amount is being made available from free foreign resources. The requirements of plant and machinery for the second phase estimated to cost Rs. 25.26 crores have been discussed with the indigenous manufacturers and the Director General of Technical Development.

Operation Flood, perhaps the most spectacular and certainly the largest development project the WFP has undertaken in any part of the world, has other enduring objectives in view. For the first time, farmers have understood their inherent strength through milk cooperatives. They depend no more on the middle man to market their milk and milk products.

In specific areas where in project is in operation, they sell their surplus milk twice daily through milk cooperatives in their respective villages and get paid for it on the same day or the next day. With an assured daily income, they are able to feed themselves and their cattle in most cases much better than they did in the past. Recent studies have shown that farmers have doubled their income in the areas where milk cooperatives are functioning effectively. Milk production will be increased through this and other programmes from 22.5 is 50 million tonnes to ensure that per capita per day availability is increased from the present 108 grammes to 280 grammes necessary for a balanced diet.

With the pressure on land, all people living in the villages cannot depend entirely on crops for sustenance. Some could easily be diverted to dairy development. It is estimated that two million people will find jobs as a result of Operation Flood, while ten million people will have additional nutrition. Since dairy development is a labour-intensive industry, it has provided a ready answer for tackling the problem of growing rural unemployment. Moreover, the drift of population to urban areas from rural areas, which is causing considerable concern to the authorities, has chances of being arrested.

In view of our ability to grow more food in much less area than in the past through high-yielding varieties of seeds, we are in a position to spare land for cattle and fodder development on an organised and scientific basis. Large scale cross-breeding of cows is expected to increase cow milk production although implementation of the programme already initiated will take some three or four years. The programme of rearing high-yielding cattle involves growing of fodder and producing other animal feed in areas close to milk sheds. About 18,75,000 milch animals will be involved in the project.

The Punjab Government has decided to set up a milk plant in every district. The output is to be doubled in the next ten years. In addition to the existing units a dairy complex is being established at Ludhiana with UNICEF assistance at a cost of Rs 2 crores with an installed capacity of 75,000 litres a day. Another milk complex is planned for Bhatinda with an investment of Rs 1.5 crores. To absorb and canalise milk produced in the rural areas, composite milk plants are proposed to be set up at Jullundur, Gurdaspur and Patiala. The Amritsar factory is to be expanded to handle 2 lakh litres of milk per day. The total cost of all these schemes is estimated to be of the order of Rs 44 crores.

Haryana's plans are no less ambitious. Its Government has provided for the commissioning of a chain of milk plants within the next three years. The State's milch animal population is nearly 75 per cent that of a premier dairy country like New Zealand. India's first wholly indigenous sweetened condensed milk plant became operative a Bhiwani in October last. A composite milk plant was commissioned at Jind in December 1970. The third is nearing completion at Ambala.

Besides Punjab and Haryana, several other States have taken steps to develop the dairy industry. The Rajasthan Government has a Rs. 6.5 crore dairy development plan. It envisages the setting up of milk banks at Jodhpur and Bikaner, each handling 100,000 litres of milk every day. Bharatpur and Alwar will have milk feeder centres, each handing 60,000 litres per day.

West Bengal's Rs 2.4 crores scheme for milk production has already been put into operation and envisages a gross output of 13.10 lakh litres per day by the end of the Fifth Plan. Kerala proposes to turn the State into a large milk-shed by the end of this decade, the aim being to provide each citizens with at least 10 ounces of milk daily. The Jawaharlal Agricultural University at Jabalpur is engaged in experiments to initiate a milk revolution in Madhya Pradesh. Its experts claim that artificial mating of even the normal village cow with an exotic breed can substantially raise milk yields of the progeny, especially if backed up with nutritious fodder.

The National Dairy Research Institute

The best for a break-through in the dairy industry is the region comprising Punjab, Haryana, Western U.P. and Rajasthan. Before partition. India's finest cattle breeding tracts were in Sind, Montgomery and Lyallpur which now from part of Pakistan. Fortunately for us, sub-continent's premier and first dairy research station, then known as the Imperial Institute of Animal Husbandry and Dairying, had been located at Bangalore in 1923. After independence it was shifted to Karnal and renamed as the National Dairy Research Institute (NDRI). Bangalore became the

Southern Regional Station of the Institute. The Western and Eastern Regional Stations are located at Aarey (Bombay) and Kalyani (near Calcutta).

The existence of the NDRI at Karnal has given a big boost to dairy development in the Punjab-Haryana region. The two States are varying with each other in raising their output of milk and its products. At present Punjab-produces 50 lakh litres a day while Haryana's daily output is 35 lakh litres. The Karnal institute has been helping dairy farmers, particularly in Punjab and Haryana, in a big way. During the 50 years of its existence, it has concentrated on research leading to the development of high yielding dairy animals and fodder cropping programme to provide nutrients necessary for milk production in the most efficient manner and at minimum cost.

The NDRI's most outstanding product is a new strain of dairy cattle, Karan Swiss. It incorporates exotic inheritance from Brown Swiss and Zebu inheritance from the famous Sahiwal and Red Sindhi. A nucleus herd of 140 Karan Swiss cows along with 187 female followers and bulls has been developed for providing germ plasm. The Karan Swiss female calves weigh about 27 kgt a birth. They are 25 to 30 per cent heavier than their contemporary pure bred Zebu. The average age at first calving for Karan Swiss is 30 months as against 40 months for Zebu. On an average Karan Swiss produces 60 per cent more milk than the Sahiwal.

Milk Chemistry

One of the successful fodder cropping programmes evolved at the institute's experimental farm consists of a mixture of jowar, bajra, sudan, teosnite and cowpea in the kharif season and a mixture of bersem, mustard and oats in the rabi season, yielding a gross of 200 tonnes of green fodder a year per hectare sufficient to support 10 cows. Another programme comprising four crops a year has a yield of 150 tonnes per hectare.

Special attention has been paid to reducing the cost of milk production. While in the Sahiwal herd averaging 2,000 litres of milk per lactation it is Rs. 1.10 per litre, for the Karan Swiss herd averaging 3,000 litres of milk per lactation it is Re 0.72. The work of the institute has clearly established that for dairying to become an industry, the cost of production of milk has to be lowered to give adequate margin to the producer.

The NDRI has worked on buffalo milk chemistry and technology to provide the base for developing processes for manufacturing milk products from buffalo milk as satisfactorily as from cow milk. Sweetened condensed milk and evaporated milk of satisfactory quality cannot be produced from buffalo milk in view of its different chemical composition and behaviour when conventional methods of manufacturing are used.

Dairy Technology

A problem of considerable importance in handling milk is its rapid transport from the place of production to the dairy. The work at the institute suggested the feasibility of concentrating milk for long distance transport and this innovation has been adopted by the Delhi Milk Scheme in getting its supplies from Gujarat.

With a view to finding new outlets for surplus milk, researches have demonstrated the feasibility of keeping Dahi over long intervals at room temperatures, preparation of cheese with various combinations of condiments and fruits, "complete" coffee, readymade tea in powder form, sterilised cream, preserving butter milk and preparing refreshing drinks from milk and sour milk.

The research programme at the NDRI is being further strengthened by providing a larger herd of buffaloes so that all aspects of the nutritional physiology and breeding behaviour of the buffalo are studied. So far, cross-breeding between the different breeds of buffaloes has not been attempted on scientific lines.

A new bio-manufacturing sector will be set up to prepare on a semi-commercial scale some of the materials which have shown promising results in the laboratories. One such material is animal and microbial 'rennet'. Animal rennet is prepared in the Western countries by slaughtering young calves. The NDRI has shown that rennet can be prepared from living calves by putting a fistula in the young animal which could be removed at a later date without any ill-effects on the subsequent growth of the animal.

At present, there is a shortage if rennet even in European countries and the finding has good commercial potentials. Process for the manufacture of microbial rennet has also been standardised and rennet from this source could also be prepared on a commercial scale.

Recently, a machine has been devised for the continuous manufacture of Khoa from milk which dispenses with considerable amount of manual labour required now. The packaging of milk products is a subject which has hardly been in India. A dynamic marketing programme is closely related to packaging. Emphasis will be given to this aspect which will also include gamma-rays for sterilising the material.

Future Approach

Complete success in the field of dairying is possible only when milk, which is an essential item of food, is made available in adequate quantities to the different sections of the population including children, expectant and nursing mothers and invalids. To reach this target, which is an increase of about 100 per cent over the present production, a comprehensive integrated programme is required to improve the productive efficiency in cattle and buffaloes through improved breeding, feeding and management practices. The recent launching of six all-India coordinated research projects in the breeding of cattle and buffaloes, and on the technology of milk and milk products by the Indian Council of Agricultural Research is the beginning of a new era in dairy research.

TABLE 23.10

Fodder Seed Production Farms

State/U.T.	*Fourth plan target*	*Established 1969-72*	*Target 1972-73*	*Remarks*
Andhra Pradesh	2	—	—	
Assam	1	1	—	
Bihar	1	—	1	
Gujarat	2	3	1	
Haryana	1	1	—	
Jammu & Kashmir	1	2	—	
Kerala	1	—	1	
Madhya Pradesh	2	3*	—	*to be expanded
Maharashtra	2	—	—	
Manipur	1	—	1	
Meghalaya	1	1	—	
Mysore (Karnatak)	2	2	—	
Orissa	1	1*	—	
Punjab	1		—	*expansion
Tamil Nadu	1	—	1	
Tripura	1	—	1	
Uttar Pradesh	—	1	—	
West Bengal	1	2	—	
Goa	1	—	1	
All India	23	18	7	

Milk trade is at present generally disorganised and only about three per cent of the milk total production is in the hands of the organised sector of the industry. The effectiveness of the process of diverting the milk from unorganised sector will depend not only on increased milk production, but also on the effectiveness of the link between dairy plants and the milk producers in their milk shed areas. This link can be established more efficiently by setting up small dairies in towns with a population ranging from 25,000 upwards. The problem of starting small dairies is closely linked up with the procurement of surplus milk from villages which are cut off from the cities due to inadequate transportation facilities and milk collection agencies and it is expected that this will be solved through a project known as Operation Flood Programme launched by the Department of Agriculture through the funds made available from World Food Programme.

Apart from the development of dairy industry, a great deal of fundamental work in dairy research is needed to improve the quality of milk products like dahi, khoa, cheese, chhenna, ghee, butter, ice cream, milk powder, etc. It is of upmost importance that there should be coordination between research institutions and the results of research should percolate into actual operations through the channels of extension and management. The speedy implementation of the dairy projects in the coming decade will, it is hoped, substantially contribute to the development and progress of dairy industry and will help India occupy a leading position in the Dairy Map of the world.

TABLE 23.11

Intensive Egg and Poultry Production-cum-Marketing Centres

State/U.T.	*Fourth plan target*	*Established 1969-72*	*Target 1972-73*
Andhra Pradesh	6	1	—
Assam	4	8	1
Bihar	1	1	2
Gujarat	3	2	—
Jammu & Kashmir	2	2	—
Kerala	1	—	—
Maharashtra	3	—	—
Manipur	—	—	1
Mysore (Karnatak)	2	2	—
Punjab	1	1	—
Rajasthan	2	1	1
Tripura	1	—	1
West Bengal	6	5	—
Goa	1	1	—
All India	33	24	6

TABLE 23.12

Establishment of Cattle Breeding Farms

State	Fourth plan target	Established 1969-72	Target 1972-73
Assam	2	1	1
Jammu & Kashmir	1	1	—
Kerala	—	—	1
Madhya Pradesh	—	—	1
Maharashtra	1	1	—
Mysore (Karnatak)	2	—	—
N.E.F.A.	1	—	—
Orissa	5	1	3
Punjab	1	3	—
Rajasthan	2	1	—
Tamil Nadu	1	—	—
All India	16	8	6

TABLE 23.13

Key Village Blocks

State/U.T.	Fourth plan target	Established 1969-72	Target 1972-73
Andhra Pradesh	32	9	—
Assam	9	11	1
Himachal Pradesh	1	1	—
Kerala	4	2	1
Madhya Pradesh	3	3	6
Maharashtra	—	2	—
Manipur	1	1	—
Meghalaya	1	1	—
Mysore (Karnatak)	1	1	6
Orissa	7	4	3
Punjab	4	2	—
Rajasthan	6	5	1
Tamil Nadu	10	4	2
Tripura	3	5	1
West Bengal	3	3	2
Delhi	3	1	3
Goa	2	4	1
Pondicherry	1	—	1
All India	91	59	28

TABLE 23.14

Progress of Intensive Cattle Development Projects—Selected Terms

State	No. of ICD projects			No. of stockmen centres			Breedable population		
	68-69	69-70	70-72	68-69	69-70	70-71	68-69	69-70	70-72 (lakhs)
A.P.	2	2	2	273	294	295	2.72	2.96	2.89
Assam	1	2	2	79	124	125	0.58	0.70	1.25
Bihar	2	2	2	164	210	164	133	2.33	2.36
Gujarat	3	4	6	268	283	354	2.45	2.78	7.41
Haryana	1	2	2	NA.	25	45	—	0.72	1.01
J. & K.	1	2	2	—	93	225	0.56	1.07	1.81
Kerala	1	1	1	100	104	100	1.06	1.40	1.40
M.P.	2	3	5	200	241	256	1.90	2.90	1.95
Maharashtra	4	4	6	245	210	327	3.20	3.94	3.47
Meghalaya	—	—	1	—	—	6	—	—	0.56
Mysore (Karnatak)	1	2	2	150	150	150	5.13	1.98	3.20
Orissa	1	1	1	100	100	100	1.19	1.20	1.33
Punjab	1	2	5	68	178	98	0.79	2.79	2.98
Rajasthan	—	—	—	—	—	—	—	—	—
Tamil Nadu	3	3	3	100*	200	—**	4.31	4.62	—**
Uttar Pradesh	2	4	5	68	182	231	2.70	4.89	6.43
West Bengal	2	2	2	124	40	177	1.35	2.07	2.24
Central Sector									
Haryana	2	2	2	109	112	159	2.32	2.40	1.67
Rajasthan	1	1	1	50	50	50	1.21	1.12	0.74
U.P.	1	1	1	73	36	128	3.51	3.51	3.51
Total	31	40	51	2171	2632	2990	37.31	42.38	46.21

* Information incomplete. "Merged with State Animal Husbandry Department.

TABLE 23.15

Number of Animals Castrated and Inoculated

(in lakhs)

	Animal breed			*No. of castration*			*No. of inoculations*		
State	*68-69*	*69-70*	*70-72*	*68-69*	*69-70*	*70-71*	*68-69*	*69-70*	*70-71*
Andhra Pradesh	0.58	0.65	0.67	0.145	0.123	0.174	12.91	15.82	13.27
Assam	0.08	0.13	0.17	0.049	0.080	0.142	0.74	1.10	0.42
Bihar	0.57	0.83	0.70	0.089	0.165	0.224	1.97	1.72	1.96
Gujarat	1.84	2.08	1.40	0.055	0.055	0.058	2.52	3.66	4.29
Haryana	—	0.01	0.04	—	0.009	0.031	—	0.08	1.07
J. & K.	0.01	0.12	0.17	0.023	0.072	0.081	0.16	2.82	1.19
Kerala	N.A.	0.85	0.79	N.A.	0.005	0.005	N.A.	0.06	0.06
Madhya Pradesh	0.19	0.26	0.28	0.112	0.163	0.333	4.71	5.71	7.78
Maharashtra	0.45	0.51	0.41	0.147	0.152	0.212	4.67	6.09	8.57
Meghalaya	—	—	neg.	—	—	0.002	—	—	0.02
Mysore (Karnatak)	0.33	0.41	0.52	0.100	0.159	0.106	—	4.23	3.69
Orissa	0.24	0.29	1.14	0.370	0.338	0.319	2.29	3.54	3.37
Punjab	—	0.70	0.26	0.059	0.760	0.098	—	2.20	2.51
Rajasthan	—	—	—	—	—	—	—	—	—
Tamil Nadu	0.75	1.51	—	0.080	0.211	—	0.68	3.93	—
U.P.	0.54	1.15	1.03	0.220	0.581	0.802	6.68	8.8	10.93
West Bengal	0.22	0.62	0.56	0.015	0.059	0.055	—	—	—
Central Sector									
Haryana	0.24	0.21	0.21	0.040	0.119	0.129	1.22	1.79	3.42
Rajasthan	0.01	0.04	0.02	0.017	0.014	0.055	—	0.23	0.45
U.P.	1.02	0.75	0.74	0.220	0.279	0.338	4.96	5.29	732
Total	7.07	11.12	9.11	1.741	3.344	3.164	43.51	67.08	70.32

* Merged with State Animal Husbandry Department.

TABLE 23.16

Foddeer Demonstrations Given and Fodder Seeds Distribution

State	*No. of fodder demonstrations*			*Quantity of fodder seeds distributed (quintals)*		
	68-69	*69-70*	*70-71*	*68-69*	*69-70*	*70-71*
Andhra Pradesh	4736	7275	30989	34.75	100.80	329.46
Assam	195	205	318	0.50	1.00	3.72
Bihar	26961	12362	10865	489.26	408.10	374.62
Gujarat	304	338	639	1281.86	2652.36	2623.82
Haryana	—	—	283	—	—	22.00
Jammu & Kashmir	155	371	107	12.10	10.27	1.24*
Kerala	—	—	148	—	—	—
Madhya Pradesh	1489	2086	1755	186.11	94.08	79.06
Maharashtra	437	530	555	66.20	50.85	105.36
Meghalaya	—	—	2	—	—	0.50
Mysore (Karnatak)	189	169	50	33.06	9.81	26.10
Orissa	185	356	292	63.79	81.04	37.70
Punjab	—	—	2*	—	16.00	N.A.
Rajasthan	—	—	—	—	—	—
Tamil Nadu	—	611	**	61.60	12.99	—**
Uttar Pradesh	813	2194	3024	1101.00	1380.48	1062.92
West Bengal	—	250	175	56.00	371.90	304.00
Central Sector						
Haryana	41	199	395	6.22	25.79	29.29
Rajasthan	—	—	—	8.80	—	—
U.P.	800	2400	800	2259.00	1099.65	917.00
Total	36305	29346	50399	5660.25	6315.20	5916.76

* Information incomplete.

** Merged with Animal Husbandry Department.

THE PROBLEM OF STRAY CATTLE IN INDIA

The magnitude of the problem of stray cattle in India is anybody's guess. According to one estimate their number in the country is 461 thousand only. *(The Dying Cow* by Bajrang Bahadur Singh Bhadri, pp. 2-3, published by Deptt, of Animal Husbandry, Govt. of Himachal Pradesh.) This estimate, however, seems to grossly err on the conservative side. On the basis of an average of five stray cattle per inhabited village in the country the number of stray cattle found in rural India alone would come to (5,64,718 inhabited villages x 5) about 28,24,000. Assuming that all male and female cattle over three years and not in use for breeding or work, all female dry cattle and the cattle not calved even once, and all female youngstock, in urban India are likely to be stray cattle, their total would come to 44,19,000 heads (Table 23.17). That means the total number of stray cattle for the entire country is likely to be in the neighbourhood of 72,43,000 heads.

The incidence of stray cattle is primarily an urban phenomenon. It is seen from Table 23.17 that the percentage of stray cattle (cow + buffaloes) to the total number of cattle (cows+buffaloes) in urban India comes to 36.2, the assumption, *inter alia,* being that none of the cattle in milk is stray. On the other hand the percentage of stray cattle to the total cattle in rural India is as low as 1.3, here the assumption being that there are no an average only five stray cattle per inhabited village. Taking the entire country (urban + rural) the percentage of stray cattle to the total number of cattle is about 2.4 assuming that (i) all male cattle over three years and not in use for breeding or work, (ii) all female cattle over three years excluding those in milk and those used for work, and (iii) all female youngstock, so astray or are likely to do so. The percentage of stray cattle to the total cattle would go down a bit if we exclude from the former all male and female cattle not used for work or breeding.

The incidence of stray cattle varies widely not only between urban and rural India, but also between different states of the Indian Union. According to Bajrang Bahadur Singh, the incidence of stray cattle is the highest in Uttar Pradesh, closely followed by Madhya Pradesh and Punjab, and the lowest in Maharashtra, Jammu & Kashmir and Delhi. But, as stated at the very out-set of this chapter, all estimates of stray cattle, whether in terms of their magnitude or incidence, are anybody's guess. There exists no authentic study on the subject. What is however, fairly obvious is that the problem of stray cattle is primarily an urban problem.

In rural India the incidence of stray cattle is, however, likely to be large during the years when, and in areas where, rains fail, droughts set in, crops are poor or fail, giving rise to a serious scarcity of *bhusa* and other roughages, and also bringing about in its wake an upward spiral of their prices. As a result being unable to feed their cattle, the more hard-up among the farmers let them stray. This also happens

TABLE 23.17

Total Cattle and Total Stray Cattle in India

(in thousands)

Region	Categories of cattle	Total no. of all cattle (buffaloes + cows)	Likely stray cattle											
			Male cattle over 3 years and not in use for breeding or work	Female stock over 3 years (excluding those used for work)						Young-stock female only	Grand Total		(12 (as) per centage of (3)	(13) as per centage of (2)
				Kept for breeding or milk production				Not in use for work or breed ing purposes	Total		(4)+(10) +(11)	(12)-(5)		
				In milk on 16th April	Dry	Not calved even once	Total							
(1)	(2)	(3)	(4)	(5)	(6)	(7)	(8)	(9)	(10)	(11)	(12)	(13)	(14)	(15)
Urban India	1. Buffaloes	2,065	12	777	324	72	1,173	18	1,191	430	1,633	856	79.1	415
	2. Cows	5,144	57	1,027	646	142	1,815	38	1,853	876	2,786	1,759	54.2	34.2
	Total	7,209	69	1,804	970	214	2,988	56	3,044	1,306	4,419	2,615	613	36.2
Rural India	1. Buffaloes	49,143												
	2. Cows	170,413												
	Total	219556										2,824*	2,824*	1.3 1.3
All-India	1. Buffaloes	51,208												
	2. Cows	175557												
	Total	226,765									72,43	5,439	3.2	14

source: Directorate of Economic and Statistics, Ministry of Food & Agriculture.

On the assumption of there being five head of stray cattle per inhabited village.

when there are widespread floods or a wave of some epidemic hitting cattle as well as those tending them.

Raison d'etre of stray cattle. In rural India a person does not normally buy food for his cattle daily. If he is a farmer, he puts aside straw, *bhusa*, etc., for his cattle's requirement of a year after he has harvested his crops. He, in fact, seldom incurs any out-of-pocket expenditure on roughages for his cattle. He certainly has not to do it every day as does the owner of a cattle in a town or a city. Besides, in rural India there are common/ *panchayat* grazing grounds where all the villagers send their cattle for the day on a nominal monthly charges. This arrangement is conspicuous by its absence in urban India. Towns and cities do not have common grazing ground over by municipality or Corporation where its cattle can graze and be left unattended every day either free of cost or at nominal charge to their owners. Besides, as stated above, their owners with urban avocations, have to buy *bhusa, kutti,* etc. every day for these cattle. This out-of-pocket expenditure irks. It irks all the more when a cattle yields no milk. There is, thus, a strong temptation for their owners to let them fend for themselves, if possible. This temptation becomes irresistable when their cattle is dry. It is this economic squeeze, abetted by the absence of common grazing grounds for cattle, that largely explains the *raison d'etre* for the presence of stray cattle in urban India.

Economic aspects of stray cattle. The cost of upkeep of stray cattle can be looked at from the point of view of the individual, *i.e.* the owner, as also from the point of view of the society, i.e. the country. From an individual's point of view, stray cattle fend for themselves. Their owners let them loose at day-break and tie them to their pegs at the end of the day. They feed them with practically nothing so long as these cattle are dry or not calved. Their cost of upkeep to them is, thus, practically nil or negligible. From the point of view of society, however, it is not so. Stray cattle, as other cattle, fill their belly. They eat mostly wild grass, rubbish and such feeds as are given to them out of charity or which they can poach upon. The monetised cost of this varied feed is a cost on the GNP. Stray cattle, thus, present the paradox of their cost of upkeep being nil or negligible to their owners but real, as of other cattle, to the country, i.e. on the GNP.

The per capita cost of upkeep of stray cattle on the GNP is, however, not of the same magnitude as of the non-stray cattle or cattle regularly maintained by a farmer or owner. It is so because stray cattle often eat up what would have in all probability otherwise gone waste or unutilised. Also stray cattle is seldom fed with concentrates. It is thus observed that the per capita cost of "upkeep" of stray cattle is in all probability likely to be much lower than the per capita cost of upkeep of cattle regularly kept and maintained. In short the per capita cost of upkeep of stray cattle even to the country is not likely to be of any significant magnitude as compared to the per capita cost of upkeep of regular cattle.

While the per capita cost of upkeep of stray cattle to the Nation is much less as compared to the per capita cost of upkeep of regularly maintained cattle, the former give the Nation almost all those benefits which the latter do. Thus, dung, meat, bones, hides, hoofs and horns of stray cattle are of as much use as are of the regularly maintained cattle. In some cases they also give as much milk as the other do. In short stray cattle are almost as useful to the Nation as are the regularly maintained cattle, and besides at a lower per capita cost, the presumption being that they cattle are normally healthy, productive, etc.

The view is held that if stray cattle does not get adequate feed for itself through foraging it is likely with every lactation its milk yield would go down and its capacity to breed and beget efficient draught bullock would suffer. ("Dairying for development of the cow," L.C. Sikka, *Building From Below—Essay on India's Cattle Economy*, I.C.A.R. Publication, pp. 20.) Also, it is often presumed that stray cattle might not be picking up adequate for its needs. Both these views, however, seem to be based apparently on commonsense only for there is not enough statistical evidence in support of these. If anything the available evidence points to the contrary. Even Sikka has observed that green grass is generally picked up by stray cattle in its most succulent state, when it has high nutrition value." *(Ibid., p. 25.)* With an abundant feeding of livestock, researchers have noticed, more of females of the species are born and *vice versa*. ("Boy or Girl?" Igar Goldman, *The Hindustan Times*, Sunday, April 28, 1968, p. 8.) If what Goldman points out is correct, and if the presumption that stray cattle must not be picking up adequate for its needs, is valid, the chances are that stray cattle would give birth to more of males of the species and the well-tended kept cattle to more of the females of the species. The whole subject of stray cattle and the issues relating to it thus remain to be explored and examined. Much work needs to be done on the subject.

24

TRACTORISATION AND ITS EFFECT ON EMPLOYMENT

Introduction

The seed-fertilizer based technology has been shown by many to have a direct favourable effect on farm employment. High opportunity cost of labour and high cost of maintaining bullocks has led to rather rapid tractorisation particularly in irrigated areas of the country. The employment effect of tractorisation, however, continues to be a subject of considerable controversy.

The higher use of human labour per unit of cropped area on mechanised farms as compared to non-mechanised farms was observed by Grewal and Kahlon[1] in Punjab, Acharya[2] in Rajasthan, Rao[3] in Ferozepur district of Punjab, and Chauhan[4] *et al* in Kanpur district of Uttar Pradesh. The differences in employment may be attributed to the differences in cropping intensity, irrigation intensity, cropping pattern, use of inputs and partly to mechanisation. Raj Krishna[5] attempted to separate out the effect of tractor technology, cropping pattern and cropping intensity on human labour employment for wheat and rice combined on Punjab Farms. He observed the decrease in employment by 95 man-hours per hectares due to use of tractor technology which was partly compensated by 64 man-hours increase in per hectare employment due to change in cropping pattern and cropping intensity. On the basis of decomposition model developed by Raj Krishna, an attempt has been made here to separate out the effect of tractor technology, cropping pattern and cropping intensity on per hectare employment of human labour.

Sample and Data

The data for the present study pertain to Ahmednagar district of Maharashtra. Three-stage stratified random sampling was used to select tehsils, villages and the sample bullock-operated and tractor-operated holdings. All the tehsils of the district were ranked on the basis of number of tractors and power threshers. Three tehsils, viz., Shrirampur, Rahuri and Kopargaon having the highest ranking performance with respect to these criteria were selected. In the selected tehsils, villages were classified into three groups on the basis of tractor and thresher population. Three villages from each group were selected randomly. All the farms of these selected villages were listed and classified as bullock-farms and tractor farms. A sample of 90 tractor-operated holdings was selected randomly. The number of tractor-operated holdings selected from each village was kept proportional to the size of tractor population in that village. To provide a proper basis for comparison of sample results, an equal size of sample of bullock-operated holdings was also selected randomly from the same selected villages following the same procedure. All the selected holdings from two categories were divided into three farm size groups, viz., below 6 hectares, 6 to 10 hectares and above 10 hectares so that the number of farms in each size group be neither too small nor too large. The selected holdings were further classified as: (1) Bullock-operated farms (BOF); (2) Bullock-operated and tractor hiring farms (BOF+ THF); (3) Tractor-operated farms (TUF); and (4) Tractor and bullock-operated farms (TOF+BOF).

The distribution of holdings, according to motive power and farm size, is given in Table 24.1.

TABLE 24.1

Distribution of Sample Holdings According to Motive Power and Farm Size

Holding size in hectares (range)	*Holding size group*	*Number of Sample holdings*				
		BOF	*BOF+THF*	*TOF*	*TOF+BOF*	*Total*
Less than 6	Small	20	18	8	7	53
6 to 10	Medium	15	20	16	18	69
Greater than10	Large	9	8	16	25	58
Total		44	46	40	50	180

The sampled farmers were personally interviewed and the Input-Output data were collected with the help of a specially designed schedule for the agricultural year 1979-80.

Empirical Model

The problem of breaking down the total difference in the labour input per hectare in crop production into the separate contribution of each factor can be handled with the help of simple algebraic decomposition equation,

Let C = Cropping intensity,

r_i = the proportion of the area of ith crop in gross cropped area,

u_i = labour per hectare (gross) in ith crop.

U = labour per hectare (net) in total crop production.

Then the gross area in crop i is Cri for every hectare of net area in all crops. Since u_i is the labour input per gross hectare in crop 1, the total employment per net hectare will be,

$$U = C\Sigma_i u_i r_i \quad \text{.............(1)}$$

and the change in U will be

$$\Delta U = \Delta C\Sigma (r_i \Delta u_i + u_i \Delta r_i + \Delta r_i \Delta u_i)$$
$$+ \Delta, C\Sigma_i (r_i u_i + r_i \Delta u_i + u_i \Delta r_i + \Delta r_i \Delta u_i) \quad \text{.............(2)}$$

where, AU is the total change in per hectare labour employment,

ΔC is the difference in cropping intensity,

Δr_i is the difference in proportion of area under ith crop,

ΔU_i is the difference in per hectare labour employment of ith crop.

We can again regroup the terms in equation (2) into

$C\Sigma_i r_i \Delta u_i$ = technology effect,

$C\Sigma_i u_i \Delta r_i$ = Crop-mix effect,

$\Delta C\Sigma_i r_i u_i$ = Cropping intensity effect on labour employment and interactions.

For decomposing the change in human labour employment between four categories of farms with three size groups each, the decomposition equation used is:

$$U = C (u_1 r_1 + u_2 r_2 + \text{........................} + u_{16} r_{16}) \quad \text{....... (3)}$$

where

$u_1, u_2, \ldots, u_{16}$ are the per hectare labour employment for different crops and

$r_1 r_2 \ldots, r_{16}$ are the proportion of area in total cropped area of each crop.

The crops considered were wheat, *jowar, bajra,* cotton (high yielding and local varieties separately), groundnut, sugarcane *(adsali:* 18 months duration, preseasonal 15 months duration and ratoon: 12 months duration), moong, gram, vegetable and lucern).

and $\Delta U =$

$$C (r_1 \Delta u_1 + r_2 \Delta u_2 + \ldots + r_{16} \Delta u_{16}) \text{— tractor technology effect}$$
$$+C (u_1 \Delta r_1 + u_2 \Delta r_2 + \ldots + u_{16} \Delta r_{16}) \text{— cropping pattern effect}$$
$$+\Delta C (u_1 r_1 + u_2 r_2 + \ldots + u_{16} r_{16}) \text{—— cropping intensity effect}$$
$$\left.\begin{array}{l} +C (\Delta u_1 \Delta r_1 + \Delta u_2 \Delta r_2 + \ldots + \Delta u_{16} \Delta r_{16}) \\ +\Delta C (u_1 \Delta r_1 + u_2 \Delta r_2 + \ldots + u_{16} \Delta r_{16}) \\ +\Delta C (r_1 \Delta u_1 + r_2 \Delta u_2 + \ldots + r_{16} \Delta u_{16}) \\ +C (\Delta u_1 \Delta r_1 + \Delta u_2 \Delta r_2 + \ldots + \Delta u_{16} \Delta r_{16}) \\ +\Delta C (\Delta r_1 \Delta u_1 + \Delta r_2 \Delta u_2 + \ldots + \Delta r_{16} \Delta u_{16}) \end{array}\right\} \quad \text{............ (4)}$$

Result and Discussion

Farm Holding and Cropping Pattern

The size of holding, intensity of cropping and cropping pattern have direct bearing on the human labour employment. Data on average farm size, cropped area and intensity of cropping for different categories and for three farm size groups are presented in Table 24.2.

TABLE 24.2

Operated Area, Cropped Area and Cropping Intensity

Farm Size	*Net cultivated area per farm (ha)*	*Cropped area per farm (ha)*	*Cropping intensity (Percentage)*	*Number of observations*
Small Farms				
Bullock Ope. farms	4.62	6.00	129.87	20
Bullock Ope. and tractor hiring farms	4.57	5.99	131.07	18
Tractor Ope. farms	5.05	7.59	150.40	8
Tractor plus bullock Ope. farms	5.49	8.14	148.44	7
Medium Farms				
Bullock Ope. farms	7.51	9.97	127.28	15
Bullock Ope. and tractor hiring farms	7.24	9.27	128.09	20
Tractor Ope. farms	8.52	11.34	131.90	16
Tractor plus bullock Ope. farms	8.79	11.41	129.89	9
Large Farms				
Bullock Ope. farms	11.89	14.15	119.09	9
Bullock Ope. and tractor hiring farms	11.65	13.48	115.67	8
Tractor Ope. farms	13.36	16.16	12154	16
Tractor plus bullock Ope. farms	13.94	16.89	121.11	25
All Farms				
Bullock Ope. farms	7.09	8.90	125.53	44
Bullock Ope. and tractor hiring farms	6.97	8.68	124.63	46
Tractor Ope. farms	9.74	12.48	128.15	40
Tractor plus bullock Ope. farms	10.91	13.69	125.55	50

The net cultivated area of pure tractor farms as well as tractor plus bullock-operated farms was more than bullock farms and bullock farms hiring tractors for all sizes of farms. The intensity of cropping varied among different categories of farms and also among different size groups of farms. On small size farms, the cropping intensity on tractor farms and tractor plus bullock farms hiring tractors. On medium and large size farms also, the cropping intensity was higher on tractor owning farms as compared to bullock farms and tractor hiring farms, but the differences are not significant. This is because of the higher proportion of area under sugarcane crop (Table 24.3) on tractor owning farms than on bullock farms and bullock farms hiring tractors which decreased the cropping intensity on tractor owning farms.

TABLE 24.3

Cropping Pattern of Sample Farms

Farm size	*Percentage to the gross cultivated area*							*Total gross*
	Cereals	*Sugarcane*	*Cotton*	*Groundnut*	*Pulses*	*Vegetables*	*Grass- and other crops*	*(ha)*
Bullock Operated Farms								
Small farms	60.11	17.11	8.26	3.26	3.00	3.09	5.18	121.00
Medium farms	61.07	17.56	7.63	2.78	3.47	2.22	5.27	144.10
Large farms	55.95	18.52	10.93	5.38	4.57	0.65	4.00	127.40
All farms	58.94	17.96	8.85	3.75	3.67	1.99	4.84	392.50
Bullock Operated and tractor hiring farms								
Small farms	48.29	24.73	10.29	6.32	1.08	4.51	4.78	106.80
Medium farms	53.68	18.29	9.36	4.95	3.01	5.16	5.54	184.70
Large farms	59.96	22.08	7.42	3.15	3.90	0.93	5.57	107.80
All farms	53.98	20.39	9.00	4.75	2.72	3.86	5.30	399.30
Tractor operated farms								
Small farms	38.77	26.50	12.84	5.80	1.99	11.10	3.02	60.70
Medium farms	50.83	25.64	11.90	5.01	1.84	1.67	3.11	179.80
Large farms	48.96	24.22	13.90	2.80	1.84	5.68	2.60	258.63
All farms	47.83	25.52	13.02	4.03	1.87	4.85	2.88	499.13
Tractor and bullock operated farms								
Small farms	47.37	27.72	11.58	7.37	1.40	1.05	3.51	57.00
Medium farms	46.85	24.88	10.84	4.47	1.79	4.25	7.23	205.44
Large farms	53.12	34.31	12.60	3.44	1.29	1.44	3.80	422.14
All farms	50.05	25.12	11.97	4.09	1.45	2.25	5.07	684.58

The cropping pattern on all farms is dominated by cereals. However, the proportion of area under cereals is higher on bullock farms than that of tractor owning farms. The reverse is the pattern in case of sugarcane, cotton and groundnut which are comparatively more labour-intensive crops. The proportion of area under sugarcane shows a increasing trend with the increase in farm size on bullock farms. On tractor owning farms the proportion of area under sugarcane decreases with the increase in farm size. The area under vegetables and pulses together occupied for about 4 to 7 per cent and fodder about 3 to 5 per cent of the total cropped area on different categories of farms.

TABLE 24.4

Employment of Human Labour (Mandays per hectare of net cropped area)

Farm size	*Bullock operated farms*	*Bullock operated & tractor hiring farms*	*Tractor operated farms*	*Tractor plus bullock farms*
Small farms	226.32	275.33	356.01	330.06
Medium farms	215.05	241.65	275.45	269.15
Large farms	187.04	207.11	241.60	240.37
All farms	207.05	240.49	265.94	256.95

Table 24.4 indicates that the employment of human labour per hectare of net cropped area on tractor owning farms is markedly higher than that on bullock farms for all farms. Also, the employment differences between tractor owning farms and bullock farms are highest on small farms followed by medium farms and lowest on large farms. From the analysis of cropping pattern and cropping intensity on different categories of farms, it is also observed that these parameters differ considerably amongst three farms size groups. Considering these points, the decomposition analysis of labour employment among different categories is attempted for each farm size group separately and also for all farms. The results obtained using the decomposition equation 4 are presented in Table 24.5.

A. Tractor Operated Farms Vs. Bullock Operated Farms

On an average, the employment of human labour per hectare of net cropped area is higher by about 59 mandays on tractor farms. The replacement of human labour due to use of tractor technology for cultivation is about 11 mandays per hectare of net cropped area. However, the differences in cropping intensity and cropping pattern had increased the employment by 7 mandays and 60 mandays per hectare respectively on tractor farms. The actual difference in per hectare employment is highest on small farms than medium and large farms. Similarly, human labour employment differences due to differences in cropping pattern and cropping intensity are higher on small farms as compared to other two size groups.

TABLE 24.5

Decomposition of Differences in Per Hectare Employment of Human Labour

Effect/Farm size	*Small farms*	*Medium farms*	*Large farms*	*All farms*
A. *Tractor operated Farms Versus Bullock Operated Farms*				
Actual change in employment(mandays)	129.69 (57.30)	60.40 (28.09)	54.56 (29.17)	58.89 (28.44)
Sources of Change				
1. Tractor technology	-19.89	-10.12	-7.52	-10.62
2. Cropping pattern	100.65	63.35	49.94	
3. Cropping intensity	47.63	6.95	10.05	7.44
4. Interactions	2.38	-0.95	1.30	3.50
Total : due to all sources (estimated)	130.77	59.23	53.77	60.04
B. *Tractor Plus Bullock Operated Farms Versus Bullock Operated Farms*				
Actual change in employment (mandays)	130.74	54.09	53.33	49.90
Sources of Change				
1. Tractor technology	-11.67	-8.80	-3.96	-6.66
2. Copping pattern	75.26	59.58	47.01	52.90
3. Cropping intensity	37.51	3.82	9.20	0.98
4. Interactions	1.35	1.59	2.65	1.93
Total : due to all sources (estimated)	102.39	56.19	54.90	49.15
C. *Tractor Operated Farms Versus Bullock Operated and Tractor Hiring Farms*				
Actual change in employment (mandays)	80.68	33.80	34.49	25.45
Sources of change'				
1. Tractor technology	-6.37	-8.13	-3.37	-8.01
2. Cropping pattern	48.23	40.65	27.03	30.43
3. Cropping intensity	42.89	5.57	11.05	5.27
4. Interactions	-2.97	-1.30	2.10	0.01
Total : due to all sources (estimated)	81.78	36.79	36.81	27.70
D. *Tractor Plus Bullock Operated Farms Versus Bullock Operated and Tractor Hiring Farms*				
Actual change in employment (mandays)	54.73 (19.86)	27.50 (11.38)	33.26 (16.06)	16.46 (6.84)
Sources of change				
1. Tractor technology	-5.91	-8.75	-3.60	-5.66
2. Cropping pattern	28.95	36.02	17.90	25.04
3. Cropping intensity	30.31	2.32	10.53	2.10
4. Interactions	-0.50	1.50	4.40	-2.13
Total : due to all sources (estimated)	52.85	31.09	29.23	19.35

Figures in parenthesis are the percentages increase in per hectare employment on tractor farms over bullock-operated farms.

B. Tractor Plus Bullock Operated Farms Vs. Bullock Operated Farms

The average per hactare employment on tractor plus bullock farms was 50 mandays more than bullock farms. This difference is less than the pure tractor farms. The negative effect of tractorisation on human labour employment was observed 7 mandays per hectare of net cropped area, while the cropping intensity and cropping pattern differences between two categories increased the employment by 1 manday and 53 mandays respectively on tractor plus bullock operated farms. The per hectare employment difference was higher on small farms than that of medium and large farms. This is again lower in comparison with pure tractor farms. Use of tractor technology has reduced employment by 12 mandays, 8 mandays and 4 mandays on small, medium and large farms respectively. The cropping pattern effect in increasing employment was observed as 75 mandays on small farms, 60 mandays on medium farms and 47 mandays on large farms. The difference in employment due to difference in cropping intensity was highest on small farms and lowest on medium farms.

C. Tractor Operated Farms Vs. Bullock Operated and Tractor-Hiring Farms

The differences in per hectare employment between these two categories were less as compared to bullock operated farms. For all farms, the employment was higher by 25 mandays per hectare of net cropped area. The negative effect of tractorisation on employment was 8 mandays. However, differences in cropping pattern and intensity of cropping increased human labour requirement by 30 mandays and 5 mandays per hectare, respectively. The actual difference in employment was higher on small farms as compared to medium and large farms. The negative effect of tractor technology on labour employment was higher on medium farms than small and large farms. The increase in employment was more on small farms than medium and large farms due to the cropping pattern difference. The increase in employment due to difference in cropping intensity was highest on small farms followed by large farms and lowest on medium farms.

D. Tractor plus Bullock Operated Farms Vs. Bullock Operated and Tractor Hiring Farms

The actual increase in per hectare employment on tractor plus bullock operated farms was 16 mandays. The estimated replacement of human labour on all farms was 6 mandays. The effect of cropping pattern and cropping intensity was observed as in increasing employment by 25 mandays and 2 mandays respectively per hectare of the net cropped area. The comparison of per hectare employment difference between three size groups indicates that actual difference was highest on small farms and lowest on large farms. The replacement of human labour due to use of tractor technology was higher on medium size farms followed by small farms and large farms. The cropping pattern dominated by more labour intensive crops had increased labour employment more on medium size farms while the cropping intensity and increased the employment more on small farms.

Conclusions

The findings of the present study are summed up below:

1. In general very close association is observed between actual and the estimated change in per hectare employment of human labour for all size groups as well as for all farms of tractor owning farms over non-tractor owning farms. This inferred that the decomposition model yielded satisfactory results.
2. Though, the use of tractor technology for cultivation decreased labour employment by 6 to 11 mandays per hectare, the differences in cropping pattern and cropping intensity between tractor owning arms and bullock farms increased the per hectare employment between 25 to 60 and 1 to 7 mandays respectively, making the tractor farms more labour-intensive than bullock farms.
3. The percentage change in employment on tractor versus non-tractor farms was highest on small farms and lowest on medium farms.
4. The differences in labour employment between tractor versus non-tractor farms have shown inverse relationship with the farm size.
5. The differences in employment due to differences in cropping intensity were higher on small farms followed by large farms and lowest on medium farms.
6. The higher cropping intensity and change in cropping pattern may be partly attributed with the use of tractor technology for cultivation and consequently increase in employment.

REFERENCES

1. Grewal, S.S. and A.S. Kahlon, 1972, Impact of Farm Mechanisation on Farm Employment in the Punjab, *Indian Journal of Agricultural Economics*, 27 (4): 214-217.
2. Acharya, S.S., 1973, Green Revolution and Farm Employment, *Indian Journal of Agricultural Economics*, 28 (3): 30-45.
3. Hanumantha Rao, C.H., 1975, *Technological Change and Distribution of Gains in Indian Agriculture*, MacMillan Company of India.
4. Chauhan, Y.S., D.S. Singh and T.R. Singh, 1981, Impact of Farm Mechanisation on Labour Use in Block Kalyanpur, District Kanpur, *Indian Journal of Agricultural Economics*, 36 (4): 56.
5. Raj Krishna, 1974, Measurement of the Direct and Indirect Employment Effects of Agricultural Growth with Technical Change, in *Employment in Developing Nations*, Ed. E.O. Edwards, Columbia University Press, New York.

25

FAMILY LABOUR UTILISATION ON FARMS

Introduction

There are, in general, two aspects of unemployment—seasonal unemployment and disguised unemployment or under-employment. Efforts have constantly been made to increase rural employment through a multiplicity of devices, but the very nature of work in agriculture is such that even though there is lesser work and much spare time for most of the period, it is never possible to spare one or more of the family members permanently. The family members, in most of the cases, are not in a position to accept a full-time employment even if it is available in the village itself. This sort of situation, renders them immobile even though they have a lot of time during which they can certainly engage themselves gainfully.[1]

This paper critically examines various facts of farms family labour availability use and its mobilisation in economic development under different agro-climatical conditions of the State by available data and personal observation of field. The specific aspects in this study are: (a) to find out farm utilisation of farm family labour and thereby quantify the surplus farm family labour in various zones of the State, and (b) to investigate into the possibilities and difficulties of its mobilisation for accelerating the process of economic development in the State.

Himachal Pradesh being a hilly State, there are hills of various heights ranging from sub-mountainous spurs and mountain ranges to the snowelad Himalayan ranges. The physical configuration of these hills makes the condition of agriculture in different parts of the State somewhat varied. Therefore, the State was categorised into three zones on the basis of elevation (height above mean sea level): (a) Low hills—elevation upto 3000 ft. (b) Mid hills—elevation from 3000 to 5000 ft. and (c) High hills—elevation above 5000 ft.

For this study, six villages were taken and out of these, two villages each from low hills, mid hills and high hills were selected on the basis of probability proportional to the net area sown in each zone. The total holding of each village was classified into two categories according to their size (i) below 2 hectares and (ii) above 2 hectares. A proportional sample of 12 householde was drawn up randomly from each village. Hence a sample of 72 households was taken, i.e., 24 each from low, mid and high hills. The month-wise information regarding labour-use pattern in different crop operations, from the selected household was obtained by the cost accounting method for the year 1978-79.[2]

Farm Utilisation and Available Surplus Labour

Because of agro-climatic difference, the cropping pattern of the three zones differs. In the low hills, wheat is the major crop followed by maize and paddy. In the mid hills, maize, paddy and wheat held equal importance. Potato is the major cash crop of the high hills, other crops being maize and wheat.

The on-farm utilisation of family labour depends upon the cropping pattern, number of animals kept on the farm and other agricultural practices used by the cultivator. On this basis, the man-hours of family labour utilisation on farms, per farm family in different size-groups in the various agro-climatic zones of the State are summarised in Table 25.1. The total annual man-hours of family labour available on farm are computed by multiplying the total available man equivalent[3] working unit in the farm family with 2400 hours (300 working days of 8 hours) i.e. 300 working days in a year is based primarily on full year round employment criterion advocated by many Economists.[4] It was justified for the purpose of quantification of surplus farm family labour because this norm is usually accepted in industrial and other sectors of economy. Family labour available for farm work, was calculated on monthly basis. The surplus man-hours in terms of actual and percentage of the total available per farm family are computed as the difference between the total working man-hours available and total working man-hours utilised on farm. The results on annual and monthly basis are given in Tables 25.1 and 25.2 respectively.

It is observed from the table that labour surplus is available in all zones and for all types of cultivators. There is a wide variation in the percentage surplus of farm labour between both the size-groups in zone as well as among various claimatical zones in a size group i.e. 20.17 to 36.86, 27.44 to 40.01 and 29.67 to 33.07 in low, mid and high hills in respect of large and small cultivators respectively.

It is further observed that utilisation of crop production-hours increases with the size of holding in each zone and same pattern is followed in the case of tending of cattle. Time spent in tending of cattle is much higher in comparison to crop production in all zones, and percentage of time spent on each of these activities increased with the size of holding. Outside employment is available in only low and mid hills and had inverse relation with elevation. If we take into account the outside employment, the percentage of labour surplus significantly changes in case of small farm holdings of low hills; it decreased from 36.86 to 17.76 per cent.

Table 25.2 shows the monthwise pattern of available surplus family labour percentage on per farm basis for different size-groups in different zones of the State. The table shows an important revealing fact that surplus family labour is much higher (about 40 per cent and above) in most of the areas during the months of November to April in general on different size- groups in all zones. The reason for this uniform pattern is that these months are off-season, and particularly, nothing can be done on the farm, leading to enforced unemployment. During busy period as July and October, farm workers of low and mid hills of large size holdings, were over-employed to finish work within a span of time, determined by climatological and biological facts. The observance of this pattern is very much important as it shows that farm family labour is surplus to the extent of about 40 per cent and above for about six to seven months in a year, in all size groups by and large in all the zones of the State. This may be of crucial importance in preparing any programme for mobilisation of surplus labour for other growth promoting purposes.

TABLE 25.1

Annual Availability, Utilisation and Surplus Family Labour Per Farm Family by Size-Group in Different Zones

(Man-hours equivalent)

Zone & Size-Group	*Available Total*	*Man-hours Utilisation on farm*			*Surplus at farm level (2-5)*	*Work available out side the farm*	*Net effective surplus time (6-7)*
		Crop production	*Animal Husbondry*	*Total Utilisation on farms (3+4)*			
1	2	3	4	5	6	7	8
Low Hills (1st Zone)							
Upto 2 hect.	5867.65	720.15	2984.70	3704.85	2162.74	1120.46	1042.28
	(100.00)	(12.27)	(50.87)	(63.14)	(36.86)	(19.10)	(17.76)
above 2 hect.	7792.86	1787.49	4433.55	6221.04	1571.82	661.23	910.59
	(100.00)	(22.94)	(56.89)	(79.83)	(20.17)	(84.8)	(11.68)
Mid Hills (2nd Zone)							
Upto 2 hect.	6960.00	1184.70	2991.00	4175.70	2784.30	411.72	2372.58
	(100.00)	(17.02)	(42.97)	(59.99)	(40.01)	(5.92)	(34.09)
above 2 hect.	8066.66	1776.30	4053.10	5329.40	2237.26	235.55	2001.71
	(100.00)	(22.02)	(50.24)	(72.26)	(27.74)	(2.92)	(24.82)
High Hills (3rd Zone)							
Upto 2 hect.	7872.50	1041.54	4227.34	5268.88	2603.62	-	2603.62
	(100.00)	(13.23)	(53.70)	(66.93)	(33.07)	-	(33.07)
above 2 hect.	17025.00	2267.29	9706.95	11974.24	5050.76	-	5050.76
	(100.00)	(13.31)	(57.02)	(70.33)	(29.67)	-	(29.67)

Note: Figures in parentheses show percentage.

Mobilisation of Labour Surplus

In estimating rural disguised unemployment[5] economists have assumed that all or most of this surplus labour could be removed from the farm without diminishing the total farm output. Kahlon[6] observed that "an effective of real labour surplus emerge only after the emergence of a product surplus (subsistence farm cannot release labour for non-farm sector where surplus making farms can do so). Thus labour surplus is the consequence of product surplus rather than causing it".

Himachal Pradesh has its own specific difficulties due to geographical and climatic limitations. Hill agriculture has small, scattered and terraced fields. Therefore, there are no possibilities for mobilisation of farm family labour towards industrial sector. Besides, the industrial sector itself is in developing stage. Without adequate agricultural (product) surplus, mobilisation is not possible for the State.

Difficulties in Mobilisation

1. The adequacy of agricultural surplus is a major macro-level constraint as without such a surplus the transfer of labour from the farm to the non-farm sector would adversely affect the wage and price in both sectors.
2. The very nature of work in agriculture is such that even though there is lesser work and much spare time for most of the period, it is never possible to spare one or more of the family members permanently. There are so many practical difficulties in transfer/migration.
3. The household sector is dominated in such an economy both in production and consumption in an integral aspect of its specialisation. The nature and composition of the output is determined upto a certain limit by the domestic or subsistence need. This forms a solid, hard core not easily amenable to change.
4. The attachment of farm labour with family and hereditary land would pose another difficulty in its migration/transfer to the industrial sector which in the absence of enough agricultural surplus, may not be able to attract it on better remunerative wages. Agricultural surplus is also essential for the capital formation which is needed for promoting industries.
5. These underemployed farm family workers can be mobilised for additional employment provided the work is compatible with their behaviour and temperament.
6. In rural areas social status hinders the family labour to work, and due to this false notion they remain underemployed.[7]
7. Small size of farm, low income, complex process to get finance, lack of credit facilities and lack of experience in the use of credit for production

TABLE 25.2

Monthwise Pattern of Surplus Labour as Percentage of its Total Monthly Available Per Farm Family

Zone & Size-Group	*MONTH*											
	July	*Aug.*	*Sept.*	*Oct.*	*Nov.*	*Dec.*	*Jan.*	*Feb.*	*Mar.*	*April*	*May*	*June*
1	2	3	4	5	6	7	8	9	10	11	12	13
Low Hills (1st Zone)												
Upto 2 hect.	1636	31.01	2559	6.76	3555	42.74	4844	59.86	57.00	4457	20.04	54.88
Above 2 hect.	-6.05	43.12	38.38	12.91	42.74	52.24	53.04	59.68	57.11	42.04	24.78	49.36
Mid Hills (2nd Zone)												
Upto 2 hect.	31.31	36.14	43.91	3.72	40.17	42.76	38.79	64.26	60.56	48.44	19.76	40.58
Above 2 hect.	-11.78	22.06	24.11	-14.12	28.42	45.83	5157	58.81	44.14	22.90	27.50	28.28
High Hills (3rd Zone)												
Upto 2 hect.	21.99	37.10	27.87	16.27	32.92	53.05	47.54	45.86	41.96	37.16	12.33	17.61
Above 2 hect.	3.88	23.22	12.62	18.99	30.79	41.53	52.06	55.04	42.29	41.51	8.88	25.85

purposes are the major difficulties for mobilisation of farm family labour into economic development. It was seen from the field survey that majority of cultivators do not utilise the full credit for the actual purpose. In fact they do not feel the importance of this credit for the production purpose. In other sectors of the economy production in credit has become so firmly established with specific type of enterprises that it is difficult to introduce new credit use.

Implication and Suggestions

Commenting on Flemings[8] assertion that the case with which labour can be transferred from agriculture to the non-agricultural industry "has frequently been exaggerated", Johnston and Mellor[9] observed that this largely ignored the significant potential that exists for rising productivity in agriculture. Thus, the mobilisation of surplus labour within agriculture/rural sector appears to be the only possible proposition for the State at present. To achieve this, the following suggestions are offered.

1. The total cropped area under cash crops should be increased not only because these crops as compared to foodgrain crops are economically more viable, but it also takes care of the problem of surplus labour as raising of cash crops needs intensive agricultural operations during the crop period.
2. Since small and marginal farmers' plight is very pathetic and most of them live in the lowest range of poverty, which ultimately affects their productivity, the State Government should impart large scale training to several individual farmers with financial assistance to enhance their knowledge of farm management, as well as to manage Cottage industries like carpentry, blacksmith, basket-making, weaving, beekeeping, poultry and dairying etc. It is probably reasonably to assume that the present credit institutions have the financial ability to serve the need of the group. These matters would be subject to review after the training programme has progressed to a certain extent to ascertain that the farmers make proper use of credit facility for increasing their farm family income.
3. Rural programme should be planned at the local or village level and work should fit with rural people's behaviour. Essentially the work should be such that it would provide employment to a large number of unskilled workers without putting much strain on resources. Several type of worksfall in this category, for instance, road construction, minor and medium irrigation, soil conservation, afforestation, etc.
4. To alleviate the problem, of surplus labour, the State Government should take effective steps for setting up rural-based small scale industries.

Seasonal industries too hold a viable claim and should be encouraged.

5 Organisation of rural people is also a primary necessity[10] for better planning to enlarge the social base of agriculture production.

REFERENCES

1. For detailed discussion see Dantwala, M.L. "Notes on Some Aspects of Rural Employment", in selected readings in Indian Journal of Agricultural Economics 1940-64, ISAE, Bombay.

2. The data were originally collected for the Comprehensive Cost of Cultivation Scheme, H.P. University, Simla-5.

3. The man equivalent family labour has been calculated by considering the members of the households engaged in agricultural occupation in the reference year. In the analysis, the attached or permanent labour within the household has been included in family labour. The conversion factors used for calculating man equivalent labour availability were as follows:
   1. Children between age 10-14 year: 0.50 man
   2. Male between age 14-60 years : 1.00 man
   3. Female between age 14-60 years: 0.75 man

4. For details see:
   (i) K.N. Raj "Employment & Unemployment in the Indian Economy", *Economic Development & Cultural Change,* Vol. VII No. 3, April-1959.
   (ii) M.L. Singh, "Unemployment in rural area of Palamau district, Bihar" : A study in Hussainabad Block, *Indian Journal of Agricultural Economics,* Vol. XXVII, No. 4, Oct-Dec. 1972.
   (iii) V.K. Pandey, S.L. Shah and A.K. Singh, "Surplus farm family labour in Uttar Pradesh and its mobilisation for Economic Development", *Indian Journal of Agricultural Economics,* Vol. XXX, No. 3, July-Sept. 1975.

5. According to Ching, the test of chronic disguised unemployment lies in the excessive supply of labour-time even at the peak of agricultural activities. The concept of disguised unemployment in the sense of zero marginal productivity of labour according to Nurkse, applies only to densly populated countries.

6. A.S. Kahlon, "Theory of Economic Growth in Over-Populated Countries", *Indian Journal of Agricultural Economics,* Vol. XXII No. 4, Oct. Dec. 1967.

7. It has been observed from some rural development programmes that sometime the workers after reporting for work for some days, do not turn up subsequently. The reason being the distance of the site of work from the place of residence.

8. J.M. Fleming, "External Economics and the Doctrine of Balanced Growth", *The Economics Journal,* Vol. LXV, June, 1955, page 254.

9. Johnston and John W. Mellor, "The Role of Agriculture in Economic Development" in the *Economics of Agriculture* edited by I Cart A Fox and D. Gale Johnson, The American Economic Association Richard D. Irwin Inc. Illinois, 1969.

10. For detailed discussion on the subject refer to J.D. Sethi, "Organising the Rural Poor". *The Economic Times,* May 18 and 19, 1981.

26

BUREAUCRACY AND DEVELOPMENT

Theoretical Aspect

Development administration, according to experts like Edward Weidner, is synonymous with an action-oriented, goal-oriented administrative system.[1] It includes the working of new agencies such as planning organisation, development corporations, the reorientation of established agencies such as department of agriculture, the delegation of administrative powers to development agencies and the creation of a cadre of administrators who can provide leadership in stimulating and supporting programmes of social and economic improvement. It is also an aspect of public administration which focuses on government-influenced change toward pro-gressive, political, economic and social objectives.

Since its main concern is with the management of change in the countries whose major national goal is development, development administration should naturally function more as an agent of change than as an instrument of stability and continuity. A bureaucracy steeped in the tradition of old style colonialism, as of India in 1947, cannot *normally* deal with development situations requiring risk-taking, quick decision-making, innovation and creativity, especially in a free democratic state committed to a socialistic pattern of society, with development as a major goal.

It is believed by all concerned that economic development, if it is to be carried out by a massive intervention of the public sector, require a breed of administrators different from those trained only in the maintenance of law and order and revenue collection. These new type of administrators would be more "freewheeling", less adhering to administrative form, less attached to the importance of hierarchy and seniority, with different kinds of values, orientation and attitudes to govern their behaviour. In developing countries it is necessary to have these cadres of administrators because there are no other social or political institutions—like in developed countries—to accomplish the economic goals.

A lot of debate and research has gone into the question whether the Indian bureaucracy, as it is, is capable of delivering the developmental goods in our conditions. Is the Indian bureaucracy static or dynamic ? Broadly, the studies show that Indian administrators have been well adjusting their behaviour and outlook to the exigencies of new situations though their reaction time has been some what longer than necessary. Perhaps, the behaviour of the individual bureaucrat is dependent upon his own cultural and value orientations, beliefs and work-ways. A study conducted in 1980-81 has put this in the following terms:

> "As all officials are human beings, possessing value judgements and biases, human values and orientations are regarded as enduring long term goals which require behaviour adaptation. Since the values of an official form part of his work environment and orientation and since [the value systems of the civil services in India have been undergoing changes due to factors like change in educational system, processes of urbanisation and industrialisation, parliamentary democracy, impact of mass media etc, the pace at which the developmental goals could be realised in India depends mainly upon its official's orientation towards certain types of values and the extent to which their own values are consistent or in contradiction with values necessary for realisation of new goals of a democratic welfare state."[2]

This very study has made the following suggestions to mould the existing bureaucratic system to serve the needs of a development administration:

(i) A three pronged policy strategy to make certain changes in the structural component of the bureaucracy by (a) reducing the number of levels in the decision-making process, especially in the organisation of field offices; (b) by visiting the officials in the field establishment with requisite authority and power to take spot decisions in the execution of development targets; and (c) making the functional division of the work between the officials such as would encourage team work with a determined leadership and direction in the implementation of developmental goals. There would need to be more open communication and professional mobility among various levels of officials.

(ii) Necessary to "make certain radical changes in the social process of the incumbent officials if their orientation towards development has to match with the reciprocal perception of the beneficiaries' confidence in their capabilities and orientation of helpfulness". Can be done by training and refresher courses.

(iii) Need to choose development administrators most carefully since they are necessarily to be an organiser, leader, a pathologist, a politician, an expediter, a specialist, a change-agent and a reformer—all rolled up in a single personality, besides wielding enormous power and prestige which is to be used for public good. These people have to be trained and reoriented in outlook and behaviour from the day they enter service. They have to develop a 'client orientation' attitude-built on a note of mutual confidence with the community they serve.

(iv) Necessary to effect changes in the officials socio-personal, socio-administrative, socio-cultural value systems, in order to effect desired changes in his behaviour towards his own colleagues, subordinates, the people and political leaders with whom he comes in daily contact. A collaborative effort of the officials, the common man and public leaders would need to permeate the developmental process.

(v) The developmental official must, by training, fully project his public image of an impartial and honest functionary and, at the same time, be closer to his subordinates and other segments of society with whom he has to work, without sacrificing his concern for public interest against the individual interest.[3]

Ultimately public bureaucracy stands in need of ethical sensitivity in order to serve public interest which can best be served when public policies are not bended towards any special interests and the various conflicts of interests are supplanted or reduced by law, organisation and process.[4]

Operational Aspect

In 1980-81, the Centre for Policy Research, New Delhi, in collaboration with the ICSSR New Delhi, did a study on Development Bureaucracy dealing essentially with the skills and capabilities of bureaucracy in development. Though the study was based on the data collected from only two districts in two different states viz., Karnal in Haryana and Gorakhpur in UP—some useful conclusions were reached for wider policy purposes.

The broad question before the study team was : how far the bureaucracy as a system, with its rules, regulations, relationships and behaviour, was capable of dealing with the various functions of development ? The team was also to identify those factors which were responsible for relative success of the efforts made at the grass roots level and their broader theoretical and policy implications of governmental systems and development. The study was confined only to agricultural development in the above two districts which had a comparable development potential but with very diverse performance record.

The Report of the study team dealt with the growth of agricultural development strategy in India since 1959 when the Ford Foundation Team of Experts had given their "Crisis Report" which led to the lADP/Package concept. The subsequent more elaborate programmes like the IAAP, HYVP, SFDA, MFAL, CADP, DPAP. and Dryland Farming Programmes were also described briefly. The improvements in agricultural programmes of production, including productivity per ha, were also noted for the period 1960-61 to 1977-78.

"The choice of the two districts-Karnal and Gorakhpur—was largely based on their natural endowments in terms of their agroclimatic conditions, especially their developmental potential *which in the view of many senior personnel was considerably similar* (emphasis added now); what then accounts for difference in their respective performance in relation to agricultural development ? In particular, are there any major administrative factors responsible for the relative success or non-success of the agricultural development programmes in the two districts ?"[5]

The main objectives of the study were framed as under :—

(i) to examine the adequacy of the various dimensions of the generic bureaucratic system of handling the task of agricultural development;

(ii) to study the attitude of the bureaucracy towards the developmental task assigned to it and towards the people whom it was supposed to service;

(iii) to assess the level of skills and capabilities of the bureaucracy implementing the programmes and;

(iv) to observe and evaluate the actual behaviour of the bureaucracy in implementing the programme including its relationship with the citizen clientele.

Focus of the Team was on the general programme. It investigated more specifically the strategy of the organisation for agricultural development in the two districts, the design of the administrative system in both and the structure of day to day administration, the skill and capabilities of the bureaucratic administrators and their attitudes and behaviour, both towards the task assigned and towards the citizens clientele.

The study particularly noted that the IADP in Karnal was better designed than the IAAP in Gorakhpur. In the former, the supply of most vital inputs like seeds, chemical fertilisers and pesticides was centrally handled by HAFED in Karnal, as against four different agencies in Gorakhpur. Karnal had 186 sales depots compared with only 150 in Gorakhpur even though the latter was twice in area and three times in population. Similarly cooperative credit was more easily available in Karnal than in Gorakhpur. 78 per cent of the net sown area in Karnal was irrigated as against only 53 per cent in Gorakhpur even though the irrigation potential there was also extensive. In Karnal all villages were electrified as against only 22.6 per cent in Gorakhpur. Network of roads was also much better in Karnal. In the matter of dissemination of information, there was a full-fledged information wing in the office of the DDA, with a plan projector, a slide projector, documentary films and slides, one audio-visual van and three jeeps. Gorakhpur had only one jeep and no other facilities for this work.

In the matter of organisational strategy, the one adopted in Karnal was found to be highly pragmatic and result-oriented, responsible for the successful implementation of the development programme. It consisted of a clear-cut division between extension functions and supply functions—the latter entrusted to autonomous corporation like HAFED and Haryana Agro-Industries Corporation. It was not so in Gorakhpur. Then there was a proper command, control and coordination of operations in Karnal-all under the DDA—to ensure a smooth and cohesive working. In Gorakhpur, on the other hand, there were separate parallel set-ups for extension, cooperation, minor irrigation and plant protection etc. No cohesion, no compactness and little coordination.

Conclusion

The study concluded that development bureaucracy did indeed have a significant share in the actual delivery of development in the field—the actual performance being related to a host of *intra* bureaucratic and *extra* bureaucratic factors. Bureaucracy included the civil service with its entire system of administrative rules and regulations, planning and implementation processes. Though the district administrative systems were comparable in both the districts, there were marked differences in performance. The two districts had a strikingly different approach to the question of administrative strategy and planning which in turn meant a different administrative design at the state level.

> "In Karnal the systematic consciousness in trying to develop a deliberate strategy for reaching the agricultural developmental programme to the farmer, administrative planning to deliver the services whether of extension or input variety and maintaining a more effective coordination between different programmes is clearly -superior to that of Gorakhpur".
>
> "In Gorakhpur, the administrative system is of the old C.D. Programme variety which became obsolete years ago.
>
> "The forte of the administrative strategy in Karnal appears to be an explicit recognition of the state of agricultural development at the local level, the level of the bureaucratic skills required to handle the task of development and the limitations of the local bureaucracy within which the programme had to be administered"[6]

Conscious efforts were made to develop an implementation machinery which was directly related to the overall administrative strategy. In Karnal it was more functional; in Gorakhpur it was quite a bit dysfunctional.

The Study concluded that —

(i) the basic parameter affecting bureaucratic performance is the bureaucratic or the administrative strategy and planning;

(ii) in other words, the administrative "how" is as important as the policy "what";

(iii) basically, bureaucracy as a system can be designed to perform developmental functions quite adequately, even in a diffused sector such as agriculture;

The most important requirement is one of deliberate and conscious development of administrative strategy—a dimension which is commonly absent in the culture of the administrative system—not a mere grafting of developmental functions on the old regulatory administrative; system will do.

(iv) to the question must it be bureaucracy to do the developmental job ? The answer is that for India the choice was made in the Constitution itself. There is no way out of the democratic option for the time being, though alternatives could be worked out.

The total upshot is that the present administrative machine in the country is quite capable of taking up developmental programmes provided it is properly trained, motivated, organised, supervised and monitored. There is need for an enlightened political leadership to inspire this machine suitably. In future recruitment, proper steps could be taken to choose persons carefully and then give them intensive in service training and motivation.

NOTES

1. Edward W. Weidner "Development Administration—A New Focus for Research" included in Paper in *Comparative Public Administration*, Ann Arbor, Michigan Institute of Public Administration 1962 p. 98.
2. R.B. Jain and P.N. Chaudhuri, *Bureaucracy and Development*, Centre for Policy Research, Chanakya Puri, New Delhi pp. 20-21 (unpublished).
3. Emmette S. Rodlord, *"Ideal and Practice in Public Administration"* (University of Alabama Press 1975).
4. Concluding Chapter of "Bureaucracy and Development" *ibid*, pp. 319-320.
5. Dr. V.A, Pai Panandikar, *"Development Bureaucracy"* (along with R.N. Bishnoi and O.P. Sharma) Centre for Policy Research, New Delhi March 1911. p. 353.
6. The above is based on "Development Democracy" *Ibid.*, (unpublished).

27

EXTENT OF ADMINISTRATIVE INVOLVEMENT IN INTENSIVE AGRICULTURE AT THE FIELD (MICRO) LEVEL

Administrative (government) involvement in intensive agriculture at the field level started in the country in the early 40s—really with the launching of the Grow More Food (GMF) Campaign in 1943. In 1950-51, this campaign was modified into an Integrated Production Programme (IPP), to include crops other than foodgrains also. The measures devised to attain the First Plan objectives aimed at securing equity/social justice in the rural areas. The agency for implementation was CD/NES through a coordinating Village Level Worker (VLW) at the bottom rung. In 1957 was brought in the Panchayti Raj (PR) set up for rural development. With this there were now three agencies in the field, viz., the Cooperatives, CD/NES and PR, to promote agricultural development. In retrospect one sees that none of them had by itself any significant impact on agricultural production. CD/NES had, however, led to an awareness of new inputs like improved seeds, fertiliser etc. but resources being inadequate, their impact on the problem did not go far. There was no new technology to be spread as such. The cooperatives were dominated by the rich farmers and the PR institutions were vitiated by political factionalism and its members pre-occupied in a race for local importance.

Whatever the agencies in name, the gut agents in the field were provided by the agriculture department the set up of which, for decades before 1947, had consisted of the Director/Deputy Directors (DA/DDA) of agriculture at the Headquarters and EADA/DAO (Extra Assistant Director Agriculture/District Agriculture Officers) in the districts, with a few agriculture inspectors in the field. Considerable expansion in these ranks took place in the 50s when the CD/NES experiment was launched. Further expansion took place when the PR structure was introduced with its Block Samities and Zila Parishads all over the country, with VLWS, each working in ten or more villages for helping agriculture and other activities, and a Block Development and Panchayat Officer (BDPO) at the block level helped by some agricultural inspectors. Hitherto, however, agricultural activity as such was considered only as a part of the overall economic developmental activity, deserving of attention as one of the subjects concerning people's welfare. All the same, the district remained the basic unit of agricultural administration.

CD/NES programme was treated as the principal vehicle of rural development. The thinking behind it was that official departmental machinery by itself could not carry out the development programmes which called for a great deal of initiative and participation on the part of the people themselves. The rural masses of people were treated as the chief producers of wealth in the country, one major resource for development, infact as the chief entrepreneurs of all local involvement. In the 50s the country was given to a lot of philosophical idealism, one recalls.

Ford Foundation Team and some other American agencies—these were the Kennedy days in early sixties—egged this process on in good faith and in good measure. Maintaining very correctly that increased agricultural production was going to be the main theme of CD/NES (and later of the P.R. programmes) they felt that the need for rapidly increasing food production could not be guaranteed under the sole direction of government administration. Democratic decentralisation, in the shape of PR, was introduced in order to decentralise responsibility, utilise and develop local area and local group initiatives. The logic was :

> "This is so partly because India is discovering through experience in her rural development programmes that neither (i) logical planning which has been done nor (ii) efficient bureaucratic administration which she possesses, accomplishes the basic task of increasing the food supply rapidly enough to greatly improve man's consumption."[1]

The Team came to the conclusion that the above two things were not adequate to engender dynamics in hundreds of millions of Indians. "Only the willing and enthusiastic endeavour of each cultivator family to farm more efficiently will create wide-spread dynamics in rural areas. The development of hundreds of thousands of largely self-directed local groups will provide the foundation stones of a basic democracy——Now India has established the framework of a CD/NES programme which provides a vehicle or machinery by which rural progress can be carried, forward by democratic methods and procedures".

Going into the reasons for the village people not taking readily to new agricultural practices, a view was expressed that little effort had been made to discover and to understand the most effective means of spreading or diffusing information about superior methods of agricultural production and not even indigenous means of spreading information had been utilised.

> "It has been due to insufficient understanding and use of social skills in the motivation of village people plus an inadequate understanding of the net work of human relationships in villages and the ways by which both technical assistants and administrators can efficiently communicate with those whom they seek to serve."[2]

Answer to these questions will be found only if Indian bureaucracy assumes responsibility for the relatively simple types of research by which these answers can be found, as is done in well-established democracies. If that is done, the Indian bureaucracy will change from the practice of *guiding* village people to an understanding of how to *motivate* village people and how to *utilise* long-established social structure of village communities as a vehicle for promoting development and democracy."[3] In this process the bureaucracy need not sacrifice its administrate integrity, even though it loses some of its elite social status and comes nearer to

This was the real crux of the matter. The administrative leaders in the field, inspired by a selfless political leadership, must be able to convince the people of their honesty of purpose, with devotion, application and personal example, apart from the logic and general acceptability of the plan proposals by the people. And while expecting the people to do their part of the job in the game of development/ agricultural production, the administration itself must deliver the goods in time, in the shape of inputs and services.

When the CD/NES and PR both more or less failed in their generalised approach to agricultural development, Intensive Agricultural Development Programme (the IDAP). called Package Programme for short, came into the picture in 1960, first on an experimental basis in 7-then in 16 districts (one district Project in each state) in pursuance of the recommendations of the Ford Foundations Food Crisis Report, 1959. In actual fact, administrative involvement in agricultural production became more intense with the lADP-followed by IAAP (Intensive Agriculture Areas Programme) and HYVP (High Yielding Varieties Programme)— programme which formed a part of the New Strategy in Agriculture, initiated and followed in the sixties and with which we have already dealt at length. Field staff was increased considerably with each programme in the covered areas; even then their inadequacy was acutely felt when the Green Revolution concept came to hold the farmer's imagination, in some states at le?st. Simultaneously the setting up of the agricultural universities in various states was of considerable help in providing the new expertise and in solving the technical problems arising locally, peculiar to the soil and natural environment, apart from their help in extension work.

Scientific agriculture is a highly specialised affair and consists of various subject matter fields even within one branch of agriculture. Competence in different agricultural disciplines was essential for tackling the numerous problem arising in the field. The subject-matter specialists were available only at the distinct level, even under the Package Programme. Difficulties arose also from the fact that, where this staff was placed under the PR institutions, their utility really depended on the interest taken in the programme by the locally elected representatives: moreover. the unified technical and administrative control at the district level suffered too. Even at the latter level, the question of coordination of various types of expert staff connected with intensive agricultural production started looming larger than before. The good old district collector-that jack of all trades was available but he was generally very much preoccupied with other work. The appointment of a wholetime CADO (Chief Agricultural Development Officer) was suggested to control and coordinate the working of district officers in various agricultural disciplines. A doubt arose whether he would enjoy enough prestige to control so many multitalented experts in the field.

With the same logic, at the block level the BDPO was to be replaced by a BADO (Block Agricultural Development Officer, directly responsible to the CADO who would act as a coordinator at the block level for agricultural production. The VLW will devote time entirely to agriculture. Other activities of development would need to be looked after separately.

Similarly at the state level, the appointment of an APC (Agricultural Production Commissioner) was suggested, next only to the Chief Secretary in rank, to plan, direct, guide and coordinate the activities of various departments connected with agricultural development. A deputy Chief Minister was to look after this important portfolio at the ministerial level.

Some states adopted ; these changes whole heartedly—others did not. As time went on, and problems or the need arose, further changes were made not only to keep up with the increased activity but also to ensure control and coordination. The strength and extension pattern of staffing in departments and directorates was, however, linked with the plan schemes and not determined by any systematic plan for strengthening the administrative structure, keeping in view the needs of agricultural development of the state as a whole. The increase in the spread of staff, therefore, lacked some sort of a rationale and coherence. Such situations arose at the level of the central government also. Sometimes food, agriculture, cooperatives, forestry, rural development and irrigation were all lumped together under one ministry, to ensure coordination. Then this system was changed and the ministry broken up into small ones, to meet political exigencies and so on.

This narrative is, by no means, intended to be a comprehensive account of all the organisational changes that were made at the centre and in the states to handle the important subject of modern agricultural production. As expansion took place and the spread of activity increased, the staff also multiplied in various fields and disciplines, throwing up problems of coordination, unified command and control. The pattern of staffing was generally uniform in all the states and yet it was only in some states that satisfactory results were forthcoming on the ground. In the four states selected for this study only Punjab and Haryana made good, UP did well only in its western region, having done only so-so in the rest of the state. Bihar did not budge much. Why ? What was wrong or had gone wrong with these two states of UP and Bihar, having almost 25 per cent of the country's population ? Was there some defect in the organisational/institutional arrangements made for ushering in the Green Revolution or was there any deficiency is the conceptual frame-work of these arrangements ? Or were there any over-whelming reasons in the sociological, historical and traditional background of these two states which stood in the way of the real effort, not only on the part of the local political leaders themselves but also the mass of the people residing there ? This could be probed further, but as it is, the conclusion is obvious that the political leadership had let down the people and failed to create conditions in which the mass of them could get down to the job of increased production and productivity.

25 years ago the PR (Panchayati Raj) Institutions were accepted in principle as the foundation of political and economic decentralisation. It was hoped that if initiative was passed on to them for planning and implemention of development programmes, people will come forward with their cooperation in right earnest and positive results would be forthcoming in quick time. Actually not much progress was made on the ground; even till today the district level planning means only a collection of centrally conceived departmental schemes broken up into districts and at the block level it is a little more than a rather disjointed exercise in implementing these schemes through a multiplicity of departmental representatives. In fact, the PR institutions have all receded into the background and the developmental activity, except in some welfare matters, has now been distributed among variously defined target programmes like SFDA, MFAL, DPAP, CADP, IRDP, and so on. The effect is that, even though these programmes in themselves are useful, a tendency towards a sort of centralisation is now growing and the concept of decentralised planning and implementation has been the sufferer, yielding place again to departmental working, however coordinated or coordinating it may be in theory. But it also remains true that departmental hierarchies would not be able to handle the area departmental plans which require a coordinated working at all levels of rural disciplines—a job as complex as a large industrial enterprise.

Let us see what the Sixth Plan document has had to say on this subject. It is stated on page 85 of it that "a clear strategy will have to be evolved to ensure that agriculture receives a very high priority in all policies and programmes.

"The postulated increase in output is expected to result from increases in area under irrigation and by a substantial increase in the consumption of chemical fertilisers and adoption of a systems approach for consolidating the gains already achieved extending the benefits of the new technology to all categories of farmers and all regions."[4]

It stated that scientific management of agriculture would require a highly orchestrated effort on policy formulation and implementation on the part of the agencies concerned with the development of packages of technology services and public policies.

> "It will be necessary to make arrangements for horizontal linkages at the block level. The task of planning and implementation for development programmes at the district level is acquiring greater complexity."

> "The planning at the block level will be further strengthened in the Sixth Plan. The programme will be made area-specific at the grass root level, based on local endownments and potential for growth and further employment. The object of these plans will be to integrate various programmes for optimal utilisation of local endowments with plan objectives and local needs."[5]

The block level has to be the primary area for planning in which case block level plan would need to be in harmony with the district and state plans. It would be necessary, apart from decentralisation of administrative machinery providing for coordination at local levels, to ensure that at every stage of planning and implementation there is full participation and involvement of the people. The PR Institutions would need to be strengthened to become institutions of democratic management of rural development, both at the district and block levels. The direct support and cooperation of the elected representatives of Panchayati Raj is vital to success.

In actual fact the PR institutions have remained ineffective, at best engaged in things like subsidy distribution. The failure to make greater progress with area development programmes could be traced, to a large extent, to serious inadequacies in the more centralised planning and implementation systems at the state level and the lack of supporting policies. Such programmes would require not only strong commitment to measures like consolidation of holdings, some land reforms but also a complex network of infrastructural investments for the development of ground and surface water resources for irrigation and drainage, integrated rural energy system and adequate communication facilities etc. In some of these tasks the overall planning may have to be at a centralised level and only implementation at the local level. It may be possible to do away with much decentralisation in the case of small states like Punjab and Haryana, but in the case of big states like UP and Bihar, local development planning will prove effective only on a decentralised basis. Whether the task is taken up at the state level or at a sub-regional level in a state, would depend on local conditions and preferences of that state. In bigger states there would certainly be a strong case of decentralisation of planning focussed on small regions, identified with regard to agroclimatic conditions, topography, density of population, level of social and economic development—above all, on the potential available for development of local resources in the broader sense.

KN Raj believes that

> "There is an essential complementarity between centralised development planning within states and the. decentralised segments of it and the need for evolving organisational arrangements that would help the technical and service personnel associated with each to establish mutually advantageous links without getting distracted from their basically different functions or coming in the way of each other".[6]

We must also seriously consider the need of giving a real political status and administrative autonomy to the PR institutions, even by going beyond the Directive Principles of the Constitution. This would imply specific statutory provisions relating to local government, definition of the area of responsibility and power, indication of sources of finance and constitutional safeguards necessary. Only then decentralised planning and implementation of development scheme at the grass-root level will take place. Government departments will not formulate such scheme at the state level, thus avoiding a fragmented approach through vertical lines of control and horizontal coordination.

An Experiment in Decentralised Planning—UP

It is gratifying to note that the UP state government has already embarked on a bold experiment in decentralised planning at the district level. In 1982-83 it started the formulation of district plans, with a view to carrying development to the people and generating development from within, as it were. People were to be involved directly in planning and decision-making on their developmental matters.

The intention was to initiate a development approach based on different physico-geographical and economic conditions, with the district as a convenient unit. In such an approach local resources, initiative, potential and capabilities could be taken into account. "The basic aim to implement district plans was to divert attention from a 'backlog problem-oriented' concept focussed on regions which had its political overtones, to a resource-cum-need based concept focussed on smaller areas."[7]

District planning was to start from below, within the frame-work of certain broad guidelines coming from the state government. (The pattern had been set by the Maharashtra government in 1972-73, as a corrective to regionalism). The districts were authorised to formulate their annual plans, keeping in view the aspirations of the people, but adhering strictly to the national and state plan priorities. Schemes under district sector included mainly programmes of agriculture and allied sectors, rural electrification, road building, small industries, drinking water etc.

On an empirical basis, the state government decided to earmark 30 per cent of the total annual plan funds/outlay for district plans on defined subjects and divided the amount among the districts on a rational basis, taking population (50 per cent) and other socio-economic indicators (like scheduled caste/tribes population, number of marginal farmers and agricultural labourers, rural village electrification and some factors) into account, with a view to evening out imbalances and backwardness etc.

There is no local taxation at present; all district plans are being financed by the state government. The district plan schemes (in the decentralised portion of the state plan) are sanctioned by the state headquarters by 15 April. There is a district level planning and monitoring committee consisting of a cabinet minister as chairman, all local MPs, MLAs, collector, DDO and district level officers as members, to finalise the district plan under the guidelines prescribed by the state government and to allocate departmental outlays between blocks on the basis of prescribed norms. The second committee is called the district plan coordination and action committee, consisting of the district magistrate, DDO, economic officer and all district level officers of development departments, to formulate the district plan, review its progress and prepare proposals for reallocation of funds. There is provision fora third committee at the district level—called the executive committee chaired by the district magistrate, with senior most member of the development department and a representative of a lead bank as its members, to decide about the provision of institutional finance for IRD and other such schemes. Then there is divisional level committee chaired by a cabinet minister, with commissioner as deputy chairman, with all districts magistrates and divisional level heads of all development departments, a representative of a lead bunk of the division as its members and joint development commissioner as its secretary, to review and approve the district plans and their implementation, to establish coordination between departments to monitor and supervise actual performance of departments against prescribed targets and lastly to draw up appropriate credit plans for mobilising maximum institutional finance for implementation of district plans.

With this type of decentralised planning the state plan is distinct from the district plans. Funds for the latter are not routed through the heads of departments but are allocated to the district directly. Considerable administrative and sanctioning powers have been delegated to the district authorities. Certain powers of the finance department for reappropriation of funds are proposed to be given to the divisional commissioner, amending the Budget Manual. Moreover, there is a proposal to further strengthen to planning cell at the district level in committee No. 2.

District plans are received and checked at the level of the divisional committee which is attended by a representative of the state planning department also. At the state government headquarters these plans are thoroughly examined further and brought into the state budget.

It will be seen that there is no statutory authority to prepare the district plans and implement them. This job is carried out informally at the executive level under the orders of the state government. So far there is not any representative of the Panchayati raj in the planning and monitoring committee as member though they individually may have adequate voice in the district planning process. Presumably this is so because all the zila parishads and block Samities too are in a state of suspended animation, having all been superseded long ago.

This is a bold experiment but it suffers from some inevitable constraints like lack of data, lack of adequate resources, lack of trained personnel and lack of coordination between departments etc. This is perhaps, a temporary phase. The biggest snag is that the local representatives are out of this experiment, for the time being, may be. A "marriage" of this planning process with the experiment of democratic decentralisation, viz. Panchayati raj institutions, in a statutory fashion will be an objective to be wished for. That will certainly bean improvement on the present scheme of things.

The above narrative, somewhat lengthy and tedious, would go to show that the job of agricultural development in the real sense is quite a complicated affair. Whereas it is easier sitting in the state secretariat, and in the departmental head offices, to draw up plans and programmes for input supplies and services, the problems arise when efforts are made to share those plans etc. with the people and to popularise them for adoption in the field. Various experiments have been tried over these two decades and modifications made in the institutional arrangements in various states in different ways; the fact emerges that decentralised planning, in keeping with the local needs, conditions and facilities etc. at the district level and later at the block level, carried out in cooperation with the elected representatives of the people and then implemented, will have a good chance of success on the ground provided all allied inputs and services have been arranged in advance by the government and the broad infrastructure created in the shape of irrigation, roads, power and consolidation of holdings etc. over a period. Above all, again the political will of the state leaders would be paramount as an input.

EPILOGUE

As said earlier, this book is intended to provoke the politician as well as the bureaucrat of UP and Bihar into thinking why the message of the Green Revolution has not reached the cultivating masses of these two states, to produce a favourable reaction of the type it did in the case of Punjab, Haryana and even Western UP. To make things clearer, the genesis of that Revolution; and its essential components, were described in detail (in Part Two of the book). It was shown that the Revolution in full vigour would be caused only by a "Package" of policy decisions relating to the components provided these were properly implemented, in the broad sequence indicated. That description was preceded by an analysis of the agricultural (foodgrains) situation in India, present and prospective, in the context of the country's demographic and economic compulsions; it was followed by (in Part III) the reactions of the states of Punjab, Haryana, UP and Bihar towards the call of the Green Revolution, showing how its effect was so varied, even contradictory interse, in its content that the impact on the ground appeared to be indicative of even a generation gap, as for example between Punjab and Bihar. It was inevitable that some analysis of these disparities between the states concerned in regard to that impact should be made, to the extent it was possible at the macro as well as the micro levels. Part IV of the book is the result of the author's effort and thinking in the matter.

Total picture, as it emerges from a study of events and acts of omission and commission of the respective four state administrations in this behalf in the last twenty years, shows that whereas Punjab and Haryana took time by the forelock and availed of the revolutionary message, in time as well as space, and in result did very well for themselves and the country, the states of UP (except for its western parts) and Bihar were not that fortunate and lost the chance by at least one generation, even though ostensibly they tried their best and even went through many motions relevant to the needs of that revolution. They made some progress in agricultural production and per ha yields of wheat and rice from the Fourth Plan onwards but the increase in population neutralised that progress and, in per capita terms, the effect was negative. As has been mentioned in an earlier chapter, the Green Revolution ideology was itself intended to neutralise the adverse effect of the demographic explosion in the Third World countries, by making a big contribution to the breadbasket in the shape of increased production of foodgrains—enough to cover the excess population and to spare for raising the nutritional standards.

That meant that the efforts made by the states of UP and Bihar were grossly inadequate in deriving the optimum benefit from the Green Revolution spread in the country from the mid sixties onwards. Why? Part IV of the Book has given some possible answers though not all. The very size of these states—unwieldy, unmanageable and over-populous—was inhibiting the success on the ground of any developmental plans, including those of agricultural production under the new technology, for the simple reason that the monitoring attention of any well meaning political or bureaucratic head of the show was not adequately available to supervise the operation at the grassroots nor was that attention of sufficient intensity and frequency to effectively deal with the delinquent officials on the spot, where necessary. When new plans are introduced—like those connected with the Green Revolution—a lot of spade work has to be done and the actual implementation has to be thoroughly supervised, with a view to producing tangible results. For such planning the states of UP and Bihar happen to be too huge to be adequately managed, administrated or developed by a single chief minister or development commissioner in their respective spheres. It bears repetition to mention here again the pertinent remark of the late Mrs Sucheta Kriplani, ex chief minister of UP, that when she addressed a conference of her (56 or 57) collectors, she felt she was addressing a public meeting. It was a polite way of conveying her helplessness in influencing directly the operative machinery in the field. New projects, to be successful, need personalised attention at the top, if they involve changing people's habits or traditional ways of doing things. Green Revolution ideology to take root in a traditional soil would require that sort of attention at many levels.

Our study showed that in Punjab and Haryana the credit for the success of the Green Revolution went to late Partap Singh Kairon in the case of Punjab and to Ch Bansi Lal in the case of Haryana. The former had prepared the ground for it unwittingly as he was broadly keen to do everything possible to improve Punjab's agriculture and economy, and he did it so well that when the Green Revolution concept swept the field in the mid-sixties, everyone in the Punjab was prepared to receive it with open arms and so willingly. The latter showed his keeness and asserted his mettle to catch up with the neighbouring state, if not to beat it in the game of agricultural production. He also did his job remarkably well. Both the leaders made full use of the local resources, exploited all political and others opportunities to make a success of their efforts to reach the goal; their objectives were so clearcut and the means to attain them so thorough that they received maximum cooperation not only from the bureaucracy but also from the masses. Do such 'go-getter' leaders arise only in small states? Not necessarily but the operational opportunities are, perhaps, more easily available in a smaller state where a strong-willed and determined leader can function more effectively in his own way from A to Z and make a thumping good success of the objectives he strives for.

Bureaucracy is an instrument in the hands of the political bosses in a free democratic country such as ours. These bosses can make use of the instrument as best as they can. If the bureaucracy suffers from inertia and other weaknesses, it can be shaken out of them overtime by a determined leadership and trained motivated in the correct direction. If the political leadership is above board, strong and determined and is capable of asserting its political will, the bureaucracy has no other option but to fall in, in the normal course. We have dealt with the weaknesses and strong points of our bureaucracy, including its capabilities, in a developmental role. There is nothing basically wrong with it except that it must be trained and motivated properly and on the right lines for that role. If there is an above-board "understanding" between the political leadership and the bureaucracy, the results in the field will be automatic since, ipsofacto, planning will be done thoroughly and the implementation of the plans carried out meticulously. The mass of our people in the raw are ever keen to do the right things and follow all useful advice, provided their economic interest is served thereby and they get a real chance to be pulled out of their penury and helplessness. A particular class of the lowest rung (economically) of the people in both the states, called the small and marginal farmers and agricultural labourers, would need particular attention of the state government concerned; any help to them, as a target group, will help the cause of the green revolution/agricultural production. We have dealt with them.

Deals with some illustrative subjects to help the planning for Green Revolution. These subjects are not exhaustive at all. There is plenty of further detail and prescription required, according to local conditions and variations, before any operational projects can be drawn up for execution by the field staff. Consolidation of fragmented holding is an instance in point. There will be many aspects of the problem, depending on the state of the revenue records and the adequacy or reliability of the revenue administration. Will separate staff be required for consolidation work, suitably trained for the job ? Could help and guidance be sought from states like Punjab and Haryana where this work was efficiently done in the fifties and sixties ? And so on.

What is the extent of the present administrative involvement in intensive agriculture in the field ? Is the system working smoothly, with discipline and efficiency or are investments in irrigation and tubwells etc. going down the drain ? What is the overall health of the bureaucracy, consisting of generalists, scientists and technocrats at various levels? Are they working in unison or at crosspurposes? Some way with the various aspects of implementation of projects on the ground, including the problems arising from dilatory, time consuming and complicated working procedures. All these numerous aspects and problems impinging on the goal we had set ourselves in the beginning—that of getting the Green Revolution on the ground, in UP and Bihar, in the manner of Punjab and Haryana—will have to ba gone into, dissolved or resolved and a working plan drawn up in a comprehensive manner, its resource and financial implications worked out, including the physical aspect of numerous input supplies and services. And so on. And then the manpower planning will need attention, to put the projects on the ground—the cadres carefully selected, trained and motivated. The further we go into the matter, the more complicated it will appear to become. This is natural but an epliogue is certainly not the place for doing it.

Coming down to brass tacks, as it were, a major feature of the Green Revolution strategy is the "package" of hyv seeds, fertilisers, pesticides, controlled water supply, some mechanical equipment like seed drills and threshers—even tractors etc. The real core is the 'miracle' seed, developed through selective breeding to be highly responsive to chemical fertilisers. To get the best out of these miracle seeds, infrastructural (some long-range) facilities like land consolidation, irrigation/controlled water supply rural electrification, rural roads, regulated markets-will have to be organised. While action on some long-range infrastructural activities gets going, a plan of action for implementation straight away will have to be prepared with whatever resources and facilities may be available, provided it fits in with the overall scheme of things. Not that any one person—poiltical or administrative—is indispensable but it will be in the fitness of an overall strategy, if the Development Commissioner is carefully chosen and retained in his job long enough to show tangible results on the ground.

Not only the basic knowledge will have to be imparted to the farmers regarding the sowing and care of the new seeds, use of proper doses of fertilisers and pesticides, watering routine, cropping rotation, essentials of double cropping etc, a major feature of the strategy would be the emphasis on organisational and institutional arrangements for the production and distribution of the package of inputs of good quality and in time. The new strategy is selective in approach as between areas and farmers. People with risk-taking capacity for investment, resourcefulness and innovativeness indispensable but it will be would get perference in the normal course. So the course. So the poor small and marginal farmers down the line would need special attention, even arrangements.

In the end it may be mentioned briefly that even though we start the Green Revolution with 'miracle' seeds, it is not automatic that miracles may happen and productivity as well as production may increase in the field. It is indeed a very long hawl between sowing of the miracle seed and the harvesting of a bumper crop. Determination, good planning efficient implementation and hard work down the line—all these things are needed in good measure. There is nothing like trying hard for any goal.

Last but not least, dear reader, let us look at the "Quotes" in the beginning of the book once again.

NOTES

1. Douglous Ensminger and Carl C. Taylor, *India's Roots of Democracy*, Orient Longman, New Delhi, 1965, p. 666.

2.& 3. Douglous Ensminger and Carl C. Taylor, *India's Roots of Democracy*, Orient Longman. New Delhi 1965, pp. 679-680.

4. *India's Sixth Five Year Plan Document*, p. 85.
5. *India's Sixth Five Year Plan Document*, pp. 92-93.

6. K.N. Raj, "Decentralisation in Perspective" *Mainstream*, New Delhi, dated 22.12.84.
7. *U.P.'s Draft Annual Plan, 1984-85*, Vol. i, p. 100.

REFERENCES

1. As Kahlon and D.S. Tyagi—"Agricultural Price Policy in India", Allied, 1933.
2. As Kahlon and others—"Dynmics of Punjab Agriculture", PAU Ludhiana, 1966.
3. As Kahlon—-Impact of New Farm Technology on Structural Changes in Farm Organisation in Punjab—A Chapter on the "Green Revolution— The Unfinished Task" Centre for Study of Social Change, Calcutta/ Delhi 1974.
4. KCS Acharya—"Food Security System in India". Concept Publishing Co., New Delhi 1983.
5. Alvin Toffler—"The Third Wave" PAN Book, Macmillan India.
6. Malcolm Darling—"Punjab Peasant in Prosperity and Debt" OUP, 1932.
7. Gilbert Etienne-"India's Changing Rural Scene", OUP, 1982.
8. Michael Chisholm—"Modern World Development—A Geographical Perspective"—Hutchison University Library, London, 1882.
9. Vadilal Dagli—Report of the Committee on Controls and Subsidies 1978-79, Govt. of India, New Delhi.
10. New Horizons—DPR Punjab, Chandigarh, 1976.
11. Biplab Das Gupta—"The New Agrarian Strategy and India", UNRI Geneva, and Macmillan, India, 1977.
12. DP Gupta and Shangari NK—"Agricultural Development in Punjab," Agricole Public Academy, 1980.

28

PROPERTY IN LAND, LAND MARKET AND TENANCY RELATIONS

The purpose of this study is essentially fourfold: first, to look critically at the theoretical categories used to characterize the nature of the pre-capitalist agrarian structure in India, in particular rural land relations; this is essential in order to judge the impact of colonialism in a proper historical perspective; second, on this basis I shall suggest alternative categories and pose the problem of agrarian transition in a way which seems useful for studying land relations in the colonial period—the period of immediate empirical concern; third, having done this I shall suggest the need to place Marx's way of posing the problem of agrarian transition for non-European agrarian societies (for their pre-colonial epoch)—namely the transition from communal to private property in land—in a proper and more concrete historical perspective. In this connection I should like to point out that the problem of the transition from communal to private property in land is only a specific derivative of the complex notion and problem of an 'Asiatic mode of production' (hereafter AMP). Notwithstanding this there have been attempts at a total negation of the notion of AMP simply on the basis of evidence of some kind of private property in land in medieval or even in ancient Indian land history. This kind of refutation we suggest is rather naive, one-sided and crudely empiricist, particularly in the light of recent reconstructions (some of which we shall attempt here in a very preliminary way in the light of the data from India history) and highly flexible interpretations of the concept of AMP. In fact the latter can be said to be still an open concept and not fully developed. The alternative categories could then be used to suggest that the supposed evidence of 'private property' in land in the pre-colonial period of Indian land history could be consistent with the notion of AMP as referring to a distinct (from feudalism in the main) socio-economic formation; fourth, we shall analyse data for a Zamindari district of British India, partly to illustrate the working out of the categories suggested.

POINTS OF DEPARTURE

There has been so much discussion about the nature of the pre-capitalist agrarian structure in India that it may seem superfluous to talk about it again. But it is important to look at the existing literature critically in order to attempt a possible reinterpretation of the existing evidence; also because much of the current discussion on agrarian transition seems to be vitiated by a kind of historical impressionism, by rigidly adhering to or taking for granted one of the two theoretical notions available for analysis, namely feudalism and AMP. There are also some 'neutral' categories suggested by those who find these two notions unacceptable, namely 'medieval Indian economy',[1] 'free peasant production',[2] and even 'peasant ownership economy'.[3]

Before looking at the views which suggest that there was a feudal epoch in India history not essentially different from the kind encountered in Western Europe and Japan, it should be made clear that the following structural elements, are in their totality taken to be the defining characteristics of feudalism, the Western European and Japanese varieties being the historical prototypes:

(1) Feudal form of property in land (noble property) and the manorial form of organization of production (demesne farming) corresponding to it; (2) fusion of economic and political power within the manor (the lord represents the 'state', so to say, with military and judicial power), i.e. existence of a landowning nobility with political power and with obligations of military service, thus marking the emergence of a fief system; (3) decentralization of political power because of (2) above and because of the fact that the extraction of the surplus from the direct producers who are in possession of their means of production has to be mediated by some form of 'extra-economic' power; (4) serfdom (in the narrow sense of labour-rent) as the characteristic though by no means the exclusive form of appropiration of the surplus, and generalization of the same (i.e. conversion at least in theory of all allodial tenures into dependent, villein status); this is not of course to deny the existence of free peasant allods and common lands in actual cases.

Now in a broad sense, the term serfdom can be said to refer to hereditary, juridical servitude—which is openly institutionalized— of the majority of the peasantry. In this broad sense serfdom would include the whole range of servile obligations besides labour-service.

(5) Low level of development of techniques of production and division of labour, both in a historically relative sense; (6) coexistence of large-scale landed property and small-scale peasant farming; (7) finally the position of the medieval Western European towns, both with respect to the feudal agrarian economy and to their internal organization, was historically distinct enough to be regarded as a crucial structural characteristic of the feudal system: they grew in opposition to the countryside

(i.e. with political and military autonomy) by virtue of a well-developed system of self-government free from direct monarchical or seigneurial control, and became vital centres of rapidly increasing merchant-manufacturing activities which were controlled by a distinct urban patriciate and tightly organized in guilds and corporations.[4]

The features listed above characterize the *internal* structure of the feudal system at any *logical* moment, and since our concern here is not movement through time, as it were, of the system, nor the relations between parts *within* the system (both of which of course depend on the contradictions arising out of the structural characteristics of the system themselves, as enumerated above), we do not need any ordered structure of determination of the system.

Nevertheless the specificity of the feudal dynamic or agrarian technological development, as shaped by the precise nature of class struggles within the structural limits of feudalism, must be firmly borne in mind. By way of summarizing a vast literature on medieval Western Europe and a few centuries of history, feudal class struggles can be classified under three broad headings: first, struggles over surplus labour/product (level of labour services, 'customary' rents from the peasant plots, arbitrary exactions like. tallages, entry fines, etc. imposed by the lords); second, struggles over land and the mobility of it (relative size of the demesne, in particular, attempts to increase it by confiscation of peasants' holdings by occupying deserted holdings or forcible seizure of peasant-reclaimed waste or of common lands; attempts to hamper inheritance/sale of land by imposing heavy fines for the same); and third, deriving from the other two kinds of struggles, struggles over the mobility and freedom of the peasantry.

In this scheme of the feudal system an analytical separation has been made between the feudal form of property in land and the characteristic form of appropriation of the surplus under it, namely labour-rent on the one hand and on the other between the latter and serfdom in its broad sense. This has been done in order to appreciate the logical possibility—which was in fact historically realized in the absolutist phase of feudalism— that even though serfdom in the narrow sense of labour-rent is abolished the feudal form of landed property may continue and the productive surplus-extraction relations could meaningfully be called feudal on that basis, i.e the feudal nobility is still the major appropriator of the social surplus. This of course is notwithstanding the historical transformation of the feudal form of property itself during the absolutist period towards a new form, namely in the direction of landlord-capitalist organization of production—the 'Prussian path'—or the triad system of landlord—capitalist tenant—agricultural labourer of English agriculture, or freehold peasant property as in France. It is important to emphasize that the elements listed here only adequately characterize the feudal system in their totality; the system cannot be derived on the; basis of any single element taken in isolation. No individual element provides a sufficient basis for feudal *productive relations* to emerge.

One cannot in general argue, in the specific case of India, that since the historical antecedents were different from those of Western European feudalism the latter could not have developed in India for that reason alone. There is the interesting case of Japan which went through a feudal epoch not essentially different from that of Western Europe, though having completely different historical antecedents. Hence the need to examine closely the existing evidence from Indian history on the basis of which an Indian feudalism is said to obtain.

It is possible to classify the 'pre-feudal' views of pre-colonial Indian history into certain categories, partly on the basis of the chronology involved in each case: (1) that feudalism was flourishing in India roughly between AD 500 and 1000, after which it started disintegrating (Sharma);[5] (2) that there were qualitatively two different stages of Indian feudalism, the second stage continuing into the Mughal period (Kosambi);[6] and (3) that there was a tendency towards feudalization from the eleventh century onwards which reached its peak during the Mughal period, before the British conquest broke off its consolidation (Kovalevsky).[7]

To begin with the third category, Kovalevsky posits a pre-feudal agrarian structure in which the most typical kind of property is communal landholding. There are five stages[8] through which the original communal form develops towards individual property in land and Kovalevsky identifies two major forces which propel this development: pressure from the tribal chiefs who wanted to claim for themselves special rights of granting unoccupied lands to individuals or groups willing to bring them under cultivation; and pressure from the priestly caste, the Brahmins, who wanted gifts of land and its revenues to sustain their religious activities, while the prevailing inalienability and indivisibility of village community and family land were a real obstacle to such gifts. The process of feudalization, according to Kovalevsky, was given a further impetus by the Muslim land policy of the granting of *iqtas*.

Now what did Kovalevsky understand by feudalism? He had actually set himself four criteria to judge what was feudal in India, namely a system of benefices, commendations, leasing of offices, and patrimonial justice (private jurisdiction). He came to the conclusion that whereas the first three clearly existed in India under the Muslims, the fourth did not. But, importantly, Kovalevsky recognizes the power of the central government in explaining the limited development of feudal relations and the persistence of 'free' peasant cultivation (rather than all-round serfdom), and popular rather than feudal justice. His qualifications are extremely important and well worth quoting in full:

> While in the West the tenure of land together with the exercise of supreme rights within the limits of individual communities and districts becomes in the end the property of the former benefice holders and local officials, where officers becomes hereditary by law, in India towards the end of Muslim sovereignty this same result was achieved in only a few districts; in the others communal and individual property remained, as before, in the hands of the native holders, but the exercise of governmental functions was entrusted to officials appointed by the Central administration ... India's process of feudalization at the time English conquered her was still far from complete....[9]

It is thus not difficult to see the distance of Kovalevsky's understanding of the term 'feudalism' from the characteristics of feudalism which I have enumerated above. Nevertheless the changes noted by Kovalevsky and their impact on communal landholding cannot be disregarded and must be conceptualized in a definite way.

The view that feudalism flourished in India between AD 500 and 1000 has been put forward by many historians, but the most formidable range of evidence has been presented by R. S. Sharma. The essential change which started off the process of feudalization has been almost generally recognized as the practice of increasingly widespread land grants mostly, though not exclusively, to Brahmins, resulting in the disintegration of state power: 'the comprehensive competence based on centralized control, which was the hallmark of the Maurya state, gave way to decentralization in the post-Maurya and Gupta periods'.[10]

Without going into the details, a review of Sharma's evidence suggests that its rather inconclusive nature forces him to make a number of qualifications in order to admit the crucial differences of Indian 'feudalism' from the European variety. The differences pointed out are: unlike Europe, in India the decentralization of political power was not the result of fiefs; foreign invasions did not play any appreciable part in the process of feudalization, as was the case in Europe, the *agraharas* or villages granted to Brahmins bore some resemblence to manors, for in some cases the beneficiaries enjoyed the right of levying forced labour of all varieties on their tenants, but, importantly, there is little evidence to suggest that most peasants were indeed subject to such intermediaries; on the contrary, the number of free peasantry seems to have been much greater. Further, the process of subinfeudation was not as extensive in India as in Europe, so that the actual tillers of the soil maintained an indirect connection with the central government, though epigrahic records for particular regions like Marwar, Saurashtra, Orissa and Bengal contain references to transfer of individuals peasants and labourers as part of grants of land (importantly, this need not necessarily imply serfdom: this could mean that already occupied as against unoccupied villages were granted the right consisting of a share in the revenue previously going to the state); in medieval Europe land was granted to the feudal barons for services rendered to the state but this secular counterpart of benefices was of a rather limited character in India; finally there is nothing conclusive known about the rights and obligations of the conquered feudatories known as *Samantas* who appeared as feudal vassals from the sixth century onwards, except that they had to furnish soldiers to their lords; further, vassals in India had to render to their lord mainly military and not administrative service as in Europe.[11] A long list of qualifications indeed. Nevertheless Sharma says, 'the main characteristics of European feudalism, the self-sufficient economy buttressed by lack of commercial intercourse and the rise of landed intermediaries leading to the subservience of the peasantry, prevailed in India'.[12] This perhaps needs no comment. In any case from our point of view it is important that there is no evidence to suggest that the various grantees possessed any right to alienate their respective rights: "Gupta grants from Bengal and Central India confer the right of enjoying the revenues from land in perpetuity on the grantee, but they do not authorize him to alienate or grant his rents or land to others.'[13]

K.S. Shelvankar also adds too many qualifications to the Western notion of feudalism: 'Indian feudalism remained fiscal and military in character, it was not manorial. There was in general none of the intermingling of peasant land with demesne land in a common village ... as marked the manorial system.... There was therefore nothing similar to the direct conflict between the manorial lord and the peasantry over the disposal and cultivation of the land and of labour services which agitated Europe from the twelfth to the eighteenth centuries.'[14] He goes on to point out that none of the major conflicts in Indian history had for its object the exercise of rights over the village. There were conflicts between overlords of various grades for the right or power to get a payment from the peasant, not to seize his land.

Finally we have Kosambi's pioneering suggestion of two stages of Indian feudalism. Kosambi implicitly postulates a stage around the beginning of the Christian era when a closed peasant village economy prevailed with practically no superior landowning class or overlords, and only limited authority and exactions were claimed in kind. Starting from such a pre-feudal social structure there was a process whereby the king alienated his rights to subordinate chiefs who came to have direct relations with the peasantry, a process which Kosambi calls 'feudalism from above'. He seems to think that it reached an advanced stage of development during the Gupta period A.D. 400-500 and Harsha (seventh century). At a later stage a class of landowners was to develop within the village, between the state and the peasantry ('feudalism from below').[15]

On the basis of Kosambi's own evidence, however, there is one feature that was common to all grants: the recipients gained at most the rights the state would normally claim, i.e. they collected the taxes already fixed by usage. No portion of the tax was to be passed on to the state or any state official but the donor had not the right to increase such taxes, nor any property rights over land and cattle. Referring to a sixth century Damodarpur plate supposedly showing purchase of land from the state, Kosambi makes the point that 'what had been bought was not the land, but the right to cultivate it in perpetuity without taxes'.[16]

After examining the whole range of evidence regarding the nature of rights in land in medieval India, Kosambi makes three general points which are of interest in the context of the notions of 'conditional' and 'absolute' private property in land:

> In the first place, most of the actual cultivators had emerged from a tribal stage where land was only territory, while primitive slash-and-burn cultivation had made individual plots useless till the day of the plough and cattle-manure fertilisation. Secondly, the holding of a field, even in the sense of mere right of cultivation, was a privilege as well as proof of membership in a community. Loss of all land would not be possible unless the individual were expelled from the peasant sub-group, usually a jati, caste. Finally, within a village community that produced virtually no commodities, land would have no purchaser, while uncleared waste or marginal land was still to be had for the cultivator.... This state of affairs continued almost to the end of the Mughal period, with local variations.[17]

These statements somehow do not square with the terms 'feudalism from above' and 'feudalism from below'. At the same time they draw attention to certain typical features of medieval Indian land rights.

Hence, to conclude our discussion of feudalism, the general tendency of those who argue that there was a feudal epoch in Indian agrarian history is to fit evidence into a far too narrow interpretation of the category of feudalism and thereby to minimize or disregard some of the crucial differences between pre-colonial India and Western Europe or Japan, with respect to the nature of property rights in land, social organization of production, state (political) power, etc. In the process the original argument in favour of the category of feudalism is ultimately burdened with too many qualifications.

This leads us to examine the other major category used to characterize the pre-capitalist/pre-colonial structure or agrarian relations in India, namely the AMP. The fundamental characteristics of the AMP as a working hypothesis according to its 'classical' formulation[18] are: (1) it is characterized by the absence of private ownership of land; (2) as a result the village community retains an essential cohesive force which has withstood the bloodiest of conquests through the ages; (3) this internal cohesion of the ancient village community is further increased by the close union of agriculture and craft industry that exists in it; (4) for reasons of climatic aridity, however, the prosperity of agriculture in these regions requires impressive hydraulic works. This irrigation requires nearly everywhere a central authority to regulate it and to undertake large-scale works; (5) for this reason the state succeeds in concentrating the greater part of the social surplus product in its own hands, which cause the appearance of privileged social strata maintained by this surplus and constituting the dominant power in society. The size of the social surplus under the AMP also allows for an advanced system of social division of labour (i.e. for the maintenance of a large body of full-time specialists not directly producting their own food).

The internal logic of a society of this kind works in favour of a very great degree of stability in basic production relations. This is because—and this is a crucial point—of the unmediated (i.e. not mediated by any class power) dominance of state power in a society of this kind, whose history is thus not properly a history of class struggles.

This scheme of an AMP is simplified and analytical and should not cause the reduction of a complex notion to one or more of certain *isolated* characteristics as if they were the determining ones or as if they were the basic premises of the entire concept as, for example, the mere absence of private property in land (thus, 'in the final analysis', some kind of state ownership of land, in which case if one finds the sovereign power 'purchasing' land from individual peasants, this constitutes a sufficient basis for denying the notion of AMP, as with Irfan Habib); or as Wittfogel emphasizes, despotic form of political rule following centralized water control and management ('hydraulic state'); or following state ownership of land, etc. Instead what the concept essentially refers to is a complex set of collective and traditional institutions and institutional constraints as well as customary and religious practices, all of which give a certain stability to basic social production relations and serve to diffuse class conflicts in a systematic way, to the point of eliminating them. Some of the more important of such institutions, institutional constraints and practices can now be discussed.[19]

To start with the village headman as a Janus-like entity was an important institution responsible for reducing and redirecting possible class oppositions. For on the one hand he was the representative of the state in the village, being an extension of the state's tax-collection machinery, the village being the unit of tax collection (i.e. the village cultivators being collectively responsible for the payment of the lump-sum tax imposed on the village). On the other hand he was the embodiment of the collective interests of the village community of which he was a member (though not always the wealthiest) and often had kinship bonds with the villagers. As a result, 'he was the village advocate against oppressive increase in rent and tax.... his position was, by its very ambivalence, an important reason for the absence or low degree of class conflict in the traditional village in Asia.'[20]

Further, there were a number of institutional constraints, both positive and negative, internal and external, operating on the villagers under the AMP, sustaining social relations in their traditional forms. On the one hand the individual as a member of the community was bound to the community as he was to the soil, by customary practices (see below). While this was an internal and positive factor of bondage, the obligations to the state in the form of rent and tax were a further positive constraint or bond, but an external one. 'The internal tradition and the external obligation served to hold each other reciprocally in place.'[21] There was a negative and external constraining factor as well, namely the lack of alternatives for employment, there being no open labour market. The individual 'cannot open up new land for cultivation save with the greatest difficulty: land is not hard to come by, but the labour, animal power and implements for clearing, watering, fertilizing the land are scarce.'[22] These constraints on the individual in the AMP 'are the opposite of the forces that push the individual off the land and pull him toward the city in modern times in Asia; they are the forces that have bound the individual to the soil in Asia in ways that are distinct from those found in European history.'[23]

One of the most important customary practices under the AMP can be identified as that of a periodic redistribution of land (both arable and homestead, and taking account of the quality of land in the case of arable) as an institutionalized means of checking inequalities within the peasantry on the basis of permanent, effective possession of land; in addition there was the practice of a periodic readjustment of the lump-sum revenue burden on the village to ensure an egalitarian sharing of this burden by the cultivating body which had a collective responsibility to meet this demand.

Again, the social bonds of the village communities in the AMP, because of such customary practices — including collective organization of work for production — went beyond mere kinship ties. They,

were at once social groups deriving their livelihood from their land. They worked upon the land because they were related as kin, as neighbours, communally and hence socially; at the same time, because they worked upon it their relations in kin were socially acknowledged; because they worked the land, its passage from one generation to the next in the same descent line was accepted by the social unity as a whole, in this case, the community.[24]

Thus the entirety of the communal relations, and hence the social relations under the AMP, could not be reduced to kinship relations.

These are some of the collective/traditional institutions, institutional constraints and customary practices under the AMP. Looked at from another angle the internal structure of societies characterized by the AMP is made up of a series of identities or non-divergences, of categories which are either opposed/diverged or not all present under a different mode of production (feudalism or capitalism); some of these inextricably linked categories are: tax and rent; community and state; public/civil law and custom; communal relations and social relations; social division of labour and village division of labour; economic and extra-economic bases of surplus appropriation; class power and state power; property relations and political relations. Though the last three identities are also particularly true of feudalism, the way in which the two sides of each identity is combined is different.

Thus, it should be recognized that on the material plane of social evolution the idea behind the introduction of the concept of AMP was to bring out the specific oppositions of this mode of production to feudalism on the one hand and capitalism on the other; in particular to account for the lack of social/technological development of 'Asiatic' societies by theoretically locating the productive forces held in check by the structure of societies resting on the AMP, and thus to provide a critique of the latter category itself. This point is not meant to imply that the 'Asiatic' societies are incapable of *any* technological development; rather that they have a tendency to get *permanently* stuck at a certain threshold of development. This level, there is some quantitative evidence to suggest, could be higher than the *average* level of slave-based societies, or even of feudalism. In order to be more precise we need comparative data of a kind which is very hard to come by. Nevertheless the clarification should be borne in mind in order to distance oneself from the orthodox thesis of a perpetual stagnation under the AMP ever since the latter's emergence.

So much for the internal theoretical structure of the AMP according to its supposedly 'classical' formulation. As regards the historical 'placing' of the concept there are two related interpretations that have been put forward. One is that 'the idea of an Asiatic mode of production is related not just to some 'primitive' Indian or Chinese society, lost in the mists of the past, but to *Indian and Chinese society as they were when European industrial capital encountered them in the eighteenth century.*'[25] The second possible interpretation, which does not necessarily exclude the first, is that the AMP constitutes one of the possible forms of transition from classless to class societies and that it contains the contradiction of this transition, i.e. the combination of communal relations of production with embryonic forms of the exploiting classes and of the state.[26] Without going into the logical problems, particularly in connection with the second 'interpretation, the essential point seems to be that there is a need for more historical research so that the process of dissolution of the Asiatic communal form of property is better understood.

As regards the first interpretation, there are a number of problems connected with its applicability to the concrete data of Indian history, some of which have been pointed out by Irfan Habib on the basis of his researches in Mughal Indian history.

I can only touch upon some of these problems and not pretend to provide any straightforward solutions.

First, the key role attributed to hydraulic and other large scale public works in the establishment of the AMP is said to be absent in Mughal India in particular, crop production being principally dependent on rainfall and village wells and tanks.[27] This is undoubtedly an important negative historical condition, all the implications of which cannot here be examined. Yet this need not be a critique of the internal logic of the model itself since the appearance of a hypertrophied despotic state is conditional on the need for large-scale hydraulic works. When these works are carried out in their essentials at the village-level, despotism does not necessarily result. If it does, even in the absence of large-scale hydraulic works—as seems to have been the case in India—the source of such despotism has to be sought elsewhere. This also seems to be true of conditions in northern Africa and particularly the Maghrib (the Arabic name for a group of countries in north-west Africa: Morocco, Algeria and Tunisia) in medieval times, and has been discovered to be a case of 'one AMP without large hydraulic works.'[28] As recent series of studies on pre-colonial Maghrib show,[29] the material basis of the monarchical state formation here was the state's monopoly of a large long-distance (trans-Saharan) trade; also that this was thus not a 'hydraulic state' as classically conceived but a 'tributary state' society with certain unique features all its own, such as the chronic instability and fragility of the state structure. The absence of a bureaucratic apparatus or standing army of any size was also revealed, as was the persistence among both peasantry and aristocracy of tribal affiliations and tribal ideology.[30]

Thus the general point which seems evident is that the need for large-scale hydraulic works is at best a sufficient and not a necessary condition for the despotic form of political rule of the classic 'Asiatic' type to emerge. Moreover only the form of political rule is the result of large-scale hydraulic works under the AMP. No other element of the concept can possibly be derived from either the existence or absence of centralized water control and management.

Therefore, partly taking cue from the situation in pre-colonial Maghrib, the data of India history would seem to fit into a weak variant of the AMP, corresponding to the prevalence of localized irrigation or rainfall farming. The distinguishing characteristic of this weak variant would be a tendency towards what has been called 'permanent feudality' (not to be confused with feudalism),[31] i.e. a recurrent tendency towards decentralization of political power, the appearance of small kingdoms, tribal aristocratic chiefships, etc. with varying degrees of subordination/autonomy, and having tributary relations with respect to the 'centralized' monarchy. This would also imply a need to accede special privileges to intermediate and village levels of power and all that needs to be done to ensure formal loyalty—which already has a shaky basis—to the 'centralized' monarchy. This would be in contrast to 'the strong internal coherence displayed by the strong variant of the Asiatic mode and its tendency to reassert itself periodically against seigneurial tendencies or at least against local separatism.'[32] This then is an important qualification to the 'classical' formulation of the notion of an AMP.

Second, again with respect to Indian history, the weight of historical evidence is said to suggest the absence of any form of communal tenure as far back as the early years of the Christian era.[33] This generalization is certainly debatable. There is evidence of such communal practices as periodic redistribution of the arable, collective organization of work for production, collective responsibility of the village peasantry for the payment of a lump-sum tax-rent to the state, periodic adjustment of the latter to ensure a certain egalitarian sharing of the same burden, etc., which in certain regions persisted even into the colonial period. What is not known and seems important to research are the conditions under which such practices fell into disuse at different points in time in different regions even in the pre-colonial period, leading in particular to inequalities within the peasantry.

Moreover, and more importantly, these so-called 'communal' practices were often enmeshed with caste factors and excluded a part of the rural population, namely the landless agricultural labour caste.

Third, we come to caste in general, and the problem of the existence of a landless agricultural labour caste in particular. I shall not deal systematically with the challenging data of Indian history in connection with rural social relations of production but will make a few tentative suggestions.

An isolated overemphasis on the caste factor, or the view that this was *the* distinguishing feature of Indian history, needs to be avoided. This is because from such a view it is but a short step towards suggesting that in the place of feudal class-based exploitation what we essentially see in Indian history is a caste-based 'exploitation' which in many cases, was indistinguishable from class-based exploitation. An extreme example of this position is in fact the suggestion of a caste-based feudalism in Indian history.[34]

By way of commenting on this view we start with the fairly straightforward observation that caste in pre-colonial India can be looked upon as an institutional means of maintaining the social division of labour in its traditional form on the one hand and the identity of the social and village divisions of labour on the other. To this extent there is clearly no contradiction between the institution of caste and the notion of an AMP. In particular, caste can be looked upon as a very important element of the internal tradition under the AMP, i.e. as an internal and positive institutional constraint.

But then is caste-based specialization necessarily and under all historical conditions exploitative, in the concrete Marxian sense of a system of privileged access to the means of production? The answer in my opinion is no. For one thing 'social division of labour means functional specification within the vocational system; but vocational groups differentiated by knowledge and skill need not per se develop opposing interests that result in differential access to the means of production'.[35] In any case it has to be shown how 'functions of domination *had* to emerge from contrast of interests rooted in vocational specialization'.[36] For example in a typical pre-colonial Asiatic society there was a social division of labour within the politically ruling elite (the priesthood, military and bureaucracy) as well as within the working population (for example between peasants and craftsmen).

Thus it would seem incorrect and one-sided to locate exploitation *within* the caste system, particularly in the context in which the state is the major appropriator of the agrarian surplus. In this connection it needs to be worked out how the unmediated dominance of state power (which, as we have seen, is a crucial structural characteristic of AMP) and the caste system relate to each other. In particular it needs to be examined how the two hold each other in balance or else exert centrifugal influences on each other, the net result in both cases being the *non-emergence* of class power in its proper sense. This is a hypothesis that sounds logically plausible but which must be subjected to rigorous testing.

Coming now to the landless agricultural labour caste: this group in the pre-colonial situation was semi-free (as compared to capitalist wage labour which is doubly free), i.e. it was free from ownership/ possession of means of production but could not enter a labour market which, in any case, was non-existent. In fact the non-existence of a labour market and the caste system can be said to be mutually supportive, they reinforce each other and hold each other in balance.

In what sense then can we say that the landless agricultural labour caste constituted a class and, if so, which class was it opposed to? I would like to hypothesize that we may expect a correspondence between the extent of unoccupied land grants to Brahmins/secular grantees and home-farm lands of Zamindars, village headmen and other revenue-collecting authorities on the one hand, and the incidence of a landless agricultural labour caste on the other. We should find some kind of employer-employee relations here, given the absence of an enserfed peasantry and the caste restrictions on manual labour. In particular, here one has to examine the non-market principles of wage determination that were followed.

Moreover here one has to see such employer-employee relations in conjunction with larger social relations and conditions in pre-colonial Indian villages, such as the existence of a peasantry— constituting the exploited majority—which was politically subordinated to the state power and not to any class power. Clearly therefore the specification of the institutional core around which the relations of production crystallized in this situation has to take into account the totality of the social relations of production, in particular the relations between the state and peasantry.

Fourth, I turn to the institution which grew up on the basis of the caste system and specified the larger social relations of exchange in pre-colonial Indian villages, namely the *jajmani* system, or the system of patron-client relations. I do not intend to engage with the various controversies around the interpretation of the jajmani system. In any case one clear-cut implication of the system is that it systematically eliminated all market exchanges and/or monetary transactions so far as intra-village exchanges were concerned, excluding those monetary transactions necessitated by the tax-rent demand of the state in cash, if at all. It could also have been a system of distribution of the rising net surplus, if the latter obtains as a result of a growth of productive forces, after the state's demand is met. In that case one has to see what principle (egalitarian or otherwise) of distribution was followed. It is also relevant to see whether the jajmani system took care of the problem of steady employment of the agricultural labour artisan caste, as it seems it did, free from the uncertainties of a market-mediated system.

Pending a systematic incorporation of these elements into an overall theory of mode of production for the pre-colonial period of Indian agrarian history (which will also lend flesh to the as yet open notion of an AMP) it is at least clear that in my opinion the Asiatic form, looking at it from the dawn of the colonial era, is historically best treated as anterior to two important processes of change in Indian history: one a 'supra-village' process of land grants (both religious and secular; revenue-free or revenue-privileged), particularly grants of unoccupied land as well as revenue assignments etc., which brought about or was later to bring about certain qualitative changes in the nature of rights over land; and the other an intra-village process of individualization of peasant cultivation and possession of land (of a *permanent nature)* as the practice of periodic redistribution of land, collective forms of organization of production, etc. tended to fall into disuse over time; and finally there are the interrelations between the two processes, all of which must be recognized and conceptualized on their own basis.

ALTERNATIVE CATEGORIES

It seems to me that a fruitful way of grappling with the existing evidence with regard to the system of land tenures in the pre-colonial period of Indian land history (without summarily rejecting the notion of AMP) is by way of posing the problem in terms of a transition from one form of private property to another form of private property in land. The two analytical categories of private property may be called *conditional,* and *absolute* private property in land, though these terms must not be taken too literally. Now there are two fundamental conditions defining absolute private property in land:

(1) The unconditional right to alienate (not only through inheritance but also through sale), where 'unconditional' is taken to mean that (a) the right of one private individual excludes a similar right of another individual (one holding a legally inferior right) to alienate the same plot of land (or a part) of which the latter individual is in possession, and (b) absence of any independent rules of alienation (those regarding whom to alienate, etc.) imposed from 'above' say by the 'state' or any form of political power.

It should be evident that the fulfilment of this right presupposes two things: that land itself is being alienated and that there has developed a market in land itself in the 'larger economy' with its own presuppositions (commoditization of the product, a certain level of productivity, a certain limitation on the state's demand on the surplus product, etc.). This right also serves to dissolve all forms of social exclusiveness (caste-based, noble lineage-based, etc.) of ownership of land, and imparts to land a social mobility, so to say.

(2) The right to regulate occupancy and appropriate and revise rent, i.e. to appropriate a given and *rising* surplus on the basis of a determinate social form of organization of production. This right implies and is implied by the condition that the absolute private property-holder in land has a complete right to eviction which he either actually exercises, or is otherwise able to change the existing terms and conditions of the contract with the direct producer to capture the (rising) surplus or a greater share of a given surplus/total output. This implication has two further presuppositions: first, a certain degree of demographic development whereby land is no more relatively abundant (relative to labour availability at a given level of technological development), and second, the direct producer has no customary occupancy right (on the basis of first clearance and/or clan/caste membership, etc.) or legal right in land (and its rising surplus) whatsoever. Thus these two together become the necessary and sufficient conditions for the right of eviction to be effectively exercised. Some qualifications to this point that the level of demographic development is a necessary condition for (2) to be satisfied must immediately be added.

I am suggesting that a low level of demographic development, i.e a relative abundance of land with respect to labour, sets negative limits within which landowners on the one hand and peasants on the other can make possible choices and can respond to the situation in predictable ways. On the one hand landowners can attempt to tie down the peasantry to the land on the basis of some form or other of political (state) military or judicial power, as under feudalism, or at least try to restrict the movement of the peasantry. One of two choices is available to the peasantry: the first is migration or movement into inferior soils by reclaiming waste. Here the level of technology, besides the availability of labour, itself sets limits to possible extensive cultivation. The second choice—the one often exercised as the possible exercise of the first choice narrows down—is collective resistance to the payment of a rising rent. The latter partly depends on the historicity of land relations itself, on the extent to which such a collectivity actually obtains, and the extent to which there are social and economic divisions within the peasantry. When there are such divisions the nature of resistance would seem to depend on whether a stratum of peasantry is to bear the increased demand for rent or more or less the whole, etc.

Absolute private property in land is different from capitalist private property, not on the basis of the first defining characteristic of the former as described above but on the basis of the second, which only implies monopoly control over land but does not necessarily imply direct, organized control over the process of production in the sense of control over the decisions regarding what to produce and how to produce. This is something which capitalist private property in land does, whatever be the specific juridical form of 'ownership'. That is to say in a capitalist enterprise such a control over production becomes an absolutely necessary condition for surplus-appropriation, thus marking the emergence of the integration of production and appropriation, unlike with pre-capitalist relations of production.[37]

Now, conditional private property in land is defined negatively when at best one of the conditions defining absolute private property is satisfied, but *not both.* I give below a few historical examples, some of which have earlier been referred to, to illustrate the nature of conditional private property in land, without implying that this was the dominant form of land relations of the respective periods to which the example belong. These examples should also help to bring out the significance of the two conditions defining absolute private property in land.

Example I: Numerous land grants to Brahmins during the Gupta period can be regarded as a form of private property in land, but as we have seen above the grantees did not have a right to alienate. Moreover it is not clear—though in some cases people along with land, i.e. inhabited tracts, were granted—whether the grantees could claim from the peasants more than that which the king would normally claim by fixed usage.

Example 2: An examination of the nature of the iqtas (the revenue assignments granted by the Delhi sultans to their military commanders) show that the *iqtadars* could not claim ownership over particular parcels of land or locality and that they could be provided with or deprived of their income at the will of the king, whereby the question of alienability of land does not arise.

Example 3: During the Mughal period the nature of 'property' rights of the crown, 'nobility' *(Jagirdars)* and the lower level 'aristocracy' (Zamindars) on the one hand and the peasantry on the other perhaps brings out what I mean by 'conditional' private property in land.

There is no denying the fact that the Mughal crown had a right over the disposal of the surplus produce, which was reflected in its rights to create revenue assignments; corresponding to this the peasantry did not have a right over its entire produce. But the crown/nobility/aristocracy did not possess (could not exercise, given the objective condition of relative abundance of land) the right of eviction of a peasant so long as he was cultivating the land. Corresponding to the absence of a right of eviction the peasant had a customary 'right' of occupancy.

On the other hand we find that the crown and the nobility had the right to restrict the movement of the peasantry. This was evidently required, in the absence of right to evict, in realizing a given revenue demand and, more importantly, in stepping up the same—as was actually done as the cost of palace establishment increased. Corresponding to this the peasantry lacked the freedom of abandoning cultivation in any given locality, i.e. the right of free disposal.

Example 4: The case of revenue-grants (known as *madad-i-ma 'ash* grants) of the Mughal period, granted mostly to men of learning, religious devotees, persons of noble lineage, etc., is an interesting instance of 'conditional' private property in land. First, the grantee could never transfer or sell the land under his possession, nor could he pass it on to his heirs, except in accordance with imperial order.

Second, the grant did not extinguish the right of the village headman *(moqaddum)* even when the grantee held the whole of the village, nor did it in any way affect the Zamindari rights established over the village, nor could the *madad-i-ma'ash* holder interfere with the occupancy rights of the peasants.

Third, 'what was granted was the right to collect the land revenue and keep it ... the grant, therefore, did not invest the grantee with any rights not claimed previously by the administration. He could not legitimately demand a larger amount of land revenue than was authorized . . .'[38]

Example 5: Without going into details it need only be pointed out that land rights in Vijayanagar society consisted only of income rights in land. Burton Stein, after examining the *nayaka* system, comes to the relevant conclusion that 'land rights do not pertain to dominion in land, but to "property" in share of income'.[39]

Example 6: Finally, we have some evidence of the nature of Zamindari and other 'property' rights in land for the immediate pre-colonial period, during Maratha rule gleaned from a few deeds of sale.

An examination of the deeds of sale of various grades of supra-village landholding interests, like that of a Choudhury's *taluk* in 1760, portion of a taluk of willaity kanungo in 1724, a Zamindari deed of sale in 1800[40] etc. brings out a number of interesting features. Apart from the fact that all these sales were in liquidation of balances of the public assessment of revenue, there is absolutely no mention of the actual quantity of land that is being sold: in the case of Zamindari sale it is said to be the *dufturry* (office?) of Zamindari that is disposed of; in the other cases the words *tussurroof malikaneh* or proprietary possession frequently occur. We are spared the mistake of calling that which is being sold, in such cases the ownership right in land itself, when we come across the following deed of sale of a moquddum 'under' a talukdar:

> I, who am Bynsee Sawant Sinnar . . . moquddum of Mouzah Isanpore, under the talooks of Juganathpersand, in the Sircar of Budruck: since I have hitherto held the moquddum of the above mauza in *full proprietary possession*. . . but am unable to pay the public revenue assessed thereon; therefore *of my own free will*, in full possession of my sense, I sell the moquddum of the above mauza . . . (dated 21st Ramzan 1178 Umlee)[41]

The implications of this are best described by Stirling, the translator of the deed of sale in 1822:

> The words, 'tusurroof malikaneh' ... are used as frequently by the selling moquddum in reference to what he sells, as by the talukdar, although avowedly subordinate to a superior holder of the latter description.... The words, I am persuaded, should be understood as descriptive of the sort of tenure of the moquddum—one enjoyed by ancient prescriptive usage, constituting a title of right, and not one which is dependent on the will of another, and cannot be transferred.... It is evident that the interest of each moquddum in the village under him was greater than that of the talookdar of the division ... although the talookdar having authority over and collecting his dues from a number of moquddums, checking their accounts, and having besides under his own management several pahikasht villages, was an officer of higher rank and more extended emoluments, as upper link in the chain of dependence.... At all events, if we consider that the thing sold always was the actual village or villages, and not merely the right of managing with the advantages incidental thereto, and thence infer a property in the soil, we are brought to this dilemma that the talukdars and moquddums were constantly proprietors of the same subject. Such a conclusion would be an absurdity, but there is none in considering them both to have possessed certain rights and official duties and emoluments in regard to the same land or villages. [42]

All in all on a descriptive plane (on an analytical plane it must be seen vis-a-vis absolute private property in land), it is possible to conceive of conditional private property in land as referring to a gradation of right 'interests' over land, one on top of another, and typically *based* on individual peasant cultivation and possession of land of a *permanent* nature. One important historical factor contributing to this seems to have been various modes of revenue-collecting arrangements introduced by successive state powers, mainly on the basis of political expediency .[43] Stirling grasped the historical nature of this well when he wrote:

> In discussing the real nature of the rights and privileges of the Zamindars, talookdars and moquddums ... it is much easier to say what they were not, than what they actually were. It is my decided opinion, that from the hereditary character pervading so remarkably all the institutions of the Hindoos, they at all times, possessed *an imperfect title of property in their offices.....* They had the right of collecting the government revenue assessed on the land *according to certain fixed rates* or determined proportion of the crop as settled by Raja Toorul Mull, whose general settlement was unquestionably the basis of all subsequent revenue management.... '[44]

Property in Land and Mode of Production

Now, one objection to the notion of conditional private property in land could be that the feudal form of landed property is also conditional in some sense, and hence the new category does not really help us in moving away from feudalism. This objection is easily disposed of. First, feudal landed property is noble property, which conditional private property in the Indian context is not. Second and more important, the feudal form of landed property is conditional because of the obligation of military service to the overlord, but conditional private property is conditional with respect to the situation 'down below'. In contrast the feudal lord had complete control over land, organization of demesne production and peasant labour (in particular, the latter's overall *reproduction) within* the manor. Thus feudal landed property may be looked upon as a particular combination of conditional and absolute private property in land, if one so likes.

Therefore in using the notion of 'conditional' private property it does not follow that a feudal *mode of production* is what we are necessarily talking about, nor that we are trying to collapse the entire socio-historical context in which private property in land was sought to be created, if at all, to mere property relations. Rather, through the notion of conditional private property in land we are implicitly contending that this larger socio-historical context can be usefully studied by means of the notion of an AMP (with suitable modifications of the latter's 'classical' formulation) and that conditional 'private property in land' and the AMP need not be inconsistent with one another.

TENANCY RELATIONS IN THE COLONIAL PERIOD: CUTTACK DISTRICT (TEMPORARILY SETTLED PORTION), 1837-1927

Having presented the historical examples of conditional private property in land in a more or less chronological order, we now arrive at the Zamindari right as created by colonial rule.

Even if it is granted that colonial rule attempted to create absolute private property in land on a generalized basis for the first time in Indian history, in the light of what bas been said it is impossible to believe that this could have been done overnight, by the stroke of a pen as it were, through Zamindari and raiyatwari tenures. By recognizing a class of private individuals as absolute proprietors of the soil the colonial state could at best have introduced an additional element of conflict, to be resolved over the whole of the colonial period.

On the other hand if we posit a tightly-knit village community before the British period, it becomes difficult to understand how an arbitrary set of private individuals could be so easily identified and superimposed on such a community. Thus it seems more plausible to suggest that such a superimposition could be facilitated rather by an already agrarian society on the basis of an appropriate alliance with the village-level dominant peasantry, i.e. by granting suitable concessions to the latter.

Thus the question should not be posed in terms of the impact of British land settlement on 'traditional' agrarian society, but in terms of a process of interaction between two structures of land tenures: the village-level landholding and rent-collection structure, including different forms of organization of production, elements of custom, etc. (the 'rent-tenure structure') on the one hand, and on the other the supra-village revenue-collection structure created by British law (the colonial 'revenue-tenure structure' consisting of those *directly* engaged with the colonial state for the purpose of revenue-collection, the colonial state itself being an important element of this structure).

Then it is possible to conceive of four possible patterns of interaction[45] in terms of long-term consequences between the two structures of land tenures which I have analytically separated, depending on particular circumstances (to be identified, as corresponding to each pattern of interaction). First, the existing landholding village elites might be completely displaced by the new holders of Zamindari titles who, by long residence, would convert the legal title into physical occupation of agricultural lands which belonged to the former village elites. Second, powerful Zamindars might succeed by the introduction of periodical survey and regular bureaucratic management, in increasing and equalizing assessment rates to such an extent that the surplus of the village elites would be wiped out, reducing them to the economic level of ordinary raiyats, even though they perhaps retain their political influence as village heads. This need not be done by the new Zamindars if the third possibility stated below is open. Third, the new Zamindars might succeed in raising assessment rates, far in excess of the amount justified by the extension of cultivation, over inferior tenants by reaching an understanding with the village-level dominant peasantry which would enable the latter to retain their margin. But in this case in a conjuncture of rising surplus, the 'alliance' may not be long-lasting or peaceful.[46] And fourth, because of the fact that under the revenue-collection arrangement of an earlier political power the village-level dominant peasantry (whose members have benefited in a number of ways from holding the offices of revenue assesement and collection, including political influence over the rest of the rural population) was vested with the responsibility of revenue collection—coupled with the facts of absenteeism, rifts between co-sharing Zamindars, scattered distribution of subdivided properties, etc.—this might so far weaken the local position of the new 'landlords' that the resident village landlords of peasant caste would raise their margin by appropriating the bulk of the increases in the economic rent of land arising from rising prices, new cash crops, increasing demographic pressure, increasing value of output due to irrigation, clearance of forests, etc.

Then it follows that the third and fourth possibilities will be reinforced if there is a considerable overlap between the two structures of land tenures discussed above. This in turn would depend on the extent to which differentiation has occurred with the cultivating peasantry and on the existence of a privileged class of village-level landholders paying a nominal rent, depending on the historicity of land relations. In particular, if the village-level dominant peasantry also happens to occupy (or attempts to do so by purchasing the newly created Zamindari rights) a place, say, in the lower rung of the 'revenure-tenure structure'. On the other hand the first and second possibilities will be reinforced in the absence of such an interpenetration of the two structures.

The particular course of development would also seem to depend on whether there was considerable reclaimable waste and the possibility of commercial agriculture, or on whether a relatively stagnant social surplus had to be shared out among different grades of landed interests.

Another contributing factor which could hasten or retard a particular line of development is the nature of intervention by the colonial state itself, the limits of which are of course set by its own interests (namely maximum and timely collection of land revenue), *within* which the kind of pressures it is subject to from different groups of landed interests would suitably modify the form and extent of intervention.

Having set out our expectations I shall now analyse data generated by concrete conjuncture. Anticipating the conclusion, it is that the third and fourth tendencies described above seem to be operating during two phases of our period for the region studied. This region, unless otherwise stated, is the temporarily settled Zamindari area of Cuttack district which constituted some 65 per cent of the then Cuttack district. This delimitation of the region has been done mainly because the relevant body of data that is comparable over time is available mostly for the temporarily settled portion.

The period of colonial rule in Orissa was preceded by about forty years of Maratha rule. The Marathas it seems followed the system of revenue collection under the management of thirty-two salaried *amils* (revenue officers of the government) and settled mostly with village-level landholding interests (head raiyats of villages), or local-level revenue assessors like *kanungoes* or revenue collectors, or like *Chaudhuris,* bypassing the collectors and managers of the revenues of large divisions, namely Zamindars and talukdars. These latter, as a contemporary said,

> had been very generally, though not universally, divested of the management of their estates for a period from 25 to 30 years by the Mahratta Soobadars, who thought that they could collect more without them, and with less annoyance to the ryots, than through their instrumentality ... I have *often* heard Mr. Ker affirm that he found the Zamindars of Cuttack in 1804, in a state of the most abject poverty and depression; they had not been for years allowed to interfere in the management of zamindaree concerns, and had lost nearly all their influence, and much of their knowledge. [47]

The immediate effect of the British policy of settling as far as possible with the existing and dispossessed Zamindars and talukdars on the Maratha 'revenue-tenure structure' must have been threefold: (1) some of the village and local-level revenue-engagers must have been completely dispossessed, perhaps sinking to the level of ordinary tenants, particularly in those cases where the dispossessed Zamindars and talukdars could make out a strong case for reinstatement; (2) some others must have managed to retain their revenue-collecting right though now subsumed under larger Zamindari estates, and thus had to pay their revenue through the Zamindars (those who came to be known as 'sub-proprietors' or 'proprietary tenure-holders'; see below); (3) but a number of Chaudhuris, kanungoes and moquddums, we know, did become 'proprietors' under the British,[48] though the extent of revenue paid and land held by then were most likely to have been much less than during the Maratha period.

More dramatic changes were in store. In the next fifteen years (1804—18), there occurred a phenomenal volume of sales of Zamindari estates for arrears of revenue and the consequent dispossession of the original proprietors; out of some 3,000 proprietors of the first settlement of 1804, only 1,449 survived till 1818. Thus 51.6 per cent of the original Zamindars paying some 78.5 per cent of the total revenue were wiped out in the three districts of the Orissa division.[49] That this was an extraordinary period is proved by the fact that, for the Orissa division as a whole, whereas for the period 1805-19 estates paying some Rs 9.5 lakhs of revenue were sold for arrears of revenue, during 1845-99 estates paying only Rs 1.5 lakhs of revenue changed hands for the same reason.

That this process of dispossession must have hit particularly hard those village-level landed interests who were recognized as 'proprietors' by the British in 1804 is attested by the following statement of Stirling, whose sympathy finally rests with these interests: 'It is my decided opinion that the operation of the British laws (in particular, the practice of public sales for recovery of arrears of revenue), has been, generally speaking, in Cuttack fatal to the rights and privileges of numerous class of middle tenantry or heads of villages... '[50] Another significant effect of this practice was likely to have been the heavy incursion of urban-based monied interests into the countryside as auction-purchasers of Zamindari rights.

Thus the relevant point which emerges is that from the very outset the body of proprietors (including the new auction purchasers) which came into being, say, by 1819, must have had to contend with a large body of formally dispossessed village-level and local-level landed interests.

The culmination of this process of 'rationalization' of the colonial revenue-collecting mechanism was the first major and long-term settlement of revenue and rent, based on a detailed survey and preparation of a record-of-rights, which came into effect in 1837 for a period of thirty years, i.e. revenue was fixed for that period. The expiry of the 1837 settlement coincided with the great Orissa famine of 1866-7 and it was therefore extended for another thirty years. The settlement carried out in 1897, known as the Provincial Settlement, was again a thirty-year settlement and thus expired in 1927 when the Revision Settlement was carried out (during 1906-12, there was a settlement for the revision of record-of-rights and certain categories of rents, but no revision of revenue). Therefore most of my data pertains to three time-points: 1837, 1897 and 1927.

III. i Dual Land Market

The practice of public auction of Zamindari estates for realizing arrears of revenue in almost all Zamindari areas of British India is widely regarded as the emergence of a market in 'land'. Though it is illegitimate to equate the purchase and sale of Zamindari rights with the emergence of a market in land as such, and with unconditional right to alienate in our sense, in view of the foregoing theoretical considerations, and as shall be clear below, what was happening in the market in Zamindari rights (in particular, the private sales of Zamindaris over a period of time and the average values at which they were being bought and sold) could be taken as an important indicator of the economic position of the Zamindars. This is particularly so when the data is juxtaposed with similar data for the market in raiyati land, which developed in the region of study comparatively late—around the 1880s—but expanded greatly thereafter. I shall start with the market in Zamindari rights.

TABLE 28.1

Revenue (Auction) Sales of Proprietary (Zamindari) Rights

Period	*No. of Sales*	*Revenue of Estates Sold (in Rs)*		*Sale Price*	
		Cuttack	*Orissa Division*	*In Multiples of Revenue Demand*	*Per* Acre (Rs)*
1805-19	N.A.	N.A.	9.5 lakhs	0.5-1 **	
1840-49	94			4.5	
1855-65	34	12,258		12	13.8
1866-75	18	1,857		20	23.0
1876-85	45	2,948		10	11.5
1886-95	109	16,233		8	9.2
1896-99	79	9,301		8	9.2
1845-99			1.5 lakhs		

Source: S.L. Maddox, *Final Report on the Survey and Settlement of the Province of Orissa (Temporarily-Settled Areas), 1890 to 1900 A.D.*, Vol. I (Calcutta: Bengal Secretariat Press, 1900), paras 275,276 (hereafter *Maddox Report).*

* Estimated by taking the average incidence of revenue per acre of assessed area to be Rs 1.15, as settled in the 1837 settlement.

** Taken from Stirling's minute of 10 October 1821, para 84; printed as an appendix to his *An Account (Geographical Statistical and Historical) of Orissa Proper or Cuttack* (Calcutta: Bengal Secretariat Press, 1822).

There are data on both revenue and private sales with respect to both revenue payable of the estates sold and aggregate sale prices realized. From this data is estimated the approximate area of the estates transferred (on the basis of the average incidence of land revenue per acre as settled in different settlements), and then the approximate price per acre of the Zamindari right thus transferred, by dividing the aggregate purchase money by the estimate of the area transferred that is already obtained.

We find in Table 28.1 that the value of zamindari right starts from an abysmally low level, only 0.5-1 times the revenue demand during the period 1805-19, and increases continuously thereafter until it reaches a peak during the period 1866-75, and starts falling off after that.

TABLE 28.2

Private Sales of Proprietary (Zamindari) Estates, 1890-9

No. of Sales of Whole or Portions	*Annual Average Revenue of Estates Sold (in Rs)*	*Annual Average Value of Estates Sold (in Rs)*	*Sale Price*	
			In Multiples of Revenue Demand	*Per acre (in Rs)*
5,161	12,107	181,619	14.68	19

Source: *Maddox Report*, para 271.

TABLE 28.3

Some Private Sales of Proprietary (Zamindari) Estates, 1900-1910

Area of Villages in which Data was Collected (in Acres)	*No. of Estates in the Selected Villages*	*No. of Sales of Whole or Portions of Estates*	*Annual Average Revenue of Estates Sold (in Rs)*	*Annual Average Value of Estates Sold (in Rs)*	*Sale Price*	
					In Multiples of Revenue Demand	*Per Acre (in Rs)*
301,440	2,383	940	3,996	46,566	11.65	15.1

Source: J.F. W. James, *Final Report on the Revision Settlement of Orissa (1906-1912 A.D.)* (Bankipore: The Bihar and Orissa Secretariat Book Depot, 1914), Appendix VIII (hereafter *James Report).*

Data on private sales of zamindari right (Tables 28.2, 28.3 and 28.4), as against revenue sales, show that the value of zamindari rights in private sales is higher than the same for revenue sales for periods which are roughly comparable. We also find that the value of zamindari right realized through private sales drops to some twelve times the revenue demand (or Rs 15 per acre) during the period 1900-10 but registers a rapid increase during the period 1922-34 to reach twenty-four times the revenue demand (or Rs 40 per acre), indicating that the zamindars were gaining some lost ground.

TABLE 28.4

Private Sales of Proprietary (Zamindari) Estates, 1922-34

No. of Sales	*Total Revenue Estates Sold (in Rs)*	*Total Value of Estates Sold (in Rs)*	*Sale Price*	
			In Multiples of Revenue Demand	*Per Acre (in Rs)*
1,259	66,635	15,84,667	23.78	39.95

Source: W.W. Dalziel, *Final Report on the Revision Settlement of Orrisa (1922-32 A.D.)* (Patna: Superintendent, Govt printing, Bihar and Orissa, 1934), Appendix VI (hereafter *Dalziel Report).*

But the movement of the value of zamindari right and its absolute magnitude must not be seen in isolation: it should be seen relative to the movement of the value of tenant right in the other segment of the land market where transfers of occupancy raiyati holdings were taking place underneath the zamindari estates of which the former were part.

The relevant data are reproduced in Tables 27.5A and 5B. Without going into all the interesting facets of the market in raiyati land partly reflected in the data (such as the volume, direction and possible causes of transfer), we find that the value of tenant right as realized in the market rises continuously till it reaches Rs 103 per acre during 1922-32. For comparable periods the values of tenant right work out to roughly three times the values of Zamindari right. Moreover these values of tenant right are average ones for the district as a whole and do not take into account three factors: (1) the case of irrigated land whose realized sale value in the market was much higher than the average within the district (2) the case of land growing high-valued crops, such as tobacco, and (3) the heavy transfer fee, amounting to 25 per cent of the purchase money, which the transferrer-raiyat must had to pay to the landlord. This must have had the effect of depressing the actual value of the tenant right by say 25 per cent.

III. ii. Land Market and the Structure of Tenancy Relations

I would like to interpret the relative movements of the values of Zamindari and tenant rights given above as reflecting the overall structure of tenancy relations and deeper changes at the level. I now turn to an elaboration of this structure and to the kind of changes on that basis which might have brought about the observed phenomenon in the 'dual land market'.

III.ii.i. *Proprietors and Sub-Proprietors:* To start with the relation between the Zamindars and sub-proprietors, Zamindars became the *de jure* proprietors of land by virtue of being directly responsible to the colonial state for the revenue demand.

It is useful to describe the incidents of Zamindari right vis-a-vis the right of various grades of sub-proprietors, on the basis of which we may expect the conflicting interests which must have emerged and the ways in which the struggle manifested itself. In a temporarily-settled estate the more important incidents of Zamindari right were as follows:[51]

1. Right to inherit according to the personal law of the proprietors.
2. Right to transfer the whole or any portion of the estate.
3. Right to partition, subject to the limitations of the Partition Laws, for the security of revenue.
4. Fixity of the revenue demand for the term of settlement subject only to

TABLE 28.5A

Transfer by Private Sale of Raiyati Holdings with Right of Occupancy

Period	*Total No. of Transactions Traders, etc.*	*Percent share in the total No. of Transactions**				*Purchase Money Rayati Area*	*Area Transferred*	*Average Price*	*Per cent of Total Occupancy Transferred*
		Zamindars	*Rayats (in Rs)*	*Money lenders, (In Acres)*	*Others (Per Acre)*				
1883-1892	347	18.4	10.5	10.5	60.5	23,673	949.2	25	
1892-1901	14,604	12.9	23.7	5.3	58.1	17,28,000	48,000	36	8.5
1903-1912	1,04,432	8.6 (9.4)	75.5 (71.6)	14.8 (18.4)	1.1 (0.5)	39,69,161	68,667	57.8	11.9
1922-1932	1,35,517	4.0 (4.8)	90.2 (85.9)	5.2 (8.2)	0.6 (0.9)	73,29,156	71,331	103.0	12.3

Source: For the first two periods, Letter No. 285-O.S., dated Cuttack, 29 October 1902, from K.G. Gupta, Commissioner of the Orissa division, to the Secretary to the Board of Revenue, L.P., Appendices A and B in Collection No. 6, File No. 257 of 1901, File-Subject: 'Transfer of Raiyati Holdings in Orissa,' Survey and Land Settlement Branch, Land Revenue Department, Board of Revenue, L.P.; for the period 1903-12, *James Report*, Appendix XV; for the period 1922-32, *Dalzial Report*, Apendix IV-A.

* Figures in parentheses represent per cent share in the total area transferred for each period.

TABLE 28.5B

Transfer by Private Sale of Raiyati Holdings with Right of Occupancy (Thani and Others), 1890-1900

	Average Annual No. of Transactions	*Purchasers*				*Annual Average of Rent Payable of Transferred Holdings (in Rs)*	*Annual Average of Area Sold (in Acres)*	*Price Paid (in Rs)*	*Sale Price*	
		Zamindars	*Raiyats*	*Money-lenders, etc.*	*Others*				*In Multiples of Rent*	*Per Acre (in Rs)*
Thani	596	141	103	40	312	1,655	774.87	26,891	16.24	34.7
Others	581	105	123	41	312	3269	606.94	31,671*	9.68	52.2
Total	1177	246	226	81	624	4924	1,381.81	58,562	11.9	42.4
Percentages	100	(20.9)	(19.2)	(6.8)	(53.0)					

Source: *Maddox Report*, para 328.

* Based on four years (1896-99) figures only.

reduction for acquisition of land by government. But no abatement or remissions can be claimed.

5. Right to the rents fixed in the settlement, and, in *villages not assigned to sub-proprietors,* to all the profits of new cultivation and to the lawful increments in rents.
6. Right to all profits from the jungle waste, and other unassessed areas and from fisheries *not assigned to sub-proprietors,* subject to the prescriptive rights of the community to fuel, pasturage, common lands, etc.
7. Right to grant leases and create intermediary tenures, though such as are *only binding for the term of settlement.*
8. Obligation not to collect unauthorized cesses *(abwabs).*
9. Obligation to pay the revenue according to the instalments *(kists)* prescribed.
10. Liability of the estate for sale in the event of failure to pay revenue before the latest day, the balance of the purchase money after satisfaction of all demand of government going to the proprietor(s).

On the other hand there was a wide variety of sub-proprietary tenures, such as those of moquddums, *Pradhans, sarbarahkars, pursethis, kharidars, shikmi* Zamindars, etc., differing from each other on the basis of the nature of their origins. Without going into details of the various incidents of their rights, it suffices to say that the sub-proprietors had the same set of rights within their tenures which Zamindars were exercising within their estates, i.e. Zamindars' rights were conditional on those of the sub-proprietors. With such a coexistence of proprietory and quasi-proprietory rights over the same lands, conflict was bound to result. That the Zamindars were constantly trying to take hold of these tenures is attested to by the fact that they were participating to a great extent as buyers of sub-proprietory tenures.

All this is not to suggest that this was indeed the principal conflict. Rather, my purpose is to show the once important position of the village-level landed interests as well as the extent of dispossession which they had suffered during the colonial period in this district. Thus they were controlling only a little more than 15 per cent of the total area assessed to revenue.

III. ii. ii. *Landlords, Tenants and Undertenants'*. Following the Zamindari settlement, those who did not engage directly with the state for the payment of land revenue or were without quasi-proprietary rights (as sub-proprietors) but were deriving an income from land by holding the same under definite terms and conditions of rent payment (i.e. liability to pay rent), became 'tenants' as legally defined. I will not here go into the nature of various grades of tenants holding lands under diverse tenures, partly reflecting their historical origins and partly the result of intervention by the colonial state in the form of rent restriction of its tenancy legislation. Rather, the

TABLE 28.6
Rent, Area and Incidence of Rent According to Class of Tenancy*

Class of Tenancy	No. of Holdings		Area (In Acres)			Incidence Per Acre (Rs Annas Paise)		
	1897	1927	1837	1897	1927	1837	1897	1927
1	2	3	4	5	6	7	8	9
Tanki Bahal	100		500 (0.08)	300 (0.03)	489	0-3-0	0-6-0	0-0-4
Khanda Jamabandi	22,600	34,501	26,900 (4.33)	30,100 (3.65)	28,007 (3.20)	1-1-0	1-8-0	2-3-0
Kamil Bazyafti	28,600		37,700 (6.07)	22,500 (2.73)		0-14-0	1-5-0	
		1,73,114			1,08,244 (12.50)			1-12-10
Nisfi Bazyafti	98,800.		87,200 (14.04)	94,800 (11.52)		0-6-0	1-1-0	
Thani	25,800		1,33,700 (21.52)	41,700 (5.06)		2-13-0	2-14-0	
Thani-Pahi Mixed	20,200	8,54,420		84,200 (10.23)	5,81,501 (67.00)		2-11-0	3-6-7
Settled and Occupancy	3,64,700			4,19,200 (50.96)			2-14-0	
Non-Occupancy	10,100	2,688	2,68,900 (43.30)	8,100 (0.98)	2,226 (0.25)	1-15-0	3-8-0	3-12-0
Chandina	22,100	22,369	10,100 (1.60)	4,800 (0.58)	3,552 (0.40)	3-15-0	4-14-0	6-6-3
Chowkidari jagir resumed	7,300	18,643		10,200 (1.24)	9,251 (1.06)		2-8-0	3-4-7

cont...

1	2	3	4	5	6	7	8	9
Other jagirs resumed	7,300			7,400 (0.89)			1-14-0	
Tenancies newly assessed	5,300	5,337		3,300 (0.40)	8,117 (0.93)		2-3-0	2-5-1
Tenancies valued for revenue	16,000	52,949		15,100	23,262 (2.70)		2-2-0	3-2-6
Lands in the possession of proprietors	26,800	44,602	42,300 (6.81)	85,800 (10.43)	88,284 (10.17)	1-12-0	2-10-0	3-5-8
Culturable			10,500	20(?)		0-7-0	5-0-0	
Biradaran Tanki			2,400 (0.38)			0-6-0		
Chakran Tanki			800 (0.12)			0-8-0		
Sairat								
Total	6,50,600	1,208,623	6,21,000 (100.0)	8,22,500 (100.0)	8,67,858 (100.0)	1-14-0	2-8-0	3-2-0

Source: Settlement Papers of the 1837 Settlement; *Maddox Report*, Appendix A.A., Vol. II; *Dalziel Report*, Appendix XII. A.

* Figures in parentheses represent per cent to total in each case.

broad purpose here is to cast doubt on the view that on the morrow of the Zamindari settlement the entire peasantry was transformed overnight into a tenantry-at-will. This is a historical in so far as it abstracts from the historical antecedents of the colonial period. With this purpose in mind one can move straight to the data reproduced in Table 28.6.

A glance at the incidence of rent per acre for different kinds of 'tenants' in this table shows that there was nothing like a uniform, competitively-determined rate of rent applied to the whole peasantry, nor did the entire peasantry become a tenantry-at-will on the morrow of the Zamindari settlement. Rather, there came into existence a privileged class of tenants whose rents could not be freely manipulated by the Zamindars. How this must have affected the profitability of 'rent ownership' (i.e. the ownership of Zamindari right) vis-a-vis that of direct cultivation on the one hand and sub-leasing of land on the other remains to be seen.

TABLE 28.7

Estimated Profitability of 'Rent Ownership'

Period	*Rental Income* Per Acre (in Rs)*	*Rental Income (of Zamindars only) Per Acre (in Rs)*
1837	0.65	0.61
1897	1.18	1.09
1927	1.38	1.27

Source: Computed from data given in Letter No. 15495, dated the 24 July 1900 from P.C. Lyon, Director of the Department of Land Records and Agriculture, Bengal, to the Secretary to the Board of Revenue, paras 12 & 15, reprinted in *Maddox Report,* Vol. II, Appendix A; *Dalziel Report,* Appendix XIII-A.

* Includes both Zamindars' and sup-proprietors' incomes.

III. ii. iii. *Relative Profitability of 'Rent Ownership' and Direct Cultivation:* We shall now see how the structure of tenancy relations given above must—along with a set of other factors like the relative movement of the price of the principal produce (relative to rent movement), predominant form of rent payment, increased value of output due to irrigation, etc.—have determined the profitability of 'rent ownership' vis-a-vis that of direct cultivation.

There are a number of crucial points to be noted about the figures presented in Table 28.7. First, these represent *gross* incomes per acre and do not exclude costs of collection. Secondly, these are the *maximum* rental incomes per acre attained *only if* the landlords are *able* to collect the entire estimated rental demand. This is so because, in any case, they had to meet the colonial state's fixed revenue demand and pocketed whatever they could collect over and above that demand. Now, there is some indirect evidence to suggest that they were not always able to collect the entire gross rental (land revenue + landlord's income), though they were able to meet the state's revenue demand: for one thing this was reflected in the fact that whereas the volume of revenue sales of Zamindaris declined over time, that of private sales increased relatively. Moreover, particularly for Zamindari estates with sub-proprietors, there were constant complaints by Zamindars about difficulties of collection and in some cases the actual per cent of the gross rental taken by the sub-proprietors was more than the prescribed; thus the figures of rental income per acre (of Zamindars only) given in Table 28.7 should be taken as upper limits which may not have been reached in many years. Thirdly, the figures given in Table 28.7 are money incomes and must be set against price movements, particularly because of the dominance of money rents received by Zamindars (see below).

We have estimated the profitability of direct cultivation for two points, 1897 and 1927, on the basis of constructing as comprehensive a cost account of farming operations as possible for each time-point, from cost of cultivation data available from different sources (for each time-point) and assuming that all wages are paid out. In any case these estimates of net income from cultivation must be regarded as rough and must be looked at in conjunction with income from 'rent ownership'. There are three points to be noted about these estimates of net profit from direct cultivation: first, the rent rate of settled and occupancy tenants is taken into account in the cost estimate, so that the net profit of *bazyaftidars* and *kharida Jamabandidars* should be higher than indicated here, since they were paying a lower rent than the settled and occupancy tenants. Secondly, the estimates are for unirrigated land. Finally, these estimates of net profit will be relevant only for farms of a small size.

With these qualifications in mind if one compares the figures in Tables 7 and 8 one finds that by 1897 profitability of direct cultivation exceeds that of 'rent ownership' by more than Rs 2 per acre, by more than Rs 4 per acre by 1927, and that the former increases faster than the latter between 1897 and 1927.

TABLE 28.8

Estimated Profitability of Direct Cultivation of Rice (Capitalistically Organized)

Year	*Net Profit Per Acre (in Rs)*
1837	—
1897	3.8
1927	5.6

Source: Computed from cost of cultivation and yield data given in N.N. Banerji, *Report on the Agriculture of the District of Cuttack* (Calcutta: Bengal Secretariat Press, 1898), pp. 37,73; *Dakziel Report*, para. 29.

Here an important word of caution is in order: by saying that the profitability of direct cultivation became more than that of 'rent ownership' I do not suggest that direct cultivation, capitalistically organized, indeed became the dominant form of the organization of production. One could argue that the rent through sharecropping or even fixed-rent sub-tenancy was greater than the profitability of direct cultivation; in any case we are not certain of the relative extent and changes over time of different forms of the organization of production.

Now to discuss the proximate causes responsible for the profitability of direct cultivation exceeding that of 'rent ownership'. There are a number of factors which seem to have worked systematically in favour of surplus-producing, cultivating 'tenants' and against purely rent-receiving 'landlords'.

A major contributing factor must have been the long-term movement of the price of rice, the principal produce (it occupied some 88 per cent of the net cropped area in 1897 and 93 per cent in 1927)',[52] relative to that of rent and wages.

The figures of annual prices given in the sources cited from which the decennial averages have been computed (see Table 28.9) are mostly taken from the price-lists prepared by the collector of the district to be submitted to the Board of Revenue to form the basis of rent revision, and hence can be regarded as prices received by farmers.

We find that the price of rice has increased two-fold between 1837/46 and 1887/96, and nearly two-fold again between 1887/96 and 1921/30.

Table 28.9

Decennial Averages of the Price of Rice

Decennial Periods	*Annual Average (Rs Per Maund)*
1837-1846	1.05
1847-1856	0.69
1857-1866	1.61
1867-1876	1.65
1877-1886	2.10
1887-1896	2.35
1901-1910	3.16
1911-1920	4.01
1921-1930	4.62

Source: Computed from average annual prices given for various years in *Maddox Report*, para. 164; N.N. Banerji, *Report on the Agriculture of the District of Cuttack*, Appendix VIII; *Dalziel Report*, para 9; L.S.S.O' Malley; *Bihar & Orissa District Gazetteers: Cuttack, District Gazetteer, Statistics, 1901-2* (Calcutta: The Bengal Secretariat Book Depot, 1905).

Referring back to Table 28.6 we find that the overall rate of rent has increased by only 33 per cent from Rs 1-14 to Rs 2-8 between 1837 and 1897 and by 25 per cent between 1897 and 1927. The settled and occupancy raiyats' rent-rate has increased by only 17 per cent during the latter period. As a result we find that by 1924 the rent of occupancy raiyats absorbs at best 10 per cent of the value of gross produce.[53]

As regards agricultural wage payments, we find that the dominant form of payment was in kind at the rate of 2 to 2.5 seers of rice per day; when paid in cash, it was 1.5 to 2 annas per day.[54] These rates are seen to obtain as early as 1814. The same kind rate seemed to have continued unchanged till the last decade of the last century.[55] But around the turn of the century there was an attempt to change over to money wage as the price of rice continued to rise: 'The real change has been in the much greater prevalence of paying labour in money ...'[56] a contemporary was writing. As a result, 'over 1875—1910 ... the cash wages of rural labourers have not kept up with the rise in price'.[57] The Settlement Officer was also writing in 1924: 'The system of paying wages in produce ... is not so prevalent as it used to be'.[58] So by 1924 we find that the money wage was increased only to 3 annas per day for agricultural day labour and that even the attached labour was paid a daily money wage besides the annual cash wage.[59]

The increasing price of the principal produce was particularly unfavourable to Zamindars primarily because of the predominance of money rents received by them: we find[60] that in 1927 Zamindars and sub-proprietors were deriving kind rent, including the command over the entire surplus produce from the lands under their cultivating possession, i.e. from only 11.6 per cent of the total assessed area, which shows the limited extent to which they could benefit from the rise in the price of the product.

III. ii. iv. *Benefits of Irrigation:* A major intervention by the colonial state in the area studied (which was also true of other temporarily-settled areas) in the latter half the last century was in the form of construction of a large network of irrigation canals in the Mahanadi delta, most of which lay within Cuttack district.

Without going into all the details, from our point of view it is relevant to see the impact of irrigation on actual rent-rates paid. It is significant to note that in 1897, after some thirty years of irrigation, from the data on pargana-wise rent-rates[61] a detailed comparison of irrigated and unirrigated groups of parganas shows that there did not occur any general enhancement of rents *on the ground of irrigation,* nor was. the level of rent higher in irrigated than in unirrigated parganas—though, as the Settlement Officer was writing in 1900, 'there is evidence that rent rates have risen more in the protected and irrigated tracts than in the unprotected and unirrigated'.[62] As a result, according to the estimate of the Settlement Officer, by

1900 for the Orissa Division as a whole the cultivators using canal water were drawing an increased value of Rs 7,90,000 (including *rabi* and sugarcane irrigated areas) plus a sum of Rs 6,00,000 saved annually owing to the protection afforded by the canals against drought. Out of these increased assets, amounting to Rs 13,90,000, they were paying to government Rs 3,15,000 annually in the shape of water-rates, keeping the remainder, Rs 10,75,000, for themselves.[63] Thus in the official discussion on the desirability of imposing an owner's rate on Zamindars in view of the very low water rate,[64] the majority opinion was against such an imposition on the basis of the conclusion that 'in Orissa the profits of irrigation have gone not to the landlords, but to the tenants. The landlords may have indirect benefits in more punctual collection of rent, but the direct benefits—the immediate profits—are those of the tenants'.[65]

III. iii. *Private Lands of Zamindars/Sub-proprietors under Direct Cultivation*

In the face of the conjuncture outlined above, a major response of rent-receiving landlords for getting round the crisis can be expected to have been in the form of attempting to extend the area under their direct cultivation, thus aiming to obtain a larger and larger share of their income from rising profits of cultivation. Here we shall see the extent to which this was happening and what limits were set to this form of response. For this one has to refer back to Table 28.5 and look at the figures there under the tenurial category shown as 'Lands in the Possession of Proprietors'.

We find that in 1837 only about 6.8 per cent of the total assessed area is under the direct cultivating possession of Zamindars (at that time, it should be noted, sub-proprietors were not allowed to hold private lands). But there does occur a phenomenal rise in the area under proprietary cultivation between 1837 and 1927, which rises at a faster rate (more than 50 per cent) than the rise in the total assessed area (32.5 per cent). In any case it still formed only 10.4 per cent of the total assessed area.

This increase seems to have taken place mostly in the estates of middle-sized/ large-sized Zamindars and is not a mere proliferation of petty proprietary cultivation. For this it is necessary to look at the data on the distribution of land revenue paid and area held across size-classes of land revenue per estate, available for two time points, 1852-3 and 1898-9 (see Tables 28.10A and 28.10B).

Unfortunately the distribution of area is not available for 1852-3.

But it is clear in any case that if we take the estates paying a revenue of Rs 50 or less as 'small-sized' estates there has not occurred any significant increase in the percentage of total number of estates lying within this category, nor in the percentage of total revenue paid. Moreover the area commanded by this category in 1898-9 works out to be some 40,768 acres (even assuming that all of this is self-cultivated), whereas the area under proprietary cultivation in 1897 was recorded as 85,800 acres.

TABLE 28.10A

Distribution of Estates and Land Revenue Across Size-Classes of Land Revenue per Estate, 1852-3

.*Size-Classes*	*No. of Estates*	*Per cent*	*Revenue paid (in Rs)*	*Per cent*
< 8 annas	5	0.2	1.3	—
8 annas—1 Re.	27	1.0	20.7	—
Re 1-10	559	22.5	2,823	0.34
Rs 10-50	696	28.0	17.156	2.06
Rs 50-250	633	25.5	74,667	9.0
Rs 250-500	216	8.7	76,758	9.25
Rs 500-1000	178	7.1	1,22,949	14.80
Rs 1,000-5,000	148	6.0	3,16,616	38.16
Rs 5,000-10,00	15	0.6	90,536	10.90
Rs 10,000-30,000	7	0.4	1,28,413	15.44
Total	2,484	100.0	8,29,570	100.00

Source: *Report on the Revenue Administration of the Lawn Provinces*, for the official year 1852-53 (Calcutta: Bengal Military Orphan Press, 1854), pp. 56-9.

Thus the significant absolute increase in the area under proprietary cultivation between 1897 and 1927 can be interpreted as an attempt mainly by middle-sized and, perhaps, large-sized Zamindars to derive a larger proportion of their income from profits of cultivation rather than from rents alone.

Table 28.10B

Distribution of Estates, Land Revenue and Area Held Across Size-Classes of Land Revenue Per Estate, 1898-9

Size-Classes	*No. of Estates*	*Per Cent*	*Revenue Paid (in Rs)*	*Per Cent*	*Area Held (in Sq. Miles)*	*Per Cent*
< 10	981	22.0	5,209	0.5	7.9	0.4
10 - 50	1,457	32.7	37,512	3.4	55.8	3.1
50 - 100	594	13.3	42,654	3.9	68.0	3.8
100- 1,000	1,204	27.0	3,82,508	34.8	603.1	33.8
1,000- 5,000	186	4.2	3,67,542	33.4	897.8	50.3
5,000-10,000	22	0.5	1,53,384	13.9		
> 10,000	5	0.3	1,09,606	9.9	150.2	8.4
Total	4,449	100.0	10,98,415	100.0	1,782.8	100.0

Source: *Maddox Report*, para 269.

Coming back to Table 28.6 we find that between 1897 and 1927 there has occurred a certain increase in the area under proprietary cultivation, but the same (2.8 per cent) is less than the rate of increase of total assessed (5.5 per cent). Thus there has occurred even a marginal decrease in the share of this area in the total to only 10.17 per cent.

This was in striking contrast to the situation obtaining in many districts of the North-Western Provinces in a more or less similar conjuncture. For example in Hoshangabad district, according to one estimate, in 1865 after deducing all expenses of cultivation the direct profits of agriculture ran at Rs 4—8 per acre as compared with a return of Rs 1—4 per acre in rental income. Correspondingly, as much as a fifth of the cultivation was held directly as *sir* by the landlords.[66] Likewise in the wheat zone, during the wheat boom which reached its high point in the 1880s, 'the *malguzars* seemed possessed of all the qualities of capitalist entrepreneurs . . . their proprietary cultivation rising by a third to become 23 per cent of the occupied area . . . what was significant was that despite a steady increase in their income from rents it still remained true in 1893 that two-thirds of their total net income (after payment of the government revenue) came from the agricultural profits reaped from their direct cultivation'.[67]

One can now set out to identify the conditions which were present in one case and not in the other in order to explain the above contrast. In general these have to be sought in the specificity of historical development of land relations in the region.

To start with, the extent to which the Zamindars could freely extend the area under their direct cultivation was limited by the historicity of tenancy relations itself. Given the fact that a privileged class of tenants had entrenched itself over a greater part of the area, the Zamindars could evict the tenants and add to their proprietary cultivation only to a limited extent; so that they must have had to fall back on waste lands that could be brought under cultivation. But, then, within this limit, there were specific disincentives or lack of incentives to proprietary cultivation, some of which were caused by the specifically colonial interests involved.

First, there was in general no preferential treatment given to Zamindars' private lands in the matter of assessment of rent. We can see this by looking at the rent per acre for 'Lands in the Possession of Proprietors' in Table 28.6.

Second, during the 1897 settlement the rent-free lands granted by Zamindars as allotment holdings to low-caste labourers in order to get a cheap source of labour for their private lands were resumed and assessed to revenue (shown under 'Tenancies Newly Assessed' and 'Tenancies Assessed for Revenue' in Table 28.6), because this meant more revenue for the colonial state, a measure which, as the Settlement Officer was writing in 1900, 'was certainly hard on the Zamindars, who . . . have complained that in consequence they have been deprived of the necessary services of the *jagirdars* who, secure in the possession of their lands, refused to work'. [68]

Finally, it is to be noted that not the entire area nominally under the cultivating possession of Zamindars was protected from the accrual of occupancy right and thus from the possibility of commutation into money rent: in 1897 only some 14 per cent of the area under cultivating possession was thus protected.[69] But between 1897 and 1927 the Zamindars were exerting pressure on the colonial revenue authorities to legally acknowledge a greater portion of the area to be free from the accrual of occupancy right.[70] At the same time the sub-proprietors, through their association (known as the Orissa Association), were urging the government to take necessary legal steps to make it possible for them to hold private lands protected from the accrual of occupancy right.[71]

As a result one of the main objects of the Orissa Tenancy Bill was stated to be 'make a definite expansion of the area held as private lands by proprietors and sub-proprietors . . .', [72] and we find that a much more liberal procedure for the record of proprietors' private lands is followed in Chapter XII of the Orissa Tenancy Act passed in 1913. Thus it is not surprising to find that by 1927 Zamindars/ sub-proprietors could claim as their absolute private property as much as 60 per cent of the lands under their cultivating possession. [73]

CONCLUDING REMARKS

We have sought to illustrate the strong historical moorings of rural land relations in British India. In the face of this historicity, even though colonial rule sought to create absolute private property in land on a generalized basis for the first time, the British could not have done this by the stroke of a pen. The tendency towards the formation of absolute private property in land was thus a halting process, but was nevertheless unfolding itself.

Because of the stage of development private property rights in land (i.e. because absolute private property in land had not yet *fully* emerged) there were two parallel markets in 'land', one in Zamindari (proprietary) right and the other in raiyati (privileged tenant) right. In the course of a century or so the value of the tenant right far exceeded the value of the proprietary right. This I have interpreted as reflecting the structure of tenancy relations and a deeper sequence of changes at that level, which has been charted.

We have sought to show that the bulk of the advantages derived from a trend increase in the price of the principal produce (and a relatively stable upward movement of the same following improved communications), a tendency for the wage rate to lag behind the increase in the price of the principal produce (reinforced by a changeover to the payment of wages in money), and increased value of output due to irrigation, increasing demographic pressure, etc. tended to flow to a privileged class of 'tenants' who had complete command over the produce or were getting a much higher money rent than they had to pay. This was possible due to the predominance of money rents received by Zamindars (which indicated the limited extent to which they could benefit from an increase in the price of the principal produce) and the sluggishness of the same relative to the rise in price, as well as because only a small percentage of area was under the direct proprietary cultivation of zamindars/sub-proprietors. There were limits to the extension of this area because of a number of factors which I have tried to identify.

The rising surplus seems to have been mostly 'invested'—in the production (i.e. non-mercantile/moneylending) sphere—in the purchase of land and later in canal irrigation. The land market seems to have sucked in a major protion of the surplus: from the figures of purchase money presented in Table 28.5A it is estimated that for thirty-one years between 1892 and 1932 some Rs 4.3 lakhs went towards the purchase of land on an *annual average*, whereas the same was some Rs 5.3 lakhs for twenty-one years between 1903 and 1932, and these do not include the heavy transfer fee of 25 per cent of the purchase money paid to the Zamindars. Truly productive investments in land (i.e. those that would increase total output) must have been so much the less.

As a result there occurred a relative weakening of the position of the purely rent-receiving landlords, and thus the profitability of 'rent ownership' was not as high as the traditional model of the 'landlord' would suppose. But it must be noted that due to a fairly high degree concentration of the ownership of Zamindari estates (see Tables 10A and 10B) the impact of the conjuncture must have been more severe on, say, the medium-sized non-cultivating Zamindars rather than on the large-sized Zamindars.

Moreover there were definite responses to this situation from the side of the Zamindars who were pressurizing the colonial authorities to make a number of concessions. These were to be embodied in the Orissa Tenancy Act of 1913, in particular regarding the protection of Zamindars' private lands, waste-land rules, etc.[74] The improving position of the Zamindars was perhaps reflected in the increased value of Zamindari right between 1900/10 and 1922/34, as well as in the decreasing rate of transfer of Zamindari rights through private sales during the latter period.

In suggesting that the profitability of 'rent ownership' was declining relative to that of direct farming for a particular period of time (during an inflationary situation), we are not trying to imply that market forces became the sole determinant of the desirability of rent income vis-a-vis profit income, but rather that the nature of the impact of such market forces and of other historical conditions (for example demographic pressure) depends on the historicity of land relations itself. Because of the coming into being of a privileged class of 'tenants' vis-a-vis Zamindars the range of choices available to the latter was rather limited in terms of how it could respond to the situation—say, by significantly increasing the area under direct cultivation, something that was happening in many districts of the North-Western Provinces, or by enhancing rents commensurate with the rise in net profit from cultivation. Neither of these was possible, in the ultimate analysis, because of the historically conditioned land relations of the region.

It is useful to look at the totally different responses of English landowners in a similar inflationary situation and increasing demographic pressure of the late sixteenth and early seventeenth particularly between 1580 and 1620,[75] to see how easily they could manipulate the *existing* tenancy arrangements to compensate for the fall in revenue from rent. This was done by deriving a much larger part of their income from direct cultivation, through alternative sources of income, say from the sale of timber, additional fines, increasing the entry fees, etc., or by creating large leasehold tenancies (which were better able to pay a higher rent) from small holdings, etc. In short there was occurring a transformation in the *structure* of manorial revenues and, correspondingly, in the organization of production, in particular the creation of capitalist tenant fanning, all of which reflected a different balance of class forces embodying a wider range of possible choices open to the landowners (even though the actual exercise of such choice met with organized peasant resistance, of course very often without success).

Such a high degree of manoeuvrability of English landowners in fact reflected the latter's structural position in English landed society undergoing a transition, in particular their absolute proprietary control over land: this the class of landowners created by colonial rule in India was only striving for.

NOTES

1. Irfan Habib, 'Economic History of the Delhi Sultanate—An Essay in Interpretation,' *Indian Historical Review*, IV, 2 (January 1978), p. 298; see also the same author's 'Problems of Marxist Historical Analysis,' *in Science and Human Progress: Essays* in *Honour of D. D. Kosambi* (Bombay: Popular Prakashan, 1974), pp. 34—48, in which the Mughal system has been called a system of centralized collection of rent/surplus—the principal contradiction being that between the king's bureaucracy and the peasantry. One wonders how this characterization by itself is inconsistent with the AMP, something which Habib seems to suggest.
2. Harbans Mukhia, 'Was there Feudalism in Indian History,'? *The Journal of Peasant Studies*, VIII, 3, (April 1981), pp. 273-310.
3. B. R. Grover, 'Nature of Land-Rights in Mughal India," *Indian Economic and Social History Review*, I, 1 (July-September 1963), 15; for Ancient India, see L. Gopal, 'Ownership of Agricultural Land in Ancient India,' *Journal of the Economic and Social History of the Orient*, IV (1961), pp. 240-63.

4. See Perry Anderson, *Passages from Antiquity to Feudalism* (1978; London: New Left Books, 1974), pp. 150, 193-93. The qualitative difference from ancient Greek city-states on the one hand and 'Asiatic' state-cities on the other, is striking.
5. R. S. Sharma, *Indian Feudalism: c300-1200* (Calcutta: University of Calcutta; 1965); see also B. P. Mazamdar, *The Socio-Economic History of Northern India (11th and 12th Centuries)* (Calcutta: Firma K. L. Mukhopadhyaya, 1960), pp. 1-28 and K. S. Shelvankar, 'Indian Feudalism: Its Characteristics,' in A. R. Desai, ed, *Rural Sociology in India* (Bombay: Popular Prakashan, 1969), pp. 150-4.
6. D. D. Kosambi, *An Introduction to the Study of Indian History* (2nd. rev.ed.; Bombay: Popular Prakashan, 1956).
7. M. M. Kovalevsky, *Communal Landholding: Causes, Course, and Results of its Disintegration* (Moscow, 1879) (in Russian). I have not been able to get hold of an English translation; instead, I have made use of Karl Marx, 'Excerpts from Kovalevsky,' in Lawrence Krader, *The Asiatic Mode of Production: Sources, Development and Critique in the Writings of Karl Marx* (The Netherlands: Van Gorcum & Co., 1975), pp. 355-84, as well as Daniel Thorner, 'Feudalism in India,' in the collection of the same author's essays, *The Shaping of Modern India* (New Delhi: Allied, 1980), pp. 283-91.
8. For details see Thorner, *The Shaping of Modern India*, pp. 377-8 and Krader, *The Asiatic Mode of Production*, pp. 346-55.
9. Cited by Thorner, 'Feudalism in India,' p. 288.
10. Sharma, *Indian Feudalism*, p. 5.
11. *Ibid.*, pp. 74-6, 271.
12. *Ibid.*, p. 271.
13. *Ibid.*, p. 5.
14. Shelvankar, pp. 151-2.
15. Kosambi, *An Introduction*, p. 295.
16. *Ibid.*, p. 320.
17. *Ibid.*, p. 323.
18. The following scheme has been adopted from Ernest Mandel, *The Formation of the Economic Thought of Karl Marx* (New York: Monthly Review Press, 1973), pp. 121-2.
19. For what follows I have relied heavily on Krader, *The Asiatic Mode of Production*, especially Ch. VII and *passim*.
20. *Ibid.*, p. 141.
21. *Ibid.*, p. 289.
22. *Ibid.*, p. 162.
23. *Ibid.*
24. *Ibid.*, p. 295.
25. Mandel, *The Formation*, p. 127; emphasis in the original.
26. See Maurice Godelier, 'The Concept of the "Asiatic Mode of Production" and Marxist Models of Social Revolution,' in David Seddon, ed., *Relations of Production: Marxist Approaches to Economic Anthropology* (London: Frank Cass, 1978); for a critique, see *ibid.*, pp. 124-7.
27. See Irfan Habib, 'An Examination of Wittfogel's Theory of "Oriental Despotism,"' *Enquiry*, 6 (1963), pp. 57—73; see also the same author's *The Agrarian System of Mughal India* (1962). I have used the typescript available at the Centre for Development Studies (Trivandrum) Library, p, 345.

28. See Umberto Melotti, *Marx and the Third World,* trans, by Pat Ransford (London: Macmillan, 1977), p. 188.
29. For an useful review see David Seddon, 'Economic Anthropology or Political Economy? (II): Approaches to the Analysis of Pre-Capitalist Formation in the Maghrib,' in John Clammer, ed., *The New Economic Anthropology* (London: Macmillan, 1978), pp. 61-109, and the references; in particular the articles of Y. Lacoste, E. Terray and R. Gallissot cited therein.
30. *Ibid.*, p. 84.
31. Melotti, *Marx and the Third World, p.* 75.
32. *Ibid.*
33. I am summarising the evidence of Sharma, *Indian Feudalism*, p. 152 and Irfan Habib 'The Social Distribution of Landed Property in Pre-British India (A Historical Survey),' *Enquiry,* Winter (1965) pp. 1-35.
34. Ramakrishna Mukherjee, *The Dynamics of a Rural Society: A Study of the Economic Structure in Bengal Villages* (Berlin: Akademic-Verlag, 1957), pp. 75-9.
35. Jurgen Habermas, *Communication and the Evolution of Society* (London: Heinemann, 1979), pp. 158-9.
36. *Ibid.*, p. 159; my emphasis.
37. I owe this clarification to E. M. Wood, 'The Separation of the Economic and Political in Capitalism,' *New Left Review*, 127 (May-June 1981), 91-2.
38. Habib, *The Agrarian System of Mughal India*, p. 400.
39 Burton Stein, *Peasant, State and Society in Medieval South India* (Delhi: Oxford University Press, 1980), p. 434.
40. Appendix to Mr Stirling's Minute on Tenures in Orissa, dated 10th October 1821; printed as an appendix to A. Stirling, *An Account (Geographical, Statistical and Historical) of Orissa Proper or Cuttack* (Calcutta: Bengal Secretariat Press, 1822), para 36.
41. *Ibid.*, para 24; my emphasis.
42. *Ibid.*, para 46.
43. See S. Nurul Hassan, 'The Position of the Zamindars in the Mughal Empire, *Indian Economic and Social History Review*, I, 4 (April-June 1964), pp. 107-19: 'there emerged not only a variety of land rights but also a kind of pyramidal structure in agrarian relations wherein rights in various kinds were superimposed upon each other', (p. 118).
44. Stirling, *An Account*, para. 59; my emphasis.
45. See Ratnalekha Ray, *Change in Bengal Agrarian Society* (New Delhi: Manohar Publications, 1979), p. 109, 274 ff., from which the broad scheme has been adopted, with modifications.
46. See Rajat Ray and Ratna Ray, 'Zamindars and Jotedars: A Study of Rural Politics in Bengal,' *Modern Asian Studies*, IX, 1 (1975), pp. 81-102, for showing how in Bengal in the latter half of the nineteenth century such an 'alliance' was breaking down.
47. Stirling, *An Account*, para 82; see also W. W. Hunter, *Orissa: Or the Vicissitudes of an Indian Province Under Native and British Rule* (2 vols.: London: Smith, Elder & Co., 1872), I, p. 256: 'The Marhattas dispossessed a number of the superior Tenure-holders, and collected dispossessed a number of the superior Tenure-holders, and collected direct from the Heads of the Village Guilds'; as well as B. C. Ray, *Orissa Under the Marathas, 1751-1803* (Allahabad: Kitab Mahal, 1960), pp. 133-4.
48. Hunter, *Orissa*, p. 260.
49. Bengal Territorial Revenue Proceedings, 17 July 1818, No. 16; Ewer's Report of 13 May 1818.

50. Stirling, *An Account*, para 77.
51. S.L. Maddox, *Final Report en the Survey and Settlement of the Province of Orissa (Temporarily - Settled Areas), 1390-1900 A.D.* (3 vols.: Calcutta: Bengal Secretariat Press, 1900), I, para 167 (hereafter *Maddox Report).*
52. *Ibid.*, para 11; W. W. Dalziel, *Final Report on the Revision Settlement of Orissa: 1922-32* AD (Patna: Superintendent, Govt. Printing, Bihar and Orissa, 1934), para. 22 (hereafter *Dalziel Report).*
53. J. F. W. James, *Final Report an the Revision Settlement of Orissa (1906-1912* AD) (Bankipore: The Bihar and Orissa Secretariat Book Depot, 1914), para. 62. (hereafter *James Report).*
54. *Maddox Report*, para. 193.
55. *Ibid.*, W. W. Hunter, *A Statistical Account of Bengal*, Vol. XVII, Cuttack and Balasore (London, 1877; rpt. Delhi: Concept Publishing Company, 1976), p. 110.
56. N. N. Banerji, *Report on the Agriculture of the District of Cuttack* (Calcutta: Bengal Secretariat Press, 1898), Appendix VIII.
57. *James Report*, para 35.
58. *Dalziel Report*, para 12.
59. *Ibid.*, para 13.
60. *Ibid.*, Appendix VII.
61. *Maddox Report*, para 111.
62. *Ibid.*
63. *Ibid.*, para 109.
64. Collection No. 6, File No. 251 of 1902; File-Subject: 'Imposition of Owner's rate-Orissa,' Survey & Settlement Branch, Land Revenue Department, Board of Revenue, Lower Provinces, *Proceedings*, 1902.
65. *Ibid.*
66. Cited by Eric Stokes, *The Peasant and the Raj: Studies in Agrarian Society and Peasant Rebellion in Colonial India* (Cambridge University Press, 1978), p. 256.
67. *Ibid.*, p. 258.
68. *Maddox Report*, para 469.
69. *James Report*, para 71.
70. *Ibid.*, para 70.
71. Sec the letters contained in File No. 80 of 1907; File Subject: 'Nij-jote Lands of Proprietary tenure-holders in Orissa', Survey & Settlement Branch, Land Revenue Department, Board of Revenue, Lower Provinces, *Proceedings*, 1907.
72. B. K. Mishra, *Land Tenure and Land Reforms in Orissa* (Cuttack: Board of Revenue, 1962), p. 10.
73. *Dalziel Report*, para 199.
74. The Orissa Tenancy Act (1913) provided that no raiyat can reclaim land without the landlord's consent, subject to the condition that the landlord's consent is to be deemed to have been given if within four-years-a long enough time-from the raiyat's beginning his reclamation he takes no steps to eject him. See Government of Orissa, Law Department, *The Orissa Tenancy Act, 1913* (Bihar and Orissa Act 2 of 1913).
75. See, for a thoroughgoing analysis of this period, Peter Bowden, 'Agricultural Prices, Farm Profits and Rents,' in *The Agrarian History of England and Wales, Vol. IV 1500-1640*, ed. by Joan Thirsk (Cambridge, Cambridge University Press, 1967), pp. 593—695, even though the causal primacy attached to the demographic factor is debatable.

29

AGRICULTURAL MARKETING

THE PRESENT STATE OF AGRICULTURAL MARKETING OF INDIA

Under subsistence cultivation, the farmer carries on agriculture mainly to satisfy his own needs. This was the position in India before the middle of the 19th century. Since then, India was gradually giving up subsistence cultivation and adopting commercial agriculture. The farmer produces rice, wheat, jowar and other food crops mainly for the requirements of his family and also produces such cash crops for the market as cotton, jute, oilseeds, tobacco, surgarcane, etc. The farmer requires the services of a well-organised marketing system backed by good transport and scientific storage facilities. For efficient agriculture, well-organised marketing is essential. For a long time, marketing conditions in India were primitive and the agriculturists were exploited by the traders and middlemen.

If repaid industrial development requires an increased amount of agricultural goods for feeding the urban population and to supply raw materials for industries, it is necessary that the government devises methods to mobilise an increasing proportion of the agricultural output. In a dictatorial system, there can be compulsory appropriation of the marketable surplus. But in a democratic set-up the best way is to induce the farmers to part with their surplus stocks through the free market mechanism. As the Foodgrain Enquiry Committee stated, the organisation of a well-regulated and controlled market would facilitate market arrivals and thus help in economic development. The rapid increase in production in wheat in Punjab and Haryana since 1967-68 has brought out clearly the need to improve facilities for agricultural goods.

Methods of Agricultural Marketing

There are many ways by which the farmer may dispose the his surplus produce. The first and the most common method is to sell away his surplus produce to the village money-lender-cum-trader. The trader may buy agricultural produce on his own or he may act as in agent of a bigger merchant or of a mercantile firm of the neighbouring 'mandi' town. It is estimated that in the Punjab, 60 per cent of wheat, 70 per cent of oils and 35 per cent of cotton are sold in the village itself.

The second method adopted by the Indian farmer is to dispose of his produce in the weekly or bi-weekly village markets, known in Hindustani as the 'hat'. Besides fairs are held once in a year or so in important villages or towns in connection with religious festivals. In 'hats' and fairs, the farmers bring their produce as well as livestock and sell them.

The third method of agricultural marketing is through the mandis in small and large towns. The mandi may be located at a distance of several miles, and, therefore the farmer has to make special effort to carry his produce to the mandi. In the mandis, there are brokers or 'dalals' who help the farmers to dispose of their produce to the wholesalers known is *'arhatiyas'*. The whole-salers many dispose of the agricultural produce which they have purchased from the farmers to retailers or flour mills and processing units. For instance, in the case of cotton, the wholesaler sells to the cotton ginning factories, and in the case of foodgrains like wheat he sells to the flour mills or to the retailer.

Basic Facilities Needed for Agricultural Marketing

In order to have best advantage in marketing of his agricultural produce the farmer should enjoy certain basic facilities:

(i) He should have proper facilities for storing his goods.

(ii) He should have been holding capacity, in the sense, that he should be able to wait for times when he could get better prices for his produce and not dispose of his stocks immediately after the harvest when the prices are very low.

(iii) He should have adequate and cheap transport facilities which would enable him to take his surplus produce to the mandi rather than dispose it of in the village itself to the village money-lender-cum-merchant at low prices.

(iv) He should have clear information regarding the market conditions as well as about the ruling prices, otherwise, he may be cheated. There should be organised and regulated markets where the farmer will not be cheated by the *dalals* and *arhatiyas*.

(v) The number of intermediaries should be as small as possible so that middlemen's profits ar reduced. This will increase the returns to the farmer.

Defect of Agricultural Marketing in India

Judging from these considerations, the position of agricultural marketing in

India is deplorable. The Indian farmer is very poor, illiterate and ignorant. First of all, he does not have facilities for storing his produce. The storage facilities which are available in the village at present are so poor that 10 to 20 per cent of the produce is eaten away by rats.

Secondly, the average farmer is so poor and indebted that he has no capacity to wait for better prices. He is forced to sell his output to the money-lender or to the trader so as to clear his debts. Such distress sales weaken the already miserable position of the average Indian farmer still further.

Thirdly, the transport conditions in rural areas are so bad that even richer farmers, who have large amounts of surplus, may not be interested in going to the mandis. Most roads are kachcha (unmetalled) and in rainy season they are unusable.

Fourthly, the conditions in the mandis are such that the farmer may have to wait for some time before he may be able to dispose of his produce. He may no have proper warehousing facilities to keep his stock while he waits. The method of transaction is generally against the interest of the farmer. In the mandis the farmer makes use of services of a dalal (broker) to sell his output to the arhatiya. The dalal and the arhatiya do not bargain openly but through a code language. The dalal is often in collusion with arhatiya and, therefore, the price which is settled is generally to the advantage of the arhatiya and not to the farmer. Moreover, through the use of false weights and measures, through unnecessary deduction on the plea that his produce is of inferior quality, the farmer often loses in going to the mandis.

Fifthly, the number of intermediaries and middlemen between the farmer and the final consumer of his produce is too many and the margin going to them is too large.

Sixthly, the farmers do not ordinarily get information about the ruling prices in the big markets. As a result, the farmers have to accept whatever price is quoted to them and have to believe whatever the traders tell them.

Steps Taken to Improve Agricultural Marketing

The Government is aware of the need for improvement and it has taken various steps to improve the situation. The All-India Warehousing Corporation has been set up to construct and manage a whole network of warehouses in all towns and mandis. To promote warehousing in villages, the cooperative societies are given necessary financial and technical facilities. To help improve the financial conditions of the farmers to save them from the clutches of the village money-lenders and to enable them to secure good prices for their products, the cooperative credit societies are giving increased credit facilities to the farmers. Further, cooperative marketing and processing societies have been started to market the agricultural goods of the farmers. Cooperative marketing would assure reasonable prices and remove all exploitation of the farmers by dalas and arhatiyas. Rural transport has been given importance under the Five-Year Plans and considerable progress has been made. Regulated markets have been started and the interests of the farmers have been very well safeguarded in these markets. Measures were being taken to give publicity to market information. Foodgrains prices are being stabilised by the Government through the recommendations of the Agricultural Prices Commissions. Finally, the Government is coming in a big way through the Food Corporation of India (FCI), the Cotton Corporation of India (CCI) etc, and is marketing agricultural produce.

REGULATED MARKETS

The purpose of a regulated market is to eliminate unhealthy market practice to reduce marketing charges and to ensure fair prices. By the beginning of the Fourth Plan, altogether 13 States had passed legislation for the establishment of regulated markets. All the states have passed legislation in this regard. In 1951, there were more than 200 regulated markets in India and by the end of the Second Five-Year Plan, i.e., in 1961, there were nearly 1,000 regulated markets. On the eve of the Fourth Plan, there were over 1,840 regulated markets in the country. At the beginning of January 1977, there were over 3,630 regulated markets.

Features of a Regulated Markets

A regulated market is started under the law either for any specific commodity or for a group of commodities. Such a market is administered by a market committee which consists of representatives of the State Government, the legal bodies (as for insurance, the district board) the traders, the commission agents or the dalals and the farmers themselves. In other words, the market committee represents all the interests. The committee is appointed by the government for a specific period and is entrusted with management of the market.

The market committee fixes the market charges, such as the commission to the charged. It ensures that no dalal represents either the buyer or the seller. It prevents unauthorised deductions from the price paid to the farmer and ensures that correct weights of measures are always used. The committee hears all the complaints and settled them. In cases of dispute, it arranges for arbitration. The committee is responsible for the licensing of brokers and weighmen. It is vested with powers to punish anyone who is fund guilty of dishonest and fraudulent practices.

The system of regulated markets has been found to be very useful in removing fraudulent practices followed by brokers and commission agents and in standardising market practices. They have helped farmers to secure fair prices for their produce and to come to the mandi without fear of being cheated. They have helped in using standard measures and weights throughout the country. Hence, it is the policy of the government to convert all markets in the country into the regulated type.

The Sixth Plan aims at the development of the marketing structure to, (a) ensure remunerative price to the producer of agricultural commodities, (b) narrow down the price spread between the producer and the consumer, (c) reduce non-functional margins of the traders and commission agents. To achieve these objectives the Government would go for comprehensive and rapid expansion of regulated marketing system. The success achieved in States like Punjab and Haryana, where regulated markets have been established in major producing areas with linked up satellite markets in the rural growth centres would be aimed at in other areas where intensive productions taken up. The regulated marketing system has also proved a good source for generating income to the marketing boards and for use for rural infrastructure. The regulated market complex will also include facilities for grading and for monitoring of prices. The development of regulated markets is proposed especially in areas where commercial crops like cotton, jute,tobacco and important non-traditional crops are produced and sold in weekly markets and 'hats'. Cooperative marketing and distribution and banking will also be linked with the regulated markets. These markets will cover all the major crops. Separate market yards are proposed for livestock, fish, fruits and vegetables.

COOPERATIVE MARKETING

Before 1954, separate cooperative marketing societies were established as distinct from the cooperative credit societies. Since 1954, however, multipurpose societies have been started with the purpose of giving both credit to the farmers and to market their surplus produce.

The cooperative marketing society functions in the following manner: The members of the society agree to sell their surplus produce to the society. As soon as they supply the produce to the society, they get an advance to carry on with their agricultural operations. The society collects the produce of all the members and those of the non-members of the village who ar willing to sell their produce, often processes the produce and then disposes it of in the mandi. It does away with many of the middlemen. If the current prices are not favourable and if it is anticipated that prices may rise in the future, the society may decide to stock the commodity. As soon as the produce is sold, the society pays the farmers the balance of the amount due to them. An important feature of the marketing society is that it is managed by paid staff. Usually a society covers a number of villages so that it may be effective and successful.

Advantages of Cooperative Marketing Societies

In some of the Western countries, cooperative marketing has been extremely successful. Denmark has been well-known throughout the world for cooperative marketing of dairy products. Many advantages are claimed for agricultural marketing on a cooperative basis. They are:

(a) The marketing society substitutes collective bargaining in place of individual bargaining. The farmer by himself is weak but the marketing society is strong. (b) It advances loans to the farmers and enables them to wait for better prices. Besides, it lends to them for their other needs. Thus, cooperative marketing societies can iink credit, farming, marketing and processing to the best advantage of the farmers. (c) It can have its own storage and warehousing facilities; it can thus remove the damage to agricultural produce through rats, ants, dampness, etc. (d) It can arrange to have quick and cheap transport and sometimes it can even have its own transport (e) It can encourage the farmers to produce graded and standardised products and discourage them from adulterating their produce. (f) It can control the flow of supplies and thus influence the prices. (g) It can eliminate many of the middlemen and thus remove their large profit margins, (h) Apart from selling the produce of the farmers it can supply them such essential goods in seeds, fertilisers, implements, etc. Thus, the cooperative marketing society is probably the best method to re-organise rural marketing and promote planned growth of our rural areas.

Scope for Development of Cooperative Marketing

There is great scope for the development of cooperative marketing societies. In the first place, there is the necessity of co-ordinate better farming, finance and marketing. At present efforts are being made to have one society which will perform all the three services for the farmers.

Secondly, the marketing societies should undertake processing of agricultural goods. Several agricultural goods can be more favourably marketed if they are processed before sale. Cotton can be ginned and pressed; oilseeds can be crushed and oil may be sold; jute can be processed and baled; and so on.

Thirdly, the cooperative marketing societies can attempt to sell agricultural goods to consumers directly (whether this is possible) and thus eliminate the middlemen and their commission.

Fourthly, the cooperative marketing societies should be made to grade their goods. The grading facilities may be made available to them by the Government through the Agricultural Marketing Government through the Agricultural Marketing Department. Grading will not only help the societies to secure better prices for their produce but also induce them to bring pressure upon their members to improve the quality of their products through the use of improved seeds, etc.

Fifthly, the cooperative marketing societies should be helped to have their own storage and warehousing facilities in the rural areas and mandis. This may be done through the provision of grants and subsidies by the Government or through cheap finance provided by the State Bank and the Reserve Bank of India. Cooperative marketing societies have built up their own storage capacity to the tune of 42 lakh tonnes in 1977-78.

Sixthly, the area of operation of the cooperative marketing society should be expanded to cover many villages (if necessary, even a tehsil) so that it may render effective marketing service to the farmers. This will also enable it to engage qualified men to manage its affairs.

Seventhly, the cooperative marketing societies should be strengthens in the matter of finance. One of the reasons why cooperative societies have not succeeded in this country have been their financial weakness. Already, the marketing societies are helped by the Reserve Bank and the State Bank of India. It may be noted that the State Bank has been asked by the Government to be specially responsive to the needs of cooperative marketing and processing societies.

Eighthly, there is very great scope for cooperative marketing societies to take up the supply of inputs also, such as fertiliser, seeds agricultural machinery and implements, pesticides, etc. In fact, between 1968-69 and 1973-74, the total value of inputs supplied by cooperative marketing societies had increased from Rs.251 crores to Rs. 450 crores.

Finally, the government should use the cooperative marketing societies, whenever it is possible and necessary. The Food Corporation of India should buy foodgrains from the cooperative marketing societies and thus eliminate the usual channels of trade. This will encourage the formation of marketing societies.

Progress in Cooperative Marketing

Under the active encouragement of the Government and the Reserve Bank of India, cooperative marketing has made great progress in Maharashtra, Andhra, Tamil Nadu, Uttar Pradesh and Bihar. The Cotton Sale Societies of Maharashtra and Gujarat gin and press cotton for members and have provided, considerable benefit to them. The cane marketing societies of U.P. and Bihar protect the interests of the members as against sugar mills and also help to improve the quality of cane; they advance funds as well as undertake welfare activities in the villages. Besides, they supply cane to the factories and finance the purchase of seeds and manures. There are cooperative societies in Maharashtra specialising in the sale of tobacco, fruit, vegetables, etc. They sell members' produce on a commission basis, but in some cases process the produce and supply manures and pure seeds. Under the recommendations of the All-India Rural Credit Survey Committee, attempts were made to link cooperative marketing with cooperative credit. On the eve of the Fourth

Plan, about 66 per cent of the primary agricultural credit societies were linked to marketing societies. The balance of 34 per cent of credit societies were also affiliated to the marketing societies subsequently.

On the eve of the Fourth Plan, there were nearly 3,300 primary marketing societies, of which about 500 were special commodity marketing societies. At the district level, there were 73 central marketing societies including 15 special commodity societies. At the state level there were 20 Apex Marketing Societies and three Marketing Federations. At the all-India level, there is one National Agricultural Cooperative Marketing Federation. Besides cooperative marketing, cooperative processing is also making headway. The cooperative marketing societies marketed agricultural produce worth Rs. 920 crores during 1972-73, Rs. 1,100 crores in 1973-74 and Rs.1,200 crores in 1975-76. The produce to be sold through the cooperative marketing societies is targeted to rise to Rs. 1,900 crores in 1978-79. The Sales of Punjab, Maharashtra, Uttar Pradesh and Gujarat accounted for 75 per cent of the total value of the agricultural produce marketed by ce-operatives. Food-grains constituted 50 per cent of the total sales of agricultural produce by cooperatives. Cooperatives are also being increasingly involved in the procurement operations of cotton and jute on behalf of the newly established cotton and jute corporation. Cooperatives have also recorded significant progress in the distribution of fertilisers and other agricultural inputs, improved seeds, agricultural machinery and pesticides. The number of agricultural processing societies organised in the cooperative sector increased from 1,755 a t the end of 1971-72 to 2,204 at the end of 1976-77.

The National Cooperative Development Corporation (NCDC) was set up under an Act of Parliament with the object of planning and promoting programmes for the production, processing storage and marketing of agricultural produce and notified commodities through cooperative societies. The Corporation supports these programmes and has initiated a number of new schemes for which assistance is provided to the State Governments. The NCDC has played an active role in the development of cooperative marketing and processing both in terms of expertise and financial assistance. The NCDC has extended its activities to the development of cooperative dairy, poultry, fisheries and minor forest produce which are basically intended to benefit the most vulnerable sections of the community. The total financial assistance provided by the NCDC since its inception (1963) up to end-March, 1978 amounted to Rs. 232 crores, of which Rs. 133 crores were outstanding.

GOVERNMENT AND AGRICULTURAL MARKETING

Let us consider the various measures which the Government has taken so far in the field of agricultural marketing.

(i) *Marketing Survey.* In the first place the government has undertaken marketing surveys of various goods and has published these surveys. These surveys have brought out the various problems connected with the marketing of goods and the suggestions for their removal.

(ii) *Grading and Standardisation*. Secondly, the government had done much to grade and standardise many agricultural goods. Under the Agricultural Produce (Grading and Marketing) Act the government has set up grading stations for commodities like ghee, flour, eggs etc. The graded goods are stamped with the seal of the Agricultural Marketing Department-AGMARK. The 'Agmark' goods have a wider market and command better prices. A Central Quality Control Laboratory has been set up at Nagpur and eight other regional laboratories in different parts of the country with the purpose of testing the quality and purity of agricultural products applying for the government's 'Agmark' have been created. During the Sixth Plan the Government is further streamlining quality control enforcement and inspection and improvement in grading. The number of testing laboratories is being increased and the program of grading at producers' level is receiving greater attention especially for commercial crops.

(iii) *Setting up of regulated markets*. A very important measure which the Government has taken to improve agricultural marketing has been the setting up of regulated markets in the country. There are now over 3,630 regulated markets. With the establishment of these regulated markets, the malpractices in mandis have disappeared and the market charges have been rationalised. As much as 70 per cent of agricultural produce is now sold in regulated markets.

In this connection the measures to standardise the weights and measures in the country should be mentioned. The government has successfully replaced the different systems of weights and measures prevalent in the country with the metric system. This is great improvement indeed as compared to the previous system under which the farmers were deceived by the traders.

(iv) *Provisions of Warehousing Facilities*. The government has done much to provide warehousing in towns and villages. The Central Warehousing Corporàtion was set up in 1957 with the purpose of constructing and running godowns and warehouses for the storage of agricultural produce. The States have set up the State Warehousing Corporation with the same purpose. At present the Food Corporation is constructing its own network of godowns in different parts of the country. The total storage capacity in the country was 131 lakh tonnes at the end of the Forth Plan. By 1977-78, a total storage capacity of 160 lakh tonnes was created.

(v) *Dissemination of Market Information.* The government has been giving attention to the broadcasting of market information to the farmers. The All India Radio broadcasts daily the closing prices of agricultural goods and gives information regarding prices, stocks and movement of staple commodities. Since most villages have radio sets, these broadcasts are actually heard by the farmers.

Finally, we should mention the setting up of the National Cooperative Development and Warehousing Board (1956).

WAREHOUSING IN INDIA

Warehousing facilities are necessary to prevent the low arising out of defective storage and also to equip the farmers with a convenient instrument of credit. Both the Agricultural Finance Sub-Committee (1945) and the Rural Banking Enquiry Committee (1950) emphasised the importance of warehousing as a method of promoting rural banking and finance in India. An important landmark in the field of warehousing was the detailed recommendation of the All-India Rural Credit Survey Committee for initiating a countrywide programme of warehousing development. The Government of India accepted the recommendations of the committee and set up the National Cooperative Development and Warehousing Board (in 1956) and the Central Warehousing Corporation (in 1957). This was followed by the setting up of state warehousing operations by all State. The Central Warehousing Corporation (CWC) was to set up central warehouses only at centres of all-India importance (as for instance in towns and cities prominent for export and import trade or inter-State trade). State Warehousing Corporations (SWCS) are to establish State warehouses in all centres of regional importance. To avoid unfair competition with the godowns of the cooperative marketing societies, the State warehousing corporations do not open warehouses at any place below the sub-divisional level.

In 1960-61, there were 10 general warehouses with a total capacity of less than 1 lakh tonnes. At the end of June, 1978 the CWG was running 203 general warehouses with a total storage capacity of about 29 lakh tonnes. The SWCs were running 854 warehouses and sub-warehouses at the end of 1975-76 with a total storage capacity of 29 lakh tonnes. The Food Corporation of India, the food and Civil Supplies Department of different States and cooperative marketing societies which work at mandi levels-all these are putting up warehouses of their own. A wide range of more than 100 commodities, such as foodgrains, pulses, oilseeds, fertilisers, sugar, gur etc. are stored in them.

From the data given in table 29.1, it may be noted that during the 4-year period (1973-74 to 1977-78), warehousing capacity increased from 119 lakh tonnes to 160 lakh tonnes-an increase of 34.4 per cent. The share of the Central Government accounts for 50.3 per cent. It is heartening to note that cooperatives have shown rapid increase in their capacity by about 13 lakh tonnes during this short span of 4 years. Cooperatives now account for 28 per cent of total storage capacity. the increase in storage capacity has been of the order of 13 lakh tones in the case of Food Corporation of India and of about 10.6 lakh tonnes in case of State Warehousing Corporations.

TABLE 29.1

Storage Capacity with Public Agencies

(Lakh tonnes)

	1973-74	*1977-78*	%
Food Corporation of India	51.3	64.6	(50.3)
Central Warehousing Corporation	11.6	15.9	
State Warehousing Corporation	6.0	16.6	(20.6)
State Governments	18.0	18.0	
Cooperatives	32.0	45.0	(28.1)
Total	119.1	160.1	(100)

Source: Draft Five Year Plan (1978-83) Revised.

Warehouses have two definite advantages— (i) to provide better and scientific storage facilities at reasonable cost, and (ii) to provide farmers and traders with a convenient instrument of credit in the form of warehouse receipt against which they can borrow from banks. Both the advantages are significant but the second one depends upon the first. In other words, the warehouses should attract adequate custom, by providing storage facilities at cheap rates consistent with local requirements. Producers and traders who are accustomed to the conventional system of storage have to be induced to use the warehousing facilities. At the same time, banks should be induced to advance loans to traders and producers on the new system of storage in warehouses. Experimental trials, grade specifications and special arrangements for the storage of such goods as gur, copra, onions etc-all these have to be arranged by the warehouses.

Advances Against Warehouse Receipts

One of the main objectives of development of warehousing in the country is to make use of warehouse receipts as an instrument to get credit from commercial banks. The latter are willing to advance loans to traders and producers on the basis of securities. It was also the intention of the Government to authorise the Reserve Bank of India to make advances on the promissory notes of any scheduled bank or State Cooperative Bank supported by documents of title to goods. The existence of licensed warehouses helps the use of such documents of title to goods. Hence, efforts have been made to develop warehousing facilities rapidly for purposes of storage and also to popularise warehouse receipts as instruments of credit. In the last few years, there has been significant expansion in the advances granted by scheduled banks on the security of warehouse receipts. The cooperative banks also extend advances against warehouse receipts, though the volume of such advances remains still insignificant. However, the warehouse receipt has not yet become a fully negotiable credit instrument in the same way as promissory notes or bills of exchange or cheques. Necessary amendments to the Negotiable Instruments Act have to be made so as to invest the warehouse receipt with all the attributes of free negotiability.

It was the intention of the All-India Rural Credit Survey Committee that there should be a close co-ordination between cooperatives and warehouses. It was always understood that warehousing facilities were primarily intended to benefit the farmers and their cooperatives. As the farmers are being organised under marketing societies, the facilities of warehouses are to be availed of by the farmers through their cooperative marketing societies. But so far cooperative societies have assumed the dominant role in agricultural marketing and accordingly they are not taking advantage of the warehousing facilities.

REFERENCES

Bhattacharjee, *Studies in Indian Agricultural Economic,* Chapter II.

All-India Rural Credit Survey, *General Report (Volume 2)* Chapters 7, 27 and 35.

30

COOPERATION AND AGRICULTURAL DEVELOPMENT

COOPERATIVE MOVEMENT: A GENERAL SURVEY

In the past, at the beginning of this century, the condition of the agriculturists was really deplorable. The joint family system was breaking down. As a result of increase in population and the absence of rapid industrialisation, more and more people depended upon agriculture resulting in too much pressure on land. The land got increasingly divided and sub-divided. Farmers who were once prosperous became petty peasants with tiny holdings of a few acres and were finally reduced to the status of landless labourers. They were in the clutches of the money-lenders. Rural debt was increasing and this increased the burden on the farmer. While his economic position was deteriorating and he was finding himself in a hopeless condition, the farmer did not have the incentive or the ability to improve his condition. The Government felt that the farmer could be helped through some type of cooperative effort.

In western countries where cooperative societies were evolved, there were three type of cooperation-consumer's cooperation, producers' cooperation and cooperative credit or finance. India started with cooperative credit for the farmers, but later developed many other forms also such as cooperative credit for others (besides farmers), producers' cooperatives, consumers' cooperatives, cooperative banking and insurance, cooperative marketing and processing, cooperative housing, and so on. Of all these forms of cooperation, the most important in India has been the agricultural credit cooperation. Naturally, we shall deal with this type of cooperation rather elaborately.

Cooperative Legislation and Progress of the Movement

The Cooperative Credit Societies, Act, 1904 was passed by the Government and rural credit societies were formed. Through the appointment of Registrars and through vigorous propaganda, the Government attempted to popularise the movement in rural areas. Within a short period, the government realised some of the weaknesses and shortcomings of the 1904 Act and therefore, passed another and comprehensive Act known as the Cooperative Societies Act of 1912. This Act recognised non-credit societies also. The progress of the rural credit movement was quite rapid in the beginning but because of difficulties in the agricultural sector particularly during 1929-33 when the country was plunged into a serious economic depression (known as the great depression as all the countries of the world suffered from it), Indian agriculture was badly affected. But during and after the Second World War agricultural prices went up and accordingly agriculturists enjoyed a good measure of prosperity; and the cooperative movement too got stimulus during this period. In spite of all this progress, only 3.1 per cent of the total credit needs of the cultivators were met by the cooperative credit societies. The most rapid progress however, has been since 1954 when the scheme suggested by the All-India Rural Credit Survey Committee was accepted and implemented by the Government and the Reserve Bank of India.

SHORT-TERM COOPERATIVE CREDIT

As indicated in the beginning of this chapter the cooperative movement was started in India largely with a view to providing agriculturists with institutional facilities for obtaining the capital required for agricultural operations at low rates of interest. The organisation of the cooperative credit for short period may be briefly out lined as follows:

Primarily Agricultural Credit Society

A cooperative credit society, commonly knows as the primary agricultural credit society (PACS) may be started with ten or more persons, normally belonging to a village. The value of each share is generally nominal so as to enable even the poorest farmer to become a member. The members have unlimited liability, that is each member is fully responsible for the entire loss of the society, in the event of failure. This will mean that all the members should know each other intimately. The Management of the society is under an elected body consisting of President, Secretary and Treasurer. The management is honorary, the only paid member being, normally, the accountant (in case the society is large and requires a paid whole-time accountant). Loans are given for short periods, normally for one year, for carrying out agricultural operations, and the rate of interest is statutorily fixed-about 6 per cent. Profit are not distributed as dividend to shareholders but are used for the welfare of the village. In the construction of a well, or maintenance of school, and so on. The position of primary societies is given in Table 30.1.

The village primary society was expected to promote thrift among the farmers and secure deposits from them. But it was found that the farmers, besides being poor, do not deposit their savings with the cooperative society.

TABLE 30.1

Primary Agricultural Credit Societies

	1950-51	*1960-61*	*1970-71*	*1978-79*
No. of Societies (thousands)	105	212	161	102
Membership (lakhs)	44	170	310	516
Loan advanced during the year (Rs. crores)	23	202	578	1,395
Loans overdue (Rs. crores)	6	44	322	895

Source: Report of the Ministry of Agriculture, Department of Cooperation 1974-75 and Reserve Bank of India, *Report on Trend and Progress of Banking in India*, 1979-80.

The usefulness of the primary credit societies has been rising steadily. In 1950-51, it advanced loans worth Rs. 23 crores; this rose to Rs. 202 crores in 1960-61, to Rs. 578 crores in 1970-71 and to Rs. 1,395 crores in 1978-79. Aggregate loans outstanding rose from Rs. 1,218. crores in 1974-75 to Rs. 1,395 crores in 1978-79. Total loans overdue fame to Rs. 895 crores in 1978-79. The primary agricultural credit societies have stepped up their advances to the weaker sections particularly the small and marginal farmers. This progress has been quite spectacular but not adequate, considering the demand for finance from farmers.

Considerable attention was given during the past few decades to build the PACS into strong institutions. Such a structure, close to the farmers, is very essential for disbursing rural credit, particularly to small farmers. However, "the primary credit society has continued to remain the weakest link in the entire cooperative structure."[1] A programme was introduced by the Government and the RBI to reorganise and revitalise the primary agricultural credit societies. It was completed in Rajasthan, Orissa, Madhya Pradesh, Kerala, Tamil Nadu and Gujarat. In other States, it has not made much headway. For example, the number of societies had come down from 2,12,000 in 1960-61 to 1,61,000 in 1970-71 and 1,02,000 in 1978-79.

To make all primary agricultural societies viable and ensure adequate and timely flow of cooperative credit to the rural areas, the Reserve Bank of India, in collaboration with State Governments, has been taking a series of steps to strengthen weak cooperative banks and to current regional imbalances in cooperative development. Steps have been taken to reorganise viable PACS and for amalgamation of non-viable societies with farmers' service societies or large sized multipurpose societies. These efforts are being intensified by providing larger funds to weak societies to write off their losses, bad debts and over-dues. The National Commission on Agriculture has recommended the organisation of Farmers' Service Societies to provide not only credit but also agriculture inputes and technical guidance to members for forming large sized multipurpose societies. These societies are being increasingly organised in all States.

The Commercial banks in India introduced in 1970 a scheme of financing PACS. The scheme is now in operation in 13 States. As at the beginning of 1980, 683 branches of 24 commercial banks had adopted 2,869 primary agricultural credit of Rs. 18 crores during 1978-79. Gradually, the funds of the commercial banks are being made available to PACS.

The scheme came in handy to commercial banks which would use PACS for disbursting agricultural loans and thus find a way out of the serious problem of not having close contacts with farmers through their own branches and field staff. The scheme, however, has not been as successful as it was anticipated. Firstly, the two systems with two unrelated culture-one commercial and the other cooperative-could not be linked effectively. Difficulties have arises from this basic incompatibility. Secondly, state cooperative banks as well as central cooperative banks have not liked successful PACS being taken away from their fold by the commercial banks. However, a good deal of scope for coordination exists between PACS and the branches of commercial banks in rural areas:

(a) The rural branches of commercial banks can assist such of those members of the PACS who are eligible for loans for who are unable to get finance from PACS for lack of funds.

(b) The rural branches of commercial banks can help the PACS with advice on management e.g. proper maintenance of books of accounts, accounting procedures, etc.

(c) The PACS can help commercial bank branches to identify eligible borrowers and to recover loans.

Cooperative Central Banks

These are federations of primary credit societies in a specified area, normally extending to the whole district (hence they are sometimes knows as district cooperative banks). These banks have a few private individuals, as shareholders who provide both finance and management. The cooperative central banks have three sources of funds-their own share capital reserves, deposit from the public and loans from the State Cooperative Bank. Their main task is to lend to village primary societies, but they were expected to attract deposits from the general public. But the expectation has not been fulfilled and many of the cooperative central banks act as intermediariers between the State Cooperative Bank on the one hand and the village primary credit societies on the other. The RBI has formulated a scheme for the rehabilitation of weak central cooperative banks. The RBI is also providing liberal assistance from the National Agricultural Credit (Long-Term Operation) Fund to the State governments for contributing to the share capital of the weak central cooperative banks selected for the purpose. The position of central cooperative banks is given in Table 30.2.

TABLE 30.2

Central Cooperative Banks

	1950-51	*1960-61*	*1970-71*	*1978-79*
Number	505	390	341	338
Share Capital (Rs. crores)	4	39	141	—
Loans advanced during the year (Rs. crores)	83	350	894	2,407

Source: Same as for Table 30.1.

State Cooperative Bank. This bank also known as the apex bank, forms the apex of the cooperative credit structure in each State. It finances and controls the working of the central banks in the State. It serves as a link between the Reserve Bank if India from which it borrows and the cooperative central banks and village primary societies on the other. The State Cooperative Bank obtains its working funds from its own share capital nd reserves, deposits from the general public and loans and advances from the Reserve Bank. The last-mentioned source is quite important, as it constitutes between 50 and 90 per cent of the working capital of State Cooperative Banks in the country. The State Cooperative Bank is not only interested in helping the cooperative credit movement but also in promoting other cooperative ventures and in extending the principles of cooperation. Table 30.3 gives the position of state cooperative banks in India.

A highly distressing fact of cooperative credit is the heavy overdues of cooperative credit institutions. The RBI appointed a "Study Team on Overdues of Cooperatives Credit Institutions" in 1972 which submitted its report in 1974. According to the Study Team. "lack of will and discipline among the cultivators to repay loans was the principal factor responsible for the prevalence of overdues in cooperatives. Defective lending policy pursued by cooperatives, the apathy of management in taking quick action against recalcitrant members and absence of favourable climate were other contributory factors."[2] In recent years, the farmers are getting organised and one of the chief demands of the farmers' union is the cancellation of their debts to cooperative societies and banks. The problem of loan overdues is thus becoming a serious one.

TABLE 30.3

State Cooperative Banks

	1950-51	*1960-61*	*1970-71*	*1978-79*
Number	15	21	25	26
Deposits (Rs. crores)	22	72	279	1,206
Loans made during the year (Rs. crores)	42	258	749	2,237

Source: Same as for Table 30.1

The cooperative credit system makes credit available to the farmers at convenient distances and has intimate knowledge of the local conditions and problems which arise. But it is organisationally and financially weak and hence, in practice, its ability to support credit to the agricultural sector is considerably limited. The All India Rural Credit Review Committee brought out the following weaknesses of the primary credit societies : (a) Cooperative credit still forms a small portion of the total borrowings of the farmers; (b) Tenants and small farmers find it difficult to satisfy their needs fully; (c) Most primary credit societies are weak and are unable to meet fully the production-oriented credit reeds; (d) Overdues at all levels are increasing alarmingly indicating and (e) They have not been able to ensure adequate and timely credit for the borrowing farmers.

By far the most salutary trend, in this context, is the continuing process of reorganisation of primary agricultural credit societies so as to make them viable. The norms for viability are a minimum short-term loan business of Rs. 2 lakhs, a gross cropped area of 2,000 hectares within a radius of 10 km and appointment of a full time paid secretary. Such reorganisation through amalgamation, according to the union Department of Rural Development in its report for 1977-78 has been more or less completed in eight States and its pace is being accelerated in others. It is expected that ultimately there will be about 90,000 viable societies, as against 1.23 lakh societies, (as of June 1977) covering 97 per cent of the villages. Though the viability of societies can thus be ensured, there will still remain the problem of uneven distribution of cooperative benefits as between different States. For instance, (in 1976-77), the loans advanced per member are as high as Rs. 769 in Gujarat, Rs. 777 in Haryana and Rs. 479 in Punjab, Rs. 114 in Orissa, Rs. 169 in Uttar Pradesh and Rs. 178 in West Bengal, against the all-India average of Rs. 278.

Viewed in another way, that is, in terms of average credit per hectare of cropped area, it is only in five States, namely, Gujarat, Haryana, Kerala, Punjab and Tamil Nadu that such credit is much higher than (double or more than double) the all-India average of Rs. 65 per hectare. The persistent problem relating to loan overdues-a major obstacle in cooperative development-also remains to be tackled. Loans overdue, it is disquieting to note, represent 42 per cent of loans outstanding in all-India, the percentage ranges from 23 in the case of Tamil Nadu to 77 in the case of Bihar. This malady afflicts, in turn, the Central cooperative banks and the State cooperative banks which lend to the primary societies.

LONG-TERM COOPERATIVE CREDIT: LAND DEVELOPMENT BANKS

The long-term requirements of the farmers were traditionally met by the money-lenders but later by other agencies also, such as the State Governments and the cooperative credit banks. But these agencies were found defective for one reason or the other. There was, thus a great need in India for an institution specially designed to cater to the long-term credit needs of the agriculturists, which would offer long-term funds at moderate rates and recover loans in annual or semi-annual instalments spread over a number of years. Land development banks (or land mortgage banks as they were previously called) were organised for the purpose of providing long-term credit to farmers.

To Punjab goes the credit for organising the first land development bank (LDB) in India in the year 1920. Madras followed suit by organising two land development banks in 1925 and Bombay followed with three banks in 1929. But the real beginning in land development banking was made by Madras with the organisation of a central land development bank in 1929 for centralising the issue of debentures and for co-ordinating the working of primary banks in the State. The progress of land development banking has been very slow and also uneven. During the great depression, land-development banks received some stimulus as agricultural prices fell considerably and the farmers needed financial assistance. But with the Second World War, the farmers experienced a good measure of prosperity and were in a position to repay their debts with the land development banks. But after Independence, land development banks have been enjoying a great degree of prosperity. However, it is important to note that whatever progress was achieved was concentrated in only a few States, viz., Andhra, Tamil Nadu, Karnataka, Maharashtra and Gujarat. The number of primary land development banks increased from 286 in 1950-51 to 856 in 1973-74 while that of Central (or State) land development banks increased from 5 to 19 during the same period. Total loans advanced by the state development banks came to Rs. 300 crores in 1979-80 as against Rs. 241 crores in the previous year; and the loans outstanding came to Rs. 1,400 crores at the end of June, 1980.

The Structure of LDBs

The long-term credit structure consists of the central land development banks (generally one for each State) and primary land development banks. In some States, there are no primary land development banks but in their place, there are branches of central land development banks. In Madhya Pradesh, the State Cooperative Bank itself functions as a central land development bank through a separate land development banking department. In Andhra, Kerala and Maharashtra, there are more than one central land development banks and efforts are being made to integrate them into a unified bank for the whole State. Similarly, there are considerable differences in the organisation of primary land development banks in the different States.

Finances of LDBs

Land development banks obtain their funds from share capital, reserves, deposits and issue of bonds or debentures,. However, the last is the most important. Debentures are long-term loans which are issued by the Central Land Development Bank, carrying fixed interest and for fixed periods, generally up to 20 years. These debentures are guaranteed by State Governments in respect of payment of interest and repayment of principal. They are subscribed by the LIC, the commercial banks, the State Bank of India and its subsidiaries and by the Reserve Bank of India. Besides ordinary debentures, the land development, banks have been floating rural debentures since 1957 for periods up to 7 years which are subscribed by farmers and panchayats and by the Reserve Bank of India (up to 66 per cent of the value of the rural debentures). In recent years, the substantial refinance facilities provided by ARDC to cooperative and development banks have helped enlarge the lending operations of these banks.

Loan Operations of LDBs

The main function of LDBs is to grant loans on the security of agricultural properties. Since they generally grant loans which run for several years, strict rules are laid down with regard to the security against which they can advance loans. Generally, these banks restrict their loans to first mortgage of agricultural property though in a few cases they may advance loans against the security of second mortgage as well. Again, they generally lend up to 50 per cent of the value of the security. In order to assess the value of land against which they lend, they employ experts who assess the value of land and are conversant with local conditions. In assessing the value of land such factors as the amount of land tax paid, the rental value of land, gross and net income from the land, sale value of the land etc., are taken into account. While granting loans the banks consider not only the value of the security offered but also examine the repaying capacity of the applicants. Fianlly, they are able to lend at fairly low rates of interest-about 9 per cent-and enable the needy farmers to secure funds for long periods.

LDBs provide credit for a variety of purposes such as redemption of old debts, improvement of land, purchase of costly agricultural equipment, construction of wells and erection of pumps and so on. As one time, the redemption of old debts was the most important and, in a sense, the only purpose for which the farmers approached the land development banks. In recent years, however, farmers have been borrowing from LDBs mainly for the purpose of land improvement and development including sinking of wells (56 per cent) and purchase of agricultural machinery (30 per cent).

Problems of LDBs

Land development banking is yet to take strong roots in India barring a few States. However, LDBs have contributed in large measure to agricultural development by lending about Rs. 3,000 crores of which about 70 per cent has gone for developing minor irrigation. The credit volume of LDBs has crossd the Rs. 300 crore mark in 1979-80. The LDBs have built up a structure which operates through about 2,100 branches and primaries located tehsil and block levels covering about 8 million members. All their loans are for productive purposes benefitting mostly the small farm holders. Though land development banking has made considerable progress in recent years, it has not really contributed much to the improvement of the financial position of the farmers through the provision of long-term credit. A large number of factors are responsible for the comparative ineffectiveness of LDBs.

Problem of Overdues

Mounting overdues in most of the LDB's have crippled the structure badly in recent years. Overdues are rising continuously crossing the containable limit. This has caused innumerable financial problems besides limiting the capacity of LDBs to lend and operate as viable units. During the 3-year period 1977-79, the loan disbursements of the banks have been stagnant at the level of about Rs. 250 crores though in the year 1979-80 there has been an appreciable increase by about Rs. 60 crores. The financial discipline imposed on the banks in the matter of eligibility to undertake fresh lending based on recovery performance has been the main limiting factor for quantitative growth of credit operations. Though to some extent, the banks themselves are to be blamed for this predicament due to faulty loaning policies, inadequate supervision, over-utilisation of loans, ineffective measures for recovery etc., contributed to the deterioration in recovering the loans. Even when the willful defaulters lands are attached and auctioned under the provision of law, the banks find few bidders coming forward to purchase such lands. Various suggestions for making the coercive measures effective have largely remained unimplemented at the level of State governments. With large overdues, restricted lending eligibility, and financial problems, the operations of the banks in a few states have come to a standstill. Unfortuately, there are neither permanent arrangements to rehabilitate borrowers and banks, nor any credit stabilisation arrangements take care of overdues effected by national calamities and other factors beyond the control of LDBs and the borrowers. Overdues, however, are not peculiar of LDBs alone. The position is equally bad in other sectors of rural credit. Some serious thinking is called for to remedy the situation at the higher levels immediately before the institutional rural credit arrangements are put out of gear.

Lack of Trained Manpower

No doubt LDBs have grown over the years, but not enough attention has been paid to build up the manpower capable of shouldering the increasing volume of business and to face the emerging challenges. To bring about qualitative changes in lending, the LDBs should be adequate technical personnel and supervisory staff. Diversification of loans for certain non-traditional purposes of agricultural development and for non-land based investment activities has to take place if the LDBs have to expand their loaning business. This requires comprehensive changes in the legal framework of respective cooperative societies, Acts and bye-laws of the LDBs.

Diversification of Loans

For the success of diversification, the banks have to change the traditional norms of landed security against loans to other forms of securities like hypothecation of assets, charge on land, government guarantee, personal surety. etc. provided investment proposals are otherwise viable. Landless labourers, village artisans and marginal farmers who could take up certain productive activities are not able to secure credit from the LDBs mainly because they do not possess land or adequate security to offer against loans. But in some states the LDBs have made a beginning in diversifying the loans which has enabled a large number of farmers and weaker sections to take loans for several activities like dairy farming, poultry farming, inland and marine fisheries, sheep rearing, sericulture, gobar gas plants and bullock carts. There is good scope for financing agro-processing industries and rural and cottage industries which have so far remained outside the purview of LDBs operation. For expansion of their business and to play the role expected of them, the LDBs have to explore the possibilities of financing various diversified activities and exploit the potential area of agricultural and rural development.

Future of LDBs

In the context of the Sixth Plan, the LDBs will be the main institutional agency to implement the minor irrigation programmes and it is expected that over 50 per cent of the target of Rs. 2,300 crores set for them will go for development of minor irrigation. Apart from minor irrigation, the LDBs are also expected to step up their credit assistance to several other agricultural development activities and for various subsidiary occupations. The integrated rural development programme will cover the entire country to improve the rural economy and institutional agencies including the LDBs are expected to involve themselves effectively to support various productive activities which will, besides supplementing the income of the people, result in creating employment potential in rural areas.

OTHER FORMS OF COOPERATION

In India, cooperation took the form of agricultural credit cooperation. In course of time, however, other forms of cooperation were also developed. A broad groups non-credit cooperation in the rural areas and another group is non agricultural cooperation. We shall see briefly these types of cooperation.

Non-credit Cooperation

In rural ares, non-credit cooperative societies were started from the earlier times to promote certain agricultural operations. For instance, non-credit societies were :

(a) *Cooperative purchase and sale societies*—these specialise in the purchase of seeds, manure, implements and machinery, etc., and for the sale of agricultural produce of the members.

(b) *Cooperative marketing societies*—these specialise in selling the surplus produce of the farmers.

(c) *Consolidation of holdings societies*—these specialise in helping members to consolidate their tiny and scattered holdings. In practice, they help the authorities in consolidation work.

(d) *Better forming societies*—these help farmers too adopt better farming methods and systems.

(e) *Manufacturing and processing societies*—The cooperative sector has moved ahead in the manufacture and processing of sugar, oilseeds, fruit and vegetables, sericulture, spinning, dairing, poultry, fishery etc. During 1977-78 a total of 129 cooperative sugar factories crushed 314 lakh tonnes of sugarcane and produced about 32 lakh tonnes of sugar. It is a matter of legitimate pride that the share of production of sugar by the cooperatives was itself about half of the total quantity produced in the country.

(f) *Retail fair pricemarketing*—1977 a country-wide infrastructure had been built consisting of 15,000 primary consumer cooperative societies, 470 central consumer cooperative stores with 4,500 branches and departmental stores. 14 State-level consumer cooperative federation and National Consumer Cooperative Federation. In all 20,000 outlets in the form of fair price shops and departmental stores have been organised.

Other non-credit societies in rural areas include societies for land reclamation, crop protection, cattle breeding, cooperative farming etc. The non-credit societies did not develop to the same extent as credit societies. This was mainly because they require better education and higher level of training which were not found in India. Besides, the credit societies themselves in many cases, undertook non-credit functions also such as purchase and sale better farming etc. However, the present tendency is the conversion of single purpose societies into multi-purpose societies which combine both credit and non-credit functions.

WEAKNESSES OF THE COOPERATIVE MOVEMENT

Though the cooperative movement has been expanding rapidly in the recent years, there were many reasons for the painfully slow growth of the movement in the first 50 years of its existence. The All-India Rural Credit Survey committee as well as many individuals and committees have brought out the various weaknesses of the movement in India and have suggested methods for improving and reorganising the movement.

(i) *Lack of spontaneity*—The cooperative movement in India did not spring from the people themselves. The movement was not voluntary and the people did not come forward to organise societies to satisfy their needs. On the other hand, the movement took the form of a government department. The villagers among whom the credit societies were started generally thought of the societies as government lending agencies. The government officials who ran the movement were ignorant of the ideals of cooperation. They were neither properly trained, nor were they aware of the needs of the farmers.

(ii) *Lack of funds*—A basic weakness of the cooperative movement was the lack of funds. It was felt that the members themselves should save and deposit their savings and thus contribute a large share to the working capital of the societies. The central and the state cooperative credit societies which were organised to help village societies could not attract as much deposits from the general public as was anticipated. This is the position even now. The Reserve Bank was willing to lend to the cooperative banks, and at concessional rates, but even this facility could not be taken advantage of by the State cooperative banks. Lack of funds was one of the basic weaknesses of the movement.

(iii) *Loans for productive purposes only*—The cooperative credit societies did not help the farmers in meeting all their credit requirements. They gave loans only for agricultural operations. But the farmers required loans to meet many of their other requirements also. For these purpose, the farmers had to depend upon the money-lenders. In other words, the cooperative credit societies should have met all the requirements of the farmers, otherwise, they would not receive full loyalty of the farmers.

(iv) *Provisions of credit only*—The cooperative movement failed to appreciate the organic connection between credit, marketing and processing. As a result, the cooperative societies provided loans to the farmers but did not help the latter in any other way. In certain areas different societies were started for different purposes. What was really needed was a cooperative society which would meet all the needs of the farmers and keep in continous touch with them throughout the year. The failure to provide credit, marketing, processing, better farming etc. to the farmers was one of the basic factors responsible for the weakness of the movement in India.

(v) *Competition from private agencies*—From the very beginning, the cooperative movement met with competition from powerful vested interests. In villages, the money-lenders and traders worked for the failure of the village societies. In urban areas, consumer societies faced bitter opposition and hostility of traders and speculators. But such opposition to the cooperative movement was not anything unique to India; such was the experience of other countries too. But on account of the financial weakness of the movement in India, cooperation could not make much headway in the face of powerful competition.

(vi) *Lack of cooperation on the part of the people*—In India, people have been largely illiterate, ignorant and extremely conservative. For one thing, most of them did not understand the real meaning and objective of cooperation. For another, the principle of unlimited liability which was the basis of the village credit societies prevented the better farmers to join the movement. Cooperation cannot succeed unless there is willing and complete cooperation from the people.

(vii) *Defective management and leadership*—The failure and liquidation of many societies in rural areas was directly due to the defective management and leadership. As mentioned earlier, the village economy is dominated by the landlords. the small farmers owe blind allegiance to the landlords. But these landlords did not care much about the success of the cooperative movement. Nor did they take genuine interest in promoting the welfare of the farmers. Besides, the work if the societies suffered from nepotism, favoritism and partiality. For instance, loans and other assistance flowed easily to rich farmers and to the relatives and friends of the office-bearers. The needy and the poor failed to get any assistance. Moreover, the office-bearers were not strict in getting repayments from those in whom they were interested. In general, the management was weak and inefficient.

(viii) *The attitude of the Government*—The Government was correct in encouraging the movement in all possible ways. The mistake it made was to convert cooperatives into a government department with all the rigidities and short sightedness associated with a government department. The tendency was to officiate the movement too much and leave little to private initiative. Moreover, the Government put more emphasis in quickening the pace of the movement instead of consolidating it. Government did not take adequate interest in improving the financial strength of the societies particularly the village primary credit societies.

(ix) *The nature of Indian rural society*—Conditions in Indian rural society did not promote cooperation among people. The average Indian farmer is illiterate and ignorant and has not been able to understand the significance and utility of cooperation. Besides, rural society is riddled with caste spirit, religious feelings and other conditions which hamper the growth of the cooperative spirit and of the cooperative movement. To a large extent, these conditions prevail even now.

We have explained a number of reasons to show the weakness of the cooperative movement and the reasons whey it has not been able to develop fast and make a mark. There were defects which were concerned with the structure of the movement and there were defect inherent in the social and economic structure of the Indian rural society. But there is nothing really wrong with the concept of cooperation. In fact, the movement has been a tremendous success in other countries-as for instance consumers' cooperation of England, credit cooperation in Germany, producers' cooperation in Denmark and Sweden and so on. If the movement has succeeded in these countries and if it has not succeeded in India it clearly means that the essential conditions which were present in England, Germany , etc. were absent in India. Unless similar conditions are created in India, there is not much chance of success here. Basically, the success depends upon the spirit of cooperation which the farmers and other groups should have and the exclusive loyalty to the principle which they should owe to it, irrespective of opposition and apparent failure at any given time.

ACHIEVEMENTS OF COOPERATION IN INDIA

The cooperative movement has been in existence for the last 76 years or so. Many have observed that the movement was an utter failure and that it should be scrapped. According to these critics of cooperation the movement has done nothing to abolish poverty of the rural masses. Nor has it contributed to increase agricultural production, to establish better marketing conditions, better living etc. It has not reduced, leave alone eliminated, the money-lender from the rural scene. The utter insignificance of the cooperative movement was indicated by the fact that in 1954 (exactly after 50 years of its existence) the cooperative institutions supplied a little more than 3 per cent of the requirements of the farmers. However, one should not get the impression that the movement has utterly failed and that, therefore, it should be scrapped. But the best assessment of the cooperative movement was given by the All-India Rural Credit Survey Committee which wrote (in 1954): "Cooperation has failed but cooperation must succeed." It is true that the movement was not much of a success before 1954, but since then with the Government and the Reserve Bank taking active interest, the movement has made great progress. The progress in the last decade is much more than the progress which has been achieved in the first 60 years of its existence.

Benefits

Cooperative societies have provided credit to the farmers at a cheap rate-about 6 per cent or so. Since 1954, the cooperative credit societies are meeting increasingly the requirements of farmers. At present nearly one-fourth to one-third of the requirements of farmers are met by the cooperative societies. The monopoly of the money lenders is being broken in the villages. While at one time of money-lenders provided more than 70 per cent of the farmers' credit needs, they are now providing less then 60 per cent. In course of time, therefore, cooperative credit would become so significant that the village money-lender will have no role to play in rural finance.

The cooperative society has led to the use of better farming methods, such as the use of improved seeds, manures etc. The marketing and processing societies have helped members to buy their requirements cheaply and sell their agricultural produce at good prices. It has also provided good storage facilities to the farmers.

Cooperative societies have also helped in improving life in rural areas. They have educated the farmers so that the latter could give up many of their bad habits like gambling, drinking. etc. They have attempted to remove undesirable social customs like reckless spending during marriages, religious ceremonies etc. They have discouraged farmers from going to the courts and indulging in costly litigation; they have attempted to solve disputes at the village level itself. In other words, they have been attempting to improve the farmers socially and morally.

The non-credit societies, like the housing cooperative societies, the consumers' cooperative societies, etc, have helped their members to raise their economic condition and have saved them from the exploitation of powerful groups. For instance, in many urban areas, housing cooperatives have enabled middle income groups to secure plots of land and also construct their own houses. Consumer cooperation have done great service to the community by helping to distribute goods in short supply equitably and in selling goods at reasonable prices; in many places they have been instrumental in preventing the private ship-keepers from taking advantage of supply shortages and charging exorbitant prices. Likewise, societies meant to help artisans like handloom weavers have helped their members with finance and marketing facilities.

Thus, the cooperative movement has been a source of great benefit to the farmers and others both economically and otherwise. Whatever be the criticism of the movement in the past, it cannot be denied that the movement is becoming increasingly useful. It is good to remember that in villages the farmers had to depend upon the money-lenders for all their requirements of funds and they had fallen victims to the machinations of the latter. The only way the farmers could be saved from the clutches of the money lenders was through the organisation of cooperative societies.

A NOTE ON SERVICE COOPERATIVES

In the past cooperatives were organised for specific purposes, i.e., for credit, for marketing, for purchase and sale, for better farming and so on. The general principle followed was: one village one society and one purpose. Single purpose societies were working with limited resources and in a small way to satisfy primarily only single need of the farmers. They did not go far enough to satisfy all the requirements of farmers and cover all their activities. Such societies could not naturally be expected to bring about an appreciable improvement in rural life. Hence, under the influence of the Reserve Bank of India, there was a shift from single purpose to multi-purpose societies which would tackle the rural problem as a whole. The multi-purpose cooperative societies have come to be knows as service cooperatives. The scope of the service cooperative is given clearly in Reserve Bank report: "Starting with credit for current needs, a society may get the old debts of its good members liquidated through a land mortgage bank, introduce better business and monetary return by inducing its members to sell their produce cooperatively, ensure their growing of improved varieties of crops, by purchasing seeds for them, save on purchases by arranging for the purchase of their needs jointly and at profitable rates on an indent system without incurring any risk or liability, save litigation expense by effecting arbitration, improve the return of crops by consolidation of holding supply pure seeds and improved implements, supplement the income of its members by inducing them to take to subsidiary industries, introducing better living measures by adopting bye-laws by common consent which will curtail ceremonial expenditure and remove insanitary habits, provide medical relief and so on."

The Case for Service Societies

The organisation of the service cooperatives to cater to all the economic needs of the villagers is advocated on the following grounds:

(a) Generally village cooperative societies have been credit societies but credit without being linked up with production and marketing is almost useless. As result the Indian peasant has not made proper use of cooperative credit provided to him. The basic defect is removed by the organisation of service cooperatives.

(b) The original principle of one-village one-society one-purpose has not succeeded. The smallness of a society with little financial resources has high propensity to fall. It is, therefore, necessary to have a society with large financial resources and which will serve a group of villages situated close by.

(c) There has been a shortage of trained personnel in villages to start and run a number of societies. It would be much better to combine the functions of different societies and entrust them of those persons with training in accounts and business.

(d) The service cooperatives will be multipurpose and thus there is a high possibility of the success of such societies. the failure of any one aspect of the society will be made up by the success of the others.

(e) The service cooperatives will bring all the farmers together. The ultimate goal of land reforms now is the organisation of cooperative joint farming which will be made possible by the organisation of service cooperatives. From service cooperative to joint farming is regarded as a necessary and logical step.

The Case Against Service Cooperatives

There are, however, a number of difficulties in the organisation of multi-purpose cooperative societies:

(i) A service cooperative needs intelligent and very hardworking organisers to run it. This is a real difficulty but this can be solved in course of time. For one thing there is rapid spread of education in rural areas and local leadership is forthcoming. For another, a multi-purpose society will have paid full time office bearers to look after its affairs.

(ii) The multipurpose society may have a high propensity to fail. It is possible that failure in one line of its business may endanger its other lines also. When once a service cooperative goes into liquidation in a particular area, it may spell disaster to cooperation itself in that area.

(iii) The service cooperative will necessarily have to serve a number of all village and therefore, its areas of operation will be large. Naturally, members will not know each other intimately, Cooperation is based on mutual knowledge and trust: this is the basis of unlimited liability. The service cooperatives cannot, therefore promote cooperation among villagers and will finally degenerate into a business proposition. There is some substance in this criticism but modern cooperators are not prepared in accept it. For proper and successful functioning single purpose cooperatives have to be replaced by service cooperatives, which necessarily have to cover a group of villages. The principle of unlimited liability will have to be given up.

(iv) Service cooperatives are said to be cumbersome and difficult to manage. It is argued that the villagers-being simple and illiterate cannot understand its working. Therefore the conclusion is that the service cooperative may not appeal to the rural folk. This type of argument has no basis at all. the major criticism against the single purpose societies is that they do not serve the farmer and do not meet all his requirements. It is to overcome this difficulty that service cooperatives are being organiser. The farmer will find that such service cooperatives will help in every way, give him credit, sell his produce, buy agricultural inputs for him, advise him on better farming etc. Besides, the farmers of these days are no more simple and illiterate as many armchair economists seem to be assuming. they are educated, alert and are thirsting the introduction of new techniques and process.

A little careful analysis will, therefore, show that many of the difficulties in the organisation of service cooperatives can be overcome with proper training of personnel and propaganda. The Reserve Bank and the State Government are now completely convinced about the idea of multipurpose service cooperatives.

COOPERATION IN FIVE-YEAR PLANS

In the pattern of development as envisaged in the Five Year Plans cooperation is expected to become progressively the principal basis of organisations in several branches of economic life, notably in agriculture, small industries, marketing, processing, distribution, construction and provision of essential services to local commuities.

Cooperation under the First Three Plans

The First Plan set a target to bring 50 per cent of villages and 50 per cent of rural population under cooperation in ten years. It was also planned that by 1955-56, the credit supplied by the cooperative movement and the Government agencies together should reach the level of Rs. 135 crores. But the actual results were consider ably short of the targets fixed. The Second Plan laid down the following targets for the cooperative movement to be achieved by the end of 1960-61: (a) supply of agricultural credit worth Rs. 225 crores through cooperatives; (b) membership of agricultural societies to be raised to 15 million persons; and (c) 10 per cent of the marketable surplus of the agricultural product to be marketed through cooperatives. To achieve these objectives, the Second Plan indicated clear targets about the number of primary societies, marketing and processing societies, cooperative warehouses etc.

The Third Plan carried forward the work of the first two plans. The primary credit societies were reorganised and their number was reduced. However, their membership rose from 17 million to 27 million and nearly 90 per cent of the villages were covered as against 75 per cent at the end of the Second Plan. Considerable attention was given to cooperative marketing of agricultural produce and a number of primary marketing societies were set up throughout the country. In addition to marketing of the produce, many of the marketing societies were undertaking distribution of fertilisers, agricultural implements, pesticides, improved seeds and the storage of agricultural produce. Some of them have been entrusted with the export of agricultural commodities to foreign countries while others are undertaking processing activities.

Cooperation in the Fourth and Fifth Plans

In the Fourth Plan more emphasis was laid on agricultural cooperatives and consumer cooperatives with a view to institutionalising the various services needed by the farmers. The object was to increase efficiency and economic viability of the cooperatives and in this direction, Government assistance was rendered to equip the cooperatives in important matters like finance, organisation and trained personnel.

In the strategy of cooperative development the Fifth Plan lays great stress on (a) structural improvement of weak central banks and a large number of non-viable primary agricultural credit societies; (b) re-organisation of a large number of marketing societies and consumer cooperative stores; (c) setting up of Farmers' Service Societies as recommended by the National Commission Agriculture; (d) reorienting the cooperatives so that they would look after increasingly the small and marginal farmers and other under-privileged sections; (e) bringing about greater cooperation between public sector banks and cooperative banks and public sector commodity corporations; and (f) taking measures for professional management of cooperation.

Cooperation in the Sixth Plan (1978-83)

The Sixth Plan accepts the Cooperative System as most widely distributed and organised credit system to perform the vast task of providing short, medium and long term credit to the rural areas. But the Plan argues that the cooperative system has been heavily biased towards more affluent sections of the rural community and that the poorer classes have generally been neglected. The Sixth Plan has outlined the following measures to reorient the cooperative system:

(a) Reorganisations of primary agricultural credit societies into strong and visible multi-purpose unit so that farmers, artisans and the self-employed get all the services at one point.

(b) Simplification of loaning procedures to facilitate the supply of credit according to their periodic requirements.

(c) Reduction of overdues in the cooperative institutions with vigorous steps taken by state governments.

(d) Greater reliance of commercial banks on financing primary credit societies; and

(e) Efficient management of the reorganised societies by professionally trained paid manager secretaries.

NOTES

Reserve Bank of India, Review of the Cooperative Movement in India; All-India, Credit Survey, Vol. II.

Government of India, Ministry of Agriculture, Department of Cooperation. *Report for latest year.*

QUESTIONS

1. What are Trade Associations? Explain their functions?
2. Briefly describe the functions of Chambers of Commerce.
3. Distinguish between Trade Association and chambers of Commerce.

REFERENCES

1. *Report of the Banking Commission,* 1972, p. 13.
2. R.B.I. *Report on Currency and Finance* 1973-74, p. 139.

31

AGRICULTURAL MARKETING

Marketing of goods normally implies their physical transfer as well as the economic terms on which these transfers are made. The efficiency in physical transfers can come about with the avoidance of wastage during transport, storage, weightment, loading and unloading.

The recent debate on taking over of wholesale trade of some important foodgrains and consequent of bold declarations of some state government in this regard bring to focus the importance of marketing. Market mechanism acts like a bridge between the producers and distributors on the one hand and consumers on the other. For a developing economy like India, surplus food supply serves the purpose of developmental plans in two ways. Self-sufficiency in food satisfies the growing demand for an increased population and surplus food significantly reduce the pressure on foreign exchange. The marketing practices in India are neither fully controlled like some of the socialist countries of eastern Europe nor fully freely run like countries in western Europe and Asia. Partially controlled market in the Indian situation is an innovation and adventure both.

With the impact of programmes for extensive and intensive development of agriculture, output as has definitely increased. Since agriculture is the sole occupation of a predominantly large percentage of rural population and the agricultural output has a marked impact on foreign exchange, it is no less important to study its final disposal though different chains of its distributive machinery.

The outlook of the farmer is now changing. He has realized the importance of adopting new techniques of production and is yearning for more income and higher standards of living. In consequence, the cropping pattern is no longer dictated by what he needs for this own personal consumption but what is responsive to the market in terms of prices received by him, yet he is not conversant with the complexities of the marketing system which is becoming more and more complicated. While that rade is very well organised, the farmers are not. There is, therefore, no wonder that many practices in vogue the markets completely ignore the interests of the farmer. By the dint of remunerative process of farm produce, agriculture is today increasingly being looked upon as a commercial proposition. Marketing, in fact, is now being considered as an integral part of production process itself. The movement of the produce from the field to the consumer in adequate quantities an of appropriate qualities at minimum incidental cost, and a reasonable margin of profit to the trader, presupposes the existence of an efficient marketing system. Thus, the final disposal of agricultural commodities produced involves the activities of real (primary) producer, small and big traders and finally consumers at different points of time and place. In between the points of origin to final distribution for consumption of such commodities there are several machineries which help the producers by mobilizing the produce to the market place on the one hand, and consumer and producer by making them available their consumption and production requirements on the other. To mention some of these market funcitonaries, they are commission agents or arhatiyas, wholesale traders, dallals or brokers, processors, mill agents, weighment, palledars, etc.

Since the produce is not immediately disposed of to the ultimate consumer, the problem of storage naturally comes in. Even the movement of such produce from the primary traders to the secondary traders or from the primary markets (hats) to the secondary markets (arhat or gola), involves a time lag, the cost of storage becomes a part of the price of be included at the point of sale. Sometimes some secondary markets do have, however, adequate storage facilities like pucca godowns and cold storage but most of the time in primary market centres miserably inadequate storage facilities become chiefly responsible for the rant nuisance or deterioration by rottening of the produce. Underdeveloped or undeveloped infra-structure of such markets is accountable for loss to the primary producer or primary trader, less price to the farmers and, what is more important, the wastage of foodgrain in situation where there is always a threat of famine almost every alternate year.

Disposal of the produce from producers to the distributors or consumers involved other important problems such as cleaning, processing, grading, and weighing. In the absence of an established and regulated market situation, middlemen in the form of distributive agencies take undue advantage of the ignorance of the farmers. Due to the lack of market intelligence services, ignorant peasants are exploited in terms of low price for their produce. Different types of weighing is always a matter of dispute between the producer and the tradesman.

From the point of view of distribution of agricultural inputs also, a developed market is an incentive to produce more. The extension and regulation of agricultural markets in thus a necessary requirement of the optimum production process.

Types of Markets

Markets for agricultural commodities can be broadly classified into three groups: *(i)* Hats and Shandies; *(ii)* wholesale markets or mandies; and *(iii)* retail markets. Hats and shandies are markets periodically held, the former usually once of twice a week and the latter either at longer intervals or on special occasions. Agricultural produce or live stock or both are sold in these primary markets, which number about 22,000 and are located mostly in the interior of the country. The areas served by a hat vary considerately. In some cases they cover only villages while in others they may extend over a radius of 100 kilometres. They can also be termed as primary markets.

Mandies, i.e. wholesale markets from the most important part of the agricultural market structure. There are about 1,700 wholesale markets in the country. Most of them command large areas and the produce is brought for sale from the primary markets, by the producers direct or by the middlemen such as village merchants, itinerant dealers, and agents of trades in the mandies.

Most mandies, particularly those in the producing areas, function as both primary and secondary markets. Primary markets are those where the bulk of the produce is brought for sale by the producer himself. The primary markets serve as feeding centres for the secondary markets which are usually wholesale markets and serve as the second point of assembling; the first being the primary market. Terminal markets are those from where the produce is exported outside. Such markets could also be a primary and secondary markets.

Bulk of marketed surplus of the farmers is usually sold in the village. The All-India Rural Credit survey shows that about 65 per cent of the total sales of crops are effected in the village. There are many reasons for the bulk of the produce being sold by the farmer in the village. The farming units are small and many farmers do not possess any of transport. Road-development is not quite adequate. Farmers are illiterate and the market information service is yet of have its remifications in the villages. The system of agricultural credit is defective. The farmers has to depend on the village merchant for loans and quite often the farmer gives an undertaking to the merchant that he will sell his produce to the latter.

In many areas keen competition among the buyers makes the village quite fair and attractive as compared with the price in mandies. It is only in backward areas that the margin of difference between the village and market price is large and the prices offered to the former are unfavourable as compared with the prices in the mandi. The conditions of marketing in the mandies are not always attractive to the small producer. He gets less attention, there are a number of unauthorised deductions and he fails to understand the complicated sale system in the mandi.

Rural Market

What should be the strategy in opening up the rural market? There are nearly 5,65,000 villages in the country. Over 60 per cent of the villages have fewer than 500 people each, accounting for 20 per cent of the rural population. Nearly half the rural population lives in villages which have a population of between 1,000 and 5,000 each and 4,000 of the bigger villages, with over 5,000 people each, account for less than 10 per cent of the country's population.

Moreover, thousands of villages are well outside the reach of railway, and roads are the only way of reaching them. The roads, being in poor shape, are unusable during the monsoon. It is a pity that against the estimated annual requirement of Rs. 300 crores for road development, only Rs. 60 crores have been allotted for it. The revenues from road transport are much greater than this. The multitude of levies, permits, and licences constitute a major impediment in the free movement of goods from villages to the urban areas and from one state to another. Surely the advantages of developing a region outweigh the extra income accruing from such levies and tolls.

Road transport and the freer movement of goods play crucial role in hastening the pace of industrialisation. On the road transport industry will fall a large share of the burden of opening up rural markets and its past record entitles it to encouragement. There is the difficult problem of ascertaining new ways of communicating with the villagers. There are, however, two hopeful factors. One, the younger generation are becoming educated and, two, in the well-to-do rural communities between 30 and 50 per cent of the villagers own radio sets. However, despite All India Radio's rural programmes, the potential of the radio as a means of educating the rural masses has barely been tapped. It has been found that one of the best ways of communicating with the illiterate villager is through the medium of films carried by mobile vans. The government should adopt this method more widely for educating and informing the farmer.

Socio-economic surveys have revealed that there is a great deal of difference in the prosperity of different districts—not all rural areas are poor as is generally believed. Agricultural production per acre varies from Rs. 1,270 in Ludhiana to Rs. 870 in Shahabad; from Rs. 730 in Tanjore to Rs. 470 in Junagadh. There are also variations in the incomes of farm families in different districts. Surveys have also revealed that the first priority on surplus income is food and the next priority is farm inputs—tubewells, pumping sets, implements, hybrid seeds, and grain storage. In other words, if out planners improve the infrastructure of development, such as roads, water, and power supply, the farmer himself will put from 25 to 60 per cent of his surplus income into the agriculture. This is turn will increase the demand for agricultural inputs and help to raise per capita income in the rural areas.

Market, Functionaries

The human agency in the marketing process is an important consideration. The functionaries of the various marketing activities at market places adopt their own techniques of operations, procedures of business, and trading practices. Technique of marketing and organization vary in accordance with the economic condition, nature of the marketed product, the area and the location of marketing, marketing agencies, and the rate of technical and commercial development.

There are several types of functionaries in the mandi such as arhatiyas or commission agent, dalals or commodity brokers, general agents, and agents for specific commodities and purposes Arhatiyas are of two kinds, namely kutcha and pucca. The kutcha one is primarily concerned with the assembling of the produce. He sometimes advances loans to the village merchants and traders on the condition that the produce purchased by them in the villages is sol in the mandi through him. Kutcha arhatiyas are more active in the vegetable and fruit markets. The pucca arhatiya acts on behalf of outstation buyers as well as on his own account. He makes advances to kutcha arhatiyas for purchases made on behalf of this clients. A dalal or broker is a per son with no establishment of his own; his function is to bring together the sell and the buyers. In addition to arhatiyas and brokers, as already mentioned, there are weighmen, sweepers, and labourers for stitching bags, loading and stocking of bags and bales.

The agencies in the chain of the marketing from the producer to the manufacturer and the consumer depend upon the nature of the crop also. Paddy usually passes from the village merchant to the mill or to the wholesale dealer in the assembling market; and, from the mill, rice passes on to the wholesale dealer in the consuming market and then through one or more retailers to the consumer. Between the wholesaler or the miller on the producing end and the wholesaler at the consuming end, a host of intermediaries exist and execute orders on the strength of samples.

In the case of groundnut, the route taken comprises the grower, the village merchant, the broker, the decorticating mill, the exporter's agent, and finally, the exporter. Similarly, in the marketing of tobacco, except in the case of cigarette tobacco which is bought by the manufacturers direct from the growers through depots established in the growing areas, many intermediaries exist between the producer and the assembling market. The existence of a long chain of intermediaries naturally reduces the share received by the actual cultivator out of the price paid by the ultimate consumer. In regulated markets there are paid marketing secretaries and other assisting staff.

Place and Time Sale

Village sales account for a major part of the marketed agricultural produce.

According to the Rural Credit Survey nearly 65 per cent of the total sales in 1951 were made in the village. This is, however, only an overall position. The marketing studies carried out by the Directorate of Marketing and Inspection, indicate considerable local variations in this respect. In Punjab, for instances 35 per cent of the paddy was estimated to have been brought to the assembling markets by cultivators as against only 3 per cent in some parts of Kerala. The direct sale by cultivators in the assembling markets depends of various factors. In some districts of Punjab with well-developed means of transport and communication, about four-fifth of the wheat crop is sold by the grower directly in the assembling markets. On the other hand, in other districts of the same state, lacking these facilities, the corresponding proportion is insignificant. A similar variation obtains among districts in Uttar Pradesh. In the western Uttar Pradesh about one-third of the local surplus is estimated to be taken to the assembling markets by the growers. This proportion is very small in the eastern districts.

In relation to time of sale, the important fact to be noted is the preponderance of sale soon after the harvest. Thus, in 1967-68, of the total market arrivals of rice and gram more than 70 per cent arrived during the two quarters immediately succeeding the harvest. The corresponding figures for wheat were 68 per cent for jowar 55 per cent. The rush of market arrivals in the post-harvest period has increased appreciably, particularly in the case of wheat as a result of the Green Revolution in recent years. It may be mentioned here that since these percentages refer to arrivals in the wholesale markets they would not fully reflect the extent to post-harvest sales in village markets owing to carry over of stocks for some period by functionaries between the village markets and the wholesale markets.

Marketing Charges

Since the bulk of the marketed produce comes to the assembling market for disposal, whether brought by cultivator the salves or by other agencies operating at the village level such as itinerant merchants, it is necessary to analyse the marketing charges for the assembling of produce. The main elements in these charges are arhat (commission), dalali (brokerage), karda (trade allowance), hamali (handling charges, tulai (weighing charges), and miscellaneous charges such as deduction for charities, market fees, fees levied by commission agents to cover the wages of his apprentices, payment to water-man, cook, sweeter, etc. Karda or trade allowance is deduction in kind payable by the seller in order to compensate the buyer for loss in weight on account of impurities, driage, etc.

A noteworthy feature of these market charges is that they differ from market to market in magnitude an may be payable by the seller in some places and by the buyer in others. In Punjab, for instance, in some of the markets, commission charges are paid partly by the seller and partly by the buyer while in other markets it is paid solely by the former Handling charges comprise of wages paid to labourers for unloading the cart, dressing the produce, and cleaning, etc. These charges are paid by sellers at some places and by buyers at other places. Deduction for charitable purposes forms a regular item in the list of customary marketing charges in almost all markets. The purposes for which this collection is made are numerous—social, educational an religious purposes being the most important.

TABLE 31.1

Marketing Margins and Producers' Share

(Percentages of consumer's price)

	Andhra Pradesh	*Madhya Pradesh*	*Tamil Nadu*	*Maha-rashtra*	*Mysore*	*U.P.*	*Average*
Producer's share	77.5	72.2	73.7	6.62	72.5	74.7	73.5
Wholesaler's margin	4.3	6.7	11.1	14.8	6.3	10.5	8.0
Retailer's margin	6.2	7.2	2.6	12.3	10.1	5.5	6.9
Cost of marketing	12.0	13.9	12.6	6.7	11.1	9.3	11.6
Consumer's price	100.0	100.0	100.0	100.0	100.0	100.0	100.0

In the post-assembling stage of marketing, marketing charges include expenses of transportation and handling from the point of delivery to the buyer's premises, the expenses of packing, commission and brokerage, etc. As a general rule all these charges are paid in cash by the wholesaler or his agent. Although there are variation in these charges from market to market, these are not so pronounced as assembling costs. The reason for this is that the contracting parties in wholesale markets are more evenly, matched than those in the assembling markets. Table 31.1 gives the producer's share and marketing margins in the price paid by the consumer for rice in different states in 1962-63.

Tow significant features of marketing of rice revealed by a study pertaining to the year 1962-63 (October-September) are that it is the wholesaler who caries the stocks from the peak marketing season to the lean season and that producer's share in the consumer price which is low during the peak season registers some increase in the lean season.

Defects in Marketing

The defects in marketing of agricultural produce can be classified into two broad groups. The first group consists of various types of malpractices prevailing in the markets such as (1) manipulation of weights and scales against the seller, (2) the practice of taking large quantities as sample, (3) arbitrary deductions from sale proceeds by the market functionaries under various pretexts, (4) sale under cover keeping the seller ignorant of price, (5) discriminatory marketing charges. The extent of these malpractices varies from market to market and it is difficult to measure it with any precision. Some typical instances of these malpractices may be noted. The arhatiyas' commission is usually higher for small cultivators as compared to village merchants who, because of their larger turnover, get better terms from the arhatiyas. This is also true of trade allowance or karda. Even in well-organised markets like Hapur and Ghaziabad, the cultivators ordinarily pay twice as much as the charges paid by village merchants.

The second group of defects in the marketing of agricultural produce arises out of inadequate facilities for transport and storage. This results in the simultaneous existence of pockets of scarcities and surpluses leading to wide inter-regional variations in prices. Another consequence flowing from preponderance of sale immediately after the harvest is wide inter-seasonal fluctuations in prices.

The impact of marketing problems on the agricultural economy was visible when the production of wheat increased by more than 100 per cent during the last few years and there was a glut in the availability of cotton during the post-harvest season after the Indo-Pakistan war in 1971. If proper marketing arrangements along with adequate storage facilities are not provided in time, this is bound to have an adverse effect on agricultural production. To avoid this phenomena there is need to have an integrated system of production, marketing and processing. The examples of sugar cooperatives in Maharashtra and cotton cooperatives in Gujarat will serve as guidelines in this respect. Besides the development of cooperatives in this sector, efforts have been made to remove other types of marketing defects by the regulating of markets.

Directorate of Marketing and Inspection

The administrative control of various regulatory measures lies with the Directorate of Marketing and Inspection. (1) It carries out surveys of marketing of agricultural produce on an all-India basis for the purpose of collecting information on which developmental work needs to be undertaken. (2) It promotes grading by fixing grade standards and organizing grading centres under the Agricultural Produce (Grading and Marketing) Act, 1937. (3) It advises state governments inestablishing regulated markets and coordinates the working of Agricultural Produce Market Acts in the states. (4) It prepares standard contract terms for adoption by the trade. (5) It carries out the quality control of the fruit products under the Fruit Products Order, 1955 and assists in the development of the Fruit preservation Industry. (6) It provides training in agricultural marketing. (7) It tenders advice on all aspects of marketing to various ministries of the Government of India, the state governments, the Indian Council of Agricultural Research, and commodity committees in the Ministry of Food and Agriculture. It keeps liaison with international organizations such as FAO and ECAFE.

Grading

At present, sales in the markets usually take place on the basis of visual evaluation of the quality by the buyers. The farmer seldom gets a price commensurate with the quality of his produce. He has, consequently, no interest in improving the quality of the produce. Marketing surveys conducted by the Directorate of Marketing and Inspection have shown that the produce brought to markets by the farmers usually contains a high percentage of refraction which could easily be reduced. In order to provide the required incentive to the farmers to improve the quality of his produce it is necessary to ensure that he gets a price commensurate with the quality of the produce. For this, steps need to be taken to introduce grading of the produce before its sale by the farmer. With this end in view, grading on the basis of standards framed by the Directorate of Marketing and Inspection was introduced in some regulated marketing societies during the Third Five Year Plan. The number of such markets at present is small. In the larger interest of the fanners, grading before sale needs to be made more common so as to cover all the assembling markets.

Regulated markets can take a leading part in introducing this improvement by insisting that all the produce be graded before sale. Side by side it is also necessary that relegated markets provide necessary facilities, such as sieves, for proper cleaning of the produce before sale. Governmental and other agencies like the Food Corporation of India, civil supplies departments in the states and the Apex Cooperative Marketing Organisation which are engaged in buying operations for building up buffer stocks can be of considerable help in popularising grading if they make purchases only on the basis of the grades prescribed by the Directorate of Marketing and Inspection.

Grading is essential not only in the interest of the farmer, but also in the interest of the consumer. In this context it may be stated that the country is slowly but steadily moving from an economy of shortages to that of surpluses. The consumer will, therefore, no longer be satisfied with whatever quality is offered to him. He is bound to become quality-conscious and ask for quality goods. Certain articles of food like ghee, butter and edible oils, are already being graded under Agmark. However, the quantities so graded from only a small part of the total quantity marketed. Grading for the consumers, therefore, needs to be intensified. The super bazars which have been established in important towns all over the country can give a lead by stocking articles of daily consumption like ghee, butter, rice, pulses, and wheat flour in consumer packs of small sizes graded under Agmark for sale to the consumers.

Market Research and Survey

A survey of regulated markets is being carried out to identify the shortcomings of such markets and making proposals for development for market yards, and for the provision of other facilities for smooth and efficient marketing. During the year

1971-72, 256 regulated markets were surveyed and the number of markets surveyed from April 1972 to November 1972 was 248. A special study for estimating marketable surplus and post-harvest losses of agricultural commodities has been taken up. Such a study on paddy is in progress. A number of *ad-hoc* inquiries and investigation into the marketing problems of different commodities were also undertaken. The Report on Regulated Markets in India (Vol. II) was published and the Report on the Marketing of Mushrooms in India is in the press.

Regulation

The marketing and inspection wing of the Union Agriculture Ministry is presently engaged in a post-harvest survey of paddy in 80 selected districts in the country, to assess the loss in storage and transport.

After the survey, expected to be concluded in a few months, a report would be prepared which would suggest improvements to storage facilities. It is also proposed to undertake a survey in respect of other major commodities as well.

Market Extension

During 1972, the Market Extension Wing of the Directorate of Marketing and Inspection, in collaboration with the Directorate of Extension, organised the exhibition "Vital Link Between Farm & Home" at Chandigarh. In addition, the Directorate of Marketing and Inspection took part in the exhibition organised by the Consumers Guidance Society at Bombay. To improve the agricultural marketing system, the Directorate of marketing and Inspection has been giving attention to grading and standardisation of agricultural produce, market research and surveys, development of regulated markets, markets extension and orderly development of cold stores.

Grading of agricultural produce is done under the provisions of Agricultural Produce (Grading and Marketing) Act, 1937. Grading activities relate to (i) compulsory grading for export and (ii) voluntary grading for internal consumption. Compulsory grading and pre-shipment inspection prior to export was in force in respect of 38 commodities. Total value of commodities graded for export amounted to Rs. 97.9 crores in 1971-72 and to Rs. 30.92 crores during the first half of 1972-73.

Grading for internal consumption comprises (a) grading for internal trade; and (b) at the producers level. The commodities graded under Agmark for internal consumption include cotton, vegetable oils, ghee, creamery butter, rice, wheat arta, pulses, gur, jaggery, potatoes, fruits, groundnuts, arecanuts, honey, eggs, bura, and whole and ground spices.

Major commodities graded before sale at the producers' level include cereals, pulses, oilseeds, fruits, vegetables, fibre crops, chillies, arecanuts, coconuts, jagger, and tobacco. The value of commodities graded under this scheme amounted to Rs.

80 crores during 1971-72 and Rs. 51.8 crores during the period January 1972 to September 1972. Grading under this scheme is undertaken in selected regulated markets, marketing cooperatives and warehouses in different states. The number of grading units increased from 411 in October 1971 to 480 in October 1972.

To ensure the purity and quality of products graded under Agmark, there is a chain of 11 Agmark laboratories in addition to the Central Agmark laboratory at Nagpur. The Central Laboratory is engaged in research and also acts as an appellate laboratory.

Market Intelligence

Weekly data on market arrivals, sales, prices, market situation, etc. are being regularly collected from over 1,300 technical reporting agencies. The 14 regional offices so far established by the Directorate of Economics and Statistics continued to exercise effective supervision over the work done by the primary reporting agencies, in close cooperation with the state governments. For studying the behaviour of retail prices an supplies of essential commodities, the Directorate collected data on retail prices for about 72 essential commodities covering both food and non-food articles from about 100 centres spread all over the country. The data on retail prices are being regularly made available to the Director-General of Civil Supplies, Ministry of Industrial Development.

On Market News side, daily prices of important agricultural commodities and the weekly reviews of food and price situation in the regions were supplied for broadcasting from all the Regional Stations of All India Radio. The Market New Agency covered over 500 markets and about 70 commodities. Display of market intelligence on notice boards was being carried on in 1,000 markets spread all over the country. Bulletins on prices and market intelligence were issued regularly by the Union and state governments for the benefit of farmers and others.

Regulated Markets

Most markets in India were unregulated till recently. Once the farmer goes with his produce to the market for any dispute arising between him and the middlemen or the purchaser, he has to depend on the decision on his broker arhatiya. The facilities for parking carts, drinking water, sanitation etc. are not good. As mentioned earlier, there is considerable room for improvement in market places and sale procedure, system of payment of price, etc.

When the sales are not done by the open auction system, the farmer's interest is often lost sight of. Even in open auctions, there may be a tacit understanding among the bidders not to bid high. For putting these malpractices to an end, marketing committees with strong representation of farmers's interest are necessary. A number of unfair practices like short weights, unwarranted deductions and higher rates of commission and other market charges prevalent in most of the mandies take away a substantial share of the price paid by the purchasers and place the farmers at a disadvantage *vis-a-vis* the more organised section of the trading class. The purpose of law and regulation is to improve competition, to strengthen bargaining power of the producer-seller and thus ensure a fair deal to the farmer. During the British regime, the regulated markets came into existence to help exports of industrial raw materials to England. But today the regulated markets have to play a positive role in the orderly development of marketing.

Legislation for regulation of markets was initiated in India in the thirties following the recommendations of the Royal Commission of Agriculture. In that decade Madras, Central Provinces, Bombay, Mysore, and Punjab enacted the necessary legislating. This process which was interrupted in the forties owing to exigencies of the World War II, gathered momentum during the last decade the legislation for regulation of marketing was enacted in Madhya Bharat, Coorg, and Orissa during the First Five Year plan period, and in Bihar and Rajasthan during the Second Five Year Plan period. By now most of the states have passed the necessary legislation. The Government of India has now formulated a scheme of assistance or the development or regulated markets in agricultural commodities in various states. Under the scheme, grants-in-aid to the tune of Rs. one crore is to be provided to Rajasthan, Bihar, Uttar Pradesh, Madhya Pradesh, Orissa, and West Bengal during the year ending March 31, 1974 for developing 48 regulated markets in paddy, wheat, jute and bajra.

Regulation of markets and market practices is now done under the Agricultural Produce Markets Act. The Directorate of Marketing and Inspection renders the necessary guidance and assistance to states in forming market legislation and its enforcement. Excepting the states of Assam, Kerala, Nagaland, Meghalaya, and Jammu & Kashmir, all states have passed legislation for market regulation. Out of 3,406 principal markets in the country, the total number of regulated markets and markets sub-yards at the end of March 1970 was 2070; this increased to 2,356 by September 1971. During 1971, study of regulated markets was undertaken with a view to assess the facilities that need to be provided to cope with the increased arrivals arising from higher levels of agriculture production.

The legislation to regulate markets generally covers all the agricultural, horticultural livestock and livestock products. The Act authorises the state government to establish a market committee for every notified area consisting of representatives of cultivators, traders, local bodies, cooperative marketing societies, and the government. The principal aim of legislation is to ensure fair marketing conditions for cultivators through elimination of malpractices and unauthorised deductions. It stipulates the sale of produce through open auction the presence of a representative of the market committee who notes down the particulars of buyers quantity sold, price, etc. It provides for licensing for market functionaries such as commission agents with provision for revocation of license if he market functionary found to follow unfair practices. It also prescribes that no allowances other than those stipulated in the legislation shall be levied. In addition to the responsibility to ensure fair practices, the market committee is also entrusted with the function of hearing the disputed between buyers and sellers and giving decisions on them which would be binding on the parties.

The regulation of markets has been helpful in securing reduction in village sales, systematisation of methods of sale and weighment, and reduction in market charges. Reduction in charges has varied from 33 to 66 per cent an unauthorised deductions completely eliminated. There has been a progressive increase in the produce brought to the regulated markets by the growers.

Apart from covering more markets by regulation, it will be necessary to expand facilities by way of market yards and other ancillaries in various markets. A recent study has shown that, in certain areas such as Punjab, during the post-harvest season of 1968-69 market arrivals increased by 150 per cent over the average of previous three year's corresponding period. The capacity for market yards proved to be inadequate for these large arrivals and considerable market transactions began to be conducted outside the market yards thus leading to malpractices. In this context, efforts are necessary to strengthen and restructure the market committees with regard to their resources and functions. A beginning has already been made by some market committees to obtain bank finance for improvement of marketing facilities. It will be necessary to enable a large number of market committees to resort to such institutional finance and thereby help in development of the market intelligence. Some of the well developed market committees are also expected to make their contribution to the development of feeder roads for which a beginning has been made in a few rural areas. Finally, it will be necessary to evolve a suitable state level machinery for supervising and coordinating the work of regulated market committees. In this context, several states are contemplating action to constitute State Agricultural marketing Boards on the lines of their functioning in Punjab and Haryana.

A stage has come when these committees should play a more dominant part in improving the marketing structure than what they have been playing so far. They should serve as focal points for initiation and development of programmes not only in respect of marketing but also in terms of the over-all development of agricultural in surrounding rural areas. In addition to providing the usual facilities and amenties, regulated markets, should earmark sited for location of shops which may stock different agricultural inputs like improved seeds, fertilizers, pesticides, etc., needed by the farmers. The shops should also stock pumping sets and other small implements and machinery required by the farmers.

Future Programmes

The government proposed to restructure and improve the existing regulated markets and greatly increase their number to cope with the rising volume of grain arrivals each year. This will afford considerable relief to farmers who now have to make do with rudimentary facilities in the mandies. As against only 2,800 regulated markets at present there is an estimated need to establish 12,000 to 14,000 additional markets centres throughout the country. As ambitious programme for augmenting and improving market facilities has already been initiated in Mysore and a similar programme is to be taken up in Uttar Pradesh, both of these with World Bank assistance. Simultaneously, the government is embarking on a programme to expand grains storage capacity over and above the 14 million tonnes expected to be available at the commencement of the Fifth Plan. Some of the proposed new storage is to be built under the auspices of the regulated markets.

The development of regulated markets is an essential part of the infrastructure that needs to be built up to stimulate modern agriculture. The traditional mandies are no more than contact points for buyers and sellers with modest or minimal facilities. The new concept of the regulated market is that it should be a growth centre providing every convenience to the farmer to bring his produce to the market, dry, clean, grade and display it in spacious yards with standard weights and measures, and facilities for auction so that buyers compete to offer the farmer a favourable price. The markets will have facilities for unloading grain that will minimise handling losses while their management will be in the hands of market committees will be expected to ensure that fraudulent practices are put down firmly. The provision of cold storages should enable the farmer to hold perishable goods till market prices firm up so that he is not compelled to make distress sales as at present. Transportation facilities and rest rooms should also enable him to move freely or to stay conveniently should he to desire.

The modern market centre is also conceived of as a place where the farmer can buy his inputs or get his consumer necessities. He will be able to purchase seeds, fertilisers, pesticide, farm equipments, spare parts, and other requirements. He should also be able to command various services such as that of agricultural machine repair stations, banks, market intelligence, and extension information. Given such facilities the modern regulated market could become nodal centres of rural development and community activity. These might also become natural centres for the establishment of agricultural processing units of various kinds and a place where men exchange not only produce but ideas. Altogether, this is programme of considerable promise an importance an one that could have a real impact on rural productivity.

Cooperative Marketing

Cooperative marketing is based on principles of cooperation as opposed to the normally business principles. Cooperative societies help the farmer in getting maximum returns for the farm products. An institutions of this type would strengthen the fibre of village life by making cultivators conscious that what used to be their weakness so far could be converted into strength, by pooling their produce, selling it through their own organisations. It could be assumed that a cooperative society will command much better bargaining position in a market than an individual cultivator can. Besides, cooperative marketing would confer on the cultivators the benefits of large-scale marketing such as pooling, grading and processing which could not be available to an individual cultivator marketing the produce on his own.

Generally speaking, prior to independence, only a few sporadic efforts were made in the development of cooperative marketing. In 1945, the Cooperative Planning Committee recommended that, within 10 years, 25 per cent of the total annual marketable surplus should be sold through cooperative organisation, and, for this purpose, a marketing society should be organised at each of the 2000 mandies in the country, it appeared that no planned efforts were made to implement this recommendation. In 1951, when the All India Rural Credit Survey was carried out, it was observed that out of 75 districts selected for the survey, there was no cooperative, marketing in as many as 63 districts. Out of the remaining 12 districts, the share of cooperatives in the marketing of agricultural produce exceeded one per cent of the total sales to all agencies in only 5 districts.

Development of cooperative marketing was given an important role in the scheme of integrated rural credit recommended in the Rural Credit Survey in 1952. According to this recommendation, primary marketing societies were to be established at important markets. During the Second Five Year plan, 1,800 primary marketing societies were set up with apex marketing societies as state level. A National Cooperative Marketing Federation composed of apex marketing societies was also constituted.

There has been a good deal of extension and diversification of the activities of the cooperative marketing and processing societies recently. The value of agricultural produce marketed by agricultural cooperatives increased from Rs. 600 crores in 1969-70 to Rs. 650 crores in 1970-71. The number of cooperative sugar factories in production went up to 75. The number of other processing units installed, including cold storages, increased to 1311 as on the 31 st March 1971. The value of fertilizers distributed by cooperatives during 1970-71 was of the order of Rs. 250 crores, as against Rs. 232 crores in the preceding year. The storage capacity created in the cooperative sector under Plan programmed went up to 2.8 million tonnes by the end of 1970-71. In addition to this, storage capacity of two lakh tonnes had been constructed by cooperatives with assistance from the Agricultural Refinance Corporation. Efforts were continued to be made for greater involvement of cooperatives in new lines of activities like production of granular fertilizers, setting up of agro-service centres etc.

The National Cooperative Development corporation, setup under the National Cooperative Development Corporation Act, 1962, with the object of planning and promoting programmes for the production, processing, storage and marketing of agricultural produce and notified commodities through cooperative societies, has continued to provide support to these programmes through various promotional measures and by providing necessary financial assistance. The Expert Committee appointed by the Government of India under the chairmanship of Shri B. Venkatappiah, to review the working of the NCDC and to examine the need for its continuance, submitted its Report in October 1971. The Committee has strongly recommended the development of cooperative programmes relating to marketing, processing and storages of agricultural produce and supply of agricultural inputs.

A wide network of 1,62,000 primary agricultural credit societies, covering 95 per cent of the villages in the country provides the main institutional base for the supply of agricultural inputs and credit to the farmers. Linked with this is an equally extensive cooperatives marketing structure comprising about 3,261 primary marketing societies covering almost all the secondary markets, 159 district marketing societies, 21 state level marketing federations and the National Cooperative Agricultural marketing Federation. In addition, there were 3 State marketing Federations for special commodities, one each in Delhi and Gujarat, for fruits and vegetables, and one in the Union Territory of Laccadives for coconut. There was also an inter-state Federation for marketing arecanut in Mysor State. The number of state partnered primary marketing societies was 2,512 representing 77 per cent of the total number of such societies. During 1971-72, attention was largely devoted to stimulated the cooperative marketing structure, intensification of activities of selected cooperatives in areas with larger potential for growth, promotion of integrated marketing system, coordination of activities between marketing consumer cooperatives etc. As a result of various measures taken to rationalise and activise the cooperative marketing structure, there was a marginal reduction in the number of primary marketing societies as some of the weak primary societies had either to be wound up or amalgamated with neighbouring societies for attaining viability. The reduction in the number of societies was mainly in the states of Andhra Pradesh and West Bengal.

Marketing Operations

A number of measures have been taken to strengthen the village cooperatives and the cooperative marketing societies to enable them to undertake diversified functions included distribution of agricultural production requisites on a larger scale, in addition to their other function such as provision of production credit by village cooperatives to farmers and marketing of agricultural produce by marketing cooperatives. These societies have been provided with governmental assistant in the form of share capital, assistance for construction of godowns, appointment of managerial staff, purchase of transport vehicles, etc.

Table 31.2 give the comparative picture of the value of different agricultural commodities marketed by cooperatives during 1969-70 and 1970-71.

TABLE 31.2

Value of Agricultural Commodities Marketed By Cooperatives 1969-70 and 1970-71

(Rs. in crores)

Commodity	*Value of Agricultural Produce marketed*	
	1969-70	*1970-71 (provisional)*
Foodgrains	202	269
Sugarcane	260	222
Other crops	138	159
Total	600	650

Cooperative Processing

The organization of processing units in the cooperative sector consisted of independent processing societies, such as sugar factories, spinning mills and solvent extraction plants and units established as adjuncts to marketing societies, such as rice mills, cotton gins, oils mills etc. Out of 1606 processing units organised upto 1970-71 (excluding a few dropped units), over 1200 were established as adjuncts to marketing societies, while the rest were independent processing societies. The cooperative sugar factories have federated themselves into eight state-level federations and one national-level federation. The cooperative spinning mills have also set up an All-India Federation. Besides, a few independent processing units have federated themselves into central processing federations for the purpose of undertaking secondary processing.

Cooperative Sugar Factories

The number of cooperative sugar factories registered/licensed at the beginning of the year 1971-72 was 117. During the year, six more sugar factories were licensed, bringing the total to 123. Out of the said 6 units, one *viz.* Saswadmali (Sholapur), having a daily cane crushing capacity of 1,000 tonnes, was taken over from the joint stock sector and cooperativised, and the remaining 5 were licensed as new sugar factories, each with a daily cane crushing capacity of 1250 tonnes. During the year, 5 existing sugar cooperatives were granted expansion in their crushing capacity. The total licensed capacity in the cooperative sector, which was 22.90 lakh tonnes out of the total of 50 lakh tonnes for the sugar industry as a whole at the beginning of the year, had increased to 24.42 lakh tonnes out of the total licensed capacity of 51.32 lakh tonnes. The total membership of the 117 cooperative sugar factories for which information is available was 5.16 lakh as on the 30th June, 1971. The grower members accounted for 4.82 lakhs representing over 93.41 per cent of the total membership. The amount of share capita collected by the above 117 cooperative sugar factories as on the 30th June, 1971 was Rs. 64.38 crores. Out of the total share capital of Rs 64.38 crores, the growers' contribution was of the order of Rs. 36.34 crores. The share of growers in the total shares capital contribution by the members, other than the state government, was 91 per cent.

Other Cooperative Processing Units

By the end of 1970-71,1584 cooperative agricultural processing units of various types (excluding sugar factories) were organised/ assisted. This included the various units like 780 rice mills of which 41 are modern ones. During 1969-70, the cooperative cotton ginnery ginned 23.36 lakh quintals of raw cotton and pressed 6.25 lakh bales. Besides Maharastra, a beginning has also been made in Madhya Pradesh. At the end of 1970-71 there were 44 cooperative jute baling units. The first cooperative jute mill set up in Assam at Naogaon went into production in January 1971. Cooperative organisations in the main jute growing areas are now to be strengthened to enable them to play their legitimate part in the working of Jute Corporation.

There were 173 oil mills and 58 groundnut decorticators in the cooperative sector by the end of 1970-71. The solvent extraction plants (six in number) with an estimated capacity of 350 types of oil cakes per day were organised/assisted during 1970-71. Every processing unit consists of fruits and vegetables and plantation crops like rubber, tea, coffee, and cashewnut.

Distribution of Agricultural Inputs

As a result of various measures taken, the cooperative have shown impressive progress from year to year in the distribution of agricultural inputs. During 1960-61, they distributed agricultural production requisites of the value of Rs. 35.7 crores, while the corresponding figure for 1970-71 was Rs. 317.2 crores. The cooperative have organised a network of retail depots for distribution of fertilisers. The number of sale points in 1966-67 was 54,603 as against 32,000 in 1961-62. This arrangement provided, on an average, one sale point for 10 villages. Cooperatives till the end of 1965-66, were handling distribution of mainly nitrogenous pool fertilisers, and some states also superphosphates, on consignment basis as agents of the state governments.

From 1966-67 onwards on the recommendations of the Sivaraman Committee, the consignment facilities have progressively been withdrawn and the cooperative wholesalers are now required to make by and large their own arrangements for distribution finance for fertilisers. The supply of poll fertilisers is also dwindling with increased availability of indigenous products for which manufacturers are free to make their own distribution arrangements. Fertilisers are, therefore, now being produced by the cooperatives by direct negotiations with indigenous manufacturers also. They are making vigorous efforts to raise adequate margin money to obtain commercial finance for increasing fertiliser distribution business. Financial assistance for raising margin money is also being provided by NCDC. In quantitative terms, cooperatives distributed about Rs. 250 crores worth of fertilisers in 1970-71 as against 32 crores in 1961-62. It is envisaged that in 1973-74, the last year of the Fourth Plan, cooperatives would account for about Rs. 450 crores worth of fertilisers out of a total estimated distribution of fertilisers valued about Rs. 900 crores.

In recent year and particularly during the last two years, cooperatives have made considerable progress in the distribution of improved seeds. As against the improved seeds valued at Rs. 3 crores distributed in 1960-61, the cooperatives have distributed seeds worth about Rs. 39 crores during 1970-71. In order to encourage the cooperatives to undertake the work of seeds production and processing as complementary to seed distribution, NCDC formulated a pilot scheme, in consolation with the National Seeds Corporation, for organisation of seed producers' cooperative processing-cum sale societies. Under this scheme, the Corporation has so fare assisted eight societies—three in Andhra pradesh. three in Mysore, and two in Maharashtra.

As in the case of all other inputs, cooperative have from year to year continued to improve their share in the distribution of pesticides/ insecticides and other production requisites. As against the distribution of this input valued at Rs. 3.8 crores in 1960-61, they have distributed these inputs worth Rs. 17 crores during 1970-71. Seven apex marketing federations have set up their own units for foumulation of pesticides/insecticides, with financial assistance from NCDC.

Besides the seeds, the cooperative have also been distribution agricultural implements. The value of such implements distributed during 1970-71 was Rs. 6.41 crores. The National Agricultural Marketing Federation also distributed 350 tractors and a large number of imported harvesters combines along with other items. Under the pilot scheme sponsored by the NCDC for the establishment of small-scale workshop for the fabrication of agricultural implements, 27 cooperatives were assisted up to the end of 1970-71.

Cooperative Storage of Foodgrains and Fertilizers

The storage capacity created in the cooperative sector under Plan programme by the end of 1970-71 worked out of about 2.8 million tonnes. In addition, some of the cooperatives which were financially strong had also constructed in earlier years storage accommodation out of their own resources and storage capacity so created was about 0.36 million tonnes. The Agricultural Refinance Corporation has made available financial assistance towards another 200,000 tonnes capacity during the year 1969-70 and 1970-71. As such, the total storage capacity, cooperatives at the end of 1970-71 was 3.36 million tonnes. In view of the inadequate owned storage capacity, cooperatives have also been hiring godowns to meet their storage requirements. According to information available, cooperatives at various levels had hired a storage capacity of about 1.25 million tonnes at the end of 1969-70.

Special Steps for Weaker Areas

Special measures have been initiated to intensify cooperative activities for processing, storage and marketing of agricultural produce in the cooperatively under-developed areas viz. The Eastern states, Rajasthan, Eastern U.P., the 199 industrially under-developed districts identified by the Planning Commission and also the tribal areas. These include liberalisation of pattern of assistance from the NCDC for these areas, assistance to cooperatives and the state governments in the preparation of specific integrated projects for production, processing and marketing of selected commodities, like fruits, vegetables, and cotton, in different areas, and the strengthening of the field organisation of the NCDC in the eastern region for a more intensive attention to the planning and implementation of the programmes.

Agricultural Marketing and Credit

Transport storage processing and marketing have a profound influence on agricultural development on any area. In fact no policy of credit can be success unless extension, credit, supplies, and marketing are thoroughly integrated. Perhaps the best way to expand the extension of farm credit is by making loans against the delivery contracts, the proceeds of which are used for our payment of loans. This is by no means a new practice. Private moneylender does this to great extent. Highly successful integration of credit and marketing has been achieved in Taiwan, Japan, and some other countries.

Long range plans for agricultural marketing have got to be developed which will meet the needs of rapidly increasing production in a fast changing technology on the one hand and rapidly expanding consumption on the other hand. This will require new marketing policies, administrative arrangements, marketing research and marketing facilities that make for more efficient assembly, transport, storage, processing, financing and distribution of agricultural products with an effective pricing system at all market levels. With the ultimate object of providing the agricultural products to the consumer at a reasonable price and at the same time ensuring remunerative returns to the cultivator, it becomes extremely necessary to reduce marketing margins. Marketing margin or price-spread can be taken as an index of the soundness of a marketing organisation. If the aim of better marketing, as pointed to long back by the Royal Commission Agriculture, is to function with greater efficiency, there is need to reduce marketing margins to the minimum. An idea about the existing position as revealed by some studies is given in Table 31.3 below:

TABLE 31.3

Middlemen's Share in the Consumer' Rupee

Products	*Wholesaler's Margin*	*Retailer 's Margin*
Sugar	5.3	9.5
Potatoes	5.4	18.6
Rice	3.1	62
Wheat	1.9	3.3
Milk	14.7	20.5

Source: M.C. Munshi, *From the farmer to the Consumer: A Study of Price Spread.*

Future Research and Other Problems

The emphasis on crop production had been greater in the fifties and the sixties but the problem of marketing and processing had been almost neglected. An awareness about the importance of this aspect has been evident more recently and the quicker the progress we make in this field in the seventies, the better will be the impact of the various crops and animal husbandry projects that are launched.

Modern agriculture requires sophisticated inputs and equipment. The supply of many of these has been the responsibility of the government agencies. This has diverted the departmental agencies from their legitimate task of helping the farmers in the new technology. A determined attempt to bring in the private sector to take over the responsibility for input supply and custom service is slowly paying results. But there is a long way to go.

In the light of what has been discussed earlier, there in an urgent need for research in exploring the possibilities of a viable marketing mechanism to minimise the cost of marketing and hence check the vicious spiral of prices. And thus, the necessity of an empirical investigation into the nature and causes of developing such centres.

To main objective of the empirical studies will be to analyse the structure of marketing in a dynamic perspective. The changing structure will be studied essentially with respect to changing pattern and importance of market functionaries, changes in volume, pattern, and the direction of trade and changes in trade and transport practices. The study should also reveal the nature and extent of the benefit to farmers (of the feeder areas), traders and consumers (in terms of quantities and price changes), benefits in terms of increased activities in the areas where new agricultural markets have been located, benefit arising out of reduction in commodities losses and impact of the expansion of organised credit consequent upon development of new agricultural markets.

For this it would be necessary to have a comprehensive survey on a sample basis of some marketing activities of feeder areas and dispersion point. Naturally, collection of data becomes the major part of the proposed study. Here is a tentative outline of the structure and scope of the study in order to compare and contrast the marketing practices of different countries.

Price is an important determinant of marketing activities. Therefore, indexing of prices commodity-wise, variety-wise, functionary-wise, and season-wise will help in evaluating the variation there of. Similarly data collection on quantity arrived in the market from farmers to the secondary market through middlemen will speak about the role, however, good or bad played by intermediaries. For analysis of cost, data collection on cost of marketing, storage, trade margin, etc. will throw light on the behaviour of prices. Nevertheless data collection in regard to transportation of the grains from farm to the market yards, handling and storage losses will aid the microscopic sphere of the study. Information about availability of market intelligence to the cultivators, functionaries and consumers will be an important aspect of the study. The evaluation of the structure and the extent of the utilisation of the available facilities in markets will display the efforts made by the governmental agencies and private institutions.

Some of the important issues on which there is the need to pay attention are:

1. Strengthening of cooperative marketing organisations.
2. (i) Extension of regulation of market practices to larger number of markets.
   (ii) Enforcement of laws concerning regulation of markets.
3. Reduction in the price spread between producer and consumer.
4. Establishment of new markets to cope with the increasing arrivals.
5. Progress in the adoption of grading and standardisation of produce.
6. Integration of production, marketing and processing.
7. Improvement in transportation and storage.

8. Marketing, transportation and packaging of perishables.
9. Training of market personnel.
10. Dissemination of market news and intelligence in the village.
11. Role of marketing in making the price support programmes of the government successful.
12. Market surveys and research to help the policy-makers— Importance of quick market surveys bringing in focus the problems arising out of the adoption of new technology.

32

AGRICULTURAL TAXATION IN INDIA

Importance of Agricultural Taxation For A Developing Economy

In a developing economy like India, agriculture constitutes the most important sector of the economy. As such, agriculture may be expected to bear a substantial portion of the cost of economic development. During the process of development, agriculture has to supply the rest of the economy with larger food surplus and raw materials. Production and marketed surplus of agricultural goods have to increase to meet the requirements of the non-agricultural sector and also to supply an increasing proportion of the community's manpower. In a developed economy, economic incentives commonly applicable to industry and trade will also help to increase the agricultural surplus. But in a developing economy with subsistence cultivation, economic incentives do not normally operate. For instance, a rapid rate of industrialisation may create a shortage of foodgrains and agricultural raw materials and may thus push up the prices; but the consequent rise in prices may not always call forth increased production and increased supply—in fact, a rise in price may actually lead to a decline in production. It is, therefore, suggested that the agricultural marketed surplus should be increased through taxation. Professor Kaldor gives clear expression to this idea when he writes: "The taxation of agriculture has a critical role to play in the acceleration of economic development since it is only the imposition of compulsory levies in the agricultural sector itself which enlarges the supply of savings for economic development." It is, thus clear that an increase in agricultural taxation is recommended and sought to be justified on the ground of raising marketable surplus from the agricultural sector.

There is yet one more reason why agricultural taxation is suggested. It is common knowledge that India's economic plans necessitate mobilisation of increasing volumes of domestic resources. Almost from the very beginning, the implementation of India' Five-Year Plans has suffered on account of inadequate domestic resources. Domestic resources for financing our plans came to 73 per cent in the First Plan, 56 per cent in the Second Plan and 59 per cent in the Third Plan. One of the reasons for the failure of the Government to raise adequate domestic resources is that is has not effectively tapped saving in agriculture. It is generally argued that, while the Government has been investing increasingly larger amounts in agriculture, it has not made a parallel effort to tax away a substantial portion of the increase in the income to the farmers. This is particularly true in the case of the large farmers whose incomes have risen considerably as a direct considerably as a direct consequence of the process of economic growth.

Present Position of Agricultural Taxation

In a abroad way, agricultural, taxation includes taxes paid by the agriculturists directly and also those borne by them indirectly. Direct taxes on agriculture consist mainly of land revenue, cesses and surcharges on land revenue, cesses on crops ad agricultural income tax. As consumers, they may be paying indirectly the excise duty, the sales tax, the import duty, the motor vehicles tax, and so on. Let us briefly describe these taxes.

Land Revenue. Traditionally, land revenue is the oldest of all taxes and, at present, the most important tax on agricultural land. The land tax is levied and collected by State Governments. In assessing land revenue, States have followed different bases, for instance, the land tax may be on net assets, or on the value of the net produce. In some states, assessment is left to the judgement of the assessing officers. At one time the land tax assessment used to be revised periodically but this has been given up in recent years. Hence the land tax has become as crude acreage tax. From Rs. 52 crores in 1951-52, the amount collected had gone up to Rs. 97 crores in 1960-61, to Rs. 113 crores in 1970-71 and to Rs. 166 crores in 1980-81 (budget). The importance of the land tax has, however, greatly declined in recent times as a result of the introduction and extension of may new levies. In 1951-52, land revenue accounted for 17 per cent of the total State tax revenue but by 1980-81 its share is less than 2 per cent. The following table illustrates the decline of the importance of land tax over the years.

As mentioned earlier, the land tax is being abolished completely or partially in most states.

Agricultural Income Tax. Agricultural income tax is levied and collected by the States. Bihar was the first to levy this tax in 1938. At present the states which have levied agricultural income tax are Assam, West Bengal, Bihar, Uttar Pradesh, Rajasthan, Madhya Pradesh, Orissa, Karnataka, Tamil Nadu and Kerala. The rates of tax have generally been lower than those applicable to the urban income tax. This tax has always occupied an insignificant role in India—a little more than 1 per cent in 1951 -52 and about 0.8 per cent in 1980-81. This can be seen from table 32.2.

TABLE 32.1

Share of Land Revenue in the State Tax Revenue

Year	*Land Revenue (Rs. crores)*	*As Percentage of the Total State Tax Revenue*
1951-52	48	17.1
1960-61	97	9.6
1980-81 (budget)	166	1.7

Source: *Reserve Bank of India Bulletins.*

Among the states only three states—Karnatka, Tamil Nadu and West Bengal—account for the bulk of the revenue from the agricultural income tax.

TABLE 32.2

Agricultural Income Tax

Year	*Amount (Rs. crores)*	*Percentage of State Tax Revenue*
1951-52	4.3	1.1
1960-61	9.5	1.0
1980-81 (budget)	83.9	0.8

Source: *Reserve Bank of India Bulletins.*

Direct taxes on agriculture accounted for about 11 per cent of the total revenue of all States taxes and duties—the range was between 3 per cent in the Punjab and 25 per cent in Assam.

Indirect taxes. Among the important indirect taxes which fall upon the farmers (as well as the urban population) special mention may be made of stamps and registration (actually a tax on transfer of property) the receipts from which have been quite substantial. Excise duties are imposed and collected by the Central Government but part of the net proceeds of the Central Government but part of the net proceeds of he Central excise is distributed among the Sates. The States too impose certain excise duties on such goods as alcoholic beverages and narcotics. The general sales tax, first levied in Tamil Nadu before World War II, has now come to occupy the most important place in State finances. Import duties, Entertainment tax, Electricity duties, Motor vehicles tax, etc. are some of the taxes which fall on the agricultural population to a certain extent. The problem with the indirect taxes in that they fall upon all people—both urban and rural—and there is no accurate method of dividing their incidence between the two sections. If such a division were possible, it would be easy to calculate the burden on the two sector.

3. Burden of Agricultural Taxation in India

Ved P. Gandhi, I.S. Gulati and E.T. Mathew have attempted to measure the gross as well as the net burden of agricultural taxation in India in recent years. According to E.T. Mathew the per capita gross burden of agricultural taxation in 1958-59 amounted to Rs. 1452 which constituted 6.8 per cent of the per capita income in the agricultural sector. The net burden of agricultural Taxation (which is gross burden minus the total share of he agricultural sector in the public expenditure on various items) was estimated to be Rs. 6.88 or 3.2 per cent of per capita income in the agricultural sector. On the other hand, the per capita gross burden of non-agricultural taxation in 1958-59 came to Rs. 46 or 9.2 per cent of the per capita income ; the per, capita net burden of non-agricultural taxation is 7.2 per cent of the per capita income. Mathew calculates the burden of individual taxes imposed on agriculture and concludes that Indian agriculture is under-taxed in an absolute sense, in a relative sense and in the context of the need for additional resources for economic development.

Gross Incidence of Agricultural Taxation

Much more exhaustive and clearer analysis of tax incidence and tax burden of agricultural taxes has been worked out by Ved P. Gandhi. The total tax incidence on agricultural and non-agricultural sectors is illustrated in table 32.3.

This table shows that in general, the agricultural sector bears relatively less taxation over the years than non-agricultural sector. In 1951-52 agricultural sector contributed Rs. 200 crores by way of direct and indirect taxes, whereas the non-agricultural sector contributed Rs. 450 crores; thus the ration between the two sectors was 44:56. By 1968-69 the gap has widened; now the agricultural sector pays over Rs. 900 crores and non-agricultural sector contributes nearly Rs. 2,700 crores; the ratio for the two sector is now 33:67.

TABLE 32.3

Total Tax Incidence on Agricultural and Non-Agricultural Sectors for Selected Years.

Year	*Agricultural Sector*	*(Rs. crores)* *Non-Agricultural Sector*
(1)	(2)	(3)
1951-52	200	450
1960-61	399	922
1962-63	619	1,750
1968-69	907	2,603

Source: (1) Ved. P. Gandhi: *Tax Burden on Indian Agriculture, p. 35.*

(2) The Economic Times Research Bureau: *"Taxable Capacity of Farm Sector.*

Thus during he planning period, the average tax in the agricultural sector, instead of rising, has actually fallen. Also during the same period while the increase in tax revenue of both the Centre and the State in the non-agricultural sector has absorbed about 25 per cent of the additional income generated in this sector, the share in the agricultural sector had not been more than 8 per cent. This implies that the non-agricultural sector carried more than two-third of the burden of additional taxation since 1951 -52.

Average Tax Rates. Let us now consider the average tax rates on the agricultural and non-agricultural sectors. Table provides some useful information.

TABLE 32.4

Average Tax Rates in the Agriculture and Non-Agricultural Sectors

	Agricultural Sector		*Non-Agricultural Sector*	
Year	*Taxes per capita* (Rs.)	*Ratio of taxes to per capitu income* (%)	*Taxes per capita* (Rs.)	*Ratio of taxes to per capita income* (%)
(1)	(2)	(3)	(4)	(5)
1951-52	8.0	3.8	41.8	9.5
1955-56	9.5	5.4	38.4	8.9
1960-61	13.4	5.6	68.9	13.0
1968-69	26.0	6.5	174.0	19.1

Source: (1) Ved P. Gandhi: *Tax Burden on Indian Agriculture,* p. 56.
(2) *Economic Times Research Bureau Taxable Capacity of Farm Sector.*

From table 32.4, we may observe that per capita taxes to on incomes are increasing in both the agricultural and non-agricultural sectors and that the ratio on taxes to per capita income has shown an increasing tendency in both the sectors from 3.8 to 6.5 per cent in the agricultural sector and from 9.5 to 19.1 per cent in the non-agricultural sector. But the more important fact from our point of view is that the agricultural sector has been contributing 6.5 per cent of its income which is only one third of the ratio of income contributed by the non-agricultural sector.

Marginal Tax Rates. In a sense, much more important than average rates of taxes are the marginal rates of taxation. Ved Gandhi estimated that during the period 1950-51 and 1960-61, the additional income which accrued to the agricultural sector was Rs. 1,717 crores but the additional taxation was Rs. 198 crores—i.e., the marginal rate of taxation was 11.5 per cent. On the other hand, during the same period, additional income in the non-agricultural sector was Rs. 2,420 crores but the additional taxation was Rs. 499 crores—the marginal rate of taxation worked out to be 20.6 per cent. It is thus clear that the marginal rate of taxation for non-agricultural sector is nearly double that of the agricultural sector. The implication is that the additional taxation imposed in recent years to finance planned economic development has been borne mainly by the non-agricultural sector.

A very interesting comparison is made here by Ved Gandhi between the marginal tax burdens of the upper and lower income groups in both agricultural and non-agricultural sectors. As mentioned in the previous paragraph, the marginal rate of taxation in the agricultural sector is 11.5 per cent; but the marginal burden on the upper classes is only 7 per cent. This means that the marginal burden of the lower classes is necessarily higher, probably about 12 or even 13 per cent. Likewise, in the non-agricultural sector, the marginal burden of taxation is 20.6 per cent; but the marginal burden borne by the upper classes is 13 per cent. This implies that the lower income group in the non-agricultural sector is left to bear a marginal burden of 23 to 24 per cent.

From our discussion of he burden of agricultural taxation in India, we may derive the following conclusions:

(a) In general, the agricultural sector bears far less tax burden than the non-agricultural sector.

(b) The upper classes of people have borne much less burden than the lower classes in both the sector, and

(c) There is a great scope for increase in agricultural taxation, specially upon the higher income groups.

The Case For Additional Taxation on Agriculture

The dimensions of taxation, direct and indirect, Central and State, have been steadily rising partly under the pressure of development expenditure and partly because of the demands of defence and civil administration. The urban sector and particularly the manufacturing and the trading sectors have had to surrender more and more of their earnings to the Government in tax payments. This process seems to have reached a stage where it is threatening to be self-defeating. It has become imperative to look for new sources—it is in this context that the taxation of agriculture is assuming great significance.

Further, thanks to the large continuous public investments made in agricultural development since the Five-Year Plans began, the generally high level of agricultural prices and the particular profitability of commercial crops, such as tobacco and sugar, not to speak of speciality farming like orchards, the earning of farmers with larger or even medium holdings have been going up. The IADP is helping agriculture to move these circumstances, the traditional system of agricultural taxation which is generally through a low fixed flat rate of land revenue, has become "out-moded, inadequate and irrelevant" and a system of progressive taxation of agricultural income in urgently called for.

Again, the direct taxes on agriculture, viz., land revenue and agricultural income tax have almost remained static since 1950-51— below 2 per cent of he income from agriculture. Agriculture has thus not contributed anything for economic development, event thought he crude index of economic welfare of the agricultural sector has been rising. Further, there is a high concentration of land in the hands of a few—and the trend is increasing in recent years. For instance, about half of he land holders own only 6 per cent of the land; on the other hand, less than half per cent of land owners own more than 11 per cent of the land. The richer sections of the agriculturists are bearing a low burden because (a) they pay land tax which is a proportional tax, and (b) they may or may not pay agricultural income tax.

The agricultural sector is undertaxed in a relative sense also, i.e., in relation to the non-agricultural sector. We have compared the burden borne by these two sectors as regards gross and net incidence of taxation and average and marginal rates of taxation. We have shown that, relatively, the agricultural sector has been under-taxed. The conclusion is that the agricultural sector can be made to bear larger burden.

Finally, as Morarji Desai stated while presenting his budget proposals (for the year 1969-70) for taxation on fertilisers and power-driven pumps, "those who benefit by the substantial investments in agricultural, including research, irrigational facilities, fertiliser plants, rural electrification credit facilities and the support prices should contribute a part of their prosperity towards the cost of development in general. This is all the more so when the benefits of improved technology cannot yet be shared by the majority of our farmers, particularly in dry regions where fertilisers and new seeds are not easy to apply. If the benefits of the new agricultural technology are to be carried progressively to a growing proportion of our farmers, the resources needed for this purpose should come at least in part from the beneficiaries of the progress."

Opposition to Additional Taxation by the Agricultural Lobby

The powerful agricultural lobby has all along opposed every move to rase additional resources from agriculture. First, the new agricultural strategy has had its direct impact so far on only a part of farm land, farming of farmer. At this stage, it is wiser to encourage it through all possible ways, particularly through a preferential treatment in the form of taxes.

Secondly, the commercialisation of agriculture in terms of modern or scientific farming of economic holdings has certainly begun, but it has not gone for enough to transform the basic situation which is that of uneconomic fragmentation and subdivision of a good part of agricultural land among millions of improvident and impecunious peasants. Consequently, it may not be in the nation's interest to tax the revenue of the developing commercial sector of agriculture and thereby discourage the shift from a peasant economy to a market economy.

Thirdly, a good part of the untaxed incomes of the more affluent farmer is certainly being invested in the accumulation of farm capital or the financing of he inputs of scientific agriculture. In other words, to the extent that the farmers could finance from their internal resources part of their own requirements of capital or working expenses, the need for the state providing for this purpose out of its resources for development finance is reduced.

Finally, the affluent sections of the farming community are coming under the net of excise taxation which covers a wide range of manufactured goods entering into the changing patterns of diversifying consumption. Consequently, the contribution of rural sector to the revenues of the state through indirect taxation is steadily growing.

Raj Committee's Proposal for Additional Taxation on Agriculture

Various proposals for additional agricultural taxation have been made by official organisations like the Planning Commission and by the economists. These tax proposals have broadly three objectives:

(a) to secure larger investible and marketable surpluses from agriculture;

(b) to provide an incentive to raise agricultural output; and

(c) to introduce an element of progression in agricultural taxation so as to make it more equitable.

In order to attain the above objective the Perspective Planning Division suggested the introduction of a progressive surcharge on land revenue, surcharge on area under commercial crops, a purchase tax on commercial crops and a tax on livestock. Some economists proposed that the land revenue rates should be raised from the present levels and that progression should be introduced. Other had suggested the elimination of the regressive land tax altogether and its replacement by a progressive agricultural income tax.

The imposition of additional taxation on agriculture has always been a source of conflict between the Centre and the States. Finally, the Chief Ministers Conference on resource mobilisation held in October, 1971, recommended the appointment of an expert committee to examine the Centre's proposal to link or merge income tax with agricultural income tax. Accordingly, the Central Government appointed the Committee on Taxation of Agricultural Wealth and Income under the chairmanship of Dr. K.N. Raj in February, 1972 to examine the question of taxation of agricultural wealth and income from all aspects. The Raj Committee submitted its unanimous report in October, 1972.

The Raj Committee's recommendations are:

(i) A progressive Agricultural Holding Tax should be imposed on agriculturists who have no other assessable income.

(ii) In the case of assesses having ono-agricultural taxable income, income from agriculture should be included in the total income for the purpose of calculating income tax.

(iii) Income from livestock, fisheries, poultry, dairy farming, etc. should be subject to tax.

(iv) An integrated taxation of agricultural property through wealth tax should be introduced.

(v) Capital gains tax on transfer of agricultural lands should be imposed.

A. Agricultural Holdings Tax (AHT)

The Raj committee has found two basic defects in the present land revenue system, viz., (a) the incidence of land revenue is relation to the productivity of land is not uniform over different parts of the country, the reason being that land revenue settlement has been done under different systems and at different times in different parts of he country; and (b) land revenue is assessed at a flat rate per hectare and hence is not progressive. It is to over come these two defects and make the agricultural sector contribute its fair and equitable share towards raising resources for further economic development that the Raj Committee has recommended the AHT. This new tax proposal takes into account the differences in the productivity of land all over the country on the basis of certain objective criteria and uniform procedures, nand reflects broadly the degree of progression applicable to other sectors of the economy.

(i) *Computation of AHT.* The AHT is a tax on the net rateable value of an agricultural land holding; or i simple terms, it is a tax on net farm business income. The simple formula for computing the actual tax liability under the AHT is:

AHT = X % of the net rateable value

2 of a holding, in which

X = Number of thousand rupees.

Net Rateable Value = gross rateable value *minus* the development allowance (which is 20 % of the retable value, subject to a maximum of Rs. 1,000)

The Committee has given the following example to illustrate the computation of AHT. Suppose the rateable value of an agricultural holding is Rs. 10,000 and the development allowance is Rs. 1,000. The AHT should be calculated as follows:

Gross Rateable value = Rs. 10,000

Development Allowance = Rs. 1,000

Net Rateable value = Rs. 9,000

X = 9

$$\text{AHT} = \frac{X}{2} \% \times \text{Rs. } 9{,}000$$

$$= \frac{9}{2} \times \frac{1}{100} \text{ Rs. } 9{,}000$$

= Rs. 405.

If the gross rateable value of agricultural holding is Rs. 20,000 and its net rateable value is Rs. 19,000, the AHT will be:

$$= \frac{19}{2} \times \frac{1}{100} \text{ Rs. } 19{,}000$$

= Rs. 1,805

As the incidence of the AHT decreases progressively on small holdings, the formula should be applied to rateable values up to Rs. 600 only. For holdings of rateable value below Rs. 600, AHT may be fixed at a flat rate of Re. 1 per holding.

(ii) *Computation of the reteable value of agricultural holding*. For the purpose of calculating the net rateable value of agricultural holdings the Raj Committee has suggested:

(a) The division of the whole county into a sufficiently large number of soil-climatically homogeneous districts/tracts.

(b) Preparation of norms of output of different crops per hectare for each year on the basis of estimates of yield for the previous 10 years, and the valuation of he norms of output at the relevant average harvest prices of the preceding three years; This will give the value of the gross output of different crops.

(c) From the above value, paid-out costs of cultivation (other than expenses on irrigation and depreciation of assets) varying between 40 and 60 per cent of the gross value should be deducted. We will get the rateable value of a hectare of land growing different crops in different districts/tracts.

(d) For each district/tract, there will be a schedule of rateable value of land per hectare under different crops. The schedule should be prepared for each year and included in the legislation of the year in question.

(e) From the above rateable value of land, we should deduct the expenses of irrigation (i.e., actual rates for irrigated crops from public sources and at 20 per cent of the rateable value of the crops irrigated from the private sources). We will then get the rateable value of the assessable agricultural holding.

To take care of the costs incurred in development and improvement of land and agricultural techniques, a development allowance at 20 per cent of rateable value (subject to a maximum of Rs. 1,000) should be given to all agricultural holdings.

(iii) *Basis of assessment and implementation.* The AHT is to be imposed on operational holdings, viz., land owned by the farmer minus any part of his land leased out or mortgaged to others plus any land leased or mortgaged by him from others. The basis of assessment will be the family and not the individual—consisting of husband, wife and minor children; this is to prevents large—scale tax avoidance.

The Raj Committee has recommended the implementation of the AHT in two phases: first, replacement of the present land revenue by the AHT on all operational holdings with rateable value of Rs. 5,000 or more; and second, extension of the AHT to all other operational holdings with rateable value below Rs. 5,000. The Raj Committee has estimated that he additional resources likely to be raised from AHT would range between Rs. 150-200 crores per years.

The salient features of the AHT may be summarised as follows:

(i) The assessment of AHT will be done on the basis of certain objective criteria and procedures that will be uniform throughout India and will ensure equity in its incidence.

(ii) The AHT takes into account the differences in productivity of land all over the country.

(iii) The tax is progressive, as it falls heavily on holdings with larger rateable value.

(iv) AHT is assessed on operational holdings (and not on owned holdings) and protects the tenancy rights on land.

(v) The AHT takes the family as the basic unit of assessment. For trusts, joint stock companies and other non-family holdings, there are special rates and procedures.

(vi) The year of assessment is the same throughout the country, viz., Ist July to 30th June.

Critical Evaluation of the AHT

The basic criticism against the AHT relates to the elaborate procedure laid down for the computation of the rateable value of agricultural holdings. The steps involved are: (i) A thorough survey of the soil with regard to its nature, quality, suitability to the various crops, irrigation requirements for different crops, etc.; (ii) A thorough survey of the climatic conditions in different districts/tracts; (iii) Division of the country into soil-climatically homogenous tracts; (iv) Cropwise production for the past 10 years for each district/tract and crop-wise prices for the last three years; (v) Preparation of farm Management studies in each soil-climatic district to compute cropwise paid-out cost of cultivation for each district/tract; and (vi) Actual assessment of the operational holdings. All these procedures will involve very heavy cost. Items (i), (ii), (iv) and (v) are particularly expensive, both initially as well as in the future. Moreover, the availability of statistics and relevant data at the right time are likely to pose a formidable problem to the implementation of the AHT.

Besides, hordes of officials will have to be engaged on a permanent basis for doing all these jobs. The tax-collection machinery will have to be strengthened as millions of farmers will be involved and provision has to be made for the setting up of separate courts as large number of disputes in the assessment of AHT will be inevitable. It is pointed to that the Raj Committees has not given proper thought to the administrative difficulties and additional costs involved.

According to C.H. Hanumantha Rao, the calculation of norms of productivity on the basis of the estimates of yield for the previous 10 years is wrong as this would definitely understate the incomes of large farmers under conditions of rapid technological progress. For example, the yield of weather per hectare in Punjab was 35 per cent more in 1971-72 when compared to 10-year average, between 1962-63 and 1971-72. Raj Committee is wrong in estimating paid-out cost at 40 to 60 per cent of he value of their gross output for all unirrigated lands, with extra deduction for expenses of irrigation. The result is that the percentage of gross output deductible as costs would be higher for irrigated crops than for unirrigated ones. But the little available evidence from farm management studies shows clearly that he unit costs are significantly lower for irrigated lands as compared to unirrigated lands. Besides, the technological revolution can be expected to reduce unit costs still further in irrigated areas. All these criticisms do not, however, invalidate the imposition of the AHT. If for administrative or non-administrative reasons, a 10-year period for calculating norms of productivity is found to be too long, a shorter period can be adopted. If annual assessment are difficult, triennial or quinquennial assessment can be tried. These are matters of administrative detail.

Some have predicted the failure of AHT on the ground of lack of records relating to occupational holdings, incorrect reporting of aggregate holdings of families, deliberate misreporting of crop patterns, problems of benami holdings and benami transfer of holding, etc. or. Raj has correctly dismissed all these criticism on the ground that they are not particular to AHT but are common to any form of direct taxation. With more experience, the tax administration will be able to overcome these difficulties to a large extent, if not eliminate them altogether.

According to some critics, the AHT contradicts the Raj committee's own guiding principle governing agricultural tax policy, viz., "The incidence of direct taxation should be broadly the same on comparable income and wealth groups irrespective of the sources of such income and the forms in which wealth is held." On the basis of this principle the Committee should have straightaway recommended that (a) all landlords getting Rs. 5,000 or more from agriculture should pay AHT roughly at the same rate as those in the non-agricultural sector pay income tax and (b) land holders getting less than Rs. 5,000 should have been exempted, as their counterparts in the non-agricultural sector. Instead, the Raj Committee has recommended the AHT for all land holders with net rateable value of Rs. 600 and above; and for all farm business incoms, below Rs. 600, the Committee has recommended uniform AHT of Re. 1 per holding. The Committee's recommendation is clearly regressive though it rationalises its recommendation by stating: "In fact, payment of land revenue or some tax on holding, is itself widely regarded by land holders as necessary evidence of their rights in their holdings." If this is true, then the present, land revenue should be retained and AHT may be imposed in addition to land revenue and not replace it.

A serious criticism against AHT is that it is based on operational holdings and, therefore, leaves out the rental income from agricultural taxation. In the words of C.H, Hanumantha Rao: "Since the case for taxing rental incomes is even stronger than the taxation of farm business respect of accepted, would introduce a fresh element of inequity in fact perversity in the system of agricultural taxation in India."

Dr. Raj accepts this as valid criticism and mentions different alternative suggestions: (a) ownership holding may be made the basis of the tax, (b) the tax liability of an operational holding may be shared between owners and tenants, and (c) the Indian Income Tax Act may be amended so that rental income derived from agriculture is treated in future as non-agricultural income.

Many have characterised the Raj Committee's estimates of additional revenue from the AHT—between Rs. 150 crores and Rs. 200 crores—as purely imaginary and speculative. The Committee confessed that it did not have reliable figures to make the estimate. D. N. Dwivedi used the AHT formula and calculated the yield of AHT for U.P. and Rs. 6 crores fore the year 1970-71—the actual yield of land revenue for the same year was Rs. 22 crores—a loss of Rs. 16 crores. Dwivedi conclusion is that the AHT may not actually bring in the additional Rs. 150 to 200 crores of revenue to the States as anticipated by the Raj Committee. The critics naturally pose the question whether it is at all worthwhile for the States to incur huge expenditure on the AHT and cancel the land revenue, when the revenue yield of AHT is highly uncertain and conjectural.

What is really necessary is not to throw out the AHT as useless and impractical but to accept the AHT in principle and make a start by covering only the top decline of family holdings in each area arranged according to the size of these holdings. Once a starts is made extension in coverage and depth can be made in course of time. The income tax and the wealth tax, it may be mentioned here, had humble beginning. Dr. Raj states: "The basic thrust of he tax could be directed at the top decile of he operational holdings and the existing land revenue retained or modified in respect of the other holdings according to the circumstances of each state: to drop altogether the proposal for a progressive tax on agricultural holdings would be not only unjustified but wholly inconsistent with the approach outlined for the Fifth Plan."

The Raj Committee may be complemented for suggesting the AHT which is sufficiently progressive form of land revenue taxation, and which, at the same time, avoids the necessity of a constitutional amendment or the Centre-State confrontation on this issue. The AHT, by and large, satisfies the specified attributes of a satisfactory system of direct taxation of agriculture. As prof. M L. Dantwala states: "No system of taxation, however, ingeniously devised, can be perfect. On the whole, the Committee appears to have made a valiant effort to adhere to the guiding principles enunciated by it at the outset. All worst, one can be sceptical about its claim to simplicity." The State Government have finally rejected the AHT on one ground or the other.

B. Partial Integration of Agricultural and Non-Agricultural Incomes

Another major recommendation of the Raj Committee is the aggregation of both agricultural and non-agricultural incomes for the purpose of income-tax. This recommendation is based on the principle that the tax burden on assessees with similar incomes does not differ sharply merely because of the fact that part of it is derived from agriculture. Integration will also help to check evasion through the device of camouflaging taxable incomes as gains from agriculture. As the proposal is not a tax on agricultural income by the Centre; it would not require any constitutional amendment.

The Raj Committee has recommended integration of agricultural and non-agricultural incomes only if any assessee has taxable income exceeding the minimum exemption limit for income-tax. In determining the rate of tax on non-agricultural income, the agricultural income and non-agricultural income should be combined in the following manner and order: (i) the initial exemption allowed out of non-agricultural income, (ii) agricultural income and (iii) balance of non-agricultural income.

A simple example will illustrate the partial integration of agricultural and non-agricultural incomes. Suppose, a tax payer is getting an agricultural income of Rs. 25,000 and non-agricultural income of Rs. 15,000. Before integration the tax-payer's normal liability is only for Rs. 15,000 (non-agricultural income) and at the present slab rates will amount to Rs. 1,350, i.e., first Rs. 5,000—exempted: income tax (I.T.) on the next Rs. 5,000-17 per cent (Rs. 500) I.T. on the next Rs. 5,000-17 per cent (Rs. 850). Under Raj Committee's proposal, the integration incomes of purpose of tax rate is given below:

(1) I.T. on the first Rs. 5,000 - Nil
(2) I.T. on the next Rs. 5,000
(farm income otherwise taxable) 10% Nil
(3) I.T. on the next Rs. 5,000
(farm income otherwise taxable) 17% Nil
(4) I.T. on the next Rs. 5,000
(farm income otherwise taxable) 23% Nil
(5) I.T. on the next Rs. 5,000
(farm income otherwise taxable) 30% Nil
(6) I.T. on the next Rs. 5,000
(farm income otherwise taxable) 40% Nil
(7) I.T, on the next Rs. 10,000
(non-agricultural income) 50% 5,000

Thus, under normal Income Tax provisions, the tax liability is only Rs. 1,350; under the Raj committee the tax liability increases to Rs. 5,000 (an extra Rs. 3,650).

But the taxpayer will pay AHT on the farm business income of Rs. 25,000 amounting to Rs. 3,125 (25/ 200x25,000). This implies that the total tax on agricultural income will be Rs. 6,775 (Rs. 3,650+3,125)

It is interesting to observe here that the Wanchoo Committee has recommended complete integration of agricultural and non-agricultural incomes. The Central Government, however, has accepted the Raj Committee's recommendation on the subject.

C. Integrated Taxation of Agricultural Property through Wealth Tax

The Raj Committee has recommended that the AHT should be supplemented with a tax on agricultural property and a tax on capital gains arising out of transaction in such property. Valuation, of farm land for wealth-tax purposes should, generally, be made through the method of income-capitalisation. A simple an adequate method would be to take 4 to 6 times the rateable value of a holding averaged over a period of years. Two radical suggestions have been made by the Raj committee which differ from the recommendations of the Wanchoo Committee on the same subject. First, the Raj committee has proposed to do away with all exemptions, except that the basic exemption should be raised to Rs. 1.5 lakhs. Second, wealth tax should be levied on family basis.

D. Integrated Taxation of Capital Gains on Agricultural Assets through Income Tax

Raj Committee has recommended that the definition of the capital asset be so widened as to permit taxation of capital gains from transfer of all agricultural land irrespective of their location. Gains from transaction in assets held for not more than a year should be treated as ordinary income and taxed accordingly.

To mitigate the inequity involved in the taxation of capital gains derived from sale of assets held for more than a year at progressive rates, a method of prorating the long-term capital gains over a number of years, without reopening any back year's assessment, may be followed.

General Assessment of Raj Committee's Proposals

Of all the recommendations of the Raj Committee the most important is the AHT. There is a general feeling that either the AHT is too radical or too cumbersome to implement. At the same time, the partial integration of agricultural income with non-agricultural income by the Centre has brought in higher agricultural income under the scope of direct taxation. States may now be encouraged to extend agricultural incomes taxation wherever they have not done so far. The higher income groups in agriculture will thus come under direct taxation. In the meantime, the lower agricultural income groups need not be disturbed through the imposition of the AHT. Again, the tendency to divide and sub-divide holdings in recent years under the impact of population pressures as well as due to fixing of land ceilings has been quite powerful throughout the country. Under these circumstances, it is suggested that the AHT cannot be significant and the proposal may, therefore, be allowed to lapse. There is no denying the fact that there is considerable opposition to the AHT proposals. The Union Government however, has accepted the Raj Committee's recommendation regarding partial integration of agricultural incomes with non-agricultural incomes by incorporating it in the 1973-74 budget proposals. If all the proposals of the Raj Committee are not accepted, the questions of equity in agricultural taxation and mobilisation of additional resources from the agricultural sector fro economic development are yet to be answered satisfactorily.

Other Proposals of Agricultural Taxation

The Perspective Planning Division proposed the following scheme of surcharge on land revenue.

TABLE 32.5

Proposed Surcharge on Land Revenue

Land revenue (Slab)			*Surcharge (per cent)*
Ist	Rs.	7	Nil
Next	Rs.	5	40
Next	Rs.	5	80
Next	Rs.	5	120
Next	Rs.	5	160
Next	Rs.	5	200
	ect.		

Source: Economic and Political Weekly, May 25,1968.

Leaving small landowners paying a land revenue of Rs. 7 or less, others are to pay a surcharge on land revenue to the extent of 40 per cent of the land revenue. The surcharge rises by 40 per cent for every slab and thus the lowest assessee will pay 40 per cent and the highest 72 per cent of the land revenue. Progression is thus introduced in land revenue. The Perspective Planning Division anticipated an annual income of about Rs. 140 crores. The system of progressive surcharge on land revenue is apparently a good source of tapping additional resources from agriculture. For one thing, land revenue system is retained. For another, the progressive principle is introduced. So long as land revenue is retained, this may prove to be a simpler method of collecting more revenue than the Agricultural Holding Tax.

Agricultural Income Tax

The Planning Commission has strongly recommended the imposition of a highly progressive agricultural income tax as a necessity. This proposal has been made in the past too but it is being pressed now on the following grounds:

(a) Government has invested large amounts of money in agriculture but very little of resources have flowed back into the exchequer from the agricultural sector;

(b) Agricultural performance has improved considerably in recent years and thus added substantially to disposable incomes in the hands of farmers; and

(c) The number of rich farmers has been increasing very rapidly as result of the government's emphasis on intensive development in selected areas. These fanners are bearing almost no burden at all.

The recommendation of the Planning Commission of the imposition of a progressive income tax has been turned down by the Chief Ministers of State. Since agricultural income tax is a state subject, the imposition of such a tax by the Centre is impossible without a constitutional amendment. It is not very difficult to show how an agricultural income tax is definitely inferior to a land tax. While the land tax is fixed and while the marginal increment of land tax is zero. The agricultural income tax is a tax on actual production and varies with production; and marginal rate of agricultural income tax is positive. As I. S. Gulati and V. N. Kothari rightly point out: "At its extreme, a tax on current agricultural produce is tantamount to taxing land use and exempting non-use of land. Thus, if the land tax is replaced by agricultural income tax, there is likely to be a transfer of land from productive to unproductive or less efficient uses and the pace of improvement is likely to slacken.... The perverse effect of the proposed change in our system of agricultural taxation cannot, therefore, be brushed aside."

The progressive agricultural income tax is sought to be justified on the ground that while thee is progression in the taxation of non-agricultural income, there is no such progression in the taxation of agricultural income. We have already shown how there is evidence to the fact that the higher income agriculturists are left free or pay a considerably small percentage of their income as tax. But it is forgotten that agricultural income tax is levied in all states and that it is highly progressive. Leaving a minimum income of Rs. 3,000 (in some states, the exemption limit is Rs. 3,500) the rate of agricultural income tax rises to as much as 45 per cent in Tamil Nadu, 50 per cent in Maharashtra and 78 per cent in Assam. From a little over Rs. 4 crores in 1951-52, agricultural income tax yielded Rs. 16 crores in 1975-76. But there is one important reason why the states have kept agricultural income tax rates low and have also not attached sufficient importance to them — this is due to the impositions of various levies and cesses along with land taxes, such as betterment levy, road and eduction cesses, water tax, etc.

In a discussion of progressive agricultural income tax in India, we should give sufficient consideration to the economic and administrative difficulties in imposing such a tax. Firstly, there is difficulty in determining farm income for purpose of levying tax on it. Normally, income is taken to be the difference between the revenue received and the expenses incurred in raising that revenue. In agriculture, revenue will be equal to sale proceeds of different crops and expenses will include the expenses of the farmers on land, hired and family labour including supervisory labour, appliances, implements, fodders, fertilisers, etc. It is extremely difficult fore an ordinary and moderately educated farmer to estimate these expenses. Nor do farmers keep any records about their income and expense. This is probably the most important difficulty in the imposition of agricultural income tax.

Secondly, there are wide fluctuations in outputs and prices of various crops. A good example is the price of chillies which has moved between Rs. 1,000 and Rs. 250 per quintal in recent years. Wide fluctuations in output and prices of various crops will result in wide fluctuations in farm income as well. Hence, the basis of taxation becomes indefinite and, therefore difficult.

Thirdly, there are different systems of land holdings and different kinds of tenancies in India. Naturally, the problem would be to locate the assessee. In many a case, the total agricultural income produced in an agricultural holding may cone under tax but when shared between the landlord and the tenants it may not he subject to tax.

Finally, the cost of collection of agricultural income tax may be very heavy since the farmer assessees are widely scattered. Besides, the number of small assessees will be exceptionally large. It is feared that the collection charges may prove to be a high proportion of the total revenue.

It is possible to argue that all these difficulties can be overcome over a period of time when the tax administration acquires more experience. This is, of course, true. However, there is the one point which should not be ignored in all our pre-occupations with concepts of equity and progression. The high income groups particularly among the industrial and treading classes have various devices of tax evasion and avoidance and hence the effective rate of income tax is considerably lower in the non-agricultural sector. These devices may not always be available to the farmers. The high rate of progression which exists only on paper as far as the non-agricultural sector in concerned, will prove, when introduced in the agricultural sector, a weapon of oppression of the agriculturists. The prosperous but unscrupulous fanner will join with the corrupt tax collector and cheat the Government. The only satisfaction would be that there is progressive agricultural income tax, similar to the progressive personal income tax, at least on paper. The Raj committee rejected the Agricultural income tax and preferred the AHT which is based on productivity and is also progressive.

Besides these two proposals, many other shave been suggested in order to tap additional revenues from agriculture. A surcharge on area under commercial crops and purchase taxes on commercial crops are other proposals. It is clear that Indian agriculture continues to be under-taxed and there is a great scope for additional taxation. This is particularly true in the case of richer farmers. However, the various proposals for additional agricultural taxation are defective and impracticable. The Union Government accepted Raj Committee's proposal to integrate agricultural and non-agricultural incomes for the purpose of calculating income tax. But the Raj Committee's major proposal of AHT has been rejected by the States. The problem of mobilisation of additional resources from the agricultural sector for purposes of development is yet to be solved satisfactorily.

33

COMMERCIALISATION AND CLASS RELATIONS

A key ingredient of Bhaduri's writings on agrarian economies has been the emphasis on agrarian economic *structure* as the most crucial factor constraining growth performance. Specifically, he has argued consistently that usurious moneylending has pre-empted accumulation for agricultural production purposes, causing stagnation. The mechanism through which the surplus is siphoned off into unproductive uses is 'forced commerce'—a phenomenon associated with sales by producers at unremunerative prices to moneylenders and traders. In his influential (if somewhat controversial) article published in 1973,[2] Bhaduri linked forced commerce with 'interlinked markets'—for product, credit, land and labour—in that the landlord was simultaneously a moneylender exploiting the same tenant cultivator through dual channels. In the essay in this volume Bhaduri views forced commerce in another context. The landowners and moneylenders/traders are two different exploiting classes and the accumulation process is governed by the pattern of interrelation between these classes. Here we shall discuss Bhaduri's hypothesis in the light of some empirical evidence on UP.

THE HYPOTHESIS

Bhaduri's statement of his hypothesis is somewhat diffuse. To reconstruct: (i) Bhaduri makes a distinction between 'forced commercialisation' and 'normal commercialisation'—the former is exploitative and the latter associated with 'gains from trade'. In the first case surplus accrues to the moneylender or trader, and in the second to the cultivator, (ii) He suggests that in any actual process of commercialisation, these two forms may coexist, but in varying proportions in different regions, (iii) The varying proportion of 'forced commercialisation' to 'normal commercialisation' is correlated closely with the relative economic power of different agrarian classes. Where 'forced commerce' has been predominant moneylending and trading classes dominate, whereas if 'normal commerce' predominates the rich agriculturists and 'progressive' landlords are the dominant agrarian class, (iv) The relative power of agrarian ruling classes determines the accumulation process in the agrarian economy. The stronger the moneylenders and traders, the more stagnant the agricultural production; conversely, agricultural dynamism is the consequence of stronger rich agriculturists and 'progressive landlords, (v) The regional difference in agricultural growth between western and eastern UP since independence can be explained by the relative significance of forced vis-a-vis normal commercialisation and its associated class configuration. Statements (iii), (iv) and (v) constitute the core of Bhaduri's hypothesis.

Before turning to the empirical part of the hypothesis, several observations could be made. First, Bhaduri's hypothesis touches upon a very fundamental question of political economy, namely the influence of *structural* or *institutional* factors on the pace of material progress in an economy. Though it is not explicitly stated, one may surmise that for Bhaduri institutional factors—particularly relations of production—are crucial in constraining agricultural growth. Even if one were to concede that forced commerce constitutes a framework in which 'productive' investments are unlikely, the question is to what extent such a framework is immutable. Once it is admitted that the structure itself changes—and not necessarily through grand or abrupt systemic ruptures—then the issue pertaining to the nature of technical, social and environmental factors which sustain a given set of production relations becomes relevant. By tracing the formation of class configurations all the way back to British revenue settlements and subsequent land reforms, and thus suggesting that post-independence agricultural performance has been conditioned by these historical forces, Bhaduri seems to imply that structural factors change very slowly. They shape, rather than are shaped by, economic conditions. It would appear that in Bhaduri's view structure changes more in response to legal mechanisms such as land reforms rather than to economic and technological forces.

Second, Bhaduri has consistently argued that agricultural backwardness in eastern and northern India is attributable to the operation of usurious moneylending. But in so far as growth performance is linked with the adoption of HYV, is it the case that moneylending (and profit from moneylending and trade through forced commerce) *blocks* the adoption of new technology? Contrary to Bhaduri's hypothesis, changes are taking place in eastern UP. According to Y.K.. Alagh: 'It is not generally recognised, for example, that in Eastern Uttar Pradesh itself wheat production has increased by about 2.4 million tonnes in the seventies and productivity levels have gone up from around 11 to 14 quintals per hectare. We may remember that a productivity level of 14 quintals was achieved in Western Uttar Pradesh in the early seventies and was regarded as one of the success stories of the Green Revolution.'[3] Should this be taken to imply that 'forced commerce' has declined in eastern UP to make this change possible? If so, what circumstances triggered this transformation? If not, it is necessary to specify the limits to agricultural progress that a given class configuration entails.

Third, in any concrete comparison between two or more regions it is very likely that features other than the class configuration may be different. For example there could be significant differences in demographic factors, agro-climatic conditions and physical infrastructure between regions. It is quite plausible to suppose that these would also exert an influence on the degree and pattern of agricultural accumulation. To what extent can structural and institutional factors be integrated with these other determinants in a theory of agrarian change? The preconditions of forced commerce include the existence of a large mass of petty tenants on the edge of economic survival and dependent on consumption loans. Obviously the existence of petty tenants depends on a number of factors which have to do with political struggle over land control, typically emphasised in Marxist literature. But it is also possible that the pressure of population on land exacerbates petty tenancy and aids in the perpetuation of forced commerce, in addition to creating other obstacles for agriculture progress. Hence, two regions which are both under the grip of forced commerce induced by heavy revenue demand but under different degrees of demographic pressure may evolve differently. One region may emerge free from the grip of forced commerce under the influence of favourable economic circumstances while the other may not. A similar argument could be made with respect to agro-climatic factors and irrigation which favour a particular crop—say wheat—in a region. If crop-specifictechnological change occurs it might be easier for this region to evade the grip of forced commerce. All this is to suggest that the evolution of class configurations may be conditioned by such factors as demographic pressure and public investments in irrigation and transport, and that class configurations do not evolve in an autonomous manner. Hence, agrarian structure and institutional framework are not unimportant, but nor are other influences.

Fourth, Bhaduri associates forced commerce with stagnation and normal commerce with expanded growth. While this may be a reasonable inference, if pushed to an extreme—(i.e. if the association becomes an identity)—this can give the subsequent empirical interpretation a tautological character. At points Bhaduri comes close to positing a rather strict relationship between the type of commercialisation and output growth. However it needs to be noted that even under normal commercialisation the distribution of 'gains from trade' depends on bargaining power. Hence a highly unequal sharing is possible, resulting in the bulk of the surplus not accruing to the direct cultivator; also that even under forced commerce it is possible that some of the surplus extracted from the cultivator may re-enter agriculture through loans tied to specific crops. Output may get a boost even though the tenant cultivators do not gain. To the extent that both normal and forced commercialisation are financed by moneylenders and traders, as seems to have been the case in UP, the difference between the two types of commercialisation may be of degree of exploitation, and not of kind.

THE EMPIRICAL EVIDENCE

The historical evidence from UP is interpreted by Bhaduri along the following lines:

(i) The impact of British policies on eastern and western UP were fundamentally different, particularly with respect to revenue settlements. The eastern region, which came under British control a little earlier, experienced the Permanent Zamindari Settlement. In the western region the land revenue settlement 'was more in the image of a ryotwari settlement combined with elements of the zamindari settlement'. High and inflexible revenue demands in the east triggered subtenancy whereby traders and moneylenders grew stronger and subdued the cultivator into forced commercialisation. By contrast the west witnessed the rise of a class of rich occupancy tenants. This phenomenon appears to be attributed by Bhaduri to temporary revenue settlements.

(ii) After 1830 the British pursued a policy of strengthening the tenants by enacting a series of legislation. However, because of the difference in 'initial conditions', this had a different impact in the two regions. In the east the moneylenders and traders who held subtenancy rights were strengthened in this process, while in the west there was a genuine development of peasant proprietorship, or at any rate more secure land control by occupancy ryots. This process culminated in the zamindari abolition, leaving behind a regionally varied agrarian class configuration at independence.

(iii) The Zamindari Abolition Act (1954) had a similarly different impact: while it permitted the tenants to purchase superior rights from the Zamindars, it meant that 'direct cultivators' in the west improved their tenurial status. But in the east tenancy rights were held by moneylenders and traders and also by the Zamindars (through subinfeudation) who consolidated their position or evaded the reforms.

(iv) This difference in class configuration meant that 'the older mix between normal and forced commerce was left largely undisturbed in eastern UP' and perhaps even strengthened forced commerce. The very opposite happened in the west.

(v) The subsequent difference in agricultural dynamism between the two regions is explainable by these differences, traceable to differences in the commercialisation process experienced by them.

TABLE 33.1

Comparative Agricultural Performance in Selected Districts of Uttar Pradesh 1950-71

Compound Annual Growth Rates of Output				*Gross output per Hectare (Rs)*
	1950-3 to 1956-9	1956-9 to 1962-5	1962-5 to 1968-71	1970-1
(1)	(2)	(3)	(4)	(5)
East UP				
Ballia	-0.99	3.82	3.25	1061
Deoria	0.29	5.03	2.43	1502
Jaunpur	-2.30	3.54	1.97	1094
Sultanpur	0.20	1.76	0.50	979
West UP				
Shahjahanpur	1.11	0.10	4.20	1177
Agra	0.88	0.71	3.12	962
Aligarh	0.75	2.01	4.41	1579
Meerut	1.24	2.29	3.66	2530

Sources: For column 2,3,4, V.N. Misra: 'Anatomy of Agricultural Growth'; T.S. Papola, V.N. Misra, H.S. Verma, R.C. Sinha and A. Joshi: *Studies on Development of Uttar Pradesh.* (1979), pp. 4-5. For column 5, R.C. Sinha: 'Agricultural Development and Rural Employment,' ibid., pp. 93—5.

In considering the empirical part of Bhaduri's hypothesis several observations may be relevant. First that while it is true in the aggregate that agricultural growth performance has generally been better in western compared with eastern UP, a closer look reveals some interesting patterns. Table 33.1 shows compound annual growth rates achieved in eight selected districts of UP between 1950—3 to 1968-71, separately for the three sub-periods. Ballia, Deoria, Jaunpur and Sultanpur are in eastern UP. Shahjahanpur, Agra, Aligarh and Meerut are from the western region.[4] From this Table certain interesting facts emerge. We notice that the eastern districts are indeed in a moribund condition during 1950-3 to 1956-9. Growth rates are negative or low. The western districts do better. Meerut, particularly, performs the best, and its agricultural prosperity is of long standing. In the second sub-period, however, the picture changes sharply. The eastern districts do much better, not only in marked contrast to their own performance during the first sub-period, but also relative to the western districts. Three out of four eastern districts show growth rates above 3.5 per cent per year and one of them higher than 5 per cent. Meanwhile two out of the four western districts show a decline in their growth rates, particularly Shahjahanpur. In view of Bhaduri's hypothesis it is worth noting that this sub-period follows closely upon the Zamindari Abolition Act of 1954. If forced commerce retained its grip in the eastern districts, evidently it did not slow down agricultural growth in that area. Bhaduri also cites a case study from Deoria district which shows that non-cultivating Brahmins and Bhumihars rather than peasant proprietors had strengthened their control over land. Yet if this is true for the entire district of Deoria it apparently did not deter 'productive' investment. As the Table shows, Deoria grew rapidly in the second sub-period and modestly in the next period. In 1970-1 it achieved a productivity level of Rs 1,502 per hectare, which compares quite favourable with three of the four western districts. Thus for the 1956-9 to 1962-5 period the data do not lend much support to Bhaduri's hypothesis.

It is in the third subperiod 1962-5 to 1968-71 that the western regions improve their growth performance substantially. All the districts from this region do very well, justifying their claim to superior dynamism. However, even in this period the growth rates in the eastern districts of Ballia and Deoria are reasonably high. Looking at the levels of agricultural productivity in 1970-1 it is also worth noting that except for Meerut the gap in productivity between the two regions is not remarkable. Hence the evidence presented, though limited, suggests that other factors may also have a bearing on relative growth performance. The variation in dynamism over successive sub-periods is notable and calls for closer analysis. Even within the broad regions there is a considerable difference in growth performance between individual districts. It is not clear how much of this difference between the performance of individual districts, some of which are contiguous, can be traced to prior differences in the historical evolution of agrarian legislation. The connection between wheat acreage and growth in the third period is evident, and is attributable to HYV wheat.

Secondly, a factor which may have some bearing on the difference in regional agricultural performance is irrigation. As is well known, in this respect the agrarian history of the two regions varied considerably.[5] The western regions received a large amount of public investment in canal irrigation in the nineteenth century and the eastern districts hardly any at all.[6] This difference in sources of irrigation remained throughout, and as Table 33.2 shows, the difference in 1948-9 is quite striking, indicating differences in infrastructural initial conditions at independence. As detailed by Whitcombe, canal irrigation triggered several changes in the agrarian economy, including a commercialisation process which entailed significant crop shifts and the emergence of wheat as a major crop in the western region. In the east paddy was, and remained, more important. There is much evidence to suggest that the commercialisation process was more vigorous in the western region and that canal irrigation, both by making water available and imposing irrigation charges, was crucial in stimulating this process.[7]

TABLE 33.2

Sources of Irrigation and Area Under Wheat in Selected Districts of UP

	Net Irrigated Area by Each source as per cent of Total Net Irrigated Area (1948-9)						*Net Irrigated area as per cent of net sown area (1949-50)*	*Area under wheat as per unit of local sown area (1949-50)*
	Total	*Canals Govt.*	*Pvt.*	*Tanks*	*Wells*	*Other sources*		
East UP								
Ballia	—	—	—	—	75.7	24.3	35.2	5.6
Deoria	—	—	—	—	74.5	25.5	37.3	10.9
Jaunpur	—	—	—	—	89.6	10.4	52.3	6.1
Sultanpur	4.7	4.7	—	—	64.5	30.8	38.4	10.2
West UP								
Shahjahanpur	57.9	57.9	—	—	14.0	28.1	12.2	23.4
Agra	60.8	60.8	—	—	36.5	2.7	26.9	13.5
Aligarh	48.1	48.1	—	—	50.6	1.4	51.6	17.6
Meerut	60.1	60.1	—	—	39.2	0.6	60.6	24.9

Note: Even though the extent of irrigation is high in east UP districts, most of this comes from traditional wells which have a low irrigating capacity.

Source: *All-India Rural Credit Survey* 1956, Vol. I, Tables 1.2, 1.3 and 1.4.

With regard to class configuration, this exogenous factor (canals were laid mainly with government funds) is likely to have induced 'productive' investments in agricultural production by landlords and 'maliks'.[8] According to Whitcombe, however, during the nineteenth century, moneylenders and traders thrived in the western sector. In fact she cites several instances of forced commerce and suggests also that the actual cultivators—even Jats—were caught in its grip.[9]

This, however, did not deter a noticeable dynamism in the commercialisation process. Why is it that in the western districts commercialisation increasingly assumed a normal form, as Bhaduri suggests, when in several other respects the 'initial conditions' were similar to the eastern districts? For example the pressure for cash to meet revenue demands and rents existed in both areas. Moneylenders and traders had extensive operations. Yet a class of tenants gradually accumulated funds. Perhaps the economic stimulus given by canal irrigation was a factor since it made possible a larger absolute volume of surplus.

Thirdly, the demographic pressure on land may have had much to do with the different evolution in the two regions. I must emphasise that the demographic factor is not an alternative explanation. However, as argued earlier, it is an additional condition which needs to be taken into account since the preconditions for 'forced' commercialisation may be more prevalent in areas of such pressure. Tables 3 and 4 show the figures for mean population density in selected districts from 1872 to 1931.

These figures clearly indicate that demographic pressure in the eastern region of UP has remained significantly higher as compared with the western region throughout the entire period. The picture is substantially unchanged in the two decades following 1951. In fact for the districts selected the density of population grew faster in eastern UP. It is true that these aggregate density figures do not accurately reflect the demographic pressure in the rural sector. This is taken into account in the figures for 1941 and 1951. The density of rural population per net sown acre is 50 per cent higher in east UP in 1951, while that for rural density is higher by 40 per cent.

TABLE 33.3

Population Density in Selected Districts of UP 1872-1931 Mean Density per Square Mile

	1872	*1881*	*1891*	*1901*	*1911*	*1921*	*1931*
East UP							
Ballia	584	784	800	794	680	675	742
Deoria*	443	574	657	649	707	721	787
Jaunpur	662	780	816	776	746	745	797
Sultanpur	607	559	628	632	612	586	614
West UP							
Shahjahanpur	551	496	532	534	548	485	513
Agra	580	525	541	572	551	500	567
Aligarh	551	525	536	617	599	545	602
Meerut	544	560	593	657	648	480	541
Indo Gangetic East	602	733	790	700	675	710	753
Indo Gangetic West	502	491	499	553	538	508	542
Indo Gangetic Central	520	519	564	571	550	528	555
Sub-Himalayan East	424	491	565	566	586	602	651

* The figure for Deoria is not available separately. The ones shown refer to Gorakhpur district of which Deoria forms a part.

Source: *The Census of India*, various years.

Notes: classified in the *Census:*

(1) Ballia and Jaunpur fall in the 'Indo Gangetic East' region.
(2) Sultanpur comes under 'Indo Gangetic Central' region.
(3) Gorakhpur (used as proxy for Deoria) belongs to the 'Sub-Himalayan East' region.
(4) All the selected west UP districts are under the 'Indo Gangetic West' region.

Admittedly the relationship between demograhic pressures and social structure is a subtle and complex one. Each influences the other. Nonetheless a significantly higher demographic pressure at comparable levels of productivity is likely to have a bearing on relations of production through its effect on bargaining power. It is partly for this reason that there is a greater incidence of family labour in the west and agricultural labour in the east, and a greater reliance on owned agricultural assets and lower rents in the west. Hence it may have proved more difficult for petty cultivators to escape the grip of forced commerce is east UP, and this led to the different impact of land reforms in the two regions.

TABLE 33.4

Population Density (Total and Rural) and Other Measures of Demographic Pressure on Land 1941-51

	Population Density per sq. mile				*Rural Population per net sown acre*	*Average size of holding*
	1941		*1951*			
	Total	*Rural*	*Total*	*Rural*	*(1951)*	*(acres)*
(1)	(2)	(3)	(4)	(5)	(6)	(7)
East UP						
Ballia	849	786	963	887	1.87	4.2
Deoria	941	921	1005	980	1.92	3.8
Jaunpur	890	849	973	927	2.01	2.3
Sultanpur	643	636	750	740	1.85	2.8
(Gorakhpur)	(818)	(770)	(919)	(858)		
West UP						
Shahjahanpur	559	477	571	489	1.08	5.3
Agra	693	503	807	554	1.17	5.1
Aligarh	704	601	792	668	1.29	7.3
Meerut	824	698	992	811	1.56	9.7

Notes: Source for columns (2)-(5), Census reports; Gorakhpur is given for comparison with Table III for Deoria.

Columns (6) and (7) are from *All India Rural Credit Surrey*, 1956, Vol. I, Table, 1.1, 1.2 and 1.5. The net sown area figure is figure is for 1951. for 1949-50 and the population

NOTES

1. Amit Bhaduri, 'Class relations and the Pattern of Accumulation in an Agrarian Economy,' *Cambridge Journal of Economics*, March 1981.
2. Amit Bhaduri, 'A Study in Agricultural Backwardness under Semi-Feudalism,' *Economic Journal*, March 1973.
3. Y. K. Alagh, interviewed in *Business India*, 3-16 January 1983, p. 132.
4. These districts have been selected because they were included in the All-India Rural Credit

Survey and they also correspond to the NSS regional classification of east and west UP. The history of British control of these districts is as follows: Deoria and Shahjahanpur were 'ceded districts.' Ballia and Jaunpur came under the Benares division. Sultanpur was a part of Oudh. Agra, Aligarh and Meerut were 'conquered districts.' See W. C. Neale, *Economic Change in Rural India* (Yale Univ. Press, 1962), p. 2.

5. See Elizabeth Whitcombe, *Agrarian Conditions in Northern India, Vol. I: The United Provinces under British Rule, 1860-1900* (California, 1972), especially Chapter II.
6. Between 1820 and 1888, the NWP received some 5,601 miles of channels and distributaries, irrigating 1,459,938 acres at a total cost of construction (excluding interest) of £4,338,384. Ibid, p. 66.
7. *Ibid.*, pp. 71—5. Whitcombe points out the ecologically deleterious consequences of the canals also. But there is no denying the stimulus they gave to commercialisation. The cropping pattern changed in favour of more 'valuable' crops such as wheat, sugarcane, indigo, opium and vegetables. Wheat cultivation was concentrated in the canal tracts. Perhaps this early existence of wheat made it easier for the western districts to adopt HYV wheat, when it became available.
8. *Ibid.*, p. 87.
9. *Ibid.*, Chapter IV, especially pp. 166-7. Also pp. 46-7 and p. 226.

ANNEXURE-I

A NOTE ON COMMERCIALISATION IN AGRICULTURE

I

In this essay we present a synoptic sketch of some ideas on the process of commercialisation. The point of view is economic-analytic, although inevitably the broad bases for my hypotheses are drawn from historical experience. It is not being denied thereby that the actual *process* of commercialisation and its historical outcome, whether or not culminating in the pervasive dominance of capitalist relations, has not been uniform, nor that it has not varied among different countries, and within India among different regions and over different historical periods.[1] Nevertheless, if an attempt is made here to delineate certain features of the process with a broad brush, it is with the belief that we may find sufficient generality in analytical interconnections to capture meaningfully the dynamics of any particular commercialisation process, or even provide some explanation for the diversity in the processes and their outcomes at different historical conjunctures. The immediate object of this note is modest: to provide a bare scaffolding.

In order to understand the kinds of questions that have been raised on the commercialisation issue we must first note the specific context in terms of which they are cast. The point of reference is Marx's depiction of history as a progressive transformation from one mode of production into another—specifically, in this case, the transition from the feudal or pre-capitalist into the capitalist mode. Two observations are pertinent here: first that Marx analysed comprehensively and systematically the mode of competitive capitalist production and investigated its laws of motion as a theoretically abstract entity—though no doubt founded on what he observed to be the dominant historical characteristics of such a formation. While-sketching the movement of history through the succession of modes he did not sufficiently elaborate the other modes, nor was the *process* of transition adequately spelt out. The capitalist mode of production was, as it were, a benchmark in relation to which characteristic features of other modes were discussed in contrast. This is not surprising for the capitalist mode is not merely *analytically* the more developed, whether in Marxist or non-Marxist literature, it is also more easily amenable to analytical abstraction, whether with regard to its conceptual characterisation or to its 'laws of motion'. With the process of general commoditisation implying an interdependent production system and, at the level of exchange, the apparent rule of Bentham—Property and Equality—the 'rules of the game' are more conveniently definable and generalizable, and hence also their interplay. In contrast the 'pre-capitalist' social formations have greatly varied from one situation to another and it cannot be decisively ascertained which of the specific features that Marx attributed to 'feudalism' should be seen as definitive and inelimin-able and which others must be allowed historical variability. The persistent debates on feudalism and the Asiatic mode of production in the context of India amply demonstrate this problem. Secondly Marx's descriptive hints on the process of transition were drawn basically from the experience of England. In that 'classic case' of capitalist development the process of commercialisation had, spreading to all spheres, acted as a dissolvent of feudal relations. In the fully articulated capitalist mode, the one Marx analysed fully, a number of developments are conceived to have fruitioned synchronously: generalised commodity production with 'labour power' becoming a commodity, realisation of surplus value in the form of profits, advancing productive forces mediated through the expanding circuit of commerce, the rise of industry drawing in the proletarianised labour of the countryside, while alongside, capitalist relations dominate agriculture as well. A number of features of this particular transition were underlined by Marx as characterising growing commercialisation: for example commutation of rent in terms of money, the displacement of crop-sharing tenancy by cash rents, a larger degree of monetisation of outputs and inputs (i.e. of the proportion of inputs and outputs purchased or sold), increased area under cash-crops or a larger proportion in output of cash-crops, the rapid increase in the number of landless labourers, etc. These were taken to reflect the dissolution of feudal forms and relations and simultaneously connote a definitive advance towards capitalist relations.

A problem of interpretation arises, therefore, in this regard. These characteristics have to be seen in terms of the specific historical *process* that Marx had in view. When severed from that process it is not clear that these isolated characteristics could be looked upon as indisputably *reflecting* a transition towards capitalist relations. A high degree of monetisation, we shall see, may indicate 'compulsive involvements' in the market without necessarily implying prevalence of *capitalist* commodity production for exchange. Similarly, the shift from crop-sharing to fixed rents or from kind wage to money wage may not necessarily reflect the establishment of capitalist relations. It is with these warnings that we look at certain aspects of the commercialisation process in India.

II

A question that has recurred in various forms in the debates on commercialisation is why the process of commercialisation has not culminated necessarily or rapidly in the pervasive dominance of capitalist relations in India, as in the 'classic case'. This question has been entwined with related controversies concerning the characterisation of the mode of production in agriculture. To my mind neither this debate, looking upon 'the mode of production *in agriculture*' statically and sectorally, nor the solution sometimes suggested of treating the coexistence of multiple modes of production in agriculture sufficiently focuses upon the *process* or dynamics of change. It is by now clear that there have been considerable variations in production and exchange relations over time and among regions, so that a reference to the pure categories of modes of production is not adequate to grasp the processes at work.

In analysing the commercialisation experience of India we may begin with the experience of the nineteenth century when commercialisation was launched and accelerated in India and on which considerable historical documentation is now available. Common to this literature are the following features:

(1) Land was made, by civil law, an alienable individual property by the colonial state. This introduction of the rule of property was considered by the British colonial government as a policy measure which would instil a spirit of capitalist entrepreneurship in agriculture.[2] (2) There was already a marked differentiation in the peasantry. (3) The imposition of a heavy revenue demand to be paid in cash and before the harvest resulted in an imperative need to raise cash. (4) The *timing* of the revenue payments coincided with a period in the productive process when the cultivators were hard of cash. The issue of *timing* of cash obligations has rightly been underlined by historians[3] as it meant unavoidable distress-borrowing and the consequential lapse into debt-traps. (5) Whereas in the case of certain cash crops like indigo and poppy direct compulsion was used to propagate cultivation, commercialisation was 'forced' upon certain sections of the cultivators in the case of other crops like jute, sugarcane and oilseeds, through seemingly free exchange (see below). (6) Exports and internal markets for cash crops developed and were made more accessible by transport and communication networks developed by the state. (7) Irrigation works were carried out by the state and other incentive inputs provided to stimulate the growth of more productive commercial crops. (8) With land becoming pledgeable property, moneylending against land mortgages became common. AH these developments did induce a remarkable increase in the degree of monetisation and led to a rising share of cash crops. The question is why this did not lead to the eventual dissolution of feudal relations and towards the proliferation of capitalist relations. Why was the pace so tardy and why in some cases did it proceed in the contrary direction—strengthening debt bondage and a servile dependence of small cultivators on moneylenders, merchants and landlords? We may try to look conceptually at the production and exchange processes to see broadly the dynamics of these relations.

The analytical framework that we outline below may be generalised to situations other than the one we have at hand, namely the nineteenth century experience. With certain modifications the framework can be adapted for the contemporary Indian situation, at least in some regions. We begin with the presence of a considerable differentiation within the peasantry. We may, roughly speaking, divide,households into categories, depending upon their access to resources (land, basically) and the level of surplus they produce on an average. Here I will not go into the intricate subtleties of the bases of categorisation wherein a number of region-specific and period-specific considerations would enter. The categories could be:

(a) Very small operators and the landless; usually chronically deficit households.
(b) Small operators who produce just enough for subsistence, i.e. they can just about even out surpluses and deficits over good and bad years.
(c) Medium operators who have a sizeable enough surplus and respond 'competitively' to market stimuli, for example by responding in terms of changes in cropping patterns to variations in relative prices.
(d) Large or dominant operators: these are households with substantial surpluses which also dominate markets by setting the terms and conditions for exchange.

We further suggest the following propositions— (i) the exchange processes are neither uniform nor equal for all participants. Not only do the quantitative terms and conditions vary, depending upon parties to the exchange, but there can be *qualitative* differences in types of exchanges and the market involvement of individual households, (ii) The exchanges are set not only in terms of 'prices' but there can be non-price factors, explicit and/or implicit, which mainly rely on personal dominance and power equations, (iii) The nature of exchange involvement as well as the terms and conditions depend largely upon the position of the participating household within the resource-status categories given above. There is thus a correspondence between the production status as a base and the concomitant exchange relations.

The exchange relations may be described in terms of a hierarchical structure. In any market—we consider here output, land, labour and credit markets as the prominent ones-there are dominant parties mostly belonging to the substantial surplus households who set the pattern as well as the terms and conditions of exchange. While the medium operators are basically price-takers and quantity-adjusters' in the sense that they have a safe enough margin of surplus to play the market game to maximise, profits, the 'chronically deficit' and 'subsistence' households are involved in exchange compulsively. The very small cultivators and the landless not possessing adequate resources for a reliable livelihood, have to enter the labour or credit market to incur consumption loans or obtain advances for circulating capital. Under the pressure of immanent cash needs they nave to sell outputs, sometimes under prior commitments, even when they have to repurchase the same under more adverse terms for their own subsistence. While the position of the chronically deficit households is one of near destitution, often encumbered with a perpetual debt burden, the subsistence households may be able to precariously defend themselves from market vagaries. But their's is also a critical marginality. Further, another predominant phenomenon characterising exchange that has received prominent attention from economists and has been noted descriptively by a number of historians, is what has been called 'interlinked markets'. A dominant party conjointly exploits the weaker parties in two or more markets by interlinking the terms of contracts. For example the land-labour interlinkage prevails quite widely, wherein the landlord stipulates, as a part of the tenancy contract, attachment of labour services which are underpaid or unpaid. The tenurial conditions also differ, depending upon the relative status of the landlord and tenant.

The land-credit linkage is present when a landlord also operates as a creditor and stipulates conditions regarding the type of crop to be grown or hypothecates part of the produce over and above rent.[5] The moneylender merchants played an extremely important role in the sugarcane and indigo belts where credit was attached to particular crops to be grown and sold to the merchant at pre-fixed prices, much lower than market prices. The continuous need to raise cash to settle revenue and other obligations compelled small peasants to accept credit on such onerous terms. Such interlinkages in markets have grave consequences on the dynamic outcome of commercialisation by shaping the exploitative process.

First, the weaker party in the exchange loses the option to participate freely in other markets due to its commitment in one. For example the tenant who has committed his labour as a part of the lease conditions cannot avail of opportunities to hire himself out at a higher wage in the open market when such opportunities arise. The producer of commercial crops cannot gain from the higher return on such crops as he loses the option of selling them on favourable terms. Unlike with pure usurious capital, the merchant-creditor intervenes directly in the production organisation of the debtor and sometimes the latter loses the option of producing any other crop that he may wish to. This closing of options also weakens the possibility of the indebted party recovering from a dependency situation.[6]

This situation also has an obverse aspect. The power of the dominant party to exploit in interlinked markets is much more than in markets taken separately. There are conventional limits to exploitation in any one market. For example the crop share is conventionally laid down, although here too variations occur. However, with interlinked markets exploitation can be spread over markets (such as intervening in an output market on the basis of a credit tie) and even across generations, when the labour of future generations is committed by the debtor or tenant. Morever, with options receding for the weaker party, the situation develops as one of dominant control over the entire livelihood of the weaker party. In proportion as the dominant party stretches the domain of exploitation, the weaker party's possibilities of redressal diminish. Commercialisation may imply then that while the exploitative element of the earlier relations is strengthened, the patronage element is eliminated gradually, as noted by Breman.[7] For example it has been observed that a shift to cash wage from kind wage has implied the renunciation of the traditional element of patronage whereby the employer provides minimum subsistence in kind. Similarly a shift to money contracts of fixed cash rent from share-cropping has also been accompanied by a shortening of leases and a quick turnover of tenants, or tenant-switching.[8]

What consequences does this structure of exchange relations have for the process of commercialisation? We have already noted the potential for greater pauperisation in certain compulsively involved sections of the peasantry. In fact historians working on commercialisation in the nineteenth century, when cash crops were introduced at an accelerated pace, have vividly described these consequences. The increasing commoditisation was for some sections of the peasantry a consequence of 'forced commercialisation'. In the classical case, production for exchange was a substantive transformation wherein the producers enhanced the possibilities of expanding their surplus through market participation, propelled by maximising net gains through exchange. In the case of compulsively involved peasantry, in essence, the producers were still concerned precariously with eking out their subsistence; their's was not a free choice always 'to commercialise', as in the case of poppy and indigo, and even if it were apparently a free choice, as in the case of other cash crops, it was an encumbered 'freedom'—for the consequences were greater dependence and servility in relation to the exploiter. Such forced commercialisation revealed itself in the fact that when the producer improved his economic position, thanks, say, to favourable prices, his market participation (i.e. share of output sold) would decline. More direct evidence was furnished by peasant struggles and revolts amongst those in dire circumstances— for example the indigo producers' rebellions.

Secondly commercialisation in output markets occurred first, stimulated by 'external' factors like transport links, pressures exerted by the colonial government and the world capitalist network to subsume peripheral countries, colonial revenue and trade policy, etc. However, this did not lead to commercialisation of other markets *pari passu*. Here an important 'external' but integrally related factor must be mentioned. As stated earlier, Marx's sketch of transition from feudalism to capitalism took the completed capitalist processes pervading all markets as a benchmark. His picture of the advance of capitalist relations towards ultimate dominance included a vigorous growth of industry along with the transformation of agriculture. In fact the two were symbiotically related. The dynamic advance of industry worked as a powerful pull factor, sucking labour from agriculture. This factor, thanks to the colonial situation, was not strong enough in India, where on the one hand supluses were being drained to the imperial country and on the other the production structure was being geared to the material requisites of a reproduction of the imperial industry. Simultaneously, with increasing impoverishment of the countryside and the colonial policy of pouring in cheap British manufactures, a process of deindustrialization was afoot, throwing artisans and handicraftsmen out of their customary occupations. The absence of a synchronous growth of industry contributed to the uneven level and pace of commercialisation in other (land, labour, credit) markets and obstructed even their full formation.

Further, the underdevelopment or muted formation of a capitalist market in one market inhibits the formation and growth of a capitalist market in another. This is a reverse gear effect, contrary to the proliferation of capitalist tendencies in all markets in the classical case. For example, as well shall see, the under-formation of capitalist markets in land and labour goes together, mutually reinforcing the negative tendencies.

Finally, the retardation of the growth of capitalist markets and the perpetuation of traditional pre-capitalist forms affects, in a private investment economy, the *rate and pattern* of reinvestment of surplus and hence, in turn, the growth of surplus itself. With the predominance of usurious, speculative and mercantile forms of capital investment, which are encouraged by the prevalent poverty and economic dependency, investment goes into unproductive channels. Productive investment, meaning: investment that enhances output growth, is at a disadvantage.[9] Where possibilities of exploiting labour become almost limitless there is less incentive to improve productive forces, i.e. undertake productive investment.

While I have emphasised above a certain kind of dynamics of the process of commercialisation based particularly on the experience of UP and eastern India in the nineteenth century, the broad framework suggested above could be seen to generate another kind of dynamics in the situation of a different structure of production and exchange relations. The retrogressive working of exchange relations—with the underdevelopment or muted formation of capitalist relations in one market reinforcing similar tendencies in others, and the operation of interlinkages diverting surpluses into unproductive channels affecting the growth and reinvestment of surplus—arises from a preponderance of the categories of 'very small operators' and 'small operators' among the differentiated peasantry. These are the chronically deficient households which have to cling to agriculture as the only source of livelihood and whose survival derives from a chronic and cumulative indebtedness. As noted above, it is also the low level of productive investment in agriculture and the weak pull-force operating from industry that perpetuates the state of backwardness. In order to push the agricultural economy out of the rut, a critical minimum level and pace of accumulation would be needed. In certain pockets where this has been achieved, 'commercialisation' has paved the way for capitalist accumulation.[10]

III

We shall elaborate here upon the observations made above regarding the under-formation of capitalist markets and the mutually reinforcing tendencies produced by such laggard and uneven commercialisation.

The first market to be commercialised, as already noted, is usually the output market. Here too the process is often distorted, with commercialisation being 'forced' on certain sections of the peasantry. With hypothecation of produce and committed distress sales it is not surprising that the usual neoclassical theoretical results regarding price quantity responses do not hold, and 'paradoxical' results obtain. It is well known for example that a rise in foodgrain prices does not affect favourably all the producers; and those who are compulsively involved in output markets and end up as net purchasers of grains are in fact affected adversely.

Further, the presence of a large mass of compulsively involved peasantry who, given the lack of alternative opportunities to raise their livelihood, are compelled to hold on to their meagre parcels of land, implies that neither labour nor land markets can form fully capitalistically.

The 'Land Market'

Let us take the land market. It has been observed by a number of historians that despite land becoming an alienable asset by law, a market involving transactions in rights to levy and appropriate a part of the revenue proceeds developed rapidly—rather than a market in land-owning rights. The enormous growth of revenue farming and subinfeudation in colonial times, following upon the legalisation of individual property in land, is well known. No doubt land transfers did occur, but these were a culmination of a protracted period of continued distress.[11] As Benoy Chowdhury observes, the market in land assets was limited to a small group and it was common to allow the debtor to continue as a share cropper on the pledged property.[12] Even with regard to transactions in revenue-appropriating rights, the extent of revenue-share appropriated rested much more on the power of dominance of the party than on any principle of competition. It has been noted that even class struggles in the agrarian economy of the time occurred mostly over the apportionment of revenue sharing and revenue levies rather than over the issue of landowning rights.[13]

Secondly, the relation between land value and actual rents paid is not deducible on any capitalist principle. Not only do rents (and rent systems) differ from party to party but the poorer tenants desperately cling to land as the only means of survival, surviving through debt-rolling or by binding future generations to labour. They pay *effectively* a rent which, if capitalised, could be many times the market value of the land.

This however does not contradict the observation that in general the market value of land is much higher than the capitalised value of rents which are nominally reported or officially documented as paid, on the average. First, the *real* rent extracted from the impoverished tenant is much more than the nominally collected, as actual rent paid assumes various implicit forms. Second, the land market is extremely imperfect and the pauperised, chronicatly-deficit tenant households which implicitly pay high rents are precisely the ones which are incapable of entering into the land market. It is thus quite likely that their capitalised value of rents exceeds the market value of land. Similarly, the petty owners would continue to hold on to their small parcels of land even when their net income accruing from land is meagre; for the alternative they face in the event of loss of their land is the precarious state of being landless labourers or helpless tenants. This state of affairs, while perpetuating the self-exploitation of family labour on small farms, also inhibits tendencies towards a concentration of landed property on the one hand and the emergence of capitalist wage labour on the other.

Further, rents on land have very little to do with the productivity of land. The lease terms depend upon the relative position of the two parties. During the colonial administration attempts were made to apply the Malthusian principle of 'differential rent. This rent determination was according to capitalist principle, Malthus having himself thought of land as comparable to machinery. That this attempt was frustrated in practice need not surprise us.[14]

While pauperisation advanced at a merciless pace, actual transfers of land did not occur commensurately. The usual expectation that with enhancing commercialisation land would be concentrated under capitalist farming and the proletarianised wage labour would be set free to be employed in agriculture and industry has not to this date been found to occur in any striking manner. Some peculiarities of landed property have to be borne in mind. First, unlike capital which is reproducible and can expand through accumulation, land is physically limited, a fixed resource. As Kautsky puts it succinctly:

> In agriculture the big landowner cannot generally increase his wealth except through centralisation, reuniting several holdings into one. In industry accumulation proceeds independently of centralisation, a big capital can form without suppressing the autonomy of the lesser enterprises. When this suppression occurs it is the *effect* of the formation of big industrial capital. Accumulation is here the starting point. On the contrary, where the land is fragmented into different properties, and where small ownership prevails, large holdings can only acquire land by centralising several smaller ones. The disappearance of the small holdings is thus the precondition for the formation of a large enterprise. But this is not enough; it is also necessary that the holdings that are expropriated should form a continuous surface. Thus the mortgage banks which in any given year acquire hundreds of small properties cannot for all that concentrate them into one big one.[15]

Even with conditions of pauperisation the attachment to land has been preserved for lack of alternative opportunities for employment. During the worst of agricultural depressions, when migration has taken place to cities, land ties are yet sought to be preserved. It is only when the depressionary state has been prolonged and the economic conditions of the petty producers has continuously declined that land transfers have occurred in sizeable magnitude. Secondly, even when land transfers occurred, land continued to be leased out in small parcels: often the indebted owner turned into a share-cropper. In recent years the policy of the state to support small operators and marginalised holdings, as also the tardy development of towns (which provide supplementary employment), has tended to stabilise the small peasantry. Land continues to remain a valued support of livelihood in the absence of any vigorous growth of alternative employment opportunities.

The 'Labour Market'

Related to this is the muted development of the wage-labour market. With the preponderance of small operators who have to depend on hiring themselves out to supplement incomes and have themselves to hire-in labour at times, the labour market shows a number of peculiarities. The non-availability of assured and continuous employment induces petty producers to cling to their tiny holdings and cultivate them intensively with the help of family labour. This is reflected in the high intensity of cultivation, relatively higher land productivity but lower labour productivity — on small rather than on large farms — and intense self-exploitation of family labour. It has also been noticed that the total (on-farm and off-farm) employment of petty operators is less than what wage labourers can sometimes obtain, and the average return per unit of time for the former may also be smaller than the casual worker's wage rate. However, the risk of not securing employment is so great with the lack of a developed wage-labour market that there is a preference to own/lease-in a parcel of land, however small.

The fact, however, that a large proportion of labour is committed to working on small farms also constrains the development of a wage-labour market.[16] Thus there appear seasonal shortages of labour which provide the rationale for attaching labour in the form of permanent servants or hypothecating labour services through interlinked contracts of land-labour or credit-labour, as discused above.

Further, a number of researchers have also observed the wide degree of segmentation in the labour market—operations-wise, sex-wise and caste-wise—and the differing terms and conditions of employment. This segmentation implies lack of mobility and renders limited the role of market forces.[17]

The 'Credit Market'

In any region a number of credit systems have been observed to prevail which have, qualitatively, very different terms attached. The monetary rate of interest does not play any formally significant role either in rendering comparable different credit systems or in posing them as options among which the debtor, party can make a choice. Nor is there a uniform rate of interest operative within any one credit system for all borrowers. The loan conditions cannot be reduced in simple terms of 'interest payable' for, in interlinked exploitation, part of the appropriation is in terms of hypothecation of produce or labour services. In fact it has sometimes been noticed that no formal interest charge is made on loans, and this has often been posed as a paradox. Historically it is known that petty borrowings have trapped the debtor into bondage over generations.[18]

The various credit systems reflect both the prevalent modes of exploitation and are relatively more severe on small borrowings. The implicit rate of interest on grain loans can be extremely heavy. Furthermore, it is the timing of the loan that also settles the terms of the loans. There are periods in the production cycle when the petty peasant needs cash desperately, either to satisfy revenue obligations or to meet the minimum requirements of subsistence or to provide circulating capital at crucial junctures in productive operations. Such loans are made particularly stringent. With very little security to offer against credit, the already indebted peasant has scarcely any option left to choose the source of credit. The situation is often one of debt-rolling, borrowing further from the same creditor to make interest payments.

The greater the predominance of the chronically deficit households the greater the diversion of reinvestible resources into usury. Not only does this offer a fertile ground for exploitation but its relative attraction over productive investment persists so long as the backward production conditions continue. Thus the notion of an expanding spectrum of avenues for investment opportunities does not apply to the decisions of the creditor, while for the debtor the ineluctable need for cash and limited access to credit reduces his options, whether with regard to the timing of the loan, the terms of credit, or even the creditor from whom to borrow. There is thus need for caution in applying capitalist logic to the market for loanable funds.

IV

We have discussed above the muted formation of markets and the way the underformation in one market reinforces similar tendencies in others, mainly with reference to the early nineteenth century commercialisation experience. However the analysis is generalizable, and although there have now emerged wide regional variations there are backward regions where the picture would still hold with only slight modifications. For example despite historical differences some similarities may be noted between that early commercialisation experience and the ongoing experiments at 'modernisation' through the launching of advanced agricultural technology (like the introduction of HYV seeds or of high-breed milch cattle). A common element is the attempt to introduce technological or commercial improvements through state policies, using primarily incentives/disincentives on the prevalent exchange network: for example making available cheaper inputs, price subsidies or supports. To the extent that market involvements of different sections of the peasantry are qualitatively different and the exchange relations are not uniform, these operations in the extant markets induce inequitous and disparate results for different sections of the peasantry; such operations also often conflict with the initial objectives of the policy.

Take the instance of a recent scheme for promoting the rearing of pedigree buffaloes for milk production. The explicit purpose was to assist small and marginal farmers by providing them with an ancillary source of income. It was observed in a field survey in UP that, first, there were present the ubiquitous corrupt practices whereby the well-to-do farmers surreptitiously availed themselves of the credit and other concessions through 'benami' deals. Second, some of the poorer farmers faced eventual consequences at variance with the objectives of the scheme. These cultivators who were earlier producing subsistence foodgrains shifted to producing fodder crops as the expensive breed of buffaloes required good quality fodder. Typically, this involved them in consumption loans borrowed from a merchant who stock a deal to purchase milk at prices stipulated by him as a part of the loan contract. Such or similar situations appear to have landed the producer in the familiar debt- trap, consequently enforcing on him the eventual sale of the buffalo or the mortgage of his land Parallel insurances may be cited in cases where, for example, under land reform treasures, small parcels of land were given to some of the landless. Not having enough circulating capital or means to subsist and carry cat production, the petty owners had soon to raise credit. mortgaging the land and eventually losing its possession.

The question at issue is not the existence at any moment of certain pre-capitalist features that have been outlined. Indeed such features did characterise the early European experience and there was even a relapse into the second.[19] Within India such features might have existed in almost all parts, including those pockets where today capitalist relations seem to be gradually acquiring dominance, as in parts of Punjab and Haryana. The question is one of the *process* of transformation of the dissolution of feudal features and the onset of a pervasive dominance et capitalist relations. I have briefly indicated the forces that work towards retarding and muting such a dynamic outcome. They can be seen, in summary, as a struggle between the forces promoting productive accumulation and those stabilising perpetuating and reinforcing unproductive investment. The balance of forces may turn one way or the other. If the forces of productive accumulation, nurtured by a set of favourable initial conditions generate a critical minimum pace, they could succeed in breaking through the tetters of the pre-capitalist remnants.

Finally the transition in agriculture cannot be discussed in isolation from developments in industry and in the overall macro-economy The formation of capitalist markets in agriculture is connected indissolubly with the advance of industry, which acts as a pull force The relationship is no doubt reciprocal: industrialisation cannot proceed at a smooth accelerated pace without the availability of adequate and stable supplies of agricultural surpluses and without the generation of an expanding effective demand for us produce.[20]

V

What are the broad lessons for policy that may be drawn from these experiences? It appears that not only does intervention through existing market channels at times reinforce the iniquitous impact of policies but intervention in *one* market produces chain reactions that frustrate the original design or, far worse, upset the precarious initial production balances. 'Commercialising' or 'modernising' one segment in the complex web of production and exchange relations, if not followed by commensurate supportive changes in others, can frustrate the policy objectives of such interventions. The policy design has to take account of the complex of activities—'the mode of livelihood'—rather than treat facets of production activity partially.

Second, the critical conditionality of backwardness appears to be, in this situation of desperate dependence, on land as the only means of livelihood for a large section of the population. It is this absence of an alternative, reliable and permanent source of livelihood that dictates its desperate clinging to land for barest survival, while entrapped in debt. Even when a succession of bad harvests and protracted distress drives members of a family to move into cities and towns in search of employment, the incomes so earned are primarily used to supplement the family's ability to retain the land (or the possibility of cultivating it). This land-lockedness has its own effects, as seen above, on the development of markets and exchange relations, and also on the pace of accumulation. In the absence of a powerful force of industrialisation lifting labour from land, state-sponsored relief schemes and tardy growth of industry succeeds only in stabilising the marginal households in agriculture. Unless an attempt is made to generate secure and regular alternative means of livelihood a breakthrough will not be possible, particularly in regions characterised by the numerical preponderance of marginal households. Diversification of employment has been looked upon in the literature on economic development as a reflection and index of development. What would be needed in situations such as discussed here is a purposive policy *directly* aiming at a *sustainable* and *productive* diversification of employment.

NOTES

1. See for example Whitcombe (1971), Benoy Chowdhury (1964, 1967 & 1970), A. Siddiqi (1973) and Shahid Amin (1982).
2. See Guha (1963) and Whitcombe (1971).
3. See Whitcombe (1971) and Amin (1982).
4. See Bharadwaj (1974) and Bharadwaj & Das (1976).
5. See Bhaduri (1973).
6. See B. B. Chowdhury (1967).
7. See Breman (1974).
8. See Bharadwaj (1974) and Bharadwaj & Das (1976).
9. See Bhaduri (1981), and his essay in this volume.
10. See Bharadwaj (1980).
11. See Pradipta Choudhury's essay in this volume.
12. See Benoy Chowdhury (1967).
13. See Shelvankar (1969).
14. See Benoy Chowdhury (1967) for a discussion of this point.
15. See Kautsky (1970).
16. See Bharadwaj (1974) for a discussion of the above points.
17. See Neeladri Bhattacharya's essay in this volume. Also see Bharadwaj (1979).
18. See Breman (1974).
19. See Dobb (1946).
20. See Bharadwaj (1980).

REFERENCES

Amin, Shahid: 'Small Peasant Commodity Production and Rural Indebtedness: The Culture of Sugarcane in Eastern UP, *c.* 1880-1920,' in *Subaltern Studies I,* ed. Ranajit Guha (New Delhi, 1982).

Bhaduri, A.: 'A Study in Agricultural Backwardness under Semi-Feudalism,' *Economic Journal,* March 1973.

———: 'The Evolution of Land Relations in Eastern India under British Rule,' *Indian Economic out Social History Review,* March 1976.

——: 'Class Relations and the Pattern of Accumulation in an Agrarian Economy,'

Cambridge Journal of Economics, March 1981.

Bharadwaj, K.: *Production Conditions in Indian Agriculture* (Cambridge, 1974).

—— & P. K. Das: 'Tenurial Conditions and the Mode of Exploitation in Some Villages of Orissa,' *Economic and Political Weekly,* July 1975.

———: 'Technical Relations in Agriculture,' in *Agricultural Development of India: Policy and Problems,* ed. C. H. Shah and C. N. Vakil (Orient Longman, 1979), pp. 265-90.

———: 'Reflections on Regional Differentiation in India,' *Economic & Political Weekly,* 1980.

Bhattacharya, N.: 'Agricultural Labour and Production' (in this volume).

Breman J.: *Patronage and Exploitation: Changing Agrarian Relations in South Gujarat* (California, 1974).

Chowdhury, B. B.: *Growth of Commercial Agriculture in Bengal* (Calcutta, 1964).

——: 'Agrarian Relations in Bengal (1859-1885)', in *History of Bengal (1757-1905),* ed. N. K. Sinha (Calcutta, 1967).

Dobb, M.: *Studies in the Development of Capitalism* (Routledge 4 Kegan Paul, 1946).

Guha, R.: *A Rule of Properly for Bengal* (Paris, 1963).

Kautsky, K.: 'La Question Agraire,' *Maspero,* 1970; translated and summarised by Jairus Bamji in *The Articulation of Modes of Production,* ed. H. Wolpe (Routledge & Kegan Paul, 1980).

Shelvankar, K. S.: 'The Problem of India,' reproduced in *Rural Sociology,* ed. A. R. Desai (New Delhi, 1969).

Siddiqi, Asiya: Agrarian *Change in a Northern Indian State: Uttar Pradesh, 1819-1833* (Oxford, 1973).

Whitcombe, E.: *Agrarian Condition in Northern India Vol. 1: The United Provinces Under British Rule, 1860-1900* (California, 1971).

34

FORESTRY AND FOOD PRODUCTION

"The well being of a people is like a tree; agriculture is its root, manufacture and commerce are it stem and branches; if the root is injured the leaves fall, the branches break away and the tree dies". The wisdom of this old Chinese adage holds good even today. In fact, from times immemorial, trees have come to symbolise the pivotal role of integrated development of individuals, societies and nations.

The use of the tree to convey important messages is not merely symbolic but it illustrates the conviction held throughout human history that it is trees and forest canopies that sustain life in a healthy condition in our planet. The relationships between forest conservation and food production are many and intimate. Some of these are: (a) protection of watersheds; (b) conservation of soil and water; (c) prevention of sedimentation in rivers and reservoirs; (d) regulation of climate, such as, temperature and wind speed; (e) conservation of the flora and fauna occurring in forest canopies; (f) prevention of atmospheric pollution; and (g) provision of a wide range of minor forest produce.

In a recent paper prepared by FAO on the role of forestry in food security, it is stated that "land use practices that lead to environmental degradation are widely recognised in developing countries. It is also known that they must put the long term future of food supplies in jeopardy. Nevertheless, they have so far received little attention in international discussions of world food security". Even when the linkages between forestry and food production are recognised and discussed, little follow-up action is forthcoming. The situation in Sub-Saharan Africa typifies the paradox arising from the growing gap between the world of words and the world of action.

In Africa, there has been widespread awareness of the ecological crisis in the offing for decades. There has also been much analysis of the important environmental maladies and possible remedies. Yet the action taken to apply the remedies has been halting and inadequate.

It may perhaps be of interest to refer to the article entitled. "The Disappearance of Tropical Forests of Africa" by Andre Marie A. Aubreville published in FAO's journal of forestry and forest industries *"Unasylva"*. The article which was originally written in 1947 warns about the adverse consequences of forest regression to African agriculture. Yet the very purpose of vast forest clearings in many cases has been to promote food production and facilitate fuel supply. While virgin forests are not easily prone to fire, the stunted secondary scrub, the savannah and the brushland become sufficiently desiccated during the dry season to be inflammable. Dr. Aubreville has described the processes which lead to the forest getting gradually thinned and becoming grassland. In the course of centuries, savannahs tend to replace the forest and even the latter would have long since vanished had it not been for the exceptional vitality of some of its component species which manage to survive in spite of all adverse circumstances.

In a recent informative and incisive analysis of methods of *"Reversing Africa's Decline"*, Lester R. Brown and Edward C. Wolf of the World Watch Institute have pleaded for a "resource-based" development strategy involving soil conservation, restoration of forests and woodlands, improvement of subsistence agriculture and stabilisation of population growth.

FAO has also developed a detailed plan for the agricultural rehabilitation of drought ravaged Africa. It is clear that ecological and agricultural rehabilitation have to proceed together. Agriculture in its broad sense is *applied ecology* involving the management of the basic life support systems comprising land, water, flora, fauna and the atmosphere on a sustainable basis. On an occasion like this, it may be appropriate to remind ourselves of the vast human dimensions of some simple statistics. Since 1967, a year which marked the beginning of the Green Revolution era in Asia and Latin America, grain output per person in Sub-Saharan Africa has fallen by nearly a third. Income per person has fallen by a fifth since 1970. About 10 million villagers have left their homes in search of food, many of them crowded into hastily erected relief camps. Norman Myers refers to such unfortunate people as "ecological refugees" since they are forced to migrate due to a breakdown in the life support systems in their original habitats.

Ecological deterioration and economic decline have led to a cereal import bill which claimed some 20 per cent of total export earnings in 1984. The Continent's debt has reached $170 billion in 1985. FAO has estimated that 21 severely affected countries would need cereal imports of over 12 million tonnes in 1985.

Africa's population is projected to roughly triple by 2025, reaching 1.5 billion. Over 40 per cent of Africa's people now live in countries where grain yields are lower than they were a generation ago. The loss of tree cover both in closed forests and in savannah settings is extensive. A few countries, such as, Mauritania and Rwanda are badly affected. In many countries wood collection for fuel and other uses exceeds the sustainable yield of remaining accessible forests. In Kenya, demand is five times the sustainable level. In Ethiopia, Tanzania and Nigeria, demand is 2.5 times to sustainable yield. In Sudan, it is roughly double. Restoring African woodland and forests is thus essential to the restoration of the hydrological cycle and to the recovery of agriculture.

The dimensions of the tree plantation programme needed are vast. If fuelwood consumption continues to rise by 3 per cent a year, Africa will need an additional 20 million hectares of fuel wood plantations by the year 2000 to cater to the increased demand without further denudation of the remaining forests. In a similar calculation, which a Fuel Wood Committee of the Government of India made in 1982, it was concluded that a minimum of three million hectares will have to be planted under quick-yielding fuelwood trees each year up to 2000 AD to meet the growing needs. Since over 40 per cent of the total area is already under agriculture, the only way this can be done in India is the planting of all *wasted lands* (often misnamed as *"waste lands"*) with leguminous shrubs and trees which can provide to the people of the area fuel, fodder, feed and fertiliser through biological nitrogen fixation. A non-Governmental Society for the Promotion of Wasteland Development was organised in 1982 for this purpose and early this year, the Government of India set up a National Wasteland Development Board for the purpose of planting trees in five million hectares of wasted lands every year.

At the UN Conference on Desertification (UNCOD) held in 1977 in Nairobi, a Plan of Action to Combat Desertification by 2000 AD was developed. A recent assessment of the implementation of this plan of action by the United Nations Environment Programme (UNEP) has revealed that the goal set up by UNCOD to stop desertification by the year 2000 is no longer feasible. Efforts undertaken to control desertification since 1978 have been inadequate. The direct impact of desertification is on food producing lands. UNEP estimates the annual losses to be six million hectares transformed into desert land. Further, every year the productivity of 20 million hectares is reduced to no economic return.

It is not in Africa alone that we face a serious threat to long term *Nutrition Security* as a result of the disappearance of the tree cover and the consequent damage to land, water, flora and fauna. Food security has been defined by FAO as the "economic and physical access to food to all people at all times". Nutrition Security is a wider concept including the availability of food of adequate quantity and quality as well as safe drinking water (Swaminathan 1984). Lack of clean drinking water in many villages in developing countries is an important cause of poor health both in children and adults. In addition, some essential vitamins and minerals may have to be supplied, particularly to pre-school children. Here again, social forestry programmes can help to provide simple botanical remedies to some of the specific nutritional disorders of an area, through the incorporation of appropriate horticultural plants in the choice of the species to be planted.

Let us cite two other examples of what is happening in continents other than Africa. Over 40 million people now live in the Himalayas. Most of them are very poor in spite of the vast natural endowments of the Himalayas. Early in this century, it was estimated that the forest cover of the Himalayas amounted to about 60 per cent on land area. Some experts calculate that it has now dwindled to only one quarter as much. Nepal's forest cover is believed to have declined from 57 to 23 per cent between 1947 and 1980. In India, the Himalayan sector accounts for one quarter of the country's forest cover. Yet it is being so rapidly reduced that several parts of the Himalayas may become barren by the beginning of the next century, unless steps are taken to provide the people with water, food, fodder, fuel, fibre and fertiliser. The local population particularly women are very conscious of the threat to their future and this resulted in the launching of the "Chipko" (hug the tree) movement largely by women. Under this movement, women hug the trees when contractors come to fell them. A very large proportion of the cattle population is maintained by the people just for meeting the manurial requirements of crops. Because of heavy animal pressure, only plants which are non-edible survive. Even these are cut for fuel wood. Not only floods and droughts and soil erosion threaten the livelihood security of the people living in the mountains, but the future of the agriculture of the plains of North India is also in jeopardy. Rising water tables resulting in the spread of soil salinity and alkalinity and frequent incidence of flash floods in periods of heavy rainfall are causing misery to the agricultural population.

The greed of the rich, the genuine needs of the poor for fuel and fodder, careless technology and the absence of a systems approach in the design and implementation of development projects in ecologically fragile areas have all contributed to the rapid decline of Himalayan forests during the last 50 years. Let me cite one example of the harmful impact of a lack of systems approach to development. The rapid progress in the planting of apple orchards in the States of Himachal Pradesh and Jammu & Kashmir of India not only resulted in clearing forest lands for horticulture but took a heavy toll on forest trees since several million tons of apples and other fruits had to be packed in wooden boxes for being sent to the plains. Later calculations showed that to meet the wood requirements for packing apples from one hectare, about ten hectares of properly managed forests will be needed, if the wood is to be provided on a sustainable basis.

A long term threat to food production as a result of forest denudation comes from the coincident extinction of genetic resources of plants and animals along with forest canopies. The tropical moist forests are veritable mines of valuable genes. In several important food crops, genes for resistance to important pests and diseases come from wild species and primitive cultivars occurring in forest areas. Some of the tropical forests have been described as pharmaceutical factories, since they contain many medicinal plans and herbs. We cannot really estimate the harm that will be done to the future of agriculture, if we do not preserve for posterity the fruits of thousands of years of natural evolution and human selection.

Thanks to an initiative by the Mexican delegation to FAO, an international undertaking for the conservation and exchange of plant genetic resources as well as FAO Commission on Plant Genetic Resources have come into existence. These developments together with the activities of International Board for Plant Genetic Resources (IBPGR) provide a more organised framework for genetic conservation activities in economic plants.

Let us cite one more example of current trends in human settlements as a result of the growth in the population of ecological refugees. Western Amazonia has large wetlands, which are in several instances not suitable for conventional agriculture. Yet in those parts of the region which lie in Colombia, Peru and Ecuador, landless labour families are moving in. In Peruvian Amazonia, such population is expected to reach about five million by the turn of the century. Extensive damage to forest eco-systems is being caused by the slash and burn system of cultivation. At the same time, there is immense scope for promoting sustainable agriculture and thriving rural communities through scientific land and water use planning. To cite an example, Amazonia is very rich in fresh-water fish fauna. One calculation reveals that the introduction of aquaculture practices would help to generate far more good quality animal protein than 40,000 square kilometers of cattle ranches. Many more examples of this kind can be cited. What is clear is the urgent need for scientific land and water use practices based on sound principles of ecology, economics, energy conservation and employment generation. How can we initiate an *economic ecology movement* in the Third World where the symbiotic relationships among forestry, food production and human happiness will be nurtured and strengthened in every village? There are numerous suggestions put forth in many conferences, books and symposia on how we can foster such a movement. Here it is proposed to refer to only a few aspects which deserve greater attention.

These aspects have been discussed briefly here below:

1. Pre-requisites for a Scientific Reorientation of Land and Water Use

Agricultural advances in recent years have been triggered by the development of high-yielding breeds of crops and livestock. The breeding of strains possessing a broad spectrum of resistance to pests and diseases and to diverse soil stresses involves extensive hybridisation followed by rigorous selection. Coupled with good management, these breeds help to raise crop and animal productivity to high levels, as is clear from the agricultural scene in most developed countries and some developing countries.

The large scale cultivation of improved genetic strains together with good soil fertility and water management helps to increase production through higher yields per hectare. Land is a shrinking resource for agriculture. The opportunities now open for increasing agricultural production through a vertical pathway of higher yields per hectare and greater intensity of cropping through multiple and intercropping make rational land use planning possible. By taking advantage of such opportunities, lands prone to erosion and which are marginal to crop husbandry can be placed under silvi-horticultural or silvi-pastoral or fuel wood farming systems.

The FAO study on "Agriculture Toward 2000" estimates that even if crop yields on land already cultivated were to increase by 72 per cent, another 200 million hectares will have to be cleared in the next 20 years for meeting the food requirements of the estimated population of over six billion people at that time. Fortunately, the untapped yield reservoir in many farming systems in most developing countries is quite high. Hence,.every country will have to accord high priority to bridging the gap between potential and actual yields in farmers' fields by identifying and removing the constraints responsible for the yield gap.

Even if farmers in some parts of a country achieve high productivity levels, farmers living in remote areas and in mountainous or desert regions lacking roads and communication facilities, will still grow only annual food crops in lands which should ideally be under tree cover or tree farming, since meeting their home needs for food and other essential commodities determines their priorities in land use. If the farming families in such areas are to be persuaded to adopt ecologically sound land use practices, it will be essential that they are assured of the supply of the staple grains they need. Governments will have to build visible stocks of the foodgrains needed by the people living in remote and ecologically endangered habitats so that the local population will have confidence that their food needs will be met. Unless such a confidence is built up, it will be futile to preach to them the harmful effects of shifting cultivation, soil erosion and loss of tree cover.

Where governments have built their own national food reserves, they should build grain stores in all areas where there is need for shifting the land use from annual to perennial crops or to agro-forestry and mixed farming systems. Before the normal sowing season, every family should be assured of the adequate supply of their staple. Where governments do not have such stocks of their own, a special *"Food for Scientific Land Use"* or *"Food for Forest Development"* or any appropriately named programme should be initiated by national governments with appropriate assistance from the international community. The World Food Programme, for example, can provide foodgrains for a programme designed strictly for bringing about desirable changes in land use, in addition to its regular programmes based on humanitarian considerations. In global terms there is abundance of food stocks today. According to FAO, these stocks are expected to reach nearly 300 million tonnes by the end of 1985. A beginning of the Food for Forest Development Programme can be made if all nations having these stocks will pledge five per cent of their stocks for such a programme. Depending on needs and circumstances, some of this foodgrain stock can go to Food for Work Programmes in the forestry sector.

In all chronically drought prone areas there is need for a *Good Weather Code,* indicating the steps that should be taken in the occasional seasons when rainfall is good. This will call for anticipatory planning and advanced preparations for extensive tree planting, if soil moisture levels prove adequate for seedling establishment. Normally both national and international funds are scarce when the rainfall is good. However, it is precisely during such seasons that much valuable work on afforestation and anti-erosion measures can be undertaken. Thus we see today a paradox. When there is drought, food and funds become readily available. But when the soil moisture conditions for tree and grass establishment are excellent, no special support becomes available. A *Good Weather Code* indicating the steps that should be taken in normal rainfall seasons will hence help to optimise the benefits from good rainfall for eco-development.

2. Harnessing New Tools of Science and Technology

We need to step up basic and applied research relating to genetic conservation, breeding better strains of locally valuable tree species and quickgrowing leguminous shrubs and trees capable of providing fuel, fodder, feed and fertiliser. Biomass utilisation is another area of research which needs great stress in developing countries. Improved post-harvest technology including recent advances in wood processing technology will help to generate additional opportunities for gainful employment in rural areas, in addition to preventing the wasteful use of wood.

In spite of all our concern and efforts, tropical rain forests as well as the tree species of arid and semi-arid areas are disappearing at alarming rates. Since cloned dioxy nitrose nucleic acid (DNA), which is the chemical substance of heredity and material having DNA in its native state can also be used for genetic conservation, we should initiate in all threatened forest eco-systems, an integrated genetic conservation strategy consisting of the maintenance of populations, individual genotypes and DNA libraries. DNA libraries could be constructed immediately at the genomic level in the case of all tree species whose populations is dwindling.

It is in the intelligent integration of classical research tools with the new tools of biotechnology including tissue culture and genetic engineering and other recent techniques resulting from advances in micro-electronics, aerial seeding, satellite imagery, computer sciences and information technology that the future acceleration in the pace of agricultural advance will depend. Developing countries with numerous small, poor farms stand to gain from advances in biotechnology that offer the hope for increasing productivity, profitability, stability and sustainability of farming systems, without adding to the cost of production.

It is obvious that in addition to biological engineering, we need a major effort in social engineering through education as also through incentives for group cooperation and action in areas like tree planting and protection, watershed management and scientific land use planning. Social fencing is the only feasible method of protecting large areas which are being reforested now in many developing countries.

This has drawn up a priority agenda for research and development action. How can all this be done? We need a large army of trained men and women who are dedicated to the cause of sustainable development and who are willing to spare a few years for working for a better future for the children as yet unborn. The organisation of an International Eco-development Corps will be of great help in marrying intellect with labour or brain and brawn, a process which Mahatma Gandhi considered to be essential of rural regeneration. This would like to indicate the broad outline of a programme for initiating such an International Eco-development Corps.

(a) Preparation of Action Plans

The first step is the preparation in each country of a National Conservation for Development Programme consisting of detailed action plans at the local level for eco-development. Where such plans exist or can be prepared soon, the kinds of professional expertise needed for implementing the plan could be articulated in fairly precise terms. Where they do not exist, one of the early tasks of the Members of the Eco-development Corps will be the preparation of detailed action plans in consultation with the local population and authorities. Thus, the corps could help in the preparation of eco-development plans as well as in the conversion of plans into accomplishments.

(b) Organisation of a Global Grid of Backup Institutions

A very important pre-requisite for the successful implementation of this programme is the availability of a global grid of outstanding support institutions. Depending on the nature of the job to be done, as for example, anti-desertification measures, production of food, fodder, fuel and fertiliser (through biological or organic sources), control of animal and human diseases, irrigation and drinking water supply, biomass utilisation, improved management, etc., the support of advanced institutions located in developed and developing countries should be enlisted for (a) providing technical help when needed throughout the duration of the project and (b) training of members of the corps. The back-up institutions will serve as an umbilical cord supporting the project until the work reaches a self-reliant and self-propelling state. The organisation of such a consortium of scientific and technical institutions for supporting the Eco-development Corps in a meaningful manner will not only help in harnessing the best available know-how, but will also generate a sense of participation among large numbers of professional institutions and experts in a programme of great human significance.

(c) Selection and Deployment of Members of Corps

This is the key element of this programme. Once the precise tasks to be performed under the National Eco-development plan are articulated, the nature of the expertise needed will be clear. There has to be a proper match between the nature of the expertise and skills needed for successful task implementation and the nature of the skills possessed by the candidate. Interest in *do-how* is more important than just interest in *know-how.*

Once the candidates have been chosen carefully according to the needs of each action plan, they should be given suitable predeployment training and orientation in appropriate institutions belonging to the back-up consortium as well as in a suitable institution in the country where they are to work.

Deployment of members of the corps will be in clusters. For example, for a project aiming at agricultural and ecological rehabilitation, there will be need for a group consisting of an agronomist, a forester, a veterinary expert, and a social scientist. In addition, every cluster should preferably have a medical graduate who can attend to human health problems. Whenever the members of the corps are unable to find solution to some of the field problems (such as new soil health and plant health problems), they should seek the assistance of the appropriate back-up institution. In the initial stages, members of the corps could help in optimising the benefits from the Food for Work programmes.

(d) Duration of the Project

In order to achieve some tangible results, this Programme should be planned until 2000 AD. Obviously, many of the young professionals agreeing to give a part of their early life for this emotionally satisfying and intellectually challenging work, may not be prepared to stay for more than three years. Each person who is leaving the project should be replaced with a person with similar expertise so that the continuity of the work is maintained. In work designed to achieve ecological rehabilitation, a 15-year period is the minimum before visible and lasting impact is achieved.

(e) Remuneration

The monthly honorarium paid to the volunteers should not exceed the amount which a national doing similar work may receive in his/her country. However, suitable dormitory and other arrangements which will help to provide free lodging and boarding and recreational facilities will be needed.

(f) Tasks to be Performed

Each country participating in this programme should have an integrated ecological and agricultural rehabilitation plan. Such a plan should consist of a portfolio of well defined tasks. Members of the Eco-development Corps will be assigned specific tasks under such an integrated development plan. Funds for executing the tasks should be available from national, bilateral and multilateral sources.

REFERENCES

1. Aubreville, Andre M.A. 1985. *The Disappearance of the Tropical Forests of Africa.* Unasylva, 37 (148): 18-27.
2. Brown, L.R. and E.G. Wolf. 1985. Reversing Africa's decline. *World Watch Paper* 65 June, 1985, 81 pp.
3. Swaminathan, M.S. 1984. Nutrition and Agricultural Development: New Frontiers, *Food and Nutrition,* 10(1) : 33-41.

35

AGRO-FORESTRY: ITS OBJECTIVES, TECHNIQUES AND IMPLEMENTATION

Social forestry pertains to those areas of forests which are man-made. Agro-forestry is conspicuously an important part of social forestry. A new scientific land management is adopted in Agro-forestry and in it the dual system of production, i.e., production of forest crops and food crops, fodders or medicinal plants, becomes possible. It meets simultaneously, at least two requirements of the participating persons. In the words of king and Chandler (1978) Agro-Forestry is "a sustainable land management system which increases the overall yield of land, combined with the production of crops (including tree crops) and forest plants and animals simultaneously or sequentially on the same unit of land and applies management practices that are compatible with the cultural practices of the local population". Thus, in Agro-Forestry, co-existence of farm and forestry is adopted on a scientific basis and consequently, the total yield of land is raised significantly.

Conventionally, there had been separation between agriculture and silviculture. Needless to say, this separation has never been complete. From time immemorial to a limited extent, a combination of food crops and forest crops had been adopted in land management by the farmers throughout the world. In the present time, the steep rise in the demand for fuelwood and food because of the rapid growth of global population, presents an urgent necessity to adopt a scientific approach on a large scale to the dual system of production or to the co-existence of forestry and farm. Greater attention is now being paid to the adoption of Agro-forestry throughout the world. The setting up of the International Council for Research in Agro-Forestry by the Food and Agricultural Organisation (FAO) in 1878 is a landmark in the developmental history of agro-forestry. The traditional system of land management consisting of a dichotomy between forestry and agriculture is now being replaced by the dual system of production. Now, on a scientific basis, the cultivation of forest crops and food crops or fodders is being adopted by the farmers.

The main objectives of agro-forestry programme in India are as follows:

(1) Its objective is to raise the supply of fuelwood in the rural areas at a convenient distance for consumers and thus meet the growing requirements of fuelwood of the society. In India, it has been estimated by the experts that the rural population use dried dung approximately 70 million tonnes as a result of which the supply of organic (natural) manures for agriculture is reduced tremendously. If by the promotion of agro-forestry, the supply of fuel-wood in rural areas is raised adequately, the misuse of cow-dung will be automatically checked. Moreover, the undue pressures put on traditional forests for obtaining fuelwood by the inhabitants of adjoining villages will be reduced to a large extent.

(2) Agro-forestry aims to raise the supply of small timbers used by the villagers for agricultural implements, house construction and some other domestic purposes. In this way, Agro-forestry can meet these requirements of the rural population. It will also reduce the pressures put on traditional forests to meet these requirements.

(3) One of the main objectives of Agro-forestry is to raise the production of food crops, legumes and tubers to meet the rapidly growing food requirements of the Indian population.

(4) Agro-forestry aims to promote the production of fruits, vegetables, pulses, milk, and meat and thus it can raise the nutritional value of food, which is urgently required for mankind. In our country an average Indian gets 2000 calories per day in food, the minimum requirements is of 3000 calories. Thus, by increasing the supply of fruits etc., the deficiency in the nutritional value of our food will be reduced to a considerable extent.

(5) Agro-forestry programme is labour-intensive. The objective of Agro-forestry programme is to create new employment opportunities among the agricultural labour and small and marginal cultivators. Their income may be largely raised by this programme and thus it can help in the formation of a harmonious social system in rural areas.

(6) Agro-forestry programme helps in obtaining an ecological balance in rural areas and thus it may be considered a matter of great significance for a country like India.

(7) This programme aims at helping in the preservation of humidity in culturable land and check soil erosion. In this way, it may be helpful in raising the productivity land.

(8) In drought-prone areas Agro-forestry aims at reducing the insecurity of the agriculturists. It has been suggested by experts that in such areas, the dual system

of production of tree and grasses or legumes ensures to stability with productivity of land.

(9) The Agro-forestry programme aims at helping in making the Indian forest policy of 1952 successful. In the forest policy of 1952 it was declared that one-third area of the country should be covered by forests and in which 60 per cent areas of the plateau and 20 per cent of the plain land are to be covered by the forests. This policy cannot be successful without the rapid development of Agro-forestry programmes.

(10) India has nearly one-fourth of livestock population of the world. For the proper feeding of this livestock, increased supply of fodder is urgently required. Agro-forestry helps in augmenting the supply of fodder and thus in the country a larger supply of milk and meat as well as better services from serviceable categories of livestock will become available.

Shortly the agro-forestry programme always fulfils the objective to raise the total yield of land considerably in comparison to the traditional system of land management consisting of a dichotomy between forestry and agriculture, horticulture, animal husbandry.

Techniques of Agro-Forestry

Agro-forestry programme can be implemented in sub-marginal marginal and super-marginal lands which can support the dual system of production. In agro-forestry, scientific analysis of the soil seems absolutely essential. Due to the presence of several varieties of soil in India, the task becomes difficult. After analysing soils it can be ascertained which kind of combinations of forest crops and agricultural crops are suitable to a particular place. In India, several research Institutes are involved in refining technologies and systems of direct relevance to agro-forestry, particularly the following:

1. Several agricultural universities scattered throughout India.
2. Forestry Research Institute, and college at Dehra Dun.
3. Grassland and Fodder Research Institute (ICFRI) in Jhansi.
4. Central Arid Zone Research Institute (CAZRI) in Jodhpur.
5. Central Plantation Crops Research Institute (CPCRI).
6. Central Soil and Water Conservation Research and Training Institute, Dehra Dun.

Experiments are to be conducted to ascertain that forest crops and agricultural crops do not compete each other for nutrients in a particular land. Moreover, measures for soil conservation are to be taken in the areas used in agro-forestry.

On the basis of experiments certain Combinations of forest crops and food crops or fodders are adopted by farmers in different parts of the country as follows:

Sinha (1984) reported that sissoo raised in the plots with intercropping has given better growth than in the control. The height of sissoo in the case of oilseeds, particularly with groundnut and soyabean, is more when grown with the food crops. Thus, raising of cash crops accelerates the rate of growth of forest species.

Some of the important species selected by BAIF for incorporation in Agro-forestry models are:

(1) Leuceana leucocephala. (2) *Acacia nilotica.* (3) *Calliandra calothyrals.* (4) *Gliricpdia maculata.* (5) *Moringa pterigosperma.* (6) Sesbania *grandiflora.*

(1) No competition with field crops. (2) Fixation of nitrogen. (3) Easy decomposition of litter. (4) Fast growth, and easy establishment. (5) Ability to regenerate. (6) High yield of wood and fodder. (7) Multiple use of wood.

In Mandar (Bihar), Mishra and Prasad (1980) conducted experiments in the newly felled forest areas under rainfed conditions. The tree crops comprised Teak *(Tectona grandis)*, sissoo *(Dalbergia sissoo)*, Subabul *(Leucaena leucocephala)* and Australian Babool (Acacia *auriculiformis)* with crops, mostly groundnut *(Arachys hypigea)*, Soyabean *(Glycine max merrill)* and till *(Sesamum indicum)* with some paddy *(Oryza sctiva)* and maize (zeamays) in the inter-rows. The experiments were quite successful and Mishra and Prasad found that they could recover the cost of afforestation through cash crops in 2-3 years, time from compatible combinations.

Based on Mishra's experiments, the tree crop combinations were extended to the farmer's fields by the Social foresty Programme of the Bihar Forest Department.

The country Report for India presented at the eighth World Forestry Congress, 1978 describes the position of agro-forestry in different states in India as practised during the last 50 years.

In Andaman and Nicobar Islands, *Ptorocarpus dalbergiodes* has been successfully raised under taungya system along with agriculture crops of sugarcane and maize.

In Assam, sal has been successfully raised along with paddy and this has even reduced jhuming, resulting in the mutual advantage of both forestry and agriculture.

In Kerala state, foodcrops are raised under the taungya system. The tree species raised are teak, soft wood and Eucalyptus. About 8,000 ha. are cleared and planted every year. In this way, regeneration of forests and cambined production of food crops is accomplished.

In Tamil Nadu, an area of nearly 2,000 ha. is taken yearly for agro-forestry, locally called kumri. Tree species being raised are teak, bamboos. Eucalyptus, sandal, tamarind, babul, wattle, cashew and rubber. Food crops raised are samai, varajin, ragi, korea gram and tapioca.

In Maharashtra, a new approach has been adopted to utilise the inter-space between the tree lines to grow fibre crops like sun-hemp, jute and mestroda and oil bearing species like sunflower castor etc.

Til was found to be a poor intercrop for teak and sissoo but with groundnut and soyabean, it was highly remunerative.

Nair (1980) reported that under the Vanlakshmi Agro-Forestry Project in Kerala, pepper is planted successfully under semal, teak and silver oak.

Shyamlal (1980) observed that the average yield of agricultured crops is more with poplars than with Eucalyptus.

Katly (1980) showed that field study in Kerala have shown that ginger is a better crop than tapocca in Eucalyptus plantation.

Chaturvedi (1981) reported that the yield of the crops with poplars decreases in the second and subsequent years as the tree shade increases but wheat can be successfully cultivated upto the fifth year. Various cash crops like turmeric, ginger, tomato, peas, onion, soyabeans, etc. are also grown successfully under poplars.

In agro-forestry for forest crops, proper attention is to be paid at various stages. Preparation of proper pits promotes the rapid growth of, tree. Digging of pits is generally completed before the month of April. Generally, 30 cm. cube pits are dug for planting the seedlings. Care is to be taken for proper spacement; lining should be from east to west to allow the percolation of sun rays. Hoening should be performed at regular intervals. Fungicides, insecticides and manures are to be used at proper time. Watering at regular intervals of seedlings and young plants also becomes essential for a few years. Arrangements are to be made for proper pruning of buds and brandies. After full growth of some kinds of trees, lopping of branches becomes helpful in growth and in getting firewood. Shortly, it may be said that for the luxuriant growth of forest crops, care in many forms is to be taken. In this connection, it may be pointed that proper cultivation of food crops in the interspaces helps the growth of forest crops automatically.

Fencing of land in Agro-forestry is essential. Fencing may be done with barbed wire and stones. Trench fencing and hedge fencing are also adopted by the farmers. Hedge fencing costs the least and may be adopted by all kinds of cultivators. Planting of subabul in live fencing is very beneficial, as by lopping, its fuel and fodders become available to farmers. Moreover, cattle-watchers are employed for protecting the forest crops and food crops and fodders.

Provision of good seedlings and inputs like seeds of food crops, manures, insecticides etc. at a concessional rate to small and marginal farmers seems at present essential. They are provided cheap credit for fencing. In this way the farmers get an incentive for greater participation in the programme of agro-forestry.

The farmers of the country require motivation for adopting the dual system of production on a large scale.

The gains of agro-forestry are explained to the farmers with the help of short length films, meetings in the villages and providing useful literatures to literate farmers. Agro-forestry programmes are implemented at several places in Government land by the personnel of forest departments. The gains of these farmers are exhibited to induce farmers to adopt similar farming on their individual lands.

Hindrances in the Developmental Programme of Agro-Forestry

(1) Presence of illiteracy, ignorance, fatalism and the philosophy of contentment among farmers of the country are hindrances in the development of the programme. Many farmers fail to understand that sometimes there may be some fall in the production of grain but the total yield of land will be greater in the dual system of production. The government is trying to reduce illiteracy and ignorance amongst rural population. The extension of educational facilities is bound to eradicate fatalism and the philosophy of contentment from the rural areas in due course. However, at present, they are adversly affecting in the adoption of any progressive programme.

(2) A steep rise in the price of foodgrains inhibits many farmers from adopting the dual system of production. These farmers think that food production alone will be more gainful on their lands.

(3) In agro-forestry, comparatively greater facilities of credit and marketing are required. Inspite of the efforts of the government, as yet institutional credit and facilities of organised marketing are not made properly available to the farmers of our country. Moreover, there remains every possibility of irregular supply of inputs like fertilizers, insecticides etc. to the farmers. These things put hindrance in the development of this programme.

(4) The success of agro-forestry programme lies in the research work done in the field. Poor allotment of funds is considered the root cause of slow progress in research. For greater and uniform facility of research work in this field throughout the country, the Central Government should make some special efforts.

(5) The success of the programme lies in the efficiency and dedication of the personnel of the forest department of the states of India. They have to motivate the farmers and provide incentives. Due to lack of dedication, success is not achieved to the extent desired.

(6) The success of the programme of agro-forestry lies in the co-ordinated activities of the departments of agriculture, horticulture, forestry and animal husbandry. Due to lack of sincere co-ordination among these departments, hindrances come in the way of development of agro-forestry.

(7) Lack of proper fencing arrangements causes enormous loss in the agro-forestry programme. At times, wild animals of the traditional forests ruin the planted seedlings/young shoots as well as the food crops. This presents a great hurdle in the developmental programme of agro-forestry. Proper care is to be taken to eliminate the chance of such loss. Attention is also to be paid to stop the loss caused by domestic cattle.

Agro-Forestry is a gift of modern science. It is designed to meet the growing requirements of mankind. There are obviously some hindrances in the implementation of the programmes of agro-forestry in an underdeveloped country like India. These hindrances are to be removed. We cannot afford to neglect the developmental programme of agro-forestry in India.

Grasses in Agro-Forestry

Permanency in agriculture is attainable only through the wise use of land. It makes imperative on every generation to undertake land use practices that ensure sustained production.

Grass among other things tends to ensure permanency to agriculture. It provides nutritious fodder to cattle which are the mainstay of agriculture in India. It also provides protection to land against erosion by water and wind. Grass is important for the preservation of productive capacity of land. Whenever the productive capacity of land has been impaired for one reason or the other, then a grass regime for a period of time will retrieve the loss considerably. Grass within the meaning of pasture and forage is not restricted to members of gramineae family; it embraces also their common associates of the legume family like cloves, lucerne, guar, cowpeas, bersum and many others.

In our country there are millions of hectares of the so-called pasture land, but these lands are wastelands.

In addition to this, there is always a retrogressive trends in the vegetation kingdom from woodland to wasteland. If we want to stop this trend, then we have to go in for Agro-forestry. Agro-forestry does not simply mean agriculture production with forestry but it is a synthesis of agriculture, horticulture, forestry and animal husbandry in suitable areas in an integrated fashion. The cattle population in India is nearly 380 million and most of them are neither stall fed, nor do they have pasture land to graze. In the rainy season, when some grasses grow, they get some fodder but again in other seasons, they remain hungry and move about here and there for getting something to chew. Those living adjacent to the forest will graze inside the forest area and cause great damage to the plantations and natural regeneration. Of course, where the regulations are a bit strict, they are stall-fed by their owners. These people collect fodder by lopping the trees and feed their cattle. But in the Western Himalaya, where the sheep graze one can hardly find any tree in those areas. This has caused ecological imbalance. And also the lands are not being properly managed.

In India we have very few perennial grasses which can be grown throughout the year, because we have a limited rainy season. We can solve the problem by incorporating this in the agro-forestry system.

Choice of Species: The forage species which could be grown along with trees etc. under rainfed conditions are many. But the choice of species depend to a great extent upon the socio-economic factors and cropping pattern.

Grasses can be divided into two groups:

1. Short duration grasses, and
2. Perennial grasses.

Among the short duration grasses are *Pennisetum-pedicellatum.*

Among the perennial grasses are the *Andropogon gayanus, Brachiaria brizantha, Panicum maximum, Chloris gayana, Dicanthium annulatum, Cenchar culiaris, Senchrus sitegous, Lusiwas zindieous etc. are important.*

Perennial Grasses can also be divided into two: (a) Grasses for open lands, (b) Grasses for shades.

Grasses for open lands are *Andropogon gayanus, Barchiaria brizantha.*

Grasses for shades are those which can grow under the canopy of trees are Guinea grass *(Panicum maximum)* Rhodes Grass *(Chloris gayana).*

These grasses generally give 400 to 500 q/ha. and the dry matter 10 to 20 per cent.

Short duration grasses gives 2 to 3 cuts and perennial grasses will give 5 to 6 cuttings.

Among the perennial legumes, *Atylosia saraboides and Stylosanthus* Species are important and among short duration legumes. Lucerne, Berseem, cowpea are important.

Grasses are usually sown after the trees are established and with the onset of the monsoon. For uniform germination or sprouting of the transplanted materials, the soil should be worked and some manure should be added. On an average, 60 Kg. of N and 50 Kg. of P_2O_5 are required per ha. Grass-legume ratio should be 3:1 and trees should be spread 10 metres apart. Legumes have a good nutritive value. These legumes are seasonal or perennial.

Under the silvipasture system of land management, suitable pasture grasses along with timber-cum-fuel-cum-fodder trees should be planted. The grasses should be able to bear the shade trees like *Dalbergia sissoo, Albizzia lebbek, Tetona grandis, Acacia arabica, Prosopis juliflora, Morns alba*, etc.

In another system, grasses and legumes should be planted with fuel-cum-fodder trees. The advantages of growing these grasses and legumes with the fodder trees are during the wet season, animals can graze on the ground vegetation, *i.e.*, grasses and legumes and during the lean period they can be fed from the fodder trees.

Grasses and legumes can also the grown with fruit trees like mango, guava, Kathal etc.

Introduction of legumes in Indian pasture is very essential in order to enrich the land and improve the quality of fodder and pastures. The species should have the following requirements:

1. The species should be perennial or should have good reseeding qualities.
2. The growing period of the legume should coincide with the growing period of the grasses.
3. It should be hardy enough to with stand the competition from grasses, and should not require much care in ploughing, seeding or irrigation, etc.
4. It should not have trailing habits, otherwise it is likely to smoother grasses.
5. It should be palatable and high-yielding.

36

ROLE OF SOCIAL FORESTRY IN SOIL CONSERVATION

Soil conservation is a process of preservation and building up of the soil resources of a nation by the application of natural principles of land use management and maintaining the process of nature under control, so as to insure maximum sustained yield from the land for the use of mankind. Soil is the weathered uppermost layer of earths crust. It is the most valuable gift given by nature. It is the base of life. No life can survive without soil, because soil gives food material, the shelter and protection to the living creature. At present, soil has deteriorated due to various causes as, for example, destruction of forests, excessive grazing, fire, siltation etc. The main cause of soil deterioration is soil erosion. Soil erosion is a process by which soil is detached from one place, transported and deposited at another place. Erosion has a very severe effect. It destroys the fertility of the soil, by the removal of the top soil. This is so specially in the hilly region because of faulty cultivation. The present system of cultivation in our country is not only defective but the land is not being managed according to its classification. It is said that one lakh acres of land are going out of cultivation every year and large areas are becoming unproductive every year due to water and wind erosion. Exhaustion of plant nutrients by leaching and excessive cropping and overgrazing are also taking place. Due to overgrazing, the land loses its protective cover and thus erosion takes place. Accumulation of toxic salts, water logging etc. causes soil deterioration. The average crop yield in India is one of the lowest in the world. This malady is an overall manifestation of various defects in our agricultural practices and socio-economic structure.

Social forestry or community forestry is nothing but a forest by the society for its own use. This will not only meet their daily demands of fuel and fodder but the adjacent forests will also be protected. In our country, the total forest area is only about 18% of the total area, and it is shrinking every year due to encroachment. The hills and steep areas which should always be covered by vegetation is also being either cultivated or grazed by a large number of cattle. Not only the tribals who depend on forests but non-tribals living below the poverty line are also using the forests for their livelihood.

The former rulers of our country took the right decision in the nineteenth century to reserve some of the forests for scientific management. They also took the decision to create village forests for the use by villagers. Unfortunately, the latter proposal was not carried out. They also left large forest areas to be controlled both by big and small landlords. The then rulers gave priority to agriculture. As a result, very little forest area came under scientific management. The areas controlled by private persons were indiscriminately cut and in some places, converted into agricultural fields.

Though the after-effects of the indiscriminate felling of trees was foreseen by the Britishers long before independence, private forests were taken over under the Private Forests Act by the different states only after independence. It look a long time for the government to manage the forests scientifically due to many reasons, e.g., dearth of manpower, unwieldy forest jurisdiction under the forest officers, extinguishing the right of the forest-dwellers and the dwellers residing in the vicinity of the forests, increase in human as well as cattle population, and lack of foresight to satisfy the daily needs of fuel to the villagers as well as urban population, and last but not the least, the greed of the forest staff as well as contractors causing heavy damage to the forests of big landlords taken over by the Government. In fact, the staff from top to bottom were trained in production forestry and looked for only revenue. In the meantime, when it was found that large scale reforestation and afforestation is required for soil conservation and protection of the dams that were coming up in the river valley projects, the Afforestation and Soil Conservation Programme was started in 1950 on a small scale. In the soil conservation programme, the Forest Department undertook only contour trenching, and afforestation works. Nothing was done to do any structural work. The Agricultural Department did only graded terracing and in the hills of H.P. and Darjeeling, some terracing was done for cultivation. As a matter of fact, no thought was given to manage the land according to its classification. Grazing was not stopped and so the programme was infructuous. Meanwhile, even the Gochar lands and community lands vanished in the densely-populated areas and the forests became grazing grounds. During the last 35 years, not only did the old forests vanish but the newly-created forests also vanished. If one goes across the Sal forests in West Bengal, some parts of Bihar and other states, he will see a large stretch of forests which are like tea bushes. These degraded forests are now called rooted wastes. Not only were the blank areas planted several times under different schemes, efforts were made to rehabilitate the degraded forests, but in vain. If the people do not get their requirements, they will not allow anything to grow, and when the Government realised this, they started the programme of social and community forestry.

Now social forestry aims not only to provide people with fuel, fodder etc. but it also helps in conserving soil. It is said that about six million tonnes of top soil are being eroded in India each year. Now let us discuss something about the role of plants in soil conservation and environmental control.

A good vegetation cover such as a thick growth of grass or a dense forest may reduce completely the effect of climate, topography and soil on earth.

The following are the major effects of vegetation on soil erosion:

(*a*) *Interception of rainfall:* A part of rainfall intercepted by the canopy of vegetation never reaches the soil but is evaporated from the leaves and stems. This part of rainfall does not contribute to the runoff. In addition, the vegetation canopy absorbs the impact of rain drops and thereby minimise the dispersal of soil by rain drops.

(*b*) *Increasing runoff velocity:* Any vegetation cover is a hindrance to runoff water. A well-distributed close-growing vegetation not only slows down the rate at which water travels down the slope but also tends to present a concentration of water. These two effects greatly reduce the erosion capacity of the runoff water.

(*c*) *Root effects:* The knitting and binding effect of the root system in the surface layer of soil aggregates the soil in the ground and increases its resistance to erosion. The roots die and decay and leave numerous cavities which increase the pore space and consequently the absorption of water into the soil. Studies have shown that a cubic meter of soil has several kilometers of root firbres.

(*d*) *Biological influences:* The soil fauna are most active in a soil having an ample vegetative cover. The soil under a thick forest cover, for example, is permeated with the channels of earthworms, beetles and other organisms. The channels increase the permeability of the soil. Vegetation in addition increases the aeration of soil.

(*e*) *Transpiration effect:* Vegetation increases the storage capacity of the soil.

(*f*) *Humus:* Due to the presence of humus on the upper layer of the soil, the infiltration and percolation capacity of the soil increases.

(*g*) *Shelterbelt:* Plant acts as a shelter belt where there is wind erosion.

How the 328.8 million ha. of land area of the country is being utilised is indicated in Table 36.1 and Table 36.2 shows the area of erosion. It would seem that the cancer of erosion has already spread to about 50 per cent of the country's land mass and that it is threatening the rest. About 21 per cent of the country's land area is in a severe stage of deterioration. It would be seen that floods and droughts affect 34 and 68 m. ha. of land area respectively. The air, land and water pollution is on a steep increase.

TABLE 36.1

Classification of Area (Thousand ha.) 1976-77

1.	Area under forest	66,808.00
2.	Area not available for cultivation	39,474.00
(i)	Area under non-agricultural uses	17,533.00
(ii)	Barren and uncultivable land	21,941.00
3.	Other uncultivated land (excluding fallowland)	33,583.00
(i)	Permanent pastures and other grazing land	12,471.00
(ii)	Land under misc. tree crops and groves not included in net area sown	3,958.00
(iii)	Culturable wasteland	17,154.00
4.	Fallow land	23,985.00
(i)	Fallow land other than current fallows	8,688.00
(ii)	Current fallows	14,297.00
5.	Net area sown	1,40,227.00
6.	Total reporting area some parts are not covered	304,000.00
7.	Area for which no return exists	24,701.00
8.	Total Geographical area	3,28,778.00

Area Subjected to Erosion

Erosion is a two-edged sword it hurts the land and the people where it takes place, so also the land and the people where it gets deposited. The following figures speak of the extent of this problem.

TABLE 36.2

Area Subjected to Erosion

1.	Total geographical area	328 m.ha.
2.	Total area subject to erosion by wind and water	150. m.ha.
3.	Area at critical stage of deterioration due to erosion	69. m.ha.
4.	Area subject to the wind erosion	32. m.ha.
5.	Area effected by gully erosion and ravines	4. m.ha.
6.	Area affected by shifting cultivation	3. m.ha.
7.	Area under rainfed farming (Non-Paddy)	70. m.ha

TABLE 36.3

Problem Lands in India

(In Million ha.)

Salt affected	6.97
Water-logged	5.99
Desert	18.79
Mined land	0.68
Shifting cultivation	2.85
Ravines area	3.97
Area affected by floods	34.00
Riverine lands	2.48
Torrent streams	0.25
Coastal sandy belts	1.63
Hill roads	88.979 km.

The preventive role of forests in the struggle against erosion, in river regulation and flood control, had been proved by experience in many countries.

TABLE 36.4

Cultivable Wasteland

Name of State	*Area Cultivable Wasteland in sq. km.*
Himachal Pradesh	639
Punjab and Haryana	3,775
Uttar Pradesh	16,360
Rajasthan	16,720
Madhya Pradesh	12,180
Bihar	5,227
West Bengal	3,000
Total	57,901

It can thus be concluded that the present floods are due to lack of sufficient forest cover in the catchment area. Dur to increase in population, it is not possible to increase the area already under agricultural use for tree planting. But in our country even the marginal wasteland, hillsides are being used for agricultural purposes. The reason is that our production of foodgrains per ha. is the lowest in the world. China with only 11 per cent of its area under agriculture is able to produce almost twice the amount of foodgrains as compared to our country with 46 per cent of the area under agriculture. So if we increase the productivity of our agricultural land, quite a large area presently under marginal cultivation will be available for afforestation.

These areas are quite open with scattered trees and shrubs and grasslands. They do not in any way serve the purpose of S.C. In order to utilise these areas as effective underground water reservoirs, systematic afforestation with hardier tree species like neem, mahua, imli, babul, shisoo, prosopis, arjun should be done. In higher altitudes, spruce, fir, deodar, oak etc. may be tried. If the approximate net area available is 4,000 sq. km. and the average expenditure per ha. is Rs. 1000/-, the total cost of planting would be Rs. 400 crores. This may be spread over to 10 years. Besides, there are areas in other catchments which need attention. The plains along the Western half of the mainland Gujarat have deep alluvial soil susceptible to accelerated erosion. Unscientific management of land over the years has caused heavy erosion and ravines formation along the banks of some of the major rivers flowing through Gujarat. It is estimated that about 0.291 million hectares of valuable land has been converted into deep ravines in Gujarat. Sabarmati is one such river whose banks have been affected with formation of deep ravines. The plantation at lakewada is a good attempt in this direction. Further erosion is arrested and the land put to productive use. Fodder production is improved.

TABLE 36.5

Sl. No.	*Expenditure*	*Rs. Per Hectare*
1.	Survey and construction of boundry trench and sowing	701.50
2.	Fencing in vulnerable points	271.80
3.	Soil working	1,800.00
4.	Inspection path	180.20
5.	Planting	1,675.00
6.	Miscellaneous	23.20
	Protection Staff	341.70
	Total	5,000.00

Chaturvedi (1985) stated that all gully heads should be filled up with brush wood from the bottom to the top. This brush wood will break the speed of the flowing water. Gully plugs should be made from top to bottom to the gully such that the base of the upper bund is in level with the top of the lower bund. The bund should be 50 cms. wide on top, 25 cms. high with a base width of 200 cms. This will give a slope angle of 18° on either side. The bunds should as far as possible be at right angles to the stream flow. The burrow pit should be dug on the upward side and should be filled up with brush wood. The sloping sides and burrow pit should be planted with branch cuttings of *Adhatoda vaisca, Ziziphus jujuba* and whatever sprouting shrubs are locally available. The object of these bunds is not to store water but to slow down its speed. The shrubs should be planted immediately after the first shower. After the first rains, some bunds may get breached while others many get filled up. The soil so collected between the two bunds is rich and fertile. Bag plants should be planted on these portions at 2m spacing. Chaturvedi reported that planting trees on degraded lands provide indirect benefits like water and soil conservation. The expenditures on Marine lands are given is Table 36.5.

It is expected that yields of about 50 tonnes of firewood will be obtained after 10 years. Present day stumpage price of firewood is about Rs. 400 per ton. Thus, a return of about Rs. 20,000 per hectare should be expected after the tenth year.

Priority of planting will have to be given in the hilly terrain. The Himalayan zone, the Vindhyan zone of U.P. and the catchment areas of Chambal in Rajasthan will have to be given very high priority. In the plains all big and small tributaries will have to be surveyed and all elevated portions of the banks should be taken up for tree planting. Bare mounds of village tanks and higher banks along smaller stream as well as should be taken up for tree planting in the initial stages. The relatively lower banks of river and areas should be taken after 5-6 years when the higher areas have been put under tree cover. Multiple row planting of trees along roads, canals and railway track should be taken up on a priority basis.

Role of Ágave Plantation in Soil Conservation

In our country, various types of soil conservation measures are taken, of which plantation is given more importance. But due to poverty, ignorance, and other reasons, it has become very difficult to protect any plantation in our country. Even the good Sal forests are being converted into rooted waste. Due to destruction of forests, long gullies and deep ravines are forming inside the forests.

As even Ipomea plants are also disappearing, it is worth the while trying Agave for soil conservation works, and it is hoped that it will be an important substitute for other plants for soil conservation measures.

The requirement of soil and moisture of this plant is very negligible. The important point in its favour is that the plantation is carried out for utilising the soil which is neither fit for agricultural crops nor for tree plantation.

Sisal plant has been found to grow on a soil depth of less than one foot and it can flourish on any type of soil. It is a drought-resistant plant and frost-hardy by nature and a rainfall of 30" to 40" is sufficient for its growth.

Technique of Planting

The plants are raised both by root-suckers and bulbils. Root suckers appear almost every year till its maturity but bulbils occur only once in its entire life cycle. These bulbils are generally of small sizes and are taken to the mother bed filled with sand, watered both in the morning and evening and when they sprout, they are transferred to the nursery beds and kept there for about twelve to fifteen months. Later on, they are transplanted at a spacing of 10 cm. x 10 cm. These bulbils are then transplanted in the fields during the plantation season when they attain the size of 25-30 cms. As for the use of suckers, these are thrown out from the roots, from the third year at the rate of 2-5 per plant on an average and are planted from the second week of June to July end. These suckers do not require any type of nursing in the nursery, but these are directly planted in the plantation areas. Normally, suckers are better planting materials.

Fencing Plants

Fencing plants do not require fencing of any type but should be protected against trampling by cattle for at least two years.

Soil Work

Soil work is done one year before and usually after the monsoon is over when the soil remains soft. Hoeing of soil in strips by pickaxes is done 15 c.m. to 20 c.m. deep, that is, soil is hoed in strips of 150 c.m. (5 ft.) wide for weathering. The space between two rows is 2.5 metres and this has an extraction path. The spacing adopted is of double rows of plants, that is, 1 m. x 1 m. and every double row of 1 m. x 1 m. is separated by 2.5 metre-wide strips to facilitate cultural operation, harvesting as well as carriage of leaves etc. 2,300 root suckers or bulbils are required for one acre plantation.

Treatment Before Planting

Before planting, the uprooted root suckers or nursery raised bulbils are subjected to pre-treatment with a view to initiate root growth. Adventitious roots are trimmed by a sharp scythe. Thus, the exterior root layers of leaves, say 1-3 leaves, are removed exposing the cream colour bulb. This process is known as root trimming. Root trimmed plants may be transplanted immediately and the entire operation from uprooting of transplanting may be completed within two days though sisal plants are prone to rough handling.

Transplanting

Root suckers and bulbils are transplanted preferably with the onset of monsoon. This operation should be completed at any cost by the end of July. Since they are draught resistant plants, pre-monsoon transplanting may also be resorted to. But the cost will be higher in this case.

Cultural Operation

Contrary to the general belief a sisal plant requires attention throughout its entire life cycle, normally ten years. It is also desirable for economic return as well.

Weeding and Hoeing

Every sisal plant is valuable because of its importance of leaves which are the only source of fibre production. Hence, as a rule, individual care must be given to all of them. Weeds may be cut and cleared at least for four years just in the monsoon season inside the strips of 150 c.m. width, where two rows of sisal plants, are transplanted and for the rest of the area, no weeds should grow up above the height of sisal plants.

The soil is loosened up to a depth of 20 c.m. in 150 c.m. wide strip except without disturbing the soil in the radius of 15-20 c.m. around the plants. This is very important for the healthy growth of plants and production of root suckers

Manuring

The plants are claciphilous. Hence, calcium fertilizers should be applied to these plants besides application of urea, single super-phosphate and potash.

The following dose per plant has been found to be satisfactory for the successful growth:

Urea - 14 gms.

Single superphosphate - 18 gms.

Potash - 4 gms.

Calcium may be added at the rate to 10 kg. per acre.

This mixture is added to each plant just after two weeks of completion of planting and it should be mixed in a radius of 15-20 c.m. from the centre of the plants. The application of fertilisers must be completed by the second week of August.

Harvesting of Leaves

Preliminary harvesting is done from the third year of planting. In this operation, all the peripheral leaves are cut, leaving behind 20 leaves in the centre. Regular harvesting is done from the fourth year.

The yield of harvested leaves is about 34,500 kg. per acre. In the subsequent operation, only 15 leaves are left, excluding the central column of the unfolded leaves. 10-15 leaves may be obtained from each plant. The leaves are cut by a curved scythe having a long handle as close to the base of possible. Care should be taken, so that the trunk of the plant is not injured. The leaves vary from 1' 6" to 3' 6" in length.

Extraction of Fibre

Sisal leaves contain only 4 per cent of fibre. These leaves are decorticated on a decortication machine run with the help of electric current. The fibre is washed at least three times and dried in the sun for three days after having hung on the specially made contrivance exactly like a towel rack and underneath sand is spread so that the colour of the fibre may not be lost in the soil. In the beginning, the fibre generally has a greenish colour but gradually, it changes into white after drying in the open for three days.

The length of the fibre is generally between 1' 6" to 3' 6". The longer the fibre, the greater is the demand in the market.

The leaves are cut from the month of October and this operation lasts upto April. The leaves must be decorticated within twenty-four hours from the time of harvesting.

The residue which is left out after decortication of fibre is also of great importance. At present, it is being used in a compost pit and after two years, it is also applied round the plants in place of fertiliser. It has a marvellous effect on the growth of sisal plants.

Normally, the decortication machine may run for eight hours at a stretch. Two labourers are required to manage one decortication machine for the production of fibre.

100 kg. of leaves yield 10 kg. of fibre with moisture and ultimately 4 kg. of dry fibre is achieved. Thus, 100 kg. of green leaves in weight will yield 4 kg. of marketable fibre.

Baling of Fibre

Dry fibre is finally baled out in baling machine made of wood and each baled out bundle of fibre is of 50 kg. to 100 kg. in weight. These bundles are stored in a closed shed on a wooden platform. The stored fibres are well protected by rain water and other injurious insects.

The following operations are done in a plantation from the 1st to 9th year:

A. First Year

(i) Soil working and bush removal in the previous year of plantation.

(ii) Digging, transport of suckers, bulbils and transplanting including alignment in the first year.

(iii) Application of fertiliser.

(iv) Weeding and hoeing.

B. 2nd Year

(i) Weeding and hoeing,

(ii) Application of fertiliser,

(iii) Shrub cutting.

C. 3rd Year

(i) Weeding, hoeing and shrub cutting.

D. 4th Year

(i) Weeding, hoeing and shrub cutting,

(ii) Application of fertilizer and shrub cutting.

E. 5th Year

(i) Shrub and climber cutting.

F. 6th Year

(i) Shrub cutting.

G. 7th Year

(i) Hoeing, weeding and shrub cutting,

(ii) Application of fertiliser.

H. 8th Year

(i) Shrub and climber cutting.

I. 9th year

(i) As above.

J. 10th Year

Collection of bulbils, etc.

The total cost of the above works, including purchase of tools, maintenance of extraction paths and protecting against cattle for nine years, comes to Rs. 2,000/-.

Harvesting and Decortication

(a) Cost of harvesting, carriage, decortication, washing and drying etc. of fibres @ 200 kg. dry fibre per acre.

(b) Harvesting of leaves.

(c) Transportation of green leaves from the plantation area to the machine site.

(d) Decortication, washing and drying etc.
(e) Preliminary harvesting in the third year, 60 kg. dry fibres.
(f) Charge of water consumption.
(g) Charge of consumption of electricity.
(h) Loss due to power failure.
(i) Cost of baling, building, stocking etc. The total cost of all these operation comes to Rs. 1,400/-.

Thus, the total expenditure of planting, maintenance, harvesting and establishment etc. per acre would be as follows:

(i)	Plantation cost	Rs. 2,000
(ii)	Harvesting and other expenditure	Rs. 1,400
	Total	Rs. 3,400
	Establishment charges 10% (approx.)	Rs. 3,40
	Jha & Sen (1988)	Rs. 3,740

Expected Yield

Preliminary harvesting is done only in the third year. Based on the experience and yield of preliminary as well as regular harvesting for seven years, it has been seen that 60 kg. and the first year 200 kg. of dry fibres from the second year onwards will be obtained per year per acre. Bulbils obtained in the 10th year are about 1,00,000.

In this connection, it is worthwhile to experiment if Eucalyptus and other fast-growing species can be planted in between the rows of Agave plants. These will also add to the profit of the plantation area. The fibres are disposed of by calling regular tenders from the intending bidders. After calculation, it has been found that a revenue of Rs. 400/- to 500/- per acre per year will be obtained without providing for interest on capital. If provision for interest on capital is taken into consideration, the revenue of Rs. 300/- per acre per year (approx.) may be obtained.

Summary

Sisal plantation will utilise unproductive soil, conserve soil from erosion and may fetch considerable amount of revenue within a short span of ten years. Thus, its plantation is worthwhile encouraging on a large scale all over India, specially in degraded forests and marginal wastelands.

37

ROLE OF COOPERATIVES IN THE PROMOTION OF SOCIAL FORESTRY

Social forestry offers enormous gains to mankind. The problems of pollution of air, scarcity of fuelwood, water conservation, irregular and scarce rainfall, soil erosion etc. may be solved to a large extent by the rapid promotion of social forestry. Moreover, it raises the production of small timber, fruits, vegetables, gains, tubers, legumes, fibres, herbs and shrubs, tassar, lac, honey etc. In short, social forestry promises variety of gains, which are urgently required to meet the growing needs of the teeming millions of India.

Rapid promotion of social forestry is not an easy task. The problem can be tackled by concerted efforts of individuals and many organisation of the society. Needless to say cooperatives can play a pivotal role in accelerating the tempo of development of social forestry both in plateau and plain regions of the country.

The rural areas of the plateau region consist of mainly tribal population and other weaker sections of Indian society. They depend largely on the produce of forests for their livelihood. Experts view that proper functioning of cooperatives would reduce their economic exploitation and raise their socio-economic status. In 1973 a study team under the chairmanship of Sri K.S. Bawa recommended the formation of large sized multipurpose primary cooperative societies to solve their various problems.

1. Such multi-purpose primaries were expected to meet short, medium and long-term credit requirements of the members.
2. Consumption credit was also to be provided to the needy members during the lean period of the year.
3. Inputs for production and essential consumer goods were to be provided by the primaries at a reasonable rate to the members.
4. Members were to be provided guidance in adopting the latest techniques of production.

5. Arrangements were to be made for the marketing of products of the members to prevent the prevalent exploitation by the intermediaries and to make the recovery of loans convenient.
6. The government was to provide share capital for the members of weaker section and some other necessary assistance.

It was expected that greater participation of the rural poor would result from these measures and ultimately, there would be rapid promotion in agriculture, social forestry and other activities in the rural areas. It was hoped that the members would imbibe the spirits of self-help, mutual aid and cultivate the habit of thrift.

Those primaries federated at each state were to form a Tribal Development Cooperative Corporation/Federation.

On the basis of the recommendation of Sri K.S. Bawa, large-sized agricultural multi-purpose societies were formed throughout the tribal belt of the country. Primary Agriculture Cooperative societies were also formed to attain multi-purpose objectives.

These multi-purpose societies helped in the promotion of social forestry, but the achievement was not up to the mark. Presently, utmost emphasis is laid for the promotion of social forestry, hence a single-purpose primary society, namely, primary social Forestry Cooperative Society may be formed throughout the tribal belt of the country to carryout efficiently the following functions, which will positively help in the speedy implementation of the social forestry programmes.

1. The requirements of the credit and inputs in the social forestry programme of members of a primary may be easily and promptly ascertained by the Executive Committee.
2. Each member will be provided timely short, medium and long-term credit. Credit will be provided for consumption purposes during the lean period of the year to eliminate exploitation.
3. Inputs of superior quality will be supplied to the member at a concessional rate without delay.
4. Proper guidance will be provided regarding the latest technique of social forestry by inviting specialists from the Forest Department.
5. Arrangement shall be made by such society for processing of the products, wherever feasible.
6. Marketing of the produce of members will be exclusively done by the primaries. Consequently, exploitation by the intermediaries will be eliminated.

7. Marketing of the produce of the members by the primaries will facilitate timely recovery of loans and hence the problem of overdues will be automatically solved.
8. The tribals and other members of weaker sections in the plateau region will be largely benefited by the smooth functioning of the primary social forestry cooperative societies and hence they will get a chance to cross the poverty line. Needless to say, social forestry programmes are labour intensive and are oriented to promote, particularly the welfare of weaker sections of the society.

The tribals and members of other weaker sections in the plateau are economically handicapped and so exploited, but it must be admitted that tney possess a special character of homogeneity and as such the successful functioning of primary social forestry societies may be expected to become a reality. The members would inculcate the habit of thrift and imbibe the spirit of self-help and mutual aid. They will lean the art of democratic management.

Primary joint social forestry cooperative societies may be formed in suitable circumstances. In such primaries, the members have to pool their individual land and consequently, the formation of a big plot or a few big plots become possible. Subsequently, production work is to be managed by the executive committees of the primaries.

Vast areas of devastated forest land and wasteland are lying idle in the tribal belt of the country. Such lands may be distributed amongst the weaker sections of the society. They may form primary joint social forestry cooperative societies and subsequently, arrangements may be made for the maximum production of suitable forest crop. A suitable example for joint social forestry cooperative society may be as follows.

In 1982-83, a sum of Rs. 1.10 crores was sanctioned by Government of India to raise tassar production in some districts of Andhra Pradesh. The government was to make block plantation of *Terminalia arjuna* in selected patches of 750 acres. After three years of tending operations and proper pollardings, the trees become bushy and fit for rearing of tassar silk worms. Those patches were to be distributed amongst 700 tribal families after three years of plantation. In such case, each patch of the area was to be organised as a joint social forestry society. Members will pool their land to make a single plot or a few plots, although the area owned by each member will stand in the name of the members in the ledger of society. The executive committee will adopt the latest techniques of production to obtain the maximum yield. It has been estimated by experts that proper rearing of silk worms can give an income of about Rs. 4,000 per ha. per rearing season. The merit of the primary joint social forestry society in such cases may be as follows:

1. The entire patch of land will be benefited by the latest techniques of production and hence the maximum yield will be possible. Merits of new techniques of production will be highlighted amongst the members.
2. Superior quality of disease-free laving and other inputs will be used in the entire patch of land in proper time by the executive committee members.
3. Leaf-sheding of plants and Pebrin disease may be controlled without delay in the entire patch.
4. In a joint social forestry cooperative society, the protection of tassar host trees in the entire patch becomes an easy task. Individuals have to devote greater energy in the protection of tassar host trees, when the planting is done on an individual basis.
5. The society will manage an adequate amount of short, medium and long-term credit required for production and hence there shall be no difficulty in adopting the latest techniques of production. Moreover, poor and innocent members will not be required to grease the palm of corrupt personnel in availing credits in this system of production.
6. The society may manage to clear the outstanding debts of members. The members may be provided consumption loan during the lean period of the year.
7. In a joint social forestry cooperative society, marketing of the products will be arranged exclusively by the society and hence exploitation by the intermediaries will be eliminated. Moreover, there will be no difficulty in the timely recovery of loan from the society.
8. In such primary societies, the yield is distributed to each factor of production on the principles decided by the members. In practice it has been found that labourers are comparatively more fairly remunerated. Thus, the income of the weaker sections may rise rapidly and they may be able to cross the poverty line.

Needless to say, there are several merits in the joint social forestry cooperative societies. We must be aware of the fact that in such primaries the members must remain self-disciplined and possess a democratic outlook. They must imbibe the habit of thrift, self-help and mutual aid. If these virtues are not found amongst the members, unfavourable results are bound to occur. Moreover, if the members of the executive committee of a primary are inefficient and dishonest, the society is bound to collapse.

Large-sized multi-purpose societies and primary Agricultural cooperative societies have been federated in the Indian states with the Tribal Development Cooperative Corporation Federation. Similarly, social forestry cooperative societies and joint social forestry cooperative societies should be federated in the different states of India with the Tribal Development Cooperative Corporation/Federation. A cell for the promotion of social forestry in the Department of Cooperation under a separate Joint Registrar may be formed in each state and requisite number of staff may be appointed at different administrative headquarters. Such arrangement will strengthen the working of primaries.

Processing of forest produce forms an integral part of social forestry. Such work can be efficiently performed by forming the primary processing cooperative societies. The Dhebar Commission in 1961 recommended that collection and processing of forest produce should be done through cooperatives. Tassar silk, silk-weaving, cane and bamboo work, bee-keeping, fruit-preservation, spinning, handloom weaving etc. can be successfully performed through the formation of primary processing cooperative societies. Subsequently, a working group, under the chairmanship of Sri M.P. Bhargava in 1962, also recommended that processing of forest produce can bc quite successfully performed through the cooperative units. In such organisation of processing work, the tribals and other poor people will be fairly remunerated for their works and thus an opportunity will be created by some poor people to cross the poverty line. Needless to say, sincere participation of workes will be available in the processing work performed through the cooperatives.

In our Forest Policy of 1952, it was clearly indicated that labour cooperative societies should be formed to eliminate the agency of contractors, who mercilessly exploits innocent tribals and other weaker sections of society of the plateau region. Labour cooperative societies should be formed and works should be allotted to them. Clearance of devastated forests lying near villages with the objective of planting saplings, levelling of land, construction of link roads, block planting of saplings near human habitation etc. are related to social forestry. These works must be carried, out through the cooperatives. The Dhebar commission opined that poor tribals remained at the mercy of contractors. Hence, the functioning of labour cooperative societies will eliminate exploitation and pave the way for raising the standard of living of exploited persons.

The role of cooperatives in the plain region of the country seems to be of a different nature. Our objective is to cover 20 per cent of land of the plain region with forests. Vast tracts of waste public or private land are rarely available. Hence, block plantation of trees in a vast area may be a rare phenomenon. Linear plantation of trees on the boundaries of high-level land and agro-forestry had been practised in the country form time immemorial. Presently, greater emphasis has been given for such plantations on scientific basis.

In the plain region of the country agro-forestry has been adopted on scientific lines on a large scale to augment the production of food, fruits, vegetables, tubers, legumes, fodders, fibres etc. Needless to say, for adopting agro-forestry on scientific lines a large amount of short, medium and long-term credits is required. Primary agricultural credit societies are functioning now in the plain region of the country and such societies can provide credit at a cheaper rate for the promotion of agro-forestry. In due course, when the primaries are capable, they may undertake the marketing of produce of the members. Thus, the participants will be able to get full price for their products. Moreover, the recovery of credit from the members will become an easy task.

In the plain region of the country, the Government is planting saplings by the sides of railway lines, roads and canals. Subsequently, tending and protection works become a problem. The help of primary agricultural credit societies of adjoining areas may be taken to solve this problem. A primary agricultural credit society may be allotted a certain portion of planted saplings for tending and protection work and it should be fairly remunerated. The advantages of such arrangement are as follows:

1. The tending and protection of saplings may be smoothly undertaken.
2. An additional source of income will be available to many primary agricultural credit societies.
3. A large number of people will be involved in the social forestry programme as members of primary agricultural credit societies.

Presently, the speedy promotion of social forestry is one of the foremost national programmes of the country. Cooperatives have been assigned the task of participating in the rapid economic development of the country. Thus, cooperatives must participate effectively in the promotion of social forestry programmes.

For making the role of cooperatives more effective in the promotion of social forestry, a few suggestions are made as follows:

1. There should be a statutory provision of. the formation of a single-purpose cooperative society throughout the country, like the Social Forestry Cooperative Society and Joint Social Forestry Cooperative Society.
2. Primary Agricultural Cooperative Societies have to play an effective role in the promotion of agro-forestry in the plain regions of the country. Hence, one of the objectives of such society must be to promote agro-forestry. Presently, in Bihar, there are twenty-two objectives of a Primary Agricultural Cooperative Society, but the objective of promoting agro-forestry/social forestry is not included. Such shortcomings must be rectified.

3. An adequate amount of fund should be allotted specifically to the cooperatives, so that they may effectively participate in the developmental programme of social forestry.
4. A special cell in the Department of Cooperation should be formed in all states of the country, so that proper supervision, guidance and control may be possible in this field.
5. There is still the idea prevalent amongst the ignorant people, particularly amongst the tribals, that cooperative is an agency of the Government. This false notion and fear is to be removed. They are to be made aware of the fact that the government assists only to promote the cooperative movement. The ultimate objective of the Government is neither to control nor to interfere in its working. The cooperatives are to be managed by their members democratically and on the principles of self-help and mutual aid.

Social forestry is oriented to promote rapidly economic welfare of the people, particularly those belonging to the weaker sections of the society. Cooperatives aim to participate in the speedy economic development of the country and help in the formation of an egalitarian social system. Thus, it may be concluded that the proper functioning of cooperatives must accelerate the tempo of development of social forestry in our country.

■ ■ ■

BIBLIOGRAPHY

A.S. Kahlon, "Theory of Economic Growth in Over-Populated Countries", *Indian Journal of Agricultural Economics,* Vol. XXII No. 4, Oct. Dec. 1967.

Acharya, S.S., 1973, Green Revolution and Farm Employment, *Indian Journal of Agricultural Economics,* 28 (3): 30-45.

Alvin Toffler—"The Third Wave" PAN Book, Macmillan India.

B.R. Grover, 'Nature of Land-Rights in Mughal India," *Indian Economic and Social History Review,* I, 1 (July-September 1963), 15; for Ancient India, see L. Gopal, 'Ownership of Agricultural Land in Ancient India,' *Journal of the Economic and Social History of the Orient,* IV (1961), pp. 240-63.

Bhaduri, A.: 'A Study in Agricultural Backwardness under Semi-Feudalism,' *Economic Journal,* March 1973.

Bharadwaj, K.: *Production Conditions in Indian Agriculture* (Cambridge, 1974).

Bhattacharya, N.: 'Agricultural Labour and Production' (in this volume).

Biplab Das Gupta—"The New Agrarian Strategy and India", UNRI Geneva, and Macmillan, India, 1977.

Breman J.: *Patronage and Exploitation: Changing Agrarian Relations in South Gujarat* (California, 1974).

Burton Stein, *Peasant, State and Society in Medieval South India* (Delhi: Oxford University Press, 1980), p. 434.

C. Subramaniam—"The New Strategy in Indian Agriculture" Vikas, New Delhi 1979, p. 23.

Chaudhary, S.P., *Proceedings of the National Institute of Sciences of India,* Vol. 26(2) Supplement.

Chauhan, Y.S., D.S. Singh and T.R. Singh, 1981, Impact of Farm Mechanisation on Labour Use in Block Kalyanpur, District Kanpur, *Indian Journal of Agricultural Economics,* 36 (4): 56.

Chowdhury, B.B.: *Growth of Commercial Agriculture in Bengal* (Calcutta, 1964).

D.D. Kosambi, *An Introduction to the Study of Indian History* (2nd. rev.ed.; Bombay: Popular Prakashan, 1956).

Dobb, M.: *Studies in the Development of Capitalism* (Routledge 4 Kegan Paul, 1946).

Dr. V.A, Pai Panandikar, *"Development Bureaucracy"* (along with R.N. Bishnoi and O.P. Sharma) Centre for Policy Research, New Delhi March 1911. p. 353.

Edward W. Weidner "Development Administration—A New Focus for Research" included in Paper in *Comparative Public Administration,* Ann Arbor, Michigan Institute of Public Administration 1962 p. 98.

Emmette S. Rodlord, *"Ideal and Practice in Public Administration"* (University of Alabama Press 1975).

Gilbert Etienne—"India's Changing Rural Scene", OUP, 1982.

Grewal, S.S. and A.S. Kahlon, 1972, Impact of Farm Mechanisation on Farm Employment in the Punjab, *Indian Journal of Agricultural Economics,* 27 (4): 214-217.

Guha, R.: *A Rule of Properly for Bengal* (Paris, 1963).

Hanumantha Rao, C.H., 1975, *Technological Change and Distribution of Gains in Indian Agriculture,* MacMillan Company of India.

Harbans Mukhia, 'Was there Feudalism in Indian History,'? *The Journal of Peasant Studies,* VIII, 3, (April 1981), pp. 273-310.

Irfan Habib, 'Economic History of the Delhi Sultanate—An Essay in Interpretation,' *Indian Historical Review,* IV, 2 (January 1978), p. 298; see also the same author's 'Problems of Marxist Historical Analysis,' *in Science and Human Progress: Essays* in *Honour of D. D. Kosambi* (Bombay: Popular Prakashan, 1974), pp. 34—48, in which the Mughal system has been called a system of centralized collection of rent/surplus—the principal contradiction being that between the king's bureaucracy and the peasantry. One wonders how this characterization by itself is inconsistent with the AMP, something which Habib seems to suggest.

J.M. Fleming, "External Economics and the Doctrine of Balanced Growth", *The Economics Journal,* Vol. LXV, June, 1955, page 254.

Kartar Singh Chela, *Crops Rotations,* (Punjab Agricultrual University Publication).

Kautsky, K.: 'La Question Agraire,' *Maspero,* 1970; translated and summarised by Jairus Bamji in *The Articulation of Modes of Production,* ed. H. Wolpe (Routledge & Kegan Paul, 1980).

Khusro, A. M. : Economics of Land Reform and Farm Size in India (1973).

Malcolm Darling—"Punjab Peasant in Prosperity and Debt" OUP, 1932.

Michael Chisholm—"Modern World Development—A Geographical Perspective"—Hutchison University Library, London, 1882.

Minhas, B.S.: Rural Poverty, Land Redistribution and Development Strategy: Facts and Policy. Indian Economic Review, Vol. V (New Series) No. 1 April, 1970.

Muni Lal, *Haryana, High Road to Prosperity,* Vikas, 1974, p. 7.

R.B. Jain and P.N. Chaudhuri, *Bureaucracy and Development,* Centre for Policy Research, Chanakya Puri, New Delhi pp. 20-21 (unpublished).

R.S. Sharma, *Indian Feudalism: c. 300-1200* (Calcutta: University of Calcutta; 1965); see also B. P. Mazamdar, *The Socio-Economic History of Northern India (11th and 12th Centuries)* (Calcutta: Firma K. L. Mukhopadhyaya, 1960), pp. 1-28 and K. S. Shelvankar, 'Indian Feudalism: Its Characteristics,' in A. R. Desai, ed, *Rural Sociology in India* (Bombay: Popular Prakashan, 1969), pp. 150-4.

Raj Krishna, 1974, Measurement of the Direct and Indirect Employment Effects of Agricultural Growth with Technical Change, in *Employment in Developing Nations,* Ed. E.O. Edwards, Columbia University Press, New York

Ramakrishna Mukherjee, *The Dynamics of a Rural Society: A Study of the Economic Structure in Bengal Villages* (Berlin: Akademic-Verlag, 1957), pp. 75-9.

Rangnekar: *Poverty and Capital Development in India*, p. 42 and 298-99.

Ruddar Datt: Giri's Plan for Co-operative Land Colonisation, Economic Times, September, 1970.

Ruddar Datt: Poor Implementation of Land Reform Programme, Economic Times. 27th November, 1973.

S.N. Sinha : *"Economics of Cropping Pattern" in AICC Economic Review,* Vol. XV, January, 1964.

Shelvankar, K.S.: 'The Problem of India,' reproduced in *Rural Sociology,* ed. A. R. Desai (New Delhi, 1969).

Shome, K.B. and Ray, (1960), "Ratings of Soils of India". In:

Siddiqi, Asiya: Agrarian *Change in a Northern Indian State: Uttar Pradesh, 1819-1833* (Oxford, 1973).

Sidhu, B.S.: Land Reforms, Welfare and Economic Growth.

V.K.R.V. Rao, *New Challenges before Indian Agriculture* (Dr. Panse Memorial Lecture).

Vadilal Dagli--Report of the Committee on Controls and Subsidies 1978-79, Govt. of India, New Delhi.

Whitcombe, E.: *Agrarian Condition in Northern India Vol. 1: The United Provinces Under British Rule, 1860-1900* (California, 1971).

INDEX

N

O

P

R

S

❑❑❑